Cold Atom

The Quantum World of Ultra-Cold Atoms and Light

Book II: The Physics of Quantum-Optical Devices

Cold Atoms

ISSN: 2045-9734

Series Editor: Christophe Salomon (*Laboratoire Kastler Brossel, École Normale Supérieure, France*)

Vol. 1 Quantum Gases: Finite Temperature and Non-Equilibrium Dynamics
edited by N. Proukakis, S. Gardiner, M. Davis and M. Szymańska

Vol. 2 The Quantum World of Ultra-Cold Atoms and Light
Book 1: Foundations of Quantum Optics
by Crispin Gardiner and Peter Zoller

Vol. 3 Quantum Gas Experiments: Exploring Many-Body States
edited by Päivi Törmä and Klaus Sengstock

Vol. 4 The Quantum World of Ultra-Cold Atoms and Light
Book II: The Physics of Quantum-Optical Devices
by Crispin Gardiner and Peter Zoller

The Quantum World of Ultra-Cold Atoms and Light

Book II: The Physics of Quantum-Optical Devices

Crispin Gardiner
University of Otago, New Zealand

Peter Zoller
University of Innsbruck, Austria

Imperial College Press

ICP

Published by

Imperial College Press
57 Shelton Street
Covent Garden
London WC2H 9HE

Distributed by

World Scientific Publishing Co. Pte. Ltd.
5 Toh Tuck Link, Singapore 596224
USA office: 27 Warren Street, Suite 401–402, Hackensack, NJ 07601
UK office: 57 Shelton Street, Covent Garden, London WC2H 9HE

Library of Congress Cataloging-in-Publication Data
Gardiner, C. W. (Crispin W.), 1942–
Physics of quantum devices / Crispin Gardiner, University of Otago, New Zealand, Peter Zoller, University of Innsbruck, Austria.
page cm. -- (The quantum world of ultra-cold atoms and light book ; II) (Cold atoms ; vol. 4)
Includes bibliographical references and index.
ISBN 978-1-78326-615-9 (hardcover : alk. paper) -- ISBN 978-1-78326-616-6 (pbk. : alk. paper)
ISBN 978-1-78326-678-4 (e-Book: individual) -- ISBN 978-1-78326-679-1 (e-Book: library)
1. Quantum optics--Mathematics. 2. Quantum computers. 3. Computer simulation. 4. Quantum systems. I. Zoller, P. (Peter), 1952– II. Title.
QC446.2.G369 2015
621.3815--dc23

2014029388

British Library Cataloguing-in-Publication Data
A catalogue record for this book is available from the British Library.

for

Helen and Hanni

Preface

It is an amazing fact that those aspects of quantum mechanics which were once seen as paradoxical, such as the EPR paradox and Schrödinger's unfortunate cat, are now seen as not only true and verifiable predictions, but also as the basis of new technologies, such as quantum computers, quantum simulators, and quantum sensors, all of which are potentially very powerful technologies. The impetus for this paradigm shift came from the field of quantum optics, where the ability to predict and control the quantum states of light and atoms provided the testing ground for the paradoxical predictions of quantum mechanics, such as the EPR paradox and Bell's inequalities, non-classical states of light (including squeezing, antibunching and quantum correlations) and even the creation of the Bose–Einstein condensate, which is a genuinely non-classical state of matter, albeit of a very different kind from EPR states. All of these situations are so far from everyday experience that no-one would have ruled out the possibility that in testing them, we might discover the boundaries of validity of quantum mechanics. But this did not happen. As investigations progressed, the quantum mechanical framework appeared to be increasingly robust—to this day no defects have been found.

So there has been a paradigm shift, and with it a need for a new orientation of the theoretical frameworks, in a way which gives more prominence to the view that the "paradoxical" aspects of quantum mechanics—most of them related to the concept of *entanglement*—are no longer located in a fringe area, to be discussed only by philosophers, but now form a central part of the theoretical toolbox needed to describe and design quantum devices. With this paradigm shift has come a new terminology; we now talk of *quantum science* or *quantum science and technology*, rather than quantum mechanics. This reflects the fact that there is now a broadening in the range of fields where progress is being made in utilizing the full power of quantum theory. There are new links between workers in the fields as quantum optics, superconductivity, solid state physics, nanotechnology and others which justify the designation of the new field of *quantum science* to encompass all of them. And central to all of the endeavours in these fields is the idea of *quantum control*, which involves the application of the techniques of each particular field to put a device into a well-defined quantum state, to process this state by some well-defined means, and to measure the resulting processed state.

Although there are many different fields within quantum science, it is becoming increasingly clear that the theoretical descriptions developed in the field of quantum optics provide the most powerful and convenient theoretical toolbox for the understanding of the devices being considered by quantum scientists. This toolbox, developed steadily over the past fifty years is based on the idea of *quantum stochastic processes*, a coherent set of techniques in which all of the main physical prerequisites of a practical quantum device are automatically taken care of.

This book, the second in our trilogy "The Quantum World of Ultra-Cold Atoms and Light", aims to present the quantum-optical toolbox in a form appropriate to the needs of those working with quantum devices. Although the quantum devices now being implemented use a number of technologies, the fundamental quantum mechanical structure underpinning them is the same as that of the quantum optics of atoms and light, and their description by means of quantum stochastic processes. The "atoms" used may be genuine atoms, or they may be based on engineered systems, such as quantum dots, nitrogen vacancy centres in diamond, rare earth dopants in YAG, nanomechanical systems, or superconducting systems; similarly the "light" may be of any wavelength between the ultraviolet and the microwave regimes.

In *Book I* we presented the basic tools needed in a relatively elementary form, which we feel should be accessible to anyone with a reasonably good grounding in quantum mechanics. This book builds on that foundation. It does not require additional supporting knowledge, although familiarity with the physics of other fields, such as superconductors, semiconductors and atomic physics will of course be helpful.

The book has nine parts, which are quite tightly connected. Part I provides an outline of the concepts of quantum optics, as well as a compact introduction to quantum information theory, in which only the very basic ideas of are presented, with emphasis on those that are most relevant to the kinds of devices considered in the book. A more extensive account can be found in the book of *Nielsen* and *Chuang*[1].

The most ideal form of optical manipulation of atoms involves only coherent fields, and for this reason in Part II we have presented the theory of this kind of manipulation for both two-level atoms and atoms with many levels, and as well we have included a chapter on the coherent manipulation of atomic motion. Quantum optics in general prefers to treat atoms as simple objects, in which only a few energy levels are relevant. This is mainly true, but we have taken some care in this part of the book to justify this point of view, and have also included a chapter on *Real Atoms*, in which we give a quite detailed explanation of those facts of atomic structure relevant to the systems we study.

The essential heart of the subject is in Part III, where we develop the full quantum stochastic formalism, a formalism that has many aspects, such as the quantum stochastic Schrödinger equation, the master equation, quantum stochastic differential equations, all of which are useful in understanding practical systems. We also include here the application of these methods to the dissipative dynamics

---

[1] M. A. Nielsen and I. I. Chuang, *Quantum Computation and Quantum Information*, 1st ed. (Cambridge University Press, Cambridge, 2000, 2010.)

of atomic systems, including systems with many atoms, a subject which is central to the application of quantum-optical systems to the construction of quantum devices.

Laser cooling is a technique which is universally used in manipulating atoms and ions, and in Part IV we give the foundations of this subject using the quantum stochastic formalism. The particular importance of ion traps in quantum devices is recognized by a chapter on the specialized techniques of cooling trapped ions, in particular, *sideband cooling*.

The measurement process in quantum-optical systems is fundamental, and Part V and Part VI are devoted to the two main aspects this subject. At the frequencies higher than those of infrared light, measurement is dominated by photon counting, which is the subject matter of Part V. In Part VI we consider phase-sensitive quantum optics, where measurement involves mixing the optical signal with a strong coherent field. Phase-sensitive quantum optics has many advantages, and in particular includes the idea of squeezing, using which quantum noise can be reduced for certain measurements. In addition, the abstract theory of quantum amplification, which can be phase-sensitive or phase-insensitive, is presented here.

The most widely used devices employ the *cavity QED* concept, of a two-level atom inside an optical cavity interacting with its quantized light modes, and the external world—this forms the subject of Part VII. The relevant Hamiltonian is that of a two-level system interacting with a harmonic oscillator, and is known as the *Jaynes–Cummings Hamiltonian.* It was already briefly introduced in *Book I*, but here we give a very detailed description of its dynamics, and how it is used. The Hamiltonian can also be emulated by the motion of a ion trapped by a harmonic potential, which we also treat in detail, since this forms the basis for the *ion-trap quantum computer*, where several ions are trapped inside a single trap, and interact with each other through their electrostatic repulsion. Consequently, there is also a chapter on the theory of this device, the first practical quantum computer to be proposed, and the first to be implemented in the laboratory.

We have emphasized that the quantum optical formalism is in our view universally applicable to quantum devices, and in Part VIII we show how superconducting devices, using Josephson junctions, can be used to make systems with all of the features of those of traditional quantum-optical systems. The radiation is in the microwave regime, and the "atoms" are Josephson junctions. These provide a potential which is sufficiently anharmonic to give rise to transitions well separated in frequency, giving rise to effective two-level systems. The microwave regime requires also different practical technologies from those used for conventional quantum-optical systems; for example, the ambient temperatures are produced by liquid helium based refrigeration, and transmission lines take the place of optical beams. Nevertheless, the theoretical formalism used is almost identical with that of quantum optics

Finally, Part IX addresses the issue of the interfacing between quantum devices and optically transmitted quantum information, for which the only currently viable technology is that of the optical fibre, the same as that currently used in telecommunication systems. The subject of interfacing is not fully mature, and what we present here are some models of how this might be done. Nevertheless, the formulation of these models brings up some very interesting techniques, and these may well outlast the particular implementations presented here.

No book of this kind can ever be complete. The choice of subject matter has been made with a view to its applicability and future usefulness, as well as its place within our own interests and expertise. The selection of subject matter follows that of lectures on quantum optics given over recent years within the Physics Department of the University of Innsbruck, but extensively augmented. It has also been strongly influenced by the research within the *Institute for Quantum Optics and Quantum Information* of the Austrian Academy of Sciences.

This book, and the trilogy as a whole, are not intended to supersede our previous book, *Quantum Noise*[2]. That book reflects the point of view of quantum optics current before the potential for quantum devices had become apparent. Hence, it concentrates more on the physical foundations of the ideas of quantum noise and dissipation, and on the production and understanding of non-classical light. However, the formulation of the quantum stochastic formalism we have made in Part III and the theory of continuous measurement in Part V, while completely consistent with that in *Quantum Noise*, have been considerably reworked and simplified into a form that we think is better adapted to the requirements of quantum device physics.

Nevertheless, the overlap between the two books is surprisingly little, and indeed we often refer the reader to *Quantum Noise* for a more extensive treatment of the material under consideration.

Otago and Innsbruck
March 25, 2015

Crispin Gardiner
Peter Zoller

[2] C. W. Gardiner and P. Zoller, *Quantum Noise*, 3rd ed. (Springer, Berlin, Heidelberg, 2004.)

Acknowledgments

This book, and its companion volumes, represent a summation of knowledge developed by ourselves and many others in the field over a period of about forty years, and during this period we have interacted with colleagues all around the world, as well as in our own countries. Our primary debt is to all of these colleagues, who have collectively created the environment within which we have worked. Particular thanks, however, are owed to those with whom we have worked most closely, namely the late Dan Walls, Ignacio Cirac, Rainer Blatt, Rob Ballagh, Matthew Collett, Jeff Kimble, Bill Phillips, Rudi Grimm, Hans Briegel, Immanuel Bloch, Scott Parkins, Ashton Bradley, Blair Blakie, Simon Gardiner and Dieter Jaksch.

During our careers, we have been supported by many institutions: the University of Otago, the Leopold Franzens University of Innsbruck, the University of Colorado, Victoria University of Wellington, the University of Southern California, and the University of Waikato.

During the time in which we have been writing we have been generously supported by New Zealand through its research funds, the *Marsden Fund* and the *New Economy Research Fund*, and by Austria through *die Österreichische Akademie der Wissenschaften, the Austrian Science Funds* and *the European Research Council.*

Contents

V CONTINUOUS MEASUREMENT AND QUANTUM TRAJECTORIES

VI Phase-Sensitive Quantum Optics

VII Quantum Processing with Atoms, Photons and Phonons

I PRINCIPLES OF QUANTUM DEVICES

PART I: PRINCIPLES OF QUANTUM DEVICES

In this introductory part we want to present our motivation for the subject matter of this book. This involves two aspects:

i) In Chap. 1 we show how the development of quantum science has provided extraordinary opportunities for quantum devices, and how we have to formulate our theoretical structures to be able to understand and design such devices.

ii) The remaining two chapters are devoted to a sketch of the abstract results of quantum information theory, which inform us about what devices we might wish to construct.

   We start with Chap. 2, which formulates the fundamental idea of a *qubit* as the fundamental unit of quantum information. This may be viewed as an idealized two-level atom, and the chapter considers what manipulations of its quantum state would be needed for quantum information processing. It guides much of the remainder of the book, which concentrates of the physical aspects of such manipulations.

   In Chap. 3 we present the ideas of quantum information, that is, the quantum states of *quantum registers*, made of assemblies of individual qubits. Most of the ideas presented here are ahead of the state of physical implementation of quantum information processing, but will play a significant role in the future.

The current situation in quantum processing is still at the "proof of principle" stage. Many devices have now been constructed, and all of the necessary protocols have been tested. Rudimentary computations can be performed. The remainder of the book is devoted to the theoretical principles needed to understand, design and improve quantum devices.

1. The Science and Technology of Quantum Optics

The success of quantum mechanics in explaining the structure of the microscopic world after its definitive formulation in the 1920s was overwhelming. The *Bohr–Rutherford* atom, while revolutionary and definitely closely related to how the microscopic world behaves, was only able to proceed by *ad hoc* methods to try to describe more complex systems than hydrogen. Similarly, the ideas of quantization of the light field formulated by *Planck, Einstein, Bose* and others, while explaining many aspects of the interaction between light and atoms, were obviously as incompatible with Maxwell's equations as the Bohr–Rutherford atom was with Newton's equations. The "new quantum theory" of *Heisenberg, Born, Dirac* and especially of *Schrödinger* dramatically swept away the "old quantum theory".

Nevertheless, the old ways of thinking survived for at least another fifty years, and during that period, probably the majority of physicists did not truly believe that the superposition principle always applies, even in macroscopic situations. The *Einstein–Podolsky–Rosen* [1.1] paradox was seen as a paradox not because it contradicted any experiment, but because it predicted results which appeared to contradict "common sense". Even today the atom is often symbolized by a small solid ball of nuclear matter encircled by point-like electrons constrained into Bohr orbits.

1.1 Quantum Mechanics—from Paradox to Paradigm

We have now made the transition to the modern era, where quantum mechanics is seen as a self-consistent, logical and apparently universal description of the world, all of whose predictions are to be taken seriously. Where we once saw paradoxes, we now see opportunities. The issue of entanglement raised by the EPR paradox is an aspect of coherence, the coherence of quantum superpositions of very different states—in this case, superpositions involving *non-local* entanglement.

After *Bell* [1.2] showed that quantum superposition states of the EPR kind were measurably different from those one would expect classically, *Clauser, Horne, Shimony* and *Holt* [1.3] pointed out that versions of Bell's inequalities were experimentally testable. A number of experimental tests[1] were made, mostly confirm-

[1] See [1.4] for a listing of these experiments.

ing quantum mechanics to some extent, culminating in *Aspect's* [1.4] quantum-optical experiment, which provided definitive verification of the existence of the kind of entanglement found so paradoxical by Einstein, Podolsky and Rosen. At this stage, *Feynman* formulated his proposal[2] [1.5] for quantum simulation and computation, which was pivotal in its recognition of quantum-mechanical Hilbert space as a way of coding information that was immensely more efficient than any classical method, and as a resource which could be very useful for information processing.

The new paradigm in quantum mechanics is founded on coherence and entanglement, the very sources of the paradoxes of the early years. Because of entanglement, quantum mechanics does things more efficiently than classical physics, and it is this efficiency which gives us the opportunities for a new world of quantum processing.

1.2 Coherence and Quantum Optics

Quantum optics started as an effort to understand the quantum properties of light, and especially those of phase, coherence and correlations of both atoms and light. Indeed, the founding work on theoretical quantum optics, that of *Glauber* [1.6], is entitled *The Quantum Theory of Optical Coherence.* Glauber's formulation of coherence made it clear that there should be *non-classical states of light*, that is, states of light exhibiting coherence or correlation properties only possible in quantum mechanics, such as antibunched light and squeezed light, as well as EPR states. Their theoretical prediction and subsequent experimental verification as states of light which were not explicable other than quantum mechanically, focused attention on quantum optics as a means of testing the validity of the quantum mechanics of states very far from any classical situation.

What this means is that the frontiers of knowledge in quantum optics are to be found in Hilbert space, involving quantum states increasingly remote from everyday experience. By constructing superpositions of very different states, and manipulating these states, quantum mechanics is tested in situations very remote from those ever originally envisaged. If these tests are passed, we will be able to construct quantum devices of enormous power and sensitivity. If not, we will have found limits on the validity of quantum mechanics, and consequently a new physics. These two possibilities are both of extraordinary interest.

1.2.1 Quantum Optics—its Scope and its Aims

Today, our understanding of the essential defining characteristic of the field of quantum optics is the existence of *quantum coherence*, and the ability to use and

[2] Feynman's proposal was apparently motivated by the definitive experimental confirmation of the reality of entanglement achieved, but although he acknowledges the existence of experimental confirmation, he makes no citations of any kind.

measure this quantum coherence. Most importantly, we include both *atomic coherence* and *optical coherence* on an equal footing. The word "optics" in the terminology "quantum optics" has its most general meaning—the science of electromagnetic radiation in the optical, ultra-violet, infra-red and microwave regimes. The "atoms" may be real atoms, or engineered systems, such as Josephson junctions, quantum dots or objects such as nitrogen vacancy centres or rare earth dopants embedded in a solid state matrix.

The field of quantum optics deals with matter at the level of individual atoms, and electromagnetic radiation at the level of individual photons. Currently, it concentrates on the description of interactions between atoms and the quantized light fields, and their use as a toolbox to build and manipulate systems of atoms and photons. Thus the current paradigm of quantum optics is that of a *composite quantum system*, made up of atoms, photons, optical cavities, and atomic traps, together with a large number of more conventional passive elements, such as mirrors and optical cavities, all manipulated using laser fields or microwaves.

1.2.2 Preparation, Control and Measurement of Composite Quantum Systems

Currently, the goal of quantum optics is to develop complete quantum control in these composite quantum systems, and this includes three fundamental aspects: The preparation of the initial quantum state of the system, the subsequent manipulation of the system, and finally the measurement of the quantum state of the system.

The extent to which the preparation, control and measurement on the single quantum level is achievable in the laboratory is extraordinary. Using a combination of optical pumping and spontaneous emission, an atom can be put into a well-defined state, starting from any initial state. Then, using coherent laser or microwave fields, one can manipulate the atom into almost any other chosen electronic state. Atoms or ions can be held in a variety of atomic traps, made using optical, electrical and magnetic techniques. Laser cooling techniques can control the motional states of trapped atoms or ions all the way down to the quantum ground state.

The interaction between atoms and photons can be realized both as a free space coupling to the continuum modes of the radiation field, or, in cavity QED, as atoms coupled strongly to a single optical mode of a high-Q cavity. Furthermore, such cavity systems can interfaced to optical fibres, which can be used to couple together different cavities, thus enabling the transmission of quantized wavepackets between one cavity system and another.

1.2.3 Quantum Devices

The impetus to develop such an extraordinary degree of control over quantum systems came from the desire to develop high precision time standards, based on atomic clocks. The capacity thus developed to control atomic systems then made it feasible to consider the experimental development of quantum information processing. The fundamental logic element of quantum information is the *qubit*—a two state quantum system, and this is straightforwardly realizable using the internal states of an atomic system. Using several atoms it is possible to make both a quantum register and, using the elements of quantum control, a quantum processor. For example, it is then possible to perform general qubit gate operations, involving one, two or many qubits, on these atoms. Finally, using the techniques of fibre coupled cavities, it is in principle possible to network such quantum computers, sending "flying qubits" between the "nodes" (that is, the individual quantum processors) of such a network.

In the last twenty years, there has been a sustained effort to construct genuine quantum devices—that is, devices which use the full power of quantum mechanics, the most prominent being a quantum computer. The idea that such a quantum computer was possible came from *Feynman* [1.7] in 1986; that it could be more powerful than classical computers for certain problems was definitively demonstrated by *Shor* [1.8] in 1994; and that a quantum computer could be engineered using quantum optical and ion trap technology was demonstrated by *Cirac* and *Zoller* [1.9] in 1995. The basic quantum gate required by the ion trap quantum computer was implemented in 2003 [1.10, 1.11], and there has since been steady development thereafter in implementing the full protocols required.

Quantum devices now exist in a number of forms, and the quantum processor is no longer merely an ambitious proposal, but is indeed a fact. The current devices are essentially embryonic, but do implement all of the necessary protocols, and can perform elementary processing based on true quantum algorithms. It can be expected that in the coming decades, technical progress will be sufficient to generate useful devices. These may not in fact be quite the devices originally envisaged by Feynman, or devices which implement Shor's pioneering algorithm, but they will be *quantum* devices, performing tasks inaccessible to conventional digital devices.

1.2.4 Quantum Information Theory

With the realization that quantum processing was potentially feasible and powerful, the field of *quantum information theory* came into existence, in which the abstract principles of quantum information and its processing were formulated and investigated. Because it is based on the principles of quantum mechanics, in particular the superposition principle and the ideas of quantum measurement theory, the field of quantum information presented totally new challenges of a mathematical and philosophical kind. The field developed with classical information theory as its model, and emphasized that abstract principles by which information processing, transfer and storage could be optimized. The classic work of *Nielsen* and *Chuang* [1.12] is in our opinion the best exposition of this subject.

1.3 Theoretical Quantum Optics

While the laser is the central enabling technology of quantum optics, it is not only as an experimental instrument that it is important, but also as an example of a quantum system of a completely new type, whose theoretical understanding initiated the development of a wholly new field of theoretical physics, that of *quantum stochastic processes*, which now dominates theoretical quantum optics. Of course quantum electrodynamics gives, in principle, a complete description of all quantum optical phenomena, and the theory of quantum stochastic processes is not a new kind of physics. It is best described as a kind quantum statistical mechanics, based on a version of quantum electrodynamics appropriately formulated for quantum optical systems, and with a very strong emphasis on the *dynamic processes.*

A particularly useful aspect of the theory of quantum stochastic processes is its natural expression of the quantum theory of measurement, especially the process of *continuous measurement*, where the light radiated by a quantum system is continuously monitored by photodetectors. This provides insights into the dynamics of the systems observed, and in addition provides a useful algorithmic computation tool to simulate the behaviour of quantum-optical systems.

The theory has developed considerably since its inception in the 1960s, and in its current form can be used to describe almost any quantum optical system. In this book we show how to formulate the theory of quantum stochastic processes, and how to use this theory. We have formulated our description of quantum optical systems using the quantum stochastic approach exclusively, in the belief that this provides the simplest, most powerful and most direct way of modelling quantum optical systems.

1.3.1 The Quantum Optical System-Environment Paradigm

The most significant aspect of quantum optical systems is that they are *open systems*, in which there is a clear separation into:

i) *The System*: This is typically microscopic, comprising perhaps a small number of atoms, and the modes of optical resonators. By “small”, we mean that a relatively small number of quantum states take part in the dynamics of interest. The system itself may actually be macroscopic, such as is the case for an optical resonator or a Josephson junction.

ii) *The Environment*: The “environment” represents, in principle, everything in the universe other than the system. In practice, when dealing with quantum optical systems, the only part of the rest of the universe which interacts directly with the system under consideration is the electromagnetic field.

The clear separation into “system” and “environment” comes about by design, and be viewed as a kind of engineering. For example, a trapped ion is held by carefully controlled magnetic and radio-frequency field in a well defined position in a very good vacuum. In this way the ion is very well isolated from most of the physical world, apart from the coupling—which is relatively weak—to the relevant resonant modes of the electromagnetic field.

1.3.2 The Electromagnetic Field as an Environment

The interaction between the electromagnetic field and an atom takes place almost exclusively in a narrow bandwidth around the frequencies determined by the internal electronic transitions, and these are so high that at ordinary temperatures there is no optical field at any of the frequencies, unless the experimenter chooses otherwise, by turning on a laser, for example. Applying an almost resonant optical field will cause the atom to change its quantum state, providing the principal experimental method of manipulating the atom.

If the atom is in an excited state, it can decay by radiating into the electromagnetic field—this is the process of *spontaneous emission.* It is most important to note that, from the point of view of an atom, this is a very weak interaction, as seen by the fact that atomic linewidths are many orders of magnitude smaller that the typical transition frequencies.

1.3.3 Inputs and Outputs

The experimenter is necessarily part of the environment, and is able to use it to set up experiments and measure results. The experimenter applies optical fields, and detects emitted light from the system, and these fields are intrinsically part of the environment. The formulation by *Gardiner* and *Collett* [1.13, 1.14] of the *input-output theory* of damped quantum systems, in which inputs and outputs to and from quantum optical systems were necessarily part of the environment, transformed quantum optics. Instead of being merely a kind of rather accurate spectroscopy, quantum optics became the study of quantum information processing, in which the main task was to determine the quantum state of the output resulting from a given quantum state of the input.

It follows therefore, that electromagnetic environment provides three principal functions, as illustrated in Fig. 1.1:

i) *Inputs*: These are the optical fields needed to set up the initial state of the system, or to control the evolution of the quantum state of the system.

ii) *Outputs*: These are the radiated optical fields, which the experimenter detects, and from the results of the detections makes inferences about the quantum state of the system.

iii) *Noise and Decoherence*: The process of spontaneous emission is the most obvious form of irreversible behaviour produced by the interaction with the environment. If we set up the system so that the applied laser field would put it in a superposition of two different quantum states, the spontaneous emission process will tend to destroy the phase relationship between the different parts of the superposition, a process known as *decoherence*. Simultaneously, it adds noise to signals which would ideally be quite precisely determined.

1.4 Quantum Control and Quantum Processing

The fundamental goal of quantum optics since its inception has been the control of quantum systems, and this has three fundamental aspects:

i) *Initializing a Quantum System*: The system of interest will be in some initial state, represented by a density operator which we shall call ρ_0. We must be able to put the system into some predetermined quantum state, represented by a density operator ρ_i, independently of the ρ_0.

   From this it follows that initialization is an irreversible process. This is the characteristic behaviour of a system coupled to an environment, and involves

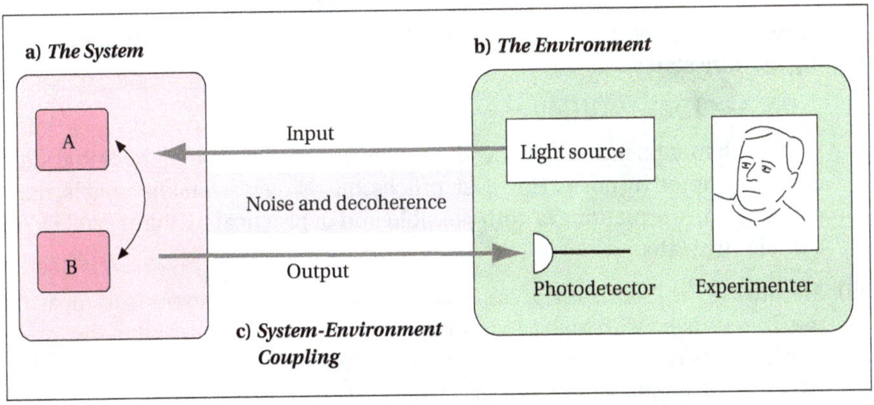

Fig. 1.1. Features of the quantum-optical system-environment model.
a) The system is composed of subsystems which interact with each other coherently. The number of such subsystems and the nature of their interconnections is arbitrary.
b) The environment comprises everything which may interact with the system. The experimenter is part of the environment, together with the experimental apparatus, such as light sources and detectors, which provide the method of controlling the system. The light sources behave as highly populated modes of the electromagnetic field; these may be coherent states, or possibly more exotic quantum states, such as squeezed or antibunched states. They may alternatively behave as carriers of quantum information.
c) The system is normally very well isolated from the environment, and the coupling between them only takes place through the electromagnetic field, and is weak. Because the environment has a very large number of modes, the interaction with the system has an irreversible nature, giving rise to decoherence and noise effects, such as spontaneous emission. However, when the light source produces a coherent driving field, this is equivalent to a coherent state of an environmental mode, and gives rise to a reversible component of the behaviour of the system.

quantum stochastic processes, such as we have already described in *Book I.*

ii) *Processing the Quantum State of a System*: Having initialized the quantum state of a system to the state ρ_i, we must be able take it to a chosen final state, represented by the density operator ρ_f. The final state is a function of the initial state, meaning that the final state can be said to be *programmable.*

In *quantum processing*, we are interested in applying a definite sequence of transformations to an initial quantum state in order to determine what the resulting final state is. These are often linear unitary transformations, but may be more general evolution operators, incorporating irreversibility.

This is similar to running a computer programme, where the output is completely determined by initial data and the programme, but is only actually known by running the programme. In quantum processing, the sequence of transformations amounts to the programme; however, the final quantum state is not the result—this is only known after some measurement process has been performed on the final quantum state.

Reversible processing, which is characterized by unitary transformations acting on pure states, such as

$$|n\rangle_i \leftrightarrow |\mathcal{F}(n)\rangle_f = \mathcal{U}(F)|n\rangle_i, \tag{1.4.1}$$

is of fundamental interest, since in this way the information in the initial state is not corrupted by noise. However, processing using irreversible (that is, non-unitary) transformations is both possible and of practical use, and can lead to a final pure state.

iii) *Measuring the Final State of the System*: The final requirement is to measure the final state ρ_f with high accuracy, so as to determine the result of the quantum processing ii). The measurement process, which we have discussed in *Bk. I: Ch.18–Ch.20*, is an irreversible process, and it necessarily involves the coupling of an environment, such as the quantized electromagnetic field, to the system. The measurements are normally performed on the electromagnetic field itself by counting photons, and from these results one makes inferences about the quantum state of the system to which the field is coupled.

1.4.1 Implementing Quantum Control Using Trapped Ions

The idealized requirements of quantum control have been put into practice in systems of trapped ions, using which elementary quantum processors have been built. These provide a very clear example of how the various abstract protocols required can actually be realized, and in which we can physically identify the above three fundamental requirements of preparation, processing and measurement.

An ion trap is a device which uses static and time-dependent electric fields to confine an ion by means of an effective potential on the ion. The main features of a typical ion trap are :

i) The ion is often $^{40}Ca^+$, and the essential electronic structure is that of a three-level system, as shown in Fig. 1.2.

ii) The lifetimes of the excited states are very different; that of the $D_{5/2}$ state is $\tau_{slow} \approx 1.1$s, while that of the $P_{1/2}$ state is $\tau_{fast} \approx 7$ns. When resonant laser light is applied to the fast transition, there is a strong fluorescent signal which appears about τ_{fast} after turning on the light—that is, essentially immediately—see Fig. 1.4.

However, this signal does not appear if the atom is in the long-lived $D_{5/2}$ state. In this case, in a time of order of magnitude τ_{slow}, the atom will make a transition to the ground state, and then the fluorescence will appear.

When initially trapped, the ion may be in any electronic state, and its motion in the harmonic trap will be best described as being thermalized, and at a relatively high temperature.

a) Initializing the Ion: There are two stages to the initialization; the ion must be put into the ground state of the harmonic trap, and its electronic state must be set equal to some chosen state.

i) *Achieving the Motional Ground State*: By applying an almost resonant optical field to the strong transition, a process of laser cooling can be induced. If the optical field is slightly *red-detuned*, to each photon absorbed by the ion, a photon of slightly higher energy is emitted, and the ion loses energy. This process can in practice be used to cool the ion to the ground state of the harmonic oscillator, as illustrated in Fig. 1.3. The process of photon emission requires the quantized electromagnetic field, and is thus the source of the irreversibility in the cooling part of the initialization process.

ii) *Achieving the Electronic Ground State*: By applying another appropriate optical field the ion can be transferred to a quantum state which has a relatively short decay time to the ground state. If the ground state is not degenerate, the atom is then in a pure electronic state. If the ground state is degenerate, it is normally possible to choose optical fields of appropriate polarization and frequency to transfer the system from all but one of the ground states to a higher

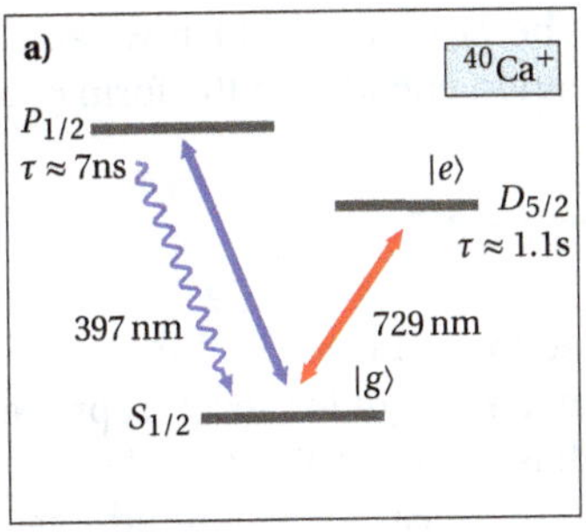

Fig. 1.2. The three-level system used in the Innsbruck quantum processor. The transition on the right has a very long lifetime, and allows the two levels it connects to be used to store quantum information. It can be manipulated using a coherent resonant field, but in the timescale of experiments spontaneous emission can be neglected. The transition on the left has a short lifetime, and when illuminated with a coherent field, produces a strong fluorescent signal as a result of atomic decay.

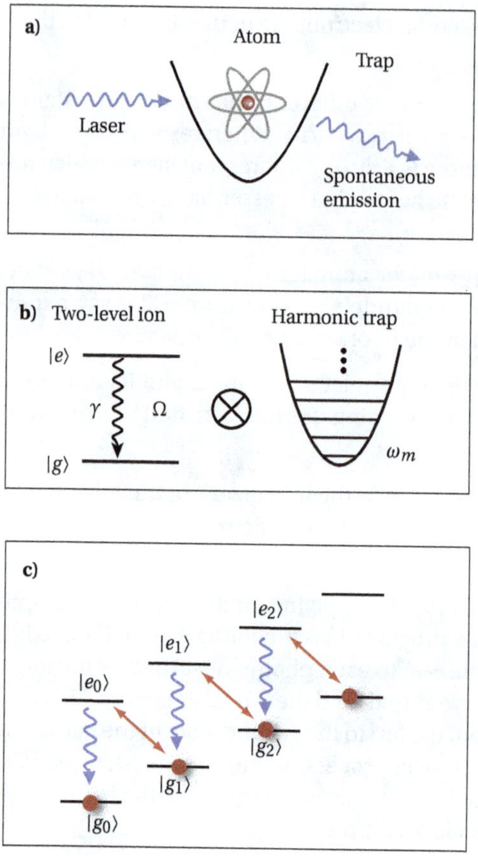

Fig. 1.3. Laser cooling principles.
a) The ion trapped in a harmonic trap is cooled by applying laser light slightly red-detuned from resonance. The photons emitted by spontaneous emission carry slightly more energy than those of the laser, making the ion lose motional and potential energy—a *cooling process results.*
b) In an ion trap both the electronic levels and the motional levels are quantized, so that the system Hilbert space is the direct product of a two-level atom Hilbert space and that of a Harmonic oscillator.
c) The energy levels of the trapped ion are those of a two-level system, reproduced for each number of motional quanta, giving a spectrum as shown. Sideband cooling is achieved by optically pumping from states $|g_n\rangle \to |e_{n-1}\rangle$, followed by a spontaneous emission yielding $|e_{n-1}\rangle \to |g_{n-1}\rangle$. This eventually takes the atom to the state $|g_0\rangle$, the electronic and motional ground state.

energy state which will then again decay to the degenerate ground state. A proportion of these decays will be to the desired ground state, the remainder are again recycled, so that eventually all of the population accumulates in the desired ground state.

b) Transforming the Quantum State of the Ion: The two-state system comprising the ground state, which we will call $|g\rangle$, and the $D_{5/2}$ state, which we shall call $|e\rangle$ can be called a *qubit*, and quantum information encoded in the form of a linear combination of the form

$$|\text{info}\rangle = a_e|e\rangle + a_g|g\rangle. \tag{1.4.2}$$

Once the atom is in the state $|g\rangle$, it can be transferred to any linear combination of the $|e\rangle$ state and $|g\rangle$ by application of a pulse 729 nm laser light for an appropriate time. Application of a pulse of 397 nm laser light has a similar effect on the $P_{1/2}$ state, but of course this is immediately destroyed by the decay to the ground state.

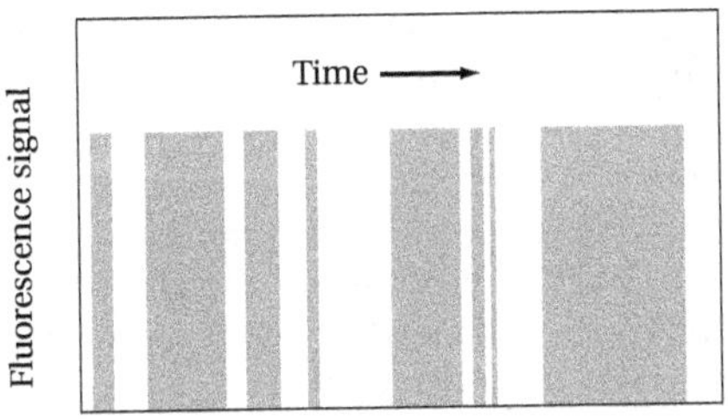

Fig. 1.4. The time-dependence of the fluorescence from the atom when it is also being driven (relatively weakly) on the transition to the $D_{5/2}$ state—one sees periods of brightness and darkness, corresponding to quantum jumps between the ground state and $D_{5/2}$ state.

c) Measurement of the Quantum State of the Ion: One applies a strong 397 nm laser field on the transition to the $P_{1/2}$ state, as discussed in Fig. 1.2. If the ion is in the state $|g\rangle$, transitions to the $P_{1/2}$ state occur within 7ns of applying the laser field, and a strong fluorescence signal is generated. This can be detected with essentially 100% efficiency. If there is no fluorescence signal, this means that the atom was in the state $|e\rangle$.

Thus the application of the 397 nm laser field produces *on demand* (or at least within 7 ns) a measurement on state $a_e|e\rangle + a_g|g\rangle$ of the two-state system.

1.4.2 Several Trapped Ions—the Quantum Processor

By trapping several ions inside a trap as in Fig. 1.5, a *composite quantum system* is created. The individual ions behave as qubits, but can communicate with each other using electrostatic forces. These forces induce collective translational modes, which are quantized, and the combined energy level spectrum which results then has a range of levels. Laser beams tuned to particular transitions can then be used to induce a wide range of unitary transformations of the kind considered in (1.4.1).

The collective vibrations of the centre of mass of the assembly of ions form a useful degree of freedom, which communicates with each individual ion in the same way. This can be used to provide a *data bus*, which can be used to transfer quantum information between the ions. Each ion can be described in terms of its electronic state and the collective vibrational mode, so that instead of the description in the form (1.4.2), we would write

$$|\text{info}, n\rangle = \big(a_e|e\rangle + a_g|g\rangle\big) \otimes |n\rangle, \tag{1.4.3}$$

where n is an index which indicates the number of "phonons", that is, vibrational quanta, in the vibrational state—this number is in practice either 0 or 1. It is then possible to tune an addressing laser to implement a unitary transformation of the form

$$\mathcal{U}|\text{info}, 0\rangle = \mathcal{U}\big\{\big(a_e|e\rangle + a_g|g\rangle\big) \otimes |0\rangle\big\}, \tag{1.4.4}$$

$$= |g\rangle \otimes \big(a_e|1\rangle + a_g|0\rangle\big). \tag{1.4.5}$$

It can be seen that the electronic excitation (whose coefficient is a_e) has been transferred to a phonon excitation, and this is equivalent to transferring the quan-

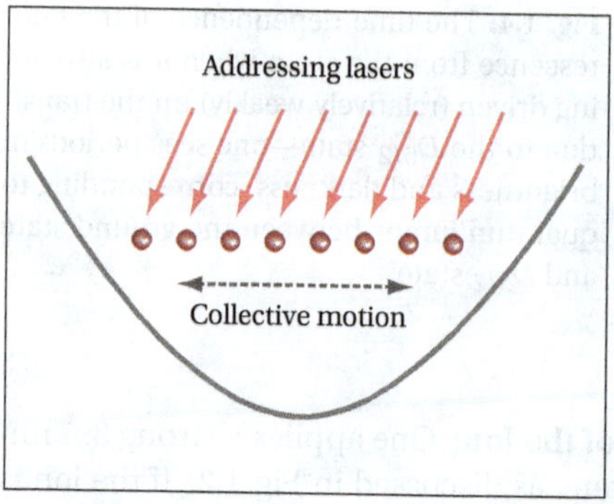

Fig. 1.5. By putting several ions in one trap, a composite many-ion system is created. Electrostatic repulsion between the ions induces collective translational motion, which enables communication between the ions, while ions can also be individually addressed by laser beams. This enables *quantum gates*, using which an ion can be unitarily transferred to a quantum state which is a function of its initial state, and that of another ion.

tum information, represented by the coefficients a_e, a_g, from the electronic excitation to the phonon bus. Since the phonon bus communicates with all of the ions, it is then possible to transfer this phononic quantum information to another ion. On this basis, a whole suite of *quantum gates* can be constructed, which transform the quantum state of a qubit in a way which depends on the quantum state of another qubit. This forms the basis of the *ion trap quantum computer* [1.9], which we will consider in detail in Chap. 24.

1.5 Distributing Quantum Information

It is clearly desirable to be able to transfer quantum information from one quantum device to another, although it must be emphasized that there are physical constraints on this which make the idea very different from the classical concept of information distribution. Quantum information is represented by the quantum state of a qubit, and there is no way of either measuring, or transferring an *unknown* quantum state without destroying some of the information in it. Thus, given a single copy of a quantum state in the form $\alpha|e\rangle + \beta|g\rangle$, there is no way of exactly determining both α and β. Given a sufficient number of copies, these coefficients can be determined as accurately as wished.

When we talk about quantum information transfer, we mean the transfer of a quantum state in the form $\alpha|e\rangle + \beta|g\rangle$ from one quantum device to another quantum device, where it may be subject to further processing. The idea is to connect one device to another by means of a direct quantum channel.

1.5.1 Quantum Networks

This leads to the concept of a *quantum network* with a set of *quantum processors*, communicating with each other using such a *quantum channel*. The principal components of a *quantum network* would be:

i) *The Quantum Processor*: The long-lived internal states of atoms and ions currently provide the best method of processing quantum information, although superconducting devices are also a viable possibility.

ii) *The Quantum Memory*: The long-lived internal states of atoms and ions currently provide the best method of storing quantum information. There are a number of options currently under experimental investigation for such quantum memories.

iii) *The Quantum Node*: It is convenient to use the terminology *quantum node* to describe any local device which contains quantum processors or quantum memory, and which can be coupled to a quantum channel by a quantum interface.

iv) *The Quantum Channel*: As noted above, the optical fibre provides the best candidate.

v) *The Quantum Interface*: The concept of a *quantum interface* represents not so much a device in itself, but a device and a choice of design of both a quantum node and a quantum channel by which an efficient transfer of quantum information can be effected.

1.5.2 Cavity QED Systems

Pellizzari, S A Gardiner, Cirac and *Zoller* [1.15] proposed a physically very different implementation of much the same Hamiltonian as for a trapped ion, which can be achieved by locating an atom inside a high finesse optical cavity, as illustrated in Fig. 1.6, and described in more detail by *Kimble* [1.16]. Here the role of the atom is the same as that of the ion in the previous section, and the cavity mode plays the role of the vibrational ion motion.

There are other ways of using both atoms (or ions) inside optical cavities, all of which provide ways of coupling the internal electronic state of an atom to a propagating optical field, producing what are known as "flying qubits", which can transfer information from one cavity to another.

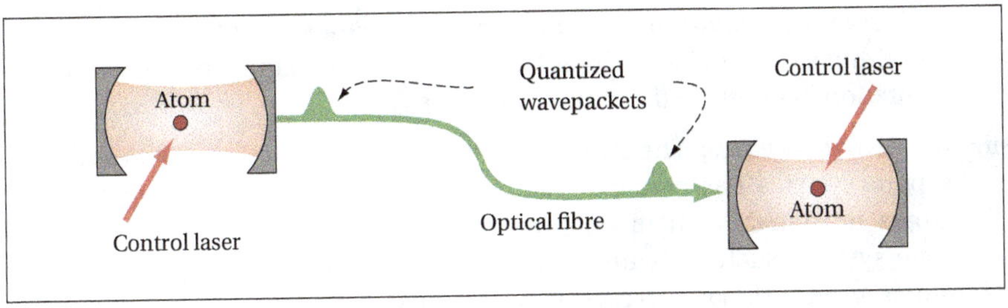

Fig. 1.6. Quantum processing of qubits can be achieved using an atom coupled to the quantized optical field in an optical cavity. Such cavities can be interfaced using optical fibres.

1.5.3 Superconducting Systems

A very promising alternative implementation of a qubit is provided by a Josephson junction operated in parameter regime where distinct resolvable energy levels exist. The nonlinearity of such a device guarantees the existence of levels which can be treated as two-level systems, thus providing qubits. The transition frequencies are in the microwave regime, so quantum processing can be provided by such systems of Josephson junctions, interacting with the microwave electromagnetic field, confined within a superconducting resonant cavity, or propagating through superconducting waveguides.

The result is called *circuit QED*, and it amounts to a microwave version of cavity QED. In Part VIII of this book we give a basic outline of the formalism required, including descriptions of both the formalism of superconductor theory and of the construction of quantum processor elements.

1.6 Technologies for Quantum Processing

There are a number of technology platforms currently under investigation, but in this book we consider in depth only the most well-developed systems, namely, those based on ion traps, those based on cavity QED and those based on superconducting systems. Among the other avenues of research, which we shall not consider in detail in this book are:

i) *Quantum Dot Systems*: *Loss* and *DiVincenzo* [1.17] proposed an implementation of a universal set of one- and two-quantum-bit gates for quantum computation using the spin states of coupled single-electron quantum dots. There are two principal kinds of quantum dot, and both have been used to construct quantum computing elements, in which the desired operations are effected by the gating of the tunnelling barrier between neighbouring dots.

   In *gate-defined quantum dots*, microscopic electrode structures are used to apply strong electric fields, producing a localized potential in which charge carriers can be trapped. For their use in quantum computing, see [1.18]

   Self-assembled quantum dots consist of small islands of semiconductor material within a matrix of a semiconductor with a larger bandgap—for the implementation of qubits in these structures, see [1.19].

ii) *Rare-Earth Dopants*: The implementation of quantum information in atomic vapour systems has certain disadvantages arising from the motion of the atoms, or from the difficulty of practical implementation compared to a solid state system, such as is universal in conventional electronics. Rare earth atoms, used as dopants in crystalline systems, provide a promising alternative approach. The rare earths differ from each other by the number of electrons in the 4f-shell, whose electrons are largely confined well within the overall charge cloud of the atom. The result is that these transitions are weakly coupled to the environment, and have very narrow linewidths. However, thermal effects are important, and experiments must be implemented at liquid helium temperatures.

A complicating feature is the existence of inhomogeneous broadening; the crystal substrate has different degrees of strain etc., so that the actual transition frequencies depend on the location of the dopant. This means that a sample of rare earth doped crystal has many different very narrow bandwidth atomic transitions at a range of frequencies.

To utilize these narrow linewidths requires appropriate strategies, such as *photon echoes*, in which an optical pulses is used to excite the dopants, which precess at their various frequencies. Another pulse reverses the motion, and the motion brings all the dopants back into phase, creating the echo of the original pulse. *Hedges, Longdell, Li* and *Sellars* [1.20] have used this technique to create a highly efficient quantum memory for light, and more recent work [1.21] has shown that a six hour coherence time can be achieved.

iii) *All-Optical Quantum Computing*: It was realized with some surprise by *Knill, Laflamme* and *Milburn* [1.22] that efficient quantum computation is possible using only beam splitters, phase shifters, single photon sources and photodetectors. Their methods exploited feedback from photodetectors and were robust against errors from photon loss and detector inefficiency. These methods have since been successfully implemented [1.23, 1.24], including an elementary implementation of Shor's factorization algorithm.

iv) *Nuclear Magnetic Resonance Systems*: Spins in the solid state can now be controlled to a degree unheard of until quite recently—single spins can be isolated, initialized, coherently manipulated and read out using both electrical and optical techniques. The application of these techniques to quantum computing shows significant promise [1.25].

1.7 Summary

In this chapter we have given a sketch of the ways in which quantum processing can be set up using quantum optical systems. The remainder of Part I provides a sketch of those results of quantum information theory which are most relevant to the construction of quantum devices.

However, the main tasks of the book are in Part II–Part IX, where we develop the theoretical concepts and techniques necessary to understand quantum-optical devices. The conceptual heart of the book is in Part III, which covers the formalism of quantum stochastic processes and Part V, which covers the theory of continuous measurement, such as is provided by a photodetector. In this book the quantum stochastic formalism is used consistently in every part. Once mastered, this formalism rapidly produces the correct theoretical description of any quantum optical system.

2. From Atoms to Qubits

When *Feynman* [2.1, 2.2] proposed the idea of a quantum computer, and a quantum simulator, he introduced the quantum version of a bit, which is now known as a *qubit.* This was chosen to be the quantum system most closely analogous to the classical concept of a bit; namely a system composed of two quantum states $|0\rangle$ and $|1\rangle$. The major difference from classical computing is the existence and possible usefulness of a quantum superposition of the form $\alpha|0\rangle + \beta|1\rangle$. The interpretation of the superposition state which results at the end of computational procedure also requires a well-defined quantum measurement procedure to generate the desired numerical results.

Quantum optics provides a natural way to implement and and manipulate a qubit, since the quantum-optical two-level system is a very well understood system, which can be manipulated using laser light. Some of the techniques have already been introduced in *Book I*, and we will give a more extensive treatment in this book. Using laser pulses of various kinds, one can transform any normalized superposition of atom-based qubits of the form $\alpha_1|0\rangle + \beta_1|1\rangle$ into any other normalized superposition $\alpha_2|0\rangle + \beta_2|1\rangle$. But more pertinently, we want to know how we can add, multiply, subtract and divide using qubits, and this means we need to design *gates*, namely, devices which will produce a qubit whose value depends on the the values of at least two qubits. In this part of this book we will outline the basic manipulations of qubits needed for quantum computation, and show how to implement them using the original proposal of *Cirac* and *Zoller* [2.3] based on cold trapped ions and laser light.

2.1 Representing and Processing Quantum Information

A computation can be considered as a physical process that transforms an *input* into an *output* according to some predetermined rules. The input and the output are both represented as patterns of bits in a storage device. In quantum computation, inputs and outputs are represented by quantum states of a system. For example, in enumerating the states of a given basis as

$$\{\,|1\rangle,\ |2\rangle,\ |3\rangle,\ |4\rangle, \ldots,\ |N\rangle\,\}\,, \tag{2.1.1}$$

the number n could be represented by the nth state of this basis.

2.1.1 Linear Unitary Evolution

If the system is isolated from the rest of the world, then its time evolution is governed by the Schrödinger equation with a Hamiltonian, which we wish to design in such a way as to generate the desired transformation of the input to the output. When implemented this way, the operation that transforms inputs into outputs is necessarily a unitary linear operator.

The requirement of unitary evolution gives rise to constraints on the possible transformations of input to output. For example, the operation that gives 1 if a number is odd and 2 if it is even, cannot be implemented unitarily in the obvious way

$$\left.\begin{array}{lll} |2n+1\rangle & \longrightarrow & |1\rangle, \\ |2n\rangle & \longrightarrow & |2\rangle, \end{array}\right\} \quad n \text{ integral.} \tag{2.1.2}$$

This operation is not unitary because:

i) It is obviously not reversible, since, for example, both $|1\rangle$ and $|3\rangle$ map to $|1\rangle$, whereas a unitary operation has an inverse, and is therefore reversible.

ii) It does not conserve the scalar product, e.g. $\langle 1|3\rangle = 0$ but the corresponding mapped states are identical, not orthogonal.

a) Use of an Auxiliary System: To solve this problem one can use an auxiliary system, consisting of two states, $|1\rangle$ and $|2\rangle$ so that the output is written in that system while keeping the unitarity of the operation. This is achieved by the unitary operation $\mathcal{U}$, defined by the mapping

$$\left.\begin{array}{lllll} |2n+1\rangle\otimes|1\rangle & \longrightarrow & \mathcal{U}\,|2n+1\rangle\otimes|1\rangle & \equiv & |2n+1\rangle\otimes|1\rangle, \\ |2n\rangle\otimes|1\rangle & \longrightarrow & \mathcal{U}\,|2n\rangle\otimes|1\rangle & \equiv & |2n\rangle\otimes|2\rangle, \\ |2n+1\rangle\otimes|2\rangle & \longrightarrow & \mathcal{U}\,|2n+1\rangle\otimes|2\rangle & \equiv & |2n+1\rangle\otimes|2\rangle, \\ |2n\rangle\otimes|2\rangle & \longrightarrow & \mathcal{U}\,|2n\rangle\otimes|2\rangle & \equiv & |2n\rangle\otimes|1\rangle. \end{array}\right\} \tag{2.1.3}$$

The requirement of unitarity forces one to specify the action of the operation on all states which can appear in either the input or the output, even if these are not used for encoding information. This leads to the inclusion of the mappings in the last two lines, both of which are necessary for the implementation of the mapping on all linear combinations of possible inputs.

This means that operations corresponding to the last two equations must be specified, even though initial states with any component of the auxiliary state $|2\rangle$ do not occur. It is straightforward to check that the operation defined by these equations is indeed unitary.

The *information* in the input is encoded in the subspace spanned by the states $|m\rangle\otimes|1\rangle$, and in contrast, no information is encoded in the subspace spanned by the states $|m\rangle\otimes|2\rangle$. This means that the most general input state is any state in this subspace, and this can be written as the linear combination

$$|\text{in}\rangle \equiv \sum_n \big(a_n|2n+1\rangle + b_n|2n\rangle\big)\otimes|1\rangle. \tag{2.1.4}$$

Under the operation (2.1.3), the input is transformed so that

$$|\text{in}\rangle \longrightarrow |\text{out}\rangle \equiv \mathcal{U}\,|\text{in}\rangle = \sum_n \left(a_n|2n+1\rangle \otimes |1\rangle + b_n|2n\rangle \otimes |2\rangle\right). \tag{2.1.5}$$

b) Evaluation of an Arbitrary Function: In general, if our algorithm consists of evaluating a given function f, we will wish to design a system with the following properties:

i) The system is spanned by states of the form $|n\rangle \otimes |m\rangle$.

ii) These states are simultaneous eigenstates of certain quantum mechanical operators which we shall call $\mathbb{N}$ and $\mathbb{M}$; that is

$$\mathbb{N}|n\rangle \otimes |m\rangle = n|n\rangle \otimes |m\rangle; \qquad \mathbb{M}|n\rangle \otimes |m\rangle = m|n\rangle \otimes |m\rangle. \tag{2.1.6}$$

iii) It will be essential to be able to measure $\mathbb{N}$ and $\mathbb{M}$ using some suitable apparatus. Usually it will be necessary that this be a *projective measurement* (see *Bk. I: Sect.18.3*), meaning that if the measured value of $\mathbb{M}$ is $\bar{m}$, then the measurement process yields

$$\sum_{n,m} C_{n,m}|n\rangle \otimes |m\rangle \longrightarrow \mathcal{N}_{\bar{m}}\left(\sum_n C_{n,\bar{m}}|n\rangle\right) \otimes |\bar{m}\rangle, \tag{2.1.7}$$

where $\mathcal{N}_{\bar{m}}$ is an appropriate normalization.

iv) We can implement an interaction Hamiltonian such that the evolution operator $\mathcal{U}_f$ transforms the input states according to the following table:

$$\left.\begin{aligned} \mathcal{U}_f|1\rangle \otimes |1\rangle &= |1\rangle \otimes |f(1)\rangle, \\ \mathcal{U}_f|2\rangle \otimes |1\rangle &= |2\rangle \otimes |f(2)\rangle, \\ \vdots \quad &\quad \vdots \quad \vdots \\ \mathcal{U}_f|N\rangle \otimes |1\rangle &= |N\rangle \otimes |f(N)\rangle. \end{aligned}\right\} \tag{2.1.8}$$

In this more general case the auxiliary space is at least N-dimensional, and not two-dimensional as in the previous example. As in that case, it is necessary to implement the transformation on all possible states in order to satisfy unitarity. If we can find a Hamiltonian which is defined on all states of the form $|n\rangle \otimes |m\rangle$, and which yields the result (2.1.8), then the full unitarity requirement will be guaranteed.

Thus, if we can devise an appropriate Hamiltonian to generate the required mapping (2.1.8), we can certainly do the same computations with quantum computers as with classical computers. All that is necessary is to start with a definite input state $|k\rangle \otimes |1\rangle$, evolve this state using $\mathcal{U}_f$ to get $|k\rangle \otimes |f(k)\rangle$, and then measure the values of $\mathbb{N}$ and $\mathbb{M}$, which will be desired results $k, f(k)$. The algorithm will be probabilistic, since we cannot be certain of exactly what value of k will be returned by the measurement, but for any such measured k, the correct value of $f(k)$ is returned.

2.1.2 Quantum Parallelism and Projective Measurements

However, the implementation of the computation by means of a linear operator means that we can use the quantum-mechanical superposition principle. Combining this with the concept of a projective measurement yields algorithms for which there is no classical analogue.

To show how we can do this, consider the following example.

i) Prepare the input state in a known superposition of all allowable input basis states, which we will write in the form

$$|\text{in}\rangle = \sum_{k=1}^{N} \alpha_k |k\rangle \otimes |1\rangle. \tag{2.1.9}$$

Here α_k form a known (both in phase and amplitude) set of normalized complex coefficients.

ii) Now perform one run of the quantum computer, so that

$$|\text{in}\rangle \longrightarrow |\text{out}\rangle = \mathcal{U}_f\, |\text{in}\rangle = \sum_{k=1}^{N} \alpha_k |k\rangle \otimes |f(k)\rangle. \tag{2.1.10}$$

All the possible values of $f(k)$ are present in this output superposition, and the coefficients α_k are preserved in both phase and amplitude.

iii) Now perform a *projective measurement* of $\mathbb{M}$, and suppose this yields the result $\bar{m}$, which is therefore the measured value of $f(k)$. Since we are measuring only $\mathbb{M}$, we do not determine the value of k apart from the knowledge we gain from the fact that

$$f(k) = \bar{m}. \tag{2.1.11}$$

If this equation has a unique solution $k_{\bar{m}}$ for k, the state after this measurement is simply $|k_{\bar{m}}\rangle \otimes |\bar{m}\rangle$, and the situation is much the same as if we had not started with a superposition state.

However, when there is *not* a unique solution, quantum mechanics continues to play a role. If we use the notation $\mathbb{S}(\bar{m})$ to be the set of all values of k such that $f(k) = \bar{m}$, then after measurement the state of the register is proportional to

$$\left(\sum_{k \in \mathbb{S}(\bar{m})} \alpha_k |k\rangle \right) \otimes |\bar{m}\rangle. \tag{2.1.12}$$

The major point here is that *both the phase and the amplitude information in the superposition have been preserved during this process.*

iv) The most interesting essentially quantum mechanical application of this result happens when the function $f(k)$ is *periodic*, that is when there is a number r such that $f(k+r) = f(k)$ for all k. If we make a projective measurement of $f(k)$ and get the value $f(\bar{k})$, then the resulting state of the system is proportional to

$$\left(\sum_{l\in\Lambda} \alpha_{\bar{k}+lr}|\bar{k}+lr\rangle\right) \otimes |f(\bar{k})\rangle, \tag{2.1.13}$$

where Λ is the set of all l such that $1 \leqslant \bar{k}+lr \leqslant N$.

Thus we obtain a quantum-mechanical superposition of periodically related states after the measurement, and this is a key property used in *Shor's* algorithm [2.4] for efficient factorization—for a tutorial exposition of the algorithm, see [2.5]. The algorithm relies on the determination of this kind of periodicity for certain functions of integers, which we shall discuss in Sect. 2.4.2.

2.2 The DiVincenzo Criteria for a Practical Quantum Computer

The model of a quantum computer we have introduced involves the availability of sets of distinct quantum states $|k\rangle$, but does not specify them in detail. The model of a quantum computer we shall describe formulates such a set more precisely as a *quantum register*, which is analogous to the registers used in classical computers, constructed from individual binary bits, which can take on two values, 0 and 1. These values can be manipulated and measured by the computer mechanism under the control of a computer program.

Correspondingly, a quantum computer needs a quantum register, built out of appropriate *quantum bits*, usually known as *qubits*, which can be manipulated and measured in a controlled way. In 2000 *DiVincenzo* [2.6] formulated a set of criteria for the implementation of a quantum computer, and these criteria are still considered as fundamental to quantum computing.

a) A Set of Qubits: These are the two-level systems which form the quantum register. They must be well-characterized, and it must be possible to create quantum superpositions of different states of an individual qubit. We want the quantum state of the qubits to be kept as pure as possible, since otherwise the power of the superpositions would not be effective. This means that the qubits must be well isolated from the environment in such a way that the process of decoherence is sufficiently slow.

We denote by $\{|0\rangle_k, |1\rangle_k\}$ two orthogonal states of the kth qubit. The quantum register consists of N qubits, and has basis states of the kind

$$|k_1, k_2, \dots k_N\rangle \equiv |k_1\rangle_1 \otimes |k_2\rangle_2 \otimes \dots \otimes |k_N\rangle_N. \tag{2.2.1}$$

A general quantum state of the quantum register can therefore be written

$$|\Psi\rangle = \sum_{k_1,k_2,\dots k_N=0}^{1} c_{k_1 k_2 \dots k_N} |k_1, k_2, \dots k_N\rangle. \tag{2.2.2}$$

We will normally use the notation $|k_1, k_2, \dots k_N\rangle$ instead of the cumbersome notation that uses tensor products, but when the tensor product notation makes for more clarity, we will use the full notation.

b) Universal Sets of Quantum Gates: The controlled manipulation of the qubits means that we can perform any unitary operation $\mathcal{U}$ on the qubits so that $|\Psi\rangle \rightarrow \mathcal{U}|\Psi\rangle$. In principle, if we want to perform general operations we should be able to engineer arbitrary interactions between the qubits. It was shown by *Deutsch* in 1989 [2.7] that any quantum computational network can be decomposed into quantum logic gates, which is similar to the situation for classical computers. Deutsch defined a *universal quantum gate* as a gate that could be used to simulate any quantum logic gate.

For classical reversible computation, the simplest universal gate has three input bits and three output bits, but for quantum computation, the situation is in fact simpler. It was shown by *Sleator* and *Weinfurter* in 1995 [2.8] that any $\mathcal{U}$ can be decomposed into a product of universal gates, and that the *universal set of gates* can be realized by using only gates with either one input and one output qubit, or two input and two output qubits. The practical task of implementing a universal set of gates is thus enormously simplified.

This means that if we are able to perform the one-qubit gates of this set on any single qubit, and the two-qubit gates on any pair of qubits, then we will be able to perform any computation (that is, any unitary operation on the register) by simply applying a sequence of appropriate gates. In Sect. 2.3 we give a detailed description of quantum gates and how to construct universal sets.

c) Measurement: One should be able to measure σ_z on each of the qubits (or, equivalently, to detect whether they are in state $|0\rangle$ or $|1\rangle$). This process necessarily requires an *irreversible* interaction with a measurement apparatus.

d) Erasure: We must be able to prepare the initial state of the system, for example the state $|0,0,\ldots,0\rangle$, and this necessarily means that we must be able to erase any information in the system. This requirement is in fact already implied by the ability to detect the state of the system. We simply detect the state of the system, and this puts all of the qubits into either the state $|0\rangle$, or the state $|1\rangle$. We then apply the single-qubit gate $\mathcal{U}_x^{(1)}(\pi)$ on all qubits with value $|1\rangle$, setting them equal to $|0\rangle$.

e) Scalability: The difficulty of performing gates, measurements, etc., should not grow exponentially as the number of qubits increases. If this is not possible, the gain in speed by using the quantum algorithms will not eventuate.

2.3 Universal Sets of Quantum Gates

There are many sets of universal gates, and of course, they are all equivalent. The most convenient set is the one that contains one two-qubit gate (an operation acting on two qubits only) plus a set of single-qubit gates. In this section we will describe some of the most important gates.

2.3.1 One-Qubit Gates

The full set of unitary transformations in two dimensions can be expressed in terms of the transformations generated by the Pauli matrices, as discussed in *Bk. I: Sect.10.7.1*. Thus, we can write any 2 × 2 unitary matrix as

$$U(\Phi, \varphi, \boldsymbol{n}) = \mathrm{e}^{\mathrm{i}\Phi} \exp(\mathrm{i}\varphi \boldsymbol{n} \cdot \boldsymbol{\sigma}). \tag{2.3.1}$$

However, we can express the second factor in terms of rotations about only two axes (conventionally the z-axis and the y-axis) using Euler angles. The overall phase transformation is irrelevant quantum-mechanically, and can be omitted.

a) Fundamental One-Qubit Gates: We are therefore able to express all one-qubit gates in terms of the two gates:

i) *Phase Gate*: $\mathcal{U}_z^{(1)}(\varphi) = \mathrm{e}^{-\mathrm{i}\varphi\sigma_z}$,

$$\left.\begin{aligned} |0\rangle &\to \mathrm{e}^{\mathrm{i}\varphi}|0\rangle, \\ |1\rangle &\to \mathrm{e}^{-\mathrm{i}\varphi}|1\rangle. \end{aligned}\right\} \tag{2.3.2}$$

ii) *Excitation Gate*: $\mathcal{U}_x^{(1)}(\theta) = \mathrm{e}^{-\mathrm{i}\theta\sigma_x}$,

$$\left.\begin{aligned} |0\rangle &\to \cos\theta\,|0\rangle - \mathrm{i}\sin\theta\,|1\rangle, \\ |1\rangle &\to -\mathrm{i}\sin\theta\,|0\rangle + \cos\theta\,|1\rangle. \end{aligned}\right\} \tag{2.3.3}$$

b) Particular One-Qubit Gates: In practice there are three particular gates which are commonly used:

i) *Hadamard Gate*:

$$H \equiv \frac{1}{\sqrt{2}}\begin{pmatrix} 1 & 1 \\ 1 & -1 \end{pmatrix} = \exp\left(-\mathrm{i}\frac{\pi}{4}\sigma_y\right). \tag{2.3.4}$$

ii) *S-Gate*:

$$S \equiv \begin{pmatrix} 1 & 0 \\ 0 & \mathrm{i} \end{pmatrix} = \mathrm{e}^{\mathrm{i}\pi/4}\exp\left(-\mathrm{i}\frac{\pi}{4}\sigma_z\right). \tag{2.3.5}$$

iii) *T-Gate*:

$$T \equiv \begin{pmatrix} 1 & 0 \\ 0 & \mathrm{e}^{\mathrm{i}\pi/4} \end{pmatrix} = \mathrm{e}^{\mathrm{i}\pi/8}\exp\left(-\mathrm{i}\frac{\pi}{8}\sigma_z\right). \tag{2.3.6}$$

2.3.2 Two-Qubit Gates

These act on pairs of qubits. The action is asymmetric, and involves a *control qubit*, whose state is not changed by the gate, and a *target qubit*, whose state is changed by the gate, in a way which depends on the state of the control qubit.

Examples of two-qubit gates are:

a) Controlled-Not Gate: $\mathcal{U}^{(2)}_{\text{CNOT}} = |0\rangle\langle 0| \otimes \mathbb{1} + |1\rangle\langle 1| \otimes \sigma_x$

$$\left.\begin{array}{lcl} |0,0\rangle & \to & |0,0\rangle, \\ |0,1\rangle & \to & |0,1\rangle, \\ |1,0\rangle & \to & |1,1\rangle, \\ |1,1\rangle & \to & |1,0\rangle. \end{array}\right\} \tag{2.3.7}$$

This gate changes the state of the second qubit conditioned on the state of the first qubit. The first qubit is therefore the *control qubit*, and the second is the *target qubit.*

b) Controlled-Phase Gate: $\mathcal{U}^{(2)}_{\pi} = |0\rangle\langle 0| \otimes \mathbb{1} - |1\rangle\langle 1| \otimes \sigma_z$

$$\left.\begin{array}{lcl} |0,0\rangle & \to & |0,0\rangle, \\ |0,1\rangle & \to & |0,1\rangle, \\ |1,0\rangle & \to & |1,0\rangle, \\ |1,1\rangle & \to & -|1,1\rangle. \end{array}\right\} \tag{2.3.8}$$

This differs from the identity operation only by the sign change in the last line; thus it changes the phase of the state conditioned on both of the initial qubits being in the state 1.

c) General Formulation: These two gates can both be written in the form of a generic two-qubit gate

$$\mathcal{U}^{(2)}(\boldsymbol{n}) \equiv |0\rangle\langle 0| \otimes \mathbb{1} + |1\rangle\langle 1| \otimes \boldsymbol{n}\cdot\boldsymbol{\sigma}, \tag{2.3.9}$$

where $\boldsymbol{n}$ is a three-dimensional unit vector. If $\boldsymbol{n}_1$ and $\boldsymbol{n}_2$ are two unit vectors related by the rotation R_{12} thus

$$\boldsymbol{n}_2 = R_{12}\boldsymbol{n}_1, \tag{2.3.10}$$

then there is a single-qubit gate $\mathcal{U}^{(1)}(R_{12})$ such that

$$\mathcal{U}^{(1)}(R_{12})\boldsymbol{n}_2\cdot\boldsymbol{\sigma}\,\mathcal{U}^{(1)}(R_{12})^{\dagger} = (R_{12}\boldsymbol{n}_1)\cdot\boldsymbol{\sigma}. \tag{2.3.11}$$

This means that the two gates characterized by $\boldsymbol{n}_1$ and $\boldsymbol{n}_2$ are related by

$$\mathcal{U}^{(2)}(\boldsymbol{n}_2) = \left(\mathbb{1} \otimes \mathcal{U}^{(2)}(R_{12})\right)\mathcal{U}^{(2)}(\boldsymbol{n}_1)\left(\mathbb{1} \otimes \mathcal{U}^{(1)}(R_{12})\right)^{\dagger}. \tag{2.3.12}$$

Since $\boldsymbol{n}_{\text{CNOT}} = (1,0,0)$ and $\boldsymbol{n}_{\pi} = (0,0,-1)$, a rotation of $-\pi/2$ about the y-axis will transform the CNOT gate to the CPHASE gate.

2.3.3 Two Different Universal Sets of Gates

These are [2.9, 2.10]:

$$\left.\begin{array}{lll} S_1 = \mathcal{U}^{(2)}_{\text{CNOT}}, & \mathcal{U}^{(1)}_x(\pi/4), & \mathcal{U}^{(1)}_z(\varphi), \\ S_2 = \mathcal{U}^{(2)}_{\pi}, & \mathcal{U}^{(1)}_z(\pi/4), & \mathcal{U}^{(1)}_x(\varphi), \end{array}\right\} \quad \varphi \in [0, 2\pi). \tag{2.3.13}$$

The two-qubit gates require different qubits to be brought into interaction with each other, and consequently are in practice more difficult to implement than single qubit gates. For example, the requirement that the operations be unitary, and therefore reversible, means that uncontrolled interaction with any part of the quantum computer other than the two qubits involved must be avoided.

2.4 Quantum Algorithms

The quantum gates of the previous section can be used to perform a wide range of computations using quantum parallelism, as we shall shortly discuss in Sect. 2.4.1, and these algorithms are *efficient*, meaning that the time taken to execute an algorithm does not grow too rapidly with the size of the problem—we define the concept precisely in (2.4.1).

However, the full power of quantum computation comes about by the use of quantum measurement theory, in the way we have discussed in Sect. 2.1.2. We shall show in Sect. 2.4.2 that an algorithm involving a quantum measurement can be used as an efficient method of determining periodicities, a capability which can be used to give an efficient algorithm for the factorization of large integers. This would yield a method for the efficient decoding of much of the encrypted information in use today for secure communication on the internet.

2.4.1 Efficient Arithmetic Operations on a Quantum Computer

The actual use of a quantum computer requires that a practical suite of algorithms for performing elementary arithmetic and certain other operations be available. Furthermore, these must be efficient algorithms.

The "size" of the problem can be specified by the number of bits (or qubits) required to specify the input; thus a number N requires $\log_2 N$ bits for its specification. The efficiency of the algorithm is measured by the number of steps that are required for its execution; this should be bounded by a polynomial in $\log_2 N$ as N becomes large. Precisely:

An efficient algorithm is one in which the number of steps n_{steps} for input N satisfies the bound

$$n_{steps} \leqslant p(\log N) \text{ for all } N, \tag{2.4.1}$$

where $p(x)$ is a polynomial in x.

Efficient quantum algorithms have been formulated [2.5,2.11] for all of the following procedures.

a) Plain Adder: The addition of two registers $|a\rangle$ and $|b\rangle$ can be written in the form given in (2.1.8), in which the result of the computation is written into one of the input registers

$$|a, b\rangle \to |a, a+b\rangle. \tag{2.4.2}$$

As one can reconstruct the input (a,b) out of the output $(a,a+b)$, there is no loss of information, and the calculation can be implemented reversibly. To prevent overflows, the second register (initially loaded in state $|b\rangle$) should be sufficiently large, *i.e.* if both a and b are encoded on n qubits, the second register should be of size $n+1$.

b) Adder Modulo N: This is similar to the plain adder; we write

$$|a,b\rangle \to |a, a+b \bmod N\rangle, \tag{2.4.3}$$

where $0 \leqslant a,b < N$. As in the case of the plain adder, there is no a priori violation of unitarity since the input (a,b) can be reconstructed from the output $(a, a+b \bmod N)$, when $0 \leqslant a,b < N$.

c) Controlled Multiplier Modulo N: The function $f_{a,N}(x) = ax \bmod N$ can be implemented in the form

$$|x,0\rangle \to |x, ax \bmod N\rangle, \tag{2.4.4}$$

and this algorithm is of the form (2.1.8).

The controlled multiplier effects this operation conditionally upon the value of some external qubit $|c\rangle$. This means we need to implement

$$|c;x,0\rangle \to \begin{cases} |c;x,ax \bmod N\rangle, & \text{if} \quad c=1, \\ |c;x,x\rangle, & \text{if} \quad c=0. \end{cases} \tag{2.4.5}$$

d) The Discrete Exponential Modulo N: Using the above gates, one can implement $f_{a,N}(x) = a^x \bmod N$ in the form

$$|x,0\rangle \to |x, a^x \bmod N\rangle. \tag{2.4.6}$$

This function is central to Shor's algorithm.

e) Discrete Fourier Transform Modulo N: This is the quantum analogue of a fast Fourier transform, and is defined with respect to the register $|0\rangle, |1\rangle, |2\rangle, \ldots, |N-1\rangle$ as

$$|x\rangle \to \mathrm{DFT}_N|x\rangle = \frac{1}{\sqrt{N}} \sum_{c=0}^{N-1} \mathrm{e}^{2\pi \mathrm{i} xc/N} |c\rangle. \tag{2.4.7}$$

2.4.2 Determination of Periodicities Using Quantum Measurements

The factorization algorithm introduced by *Shor* [2.4] depends on the close connection between factorization and the determination of the periodicity of the discrete exponential function function introduced in Sect. 2.4.1d. More precisely, if we know the periodicity of $a^x \bmod N$ as a function of x for some appropriately chosen values of a, we can efficiently determine the factors of N. This is a mathematical result, and is explained very concisely and in the context of quantum computation by *Ekert* and *Josza* [2.5].

a) Physics of the Algorithm: This comes about from quantum measurement theory, in the form which we have already mentioned in Sect. 2.1.2. For this case we proceed (following the exposition in [2.5]) thus:

i) To start with, we need a quantum register of length L, such that $N^2 < 2^L < 2N^2$.

ii) Prepare the quantum register in the state $|0\rangle$; that is, put all L qubits into the state $|0\rangle$.

iii) Let $q = 2^L$, and apply the discrete Fourier transform modulo q to the register, giving

$$|0\rangle \to \mathrm{DFT}_q|0\rangle = \frac{1}{\sqrt{q}}\sum_{a=0}^{q-1}|a\rangle. \tag{2.4.8}$$

iv) Now choose a random number y and compute $y^a \bmod N$, storing the result in a second, auxiliary, register, giving

$$\frac{1}{\sqrt{q}}\sum_{a=0}^{q-1}|a\rangle\,|y^a \bmod N\rangle. \tag{2.4.9}$$

v) We want to determine the periodicity of $y^a \bmod N$ as a function of a. Suppose the period is r. Do a projective measurement on the second register, getting a result z, which we can write as $z = y^l \bmod N$, where l is the *minimum* value of l giving this value z. Supposing the period is r, then for all j, the states $|y^{l+jr} \bmod N\rangle$ *are all the same,* so that after the measurement the state of the first register is

$$|\varphi_l\rangle = \frac{1}{\sqrt{A}}\sum_{j=0}^{A}|l+jr\rangle, \tag{2.4.10}$$

where A is the number of states in the sum, that is, the largest value of j such that $l + jr$ is less than the register size $q = 2^l$.

vi) The basic procedure to follow now is clear. A superposition of the kind we have produced is very common in physics; it is very like the signal from a diffraction grating. There, the periodic nature of the waveform produced by the grating can be seen in the fact that at large distances we see propagation in definite directions, and these directions are determined by the periodicity. To measure these directions is quite straightforward physically, yielding the periodicity of the grating.

The distinct propagation directions, mathematically regarded, arise from a Fourier transform. Here we can use the discrete Fourier transform on the state (2.4.10) to determine the periodicities. The only snag is that the state (2.4.10) is not quite a periodic superposition. If we were lucky, and we found that $(A+1)r + l = l \bmod q$, then the states could be regarded as cyclically periodic, and applying DFT_q would yield a state proportional to $|r\rangle$. Of course we are not normally lucky in this sense, but in general we have a periodic superposition of a large number of states, and the DFT_q can determine the periodicity sufficiently accurately for us to distinguish between a value r and the neighbouring values $r \pm 1$.

The situation is analogous to that found in spectroscopy, where a very accurate determination of wavelength is possible from a finite wavetrain of sufficient length.

b) Summary: The details of the process are complicated, and require a probabilistic estimation of the result, as detailed in [2.5]. But the essence of the power of the method of quantum parallelism lies in the fact that a quantum superposition is characterized by amplitudes (which yield probabilities) as well as phases. The power of the algorithm for finding a periodicity we have just outlined is directly related to the existence of the phases, and their preservation in projective measurements.

2.5 The Universal Quantum Simulator

When real physical systems evolve over time, the governing quantum mechanical laws evolve a quantum state of enormously high dimension with fidelity and rapidity. Nevertheless, even for a relatively small system, the number of Hilbert space dimensions can be so huge that there is no hope of containing it within any conventional computer. For example, a system consisting of 100 qubits requires a Hilbert space of dimension d of $2^{100} \approx 10^{30}$, which would require a computer memory of about 10^{20} Gb. Since Avogadro's number is only about 6×10^{23}, getting sufficient memory for this problem would technically very challenging. Moreover, to predict the value of a physical quantity, we would have to add, multiply or otherwise combine all of those coefficients, something that will require a time that scales exponentially with d as well.

2.5.1 Efficient Quantum Simulation

Of course quantum mechanics does things differently—in fact all that is needed to simulate such a system is the 100 qubit register, which can be created with 100 two-level atoms, and a suitable way of emulating the Hamiltonian. This idea was first proposed by Feynman [2.1] as a possible way of overcoming the clear impossibility of using classical computational methods. If we had a system at our disposal, which we could manipulate at will, then we would be able to engineer the desired interactions between those particles, and thus predict the values of physical quantities by simply performing the appropriate measurements on our system.

That this is able to be done efficiently was demonstrated by *Lloyd* [2.12] in 1996. The key to this result is that *physics is local*. It turns out that efficient simulation can only be achieved if the interactions between the qubits are almost local; for example, if $|i\rangle$ and $|j\rangle$ are two qubits, then the Hamiltonian $\mathcal{H}$ must only have nonvanishing matrix elements $\langle i|\mathcal{H}|j\rangle$ when $i = j$ or when i and j are adjacent (or almost adjacent). Since this appears to be the case for all known theoretical descriptions of physics, this restriction is not a problem—it is perhaps even a welcome restriction.

The most general kind of problem we would want to consider would involve the time evolution generated by time-dependent $\mathcal{H}(t)$. This describes interactions between the particles on a lattice, which we represent by a quantum register. We specify the initial state $|\Psi,0\rangle$ and a final time t. The goal is to determine certain physical properties of the final state $|\Psi,t\rangle$. Lloyd's work demonstrated that evolving in small time steps would allow efficient simulation of any many-body quantum Hamiltonian containing local or almost local few-particle interactions; that is, the overall time needed for the simulation would be polynomial bounded in the number of particles, and thus would not grow exponentially with the number of particles.

a) Local Hamiltonians: The main idea is to write the Hamiltonian as a sum of terms, each of which can be efficiently exponentiated. For example we might write a Hamiltonian which is the sum of a "potential term" $\mathcal{V}$ and a "kinetic term" $\mathcal{T}$ in the form

$$\mathcal{H} = \hbar(\mathcal{V}+\mathcal{T}), \tag{2.5.1}$$

where the operators $\mathcal{T}$ and $\mathcal{V}$ have matrix elements between the qubits $|i,\alpha\rangle\ |j,\beta\rangle$ (where α,β take on the qubit values $0,1$, and i describes the location of the qubit in the register) of the form

$$\langle i,\alpha|\mathcal{V}|j,\beta\rangle = \delta_{i,j}\,(\alpha|V_i|\beta), \tag{2.5.2}$$

$$\langle i,\alpha|\mathcal{T}|j,\beta\rangle = \left(\delta_{i,j+1}+\delta_{i,j-1}-2\delta_{i,j}\right)(\alpha|T_i|\beta). \tag{2.5.3}$$

A Hamiltonian of this kind, with interactions extending no further than to neighbours, is said to be a *local Hamiltonian.* The exponentiation of $\mathcal{V}$ reduces to the direct product of the exponentials of 2×2 matrices, while we can use the discrete Fourier transform to diagonalize $\mathcal{T}$ in i,j, so that its exponentiation also reduces to the direct product of the exponentials of 2×2 matrices. As Lloyd pointed out, these are efficient exponentiation algorithms.

b) The Lie–Trotter–Kato Product Formula: This formula provides the algorithm for exponentiating the full Hamiltonian; it is given by

$$\mathrm{e}^{\mathrm{i}(\mathcal{V}+\mathcal{T})t} = \lim_{n\to\infty}\left(\mathrm{e}^{\mathrm{i}\mathcal{V}t/n}\mathrm{e}^{\mathrm{i}\mathcal{T}t/n}\right)^n. \tag{2.5.4}$$

This algorithm was invented in the 19th century by *Lie* for the case when the system has finite dimensions, and thus $\mathcal{V}$ and $\mathcal{T}$ are both finite matrices, as is the case here, and is therefore more correctly known as the *Lie product formula.* The generalization to the case that $\mathcal{V}$ and $\mathcal{T}$ are more general operators in Hilbert space (or other mathematical spaces) was made by *Trotter* [2.13] and *Kato* [2.14], and this generalization is known as the Trotter–Kato formula.

Lloyd was able to show that, for local interactions, the Lie–Trotter–Kato formula provides an accurate efficient algorithm for exponentiating the full Hamiltonian, and thus that a quantum simulator could indeed be constructed.

2.5.2 The Tasks for a Quantum Simulator

Feynman envisioned a quantum simulator to be "a quantum machine that could imitate any quantum system, including the physical world" [2.1]. Such a general device would be as difficult to build as a quantum computer. However, if we are more modest and demand only that our simulator imitate certain physically interesting systems, a quantum simulator might be easier to construct than a quantum computer. It would nevertheless still be an important and useful device.

Normally, when we do simulations, we are not interested in the faithful representation of a complete many-body wavefunction. Rather, the main interest lies in certain physical properties, typically related to intensive quantities—such as densities, magnetization per lattice site, or few-body correlations—which in turn determine the phase diagram.

Moreover, a quantum simulator would be expected to be more robust against imperfections than a quantum computer. Roughly speaking, if one of the lattice sites undergoes an undesired perturbation, this may affect expectation values of intensive quantities by only a few per cent, as long as the lattice is sufficiently large. In fact, in many real materials that we would like to imitate with the simulator, such imperfections exist naturally, without destroying the main physical properties of interest.

3. Quantum Information

In the previous chapter we have formulated an abstract model of a quantum computer, and in this model the central object is the *quantum register*, composed of a set of qubits. The quantum register thus constructed obeys the laws of quantum mechanics, and with these come the counterintuitive predictions of quantum mechanics, such as the EPR paradox, and the Schrödinger cat paradox, which we have already mentioned in *Bk. I: Sect.1.2.* Most of the counterintuitive predictions of quantum mechanics are related to the superposition principle, and the basic ingredient involved in such paradoxes is that of *entanglement*, which we have already introduced in *Book I*.

However, apart from its fundamental interest, entanglement plays a central role in the field of quantum information. The importance of entanglement comes directly from the way in which a quantum computer works. The quantum gates introduced in the previous chapter inevitably produce entangled states from any initial state. For example the state (2.4.9), produced by the periodicity algorithm, entangles the two quantum registers, and the production of the periodic superposition state (2.4.10) in the first register depends on this entanglement. *Entanglement is absolutely central to the working of the algorithm.*

Kimble [3.1] has suggested a *quantum internet*, in which quantum information is transmitted all around the world—a very ambitious concept. But even the issue of transmitting the contents of quantum registers within a single device of realistic power raises the question of the fidelity of the transfer of the information. Since entangled states play such an important role in quantum algorithms, preservation of entanglement will be vital when information transmission is required. The characterization of entanglement is one of the central theoretical issues in quantum information theory, since without reliable ways of quantifying entanglement, we cannot assess the quality of any protocol for the transmission of quantum information.

In fact, there are still many open questions regarding the entanglement properties of quantum systems. *Bipartite entanglement*—that is, the entanglement of the states of two systems—is well understood when we are dealing with pure states. *Multipartite entanglement* occurs when there are more than two systems whose states are entangled, and is much more difficult to quantify satisfactorily. The situation becomes even more difficult if the state of the system is mixed, which will inevitably happen as the result of interactions with the environment, or because of imperfections in the experimental apparatus. In that case a density operator description is necessary, and there is in general no straightforward way of determining whether two systems are entangled or not.

In this chapter we will first review the description of entanglement, both for pure and mixed states and will give some of the known criteria to determine whether a mixed state is entangled or not. We will then introduce the idea of *teleportation*, in which an arbitrary quantum state $|S\rangle$ can be transferred from place A to place B without it being necessary to know what state is represented by $|S\rangle$. This leads to the concept of *quantum communication*, where quantum information contained in qubits is transferred from point to point, but is never measured at any stage in the process.

Quantum information is not truly digital information—it is necessary to consider and be able to preserve all quantum superposition states, for example $\cos\theta|0\rangle + \sin\theta e^{i\phi}|1\rangle$. But this means that the information which must be preserved is specified by the two real numbers θ and ϕ. If the information is contained in a mixed state, these parameters then become random variables—this is equivalent to adding noise to these variables. This is really analogue information, whose accurate transmission even classically is not nearly as reliable as that used for digital transmission. The situation for quantum information is analogous, but more difficult. Classically, to get good transmission of an analogue signal we must:

i) Make the input signal large.

ii) Ensure that sources of noise during the transmission are as small as possible.

iii) Provide accurate low-noise amplification during the transmission. In this way, the ratio of the signal to the noise added by the transmission process is maximized.

Quantum mechanically, the situation is very much more delicate. We cannot amplify a state such as $|I\rangle \equiv a|0\rangle + b|1\rangle$, in fact the *quantum no-cloning theorem* states that we cannot even make a single copy of the state $|I\rangle$ unless we destroy $|I\rangle$. However, we can prepare in advance many copies of the same state by repeating the procedure to produce the first copy—this kind of procedure is done all the time in a quantum optics laboratory when we prepare an atom in a definite quantum state. Procedures to encode quantum information using this kind of system and transmit it reliably—known as *purification protocols*—have been developed, and these allow, in principle, quantum communication over arbitrarily long distances.

3.1 Entanglement

Entanglement is a property of composite systems, such as, for example, two subsystems A and B whose states belong to two Hilbert spaces, H_A and H_B, respectively. It is usual to think of these systems as atoms located at different places, for example, two two-level atoms located at positions $\boldsymbol{x}$ and $\boldsymbol{y}$. But one could also consider a single atom, in which system A is described by the spin angular momentum wavefunction and system B is described by the orbital angular momentum wavefunction.

In this section we will give only the most elementary description of entanglement. For a deeper understanding, the reader is referred to classic work of *Nielsen* and *Chuang* [3.2], and to the review by the four *Horodeckis* [3.3].

3.1.1 Entanglement of Pure States

Let us consider states of the composite system in which one subsystem is in a certain state $|i\rangle_A$, and the other in $|j\rangle_B$. One denotes those states as $|i\rangle_A \otimes |j\rangle_B \in H = H_A \otimes H_B$, or simply $|i, j\rangle \in H$.

The central issue in entanglement is the superposition principle—*any* superposition of these states, such as $|\Psi\rangle \equiv \alpha|0,0\rangle + \beta|1,1\rangle \in H$, is a valid quantum state, and thus represents a physically realizable situation. However, a state of this form cannot be described as the direct product of state for system A and a state for system B. That is, when neither α nor β is zero, there exists no pair of vectors

$$|\phi_1, A\rangle \in H_A, \quad |\phi_2, B\rangle \in H_B, \tag{3.1.1}$$

such that

$$|\Psi\rangle = |\phi_1, A\rangle |\phi_2, B\rangle. \tag{3.1.2}$$

Definition of an Entangled State: An entangled state is defined as a state which cannot be written in the product form (3.1.2).

3.1.2 Bipartite Entangled States

In the following, we will use the notation

$$\left.\begin{array}{ll} |k\rangle_A\,, & k = 0, 1, 2, \ldots, d_A\,, \\ |k\rangle_B\,, & k = 0, 1, 2, \ldots, d_B\,, \end{array}\right\} \tag{3.1.3}$$

for an orthonormal basis in H_A and H_B respectively.

a) Entangled States and Correlations: A pure state in which there are correlations between the observable in the two systems is necessarily a correlated state—that is, if F_A and F_B are observables in the two Hilbert spaces, then for all choices of F_A and F_B it will be true that

$$\langle\Psi|F_A \otimes F_B|\Psi\rangle = \langle\phi_1, A|F_A|\phi_1, A\rangle\langle\phi_2, B|F_B|\phi_2, B\rangle, \tag{3.1.4}$$

$$= \langle\Psi|F_A|\Psi\rangle\,\langle\Psi|F_B|\Psi\rangle, \tag{3.1.5}$$

only if $|\Psi\rangle$ has a factorized form such as (3.1.2). Thus, for an entangled state, one can always choose a pair of operators F_A and F_B for which the factorized form (3.1.4) is not true. Such a state is therefore correlated.

b) Classical Correlations: The existence of correlations, by itself, is not a property of entangled states. However, classical correlations always arise from mixed states, which can only be described using a density operator. Such classical correlations are restricted by the Bell inequalities [3.4], whereas correlations arising from entangled states may violate these inequalities.

c) Interactions are Essential for Entanglement: To produce an entangled state from a product state, interactions between the two systems are essential. Without interactions, the Hamiltonian describing the evolution of systems A and B can be written as $H = H_A \otimes \mathbb{1}_B + \mathbb{1}_A \otimes H_B$, leading to an evolution operator of the form $U(t) = U_A(t) \otimes U_B(t)$, which clearly transforms a product state into another product state.

Operators, such as this evolution operator, which factorize into the product of operators for each subsystem, are known as *local operators*. Entanglement cannot be created using only local operators.

3.1.3 Bipartite Entanglement of Qubits

Although the definitions and results apply for general dimensions, the simplest entangled system is that of two sets of qubits, that is, two systems where $d_A = d_B = 2$. In that case we will write the basis states as $|0\rangle$ and $|1\rangle$.

a) Notation for the Pauli Matrices in Quantum Information: It is very convenient to use a notation for the Pauli matrices (see *Bk. I: Sect.10.7*) which does not involve subscripts—the notation conventionally chosen is

$$\left.\begin{aligned} X &\equiv \sigma_x = |1\rangle\langle 0| + |0\rangle\langle 1| = \begin{pmatrix} 0 & 1 \\ 1 & 0 \end{pmatrix}, \\ Y &\equiv \sigma_y = -\mathrm{i}\big(|1\rangle\langle 0| - |0\rangle\langle 1|\big) = \begin{pmatrix} 0 & -\mathrm{i} \\ \mathrm{i} & 0 \end{pmatrix}, \\ Z &\equiv \sigma_z = |1\rangle\langle 1| - |0\rangle\langle 0| = \begin{pmatrix} 1 & 0 \\ 0 & -1 \end{pmatrix}, \\ \boldsymbol{X} &\equiv (X, Y, Z). \end{aligned}\right\} \tag{3.1.6}$$

b) Bell States: The entangled states introduced by *Bell* [3.4] in his analysis of the EPR paradox play a very important role in quantum information—they are defined by

$$|\Psi^\pm\rangle = \frac{|0,1\rangle \pm |1,0\rangle}{\sqrt{2}}, \qquad |\Phi^\pm\rangle = \frac{|0,0\rangle \pm |1,1\rangle}{\sqrt{2}}. \tag{3.1.7}$$

The Bell states have the following properties:

i) The four Bell states form a complete orthonormal set, and all of them are what one might call "fully entangled", meaning that the non-zero contributions from factorized components are equal.

ii) Each of the Bell states can be transformed into any other Bell state by a unitary local operation. For example,

$$X_1|\Psi^\pm\rangle = -|\Phi^\pm\rangle, \qquad Z_2|\Psi^\pm\rangle = |\Psi^\mp\rangle, \tag{3.1.8}$$

and the Pauli matrix properties ensure that X_1 and Z_2 etc., are unitary matrices.

iii) The Bell states can be characterized as eigenstates of the two commuting Hermitian operators

$$\mathcal{Q} \equiv Z_1 Z_2, \qquad \mathcal{S} = X_1 X_2. \tag{3.1.9}$$

The eigenvalue equations are

$$\mathcal{Q}|\Phi^\pm\rangle = |\Phi^\pm\rangle, \qquad \mathcal{Q}|\Psi^\pm\rangle = -|\Psi^\pm\rangle, \tag{3.1.10}$$

$$\mathcal{S}|\Phi^\pm\rangle = \pm|\Phi^\pm\rangle, \qquad \mathcal{S}|\Psi^\pm\rangle = \pm|\Psi^\pm\rangle. \tag{3.1.11}$$

Hence, when we measure $\mathcal{Q}$,we identify which *kind* of Bell state (Ψ or Φ), and when we measure $\mathcal{S}$, we identify which of the signs $\pm$ the state has.

iv) A unitary transformation which will generate the Bell states is

$$\mathcal{U}_{\text{Bell}} = \exp\left(\mathrm{i}(Z_1+1)\pi/4\right)\frac{1+\mathrm{i}\mathcal{S}}{\sqrt{2}}, \tag{3.1.12}$$

namely

$$\mathcal{U}_{\text{Bell}}\begin{pmatrix}|0,1\rangle\\|1,0\rangle\\|0,0\rangle\\|1,1\rangle\end{pmatrix} = \begin{pmatrix}|\Psi^-\rangle\\\mathrm{i}|\Psi^+\rangle\\|\Phi^-\rangle\\\mathrm{i}|\Phi^+\rangle\end{pmatrix}. \tag{3.1.13}$$

This is a product of a local operation, and a term involving interactions arising from $\mathcal{S}$, which by its definition (3.1.9) is an interaction between the two systems.

Exercise 3.1 Correlations in Bell States: Using the notation of (3.1.6), choose $F_A = \boldsymbol{X}_A\cdot\boldsymbol{n}$ and $F_B = \boldsymbol{X}_B\cdot\boldsymbol{n}$, where $\boldsymbol{n}$ is a unit vector. For any $\boldsymbol{n}$ show that

$$\langle\Psi^-|\boldsymbol{X}_A\cdot\boldsymbol{n}\,\boldsymbol{X}_B\cdot\boldsymbol{n}|\Psi^-\rangle = -1, \tag{3.1.14}$$

$$\langle\Psi^-|\boldsymbol{X}_A\cdot\boldsymbol{n}|\Psi^-\rangle = \langle\Psi^-|\boldsymbol{X}_B\cdot\boldsymbol{n}|\Psi^-\rangle = 0, \tag{3.1.15}$$

so that this Bell state is correlated.
Note: The state $|\Psi^-\rangle$ is a state of zero total pseudospin, and is thus rotationally invariant. Hence $\boldsymbol{n}$ can be chosen in any convenient direction for the evaluation of the means.

Exercise 3.2 Bell Operator Properties: Check the properties of the Bell operators $\mathcal{Q}$ and $\mathcal{S}$ and the Bell states given in iii) above.

Exercise 3.3 Generating Bell States: Check the truth of (3.1.13). Note that the second factor generates an entangled state with an imaginary coefficient, while the first factor adjusts the the relative phase of the two components.

Modify the transformation to avoid the imaginary coefficients in (3.1.13).

3.1.4 The Schmidt Decomposition

Let us show now how we can determine whether a state is entangled, and how to give a quantitative measure of the amount of entanglement. We consider a state of the form

$$|\Psi\rangle = \sum_{i=1}^{d_A}\sum_{j=1}^{d_B} c_{i,j}|i\rangle_A \otimes |j\rangle_B. \tag{3.1.16}$$

Here the states $|i, j\rangle$ are an arbitrary orthonormal basis, and in this basis, all information about the state is determined by the coefficients $c_{i,j}$, which form a $d_A \times d_B$ matrix, which we shall call C. This matrix determines how the states in the two different spaces are connected to each other in the state $|\Psi\rangle$ in the product space.

a) Singular Value Decomposition of a Matrix: This theorem states that for any $m \times n$ matrix C, we can write

$$C = UDV, \tag{3.1.17}$$

in which

$$\left.\begin{array}{ll} U & \text{is an } m \times m \text{ unitary matrix,} \\ V & \text{is an } n \times n \text{ unitary matrix,} \\ D & \text{is an } m \times n \text{ diagonal matrix with positive non-zero elements } d_k, \\ \multicolumn{2}{l}{\text{The non-zero eigenvalues of } C^\dagger C \text{ and } CC^\dagger \text{ are given by } d_k^2.} \end{array}\right\} \tag{3.1.18}$$

The decomposition is unique apart from:

i) The ordering of the diagonal elements d_k of D is arbitrary.

ii) If there are sets of identical diagonal elements, there is an ambiguity related to the possibility of arbitrary unitary transformations in the corresponding subspace.

b) Application to the Schmidt Decomposition: Using the singular value decomposition, we can write

$$|\Psi\rangle = \sum_{k=1}^{d} d_k|u_k\rangle_A \otimes |v_k\rangle_B, \tag{3.1.19}$$

in which:

- $d_k \geqslant 0$ are the non-zero diagonal elements of D, (3.1.20)
- $d \leqslant \min(d_A, d_B)$ is the number of non-zero d_k , (3.1.21)
- because $\langle\Psi|\Psi\rangle = 1$, it follows that $\sum_{k=1}^{d} d_k^2 = 1$, (3.1.22)
- $|u_k\rangle_A = \sum_{i=1}^{d_A} U_{i,k}|i\rangle_A$, (3.1.23)
- $|v_k\rangle_B = \sum_{j=1}^{d_B} V_{k,j}|j\rangle_B$, (3.1.24)

Both $|u_k\rangle_A$ and $|v_k\rangle_B$ form an orthonormal basis in their respective spaces. The d_k are called *Schmidt coefficients* and the bases $|u_k\rangle_A$ and $|v_k\rangle_B$ are called *Schmidt bases.*

c) Reduced Density Operators of the Subsystems: The Schmidt decomposition describes the entanglement relationship between the two spaces with great clarity. For example:

i) Measurements performed on the subsystems are described by the reduced density operators

$$\rho_A \equiv \mathrm{Tr}_B\{|\Psi\rangle\langle\Psi|\} = \sum_{k=1}^{d} d_k^2 |u_k\rangle\langle u_k|, \tag{3.1.25}$$

$$\rho_B \equiv \mathrm{Tr}_A\{|\Psi\rangle\langle\Psi|\} = \sum_{k=1}^{d} d_k^2 |v_k\rangle\langle v_k|. \tag{3.1.26}$$

These are essentially isomorphic density operators, differing only by the choice of orthogonal basis.

ii) The Schmidt coefficients can be found by diagonalizing either reduced density operator, but to find the corresponding bases one must diagonalize both reduced density operators. Thus, if the two reduced density operators do arise from a pure state $|\Psi\rangle$, then they contain all the information needed to construct $|\Psi\rangle$.

iii) A product state $|\phi_1\rangle_A \otimes |\phi_2\rangle_B$ is already written in the Schmidt basis form, with only one $d_k = 1$. Conversely, if we have a state with only one Schmidt coefficient then it must be a product state. Equivalently, $|\Psi\rangle$ is a product state if and only if the corresponding reduced density operators correspond to pure states.

iv) This means that if $|\Psi\rangle$ is an entangled state, the corresponding reduced density operators must correspond to mixed states, or, equivalently, that there must be more than one non-zero Schmidt coefficient.

d) Entropy of Entanglement: Thus, we see that the entanglement of a state is directly related to the mixedness of the reduced density operators. This is intuitively clear since, as we mentioned above, entangled states give rise to correlations and if we only observe one of the systems we lose information about these correlations which results in the fact that we will effectively have a mixed state. This suggests that we can measure the degree of entanglement by the degree of mixedness of the reduced density operators. There are several measures of mixedness of density operators; perhaps the most popular one is the *von Neumann entropy*

$$S(\rho) = -\mathrm{Tr}\{\rho \log_2 \rho\}. \tag{3.1.27}$$

For a pure state this entropy is zero, whereas for a maximally mixed state (described by the identity operator, properly normalized) it gives $\log_2 d$, where d is the dimension of the Hilbert space.

The von Neumann entropy is a convex function of ρ, that is, for $p \in [0,1]$,

$$S(p\rho_1 + (1-p)\rho_2) \geqslant pS(\rho_1) + (1-p)S(\rho_2). \tag{3.1.28}$$

This means that the entropy increases when two states are mixed, corresponding to a loss of information.

This motivates the following definition: Given a state $|\Psi\rangle$, we define the *entropy of entanglement*, $E(\Psi)$ as the von Neumann entropy of the reduced density operator [3.5]. Thus, we have

$$E(\Psi) = S(\rho_A) = S(\rho_B) = -\sum_{k=1}^{d} d_k^2 \log_2(d_k^2). \tag{3.1.29}$$

States $|\Psi\rangle \in H_{AB} = C^d \otimes C^d$ for which $E(\Psi) = \log_2(d)$ are called *maximally entangled states* in d dimensions.

The entropy of entanglement depends only on the Schmidt coefficients, but not on the corresponding basis. This means that it is invariant under local unitary operations. That is, if $|\Psi'\rangle = (U_A \otimes U_B)|\Psi\rangle$, then $E(\Psi') = E(\Psi)$.

On the other hand, one can show that the entropy of entanglement cannot increase in average by local operations [3.6]. That is, if we perform independent measurements in A and B and obtain the state $|\Psi_k\rangle$ after the measurement with probability p_k, then

$$E(\Psi) \geqslant \sum_k p_k E(\Psi_k). \tag{3.1.30}$$

Note, however, that the previous inequality does not imply that none of the $E(\Psi_k)$ can be larger than $E(\Psi)$, or even the maximum allowed $\log_2 d$.

3.1.5 Multipartite Entanglement

The quantification of multipartite entanglement is a very complex issue, which we will only briefly mention here—for a comprehensive treatment see [3.3,3.7].

a) Entanglement Partitions: When we consider several systems $A_1, A_2, \ldots, A_N$, entanglement can be exist within any or all of the possible partitions of the individual systems. For example, we can have a state of the form $|\Psi\rangle = |\phi_1\rangle_{A_1A_3} \otimes |\phi_2\rangle_{A_2A_5A_6} \otimes |\phi_3\rangle_{A_4}$, where none of $|\phi_1\rangle$, $|\phi_2\rangle$ or $|\phi_3\rangle$ can be written as product states. In such a state the parties A_1 and A_3 are entangled with each other, but not with the others; similarly, the parties A_2, A_5 and A_6 are entangled with each other, and the party A_4 is completely disentangled.

To give a description of entanglement, we must partition the system, for example, we can consider the partition $(A_1A_3)(A_4)(A_2A_5A_6)$. We can classify the entangled states according to the different partitions. That is, a state is entangled according to some partition if it can be written as a product state of the corresponding disjoint elements of the groups, but not within each of the groups. In order to determine the partition corresponding to a particular state we can calculate all possible reduced density operators and determine whether they correspond to mixed states or not.

b) Tripartite Entanglement: *Dür*, *Vidal* and *Cirac* [3.8] showed that there are exactly two kinds of tripartite entanglement, based on the states

$$|\mathrm{GHZ}\rangle = \frac{1}{\sqrt{2}}\big(|0,0,0\rangle + |1,1,1\rangle\big), \tag{3.1.31}$$

$$|\mathrm{W}\rangle = \frac{1}{\sqrt{3}}\big(|0,0,1\rangle + |0,1,0\rangle + |1,0,0\rangle\big). \tag{3.1.32}$$

They were able to show that:

i) All tripartite states can be converted, using only local operations and classical communication, with non-zero probability, into either the W state or the GHZ state.

ii) There is no way, using only local operations and classical communication, to convert a W state into a GHZ state.

Thus these two states define two equivalence classes of tripartite entanglement. The state $|\mathrm{GHZ}\rangle$ had already been introduced by *Greenberger, Horne* and *Zeilinger* [3.9] (after whom it is named) as a remarkable extension of the Bell states (3.1.7), whose properties could be shown to be definitively incompatible with classical physics.

3.1.6 Decoherence, Mixed States and Entanglement

The states that we have considered in the previous section are idealized, but in reality, all systems interact with some sort of environment, and undergo a process of *decoherence*, such as we have already treated in *Bk. I: Sect.19.4.* There it was seen that the process of decoherence is intimately connected with the entanglement of the system and the environment—in fact, the interaction between system and environment will almost always generate entanglement even if the initial state is a product state:

$$|\Psi_{\mathrm{Sys}}, 0\rangle_{\mathrm{Sys}} \otimes |\Psi_{\mathrm{Env}}, 0\rangle_{\mathrm{Env}} \to |\Psi, t\rangle_{\mathrm{Sys,\,Env}}. \tag{3.1.33}$$

For the system itself, the only information that we can acquire without performing measurements in the environment is contained in the reduced density operator

$$\rho_{\mathrm{Sys}}(t) = \mathrm{Tr}_E\left\{|\Psi, t\rangle_{\mathrm{Sys,\,Env}}\langle\Psi, t|_{\mathrm{Sys,\,Env}}\right\}, \tag{3.1.34}$$

which will correspond to a mixed state.

a) Defining Entanglement for Mixed States: We define an entangled mixed state as follows.

i) A density operator ρ is *separable* if and only if there exist $p_k \geqslant 0$, a set of $|a_k\rangle\} \in H_A$ and a set of $|b_k\rangle \in H_B$ such that

$$\rho = \sum_k p_k |a_k, b_k\rangle\langle a_k, b_k|. \tag{3.1.35}$$

ii) A density operator is is *entangled* when it is not separable.

b) Interpretation of the Definition: This definition of entanglement is equivalent to the physical requirement that a state is called entangled if it cannot be prepared by local operations (and classical communication) out of a product state.

Let us give some examples for the case of two qubits:

i) The state described by $\rho = |0,0\rangle\langle 0,0|$ is separable since it is already a product state.

ii) The state

$$\rho = \tfrac{1}{2}\left(|0,0\rangle\langle 0,0| + |1,1\rangle\langle 1,1|\right), \tag{3.1.36}$$

is separable since it can be locally prepared as follows. We choose either 0 or 1 randomly, and with equal probability. If the result is 0, we prepare both A and B in state $|0\rangle$, if the result is 1 we prepare them in state $|1\rangle$. Obviously, in this way we do not need any interaction between the systems. We just need classical communication between the location of A and B so that the corresponding preparers can agree on the state they prepare.

iii) In general a density operator can be always be written in the form

$$\rho = \sum_k p_k |\phi_k\rangle\langle\phi_k|, \tag{3.1.37}$$

where the p_k are positive and add up to one. One particular decomposition of this form is the spectral decomposition, in which, additionally, the vectors $|\phi_k\rangle$ form an orthonormal basis. In general, except for pure states (rank one density operators) there are infinitely many decompositions. Note that a decomposition like (3.1.37) tells us one way of creating a state described by ρ. We simply have to prepare the system in state $|\phi_k\rangle$ with probability p_k. The fact that there exist infinitely many decompositions of a state means that it can be prepared in infinitely many different forms.

iv) Any density operator of the form $\rho = \rho_A \otimes \rho_B$ is separable since the states ρ_A and ρ_B can be prepared locally out of the state $|0\rangle$. To see this, if we write ρ as in (3.1.37) we can simply transform the state $|0\rangle$ into $|\phi_k\rangle$ with probability p_k.

c) The Peres–Horodecki Criterion: It is very hard in practice to determine whether a given state is entangled or not. However, the *Peres–Horodecki* criterion [3.10, 3.11] provides a test which is valid for bipartite entanglement between states in two Hilbert spaces $\mathcal{H}_A$ and $\mathcal{H}_B$.

To state the criterion, we need to define the *partial transpose* ρ^{T_A} of a bipartite density operator ρ with respect to the first system in the basis:

$$\langle k,q|\rho^{T_A}|k',q'\rangle \equiv \langle k',q|\rho|k,q'\rangle. \tag{3.1.38}$$

The *Peres–Horodecki* criterion states: *If $d_A \times d_B \leqslant 6$ then ρ is entangled if and only if ρ^{T_A} has a negative eigenvalue.*

For higher dimensions, the criterion is a sufficient condition; if it is satisfied, the density operator is entangled. However, there are density operators not satisfying the criterion which are nevertheless entangled.

d) Fidelity: It is important to have a measure of how well a state created in the laboratory, and corresponding to the density operator ρ, represents a given pure state $|\Phi\rangle$. The *fidelity* is defined to achieve this aim, and is defined by

$$F = \langle\Phi|\rho|\Phi\rangle \leqslant 1. \tag{3.1.39}$$

Notice that:

i) $F = 1$ if and only if $\rho = |\Phi\rangle\langle\Phi|$.

ii) The fidelity is a valid measure even if ρ corresponds to a pure state, i.e. if $\rho = |\Psi\rangle\langle\Psi|$, in which case $F = |\langle\Phi|\Psi\rangle|^2$.

iii) For a completely random state in a Hilbert space of dimension d, the density operator would be $\rho = 1/d$, in which case the fidelity is $F = \langle\Psi|\Psi\rangle/d = 1/d$.

Experimentally, one aims for $F \approx 1$, and currently fidelities of the order of 70% to 99% are achievable, depending on the system and the desired state—see, for example [3.12].

3.2 Transferring Quantum Information

When we transfer classical information, we normally copy the information at the same time, so that both the sender and the recipient have a copy of the same information. This process also allows one to check that the information has been transferred correctly. Quantum information transfer is subtler, and is fundamentally limited by the no-cloning theorem, which forbids copying. However, provided the sender does not require a copy, quantum information can be transferred. One method is by *teleportation*, which we will deal with in Sect. 3.2.2, and reasonably straightforward protocols using optical fibres, as discussed in Part IX, are also feasible.

3.2.1 The No-Cloning Theorem

The no-cloning theorem was first proved by *Wooters* and *Zurek* [3.13], and has been significantly generalized by *Barnum, Caves, Fuchs, Josza* and *Schumacher* [3.14]. We will give here a proof of the simplest version of the theorem.

To clone a quantum state would be a physical process, so it must be achieved as a result unitary time development operation. The simplest model is to implement the transformation

$$|\psi\rangle \otimes |e\rangle \longrightarrow \mathcal{U}|\psi\rangle \otimes |e\rangle = |\psi\rangle \otimes |\psi\rangle. \tag{3.2.1}$$

By this equation, we mean that we take an *arbitrary* quantum state $|\psi\rangle$ within a subspace, and map a copy of $|\psi\rangle$ into an auxiliary subspace, at the same time preserving the original state. The state $|e\rangle$ is an assumed initial setting of the state in the auxiliary subspace.

The defect in this idea is fundamental; $\mathcal{U}$ is a linear transformation, but this mapping is not linear. If the initial state is of the form $\alpha|a\rangle+\beta|b\rangle$, then the cloning should be of the form

$$\big(\alpha|a\rangle+\beta|b\rangle\big)\otimes|e\rangle \longrightarrow \big(\alpha|a\rangle+\beta|b\rangle\big)\otimes\big(\alpha|a\rangle+\beta|b\rangle\big). \tag{3.2.2}$$

On the other hand, since $\mathcal{U}$ is a linear unitary operator, we should get

$$\mathcal{U}\big\{(\alpha|a\rangle+\beta|b\rangle)\otimes|e\rangle\big\} = \alpha|a\rangle\otimes|a\rangle+\beta|b\rangle\otimes|b\rangle. \tag{3.2.3}$$

Thus it is not possible to execute a cloning of this kind. The more general results of [3.14] confirm the no-cloning theorem for any conceivable model.

3.2.2 Teleportation

By teleportation we define the transfer of an intact quantum state from one place to another, by a sender who knows neither the state to be teleported nor the location of the intended receiver [3.15]. Consider two partners, Alice and Bob, located at different places. Alice has a qubit in an unknown state $|\phi\rangle$, and she wants to teleport it to Bob, whose location is not known. Prior to the teleportation process, Alice and Bob share two qubits in a Bell state

$$|\Psi^-\rangle = \frac{|0,1\rangle-|1,0\rangle}{\sqrt{2}}\,. \tag{3.2.4}$$

The idea is that Alice performs a joint measurement of the two-level system to be teleported and her particle. Due to the nonlocal correlations contained in the Bell state, the effect of the measurement is that the unknown state appears instantaneously in Bob's hands, except for a unitary operation which depends on the outcome of the measurement. If Alice communicates to Bob the result of her measurement, then Bob can perform that operation and therefore recover the unknown state (for experiments, see [3.16–3.18]).

a) The Procedure: Let us call particle 1 that which has the unknown state $|\phi\rangle_1$, particle 2 the member of the EPR that Alice possesses and particle 3 that of Bob. We write the state of particle 1 as $|\phi\rangle_1 = a|0\rangle_1+b|1\rangle_1$ where a and b are (unknown) complex coefficients. The state of particles 2 and 3 is the Bell state $|\Psi^-\rangle$. The complete state of particles 1, 2 and 3 is therefore

$$|\Psi\rangle_{123} = \frac{a}{\sqrt{2}}(|0,0,1\rangle-|0,1,0\rangle)+\frac{b}{\sqrt{2}}(|1,0,1\rangle-|1,1,0\rangle). \tag{3.2.5}$$

In order to teleport the state, Alice and Bob follow this procedure:

i) *Alice Makes her Measurement*: Alice makes a joint measurement of her particles (1 and 2) in the Bell basis

$$|\Psi^\pm\rangle = \frac{1}{\sqrt{2}}(|0,1\rangle\pm|1,0\rangle) \tag{3.2.6}$$

$$|\Phi^\pm\rangle = \frac{1}{\sqrt{2}}(|0,0\rangle\pm|1,1\rangle) \tag{3.2.7}$$

ii) *Alice Broadcasts her Result*: Then she broadcasts (classically) the outcome of her measurement. That is, she has to send Bob two bits of classical information which indicate the outcome of the measurement.

iii) *Bob Restores his State*: Bob then applies a unitary operation to his particle to obtain $|\phi\rangle_3$.

b) Outcomes of the Procedure: According to the state of the particles the possible outcomes are:

i) With probability 1/4, Alice finds $|\Psi^-\rangle_{12}$. The state of the third particle is automatically projected onto $a|0\rangle_3 + b|1\rangle_3$. Thus, in this case Bob does not have to perform any operation.

ii) With probability 1/4, Alice finds $|\Psi^+\rangle_{12}$. The state of the third particle is automatically projected onto $-a|0\rangle_3 + b|1\rangle_3$. Teleportation occurs if Bob applies σ_z to his particle.

iii) With probability 1/4, Alice finds $|\Phi^-\rangle_{12}$. The state of the third particle is automatically projected onto $a|1\rangle_3 + b|0\rangle_3$. Teleportation occurs if Bob applies σ_x to his particle.

iv) With probability 1/4, Alice finds $|\Phi^+\rangle_{12}$. The state of the third particle is automatically projected onto $a|1\rangle_3 - b|0\rangle_3$. Teleportation occurs if Bob applies σ_z to his particle.

c) Remarks: Notice that:

i) Alice ends up with no information about her original state so that no violation of the no-cloning theorem occurs. In this sense, the state of particle 1 has been transferred to particle 3.

ii) On the other hand, there is no instantaneous propagation of information. Bob has to wait until he receives the (classical) message from Alice with her outcome. Before he receives the message, his lack of knowledge prevents him from having the state.

iii) No measurement can tell Bob whether Alice has performed her measurement or not.

iv) Since teleportation is a linear operation applied to a state, it will also work for statistical mixtures, or in the case in which particle 1 is entangled with other particles.

II Coherent Optical Manipulation of Atoms

Part II: Coherent Optical Manipulation of Atoms

The universal tool which underpins all of quantum optics is the laser, whose development into a reliable source of controlled, intense and coherent light promoted optical science into one of the key enabling technologies of modern society. The kind of physics we want to study in this book is dominated by techniques of the manipulating quantum states—mainly, but not exclusively, states of atoms—by the careful application of precisely designed pulses of electromagnetic radiation. In practice, laser light, microwaves and radio-frequency fields are all used, often in conjunction with each other.

Mostly, we are interested in pulses which are almost resonant with particular transitions in atoms, and the kinds of systems which can be studied experimentally are often those for which suitable lasers are readily available. This nevertheless supplies a very wide variety of possible systems.

The strategy in this part of the book will be to present the techniques of manipulation in their most elementary form. By this we mean that we consider idealized systems, in which the following simplifications are made.

i) *The Two-Level Atom Approximation*: Nature does not present us with two-level atoms, but rather with real atoms, with very large numbers of energy levels. However, in Chap. 4 we will show that when a laser field is very close to resonance with a transition between two energy levels, the system behaves as a two-level atom with corrections arising from the effects of the other energy levels, which can be calculated perturbatively.

ii) *The Electromagnetic Field is Coherent and Intense*: Under these conditions, quantum effects and especially quantum fluctuations can be neglected. The optical field is therefore modelled as a classical field—that is, the kind of field produced by a highly stabilized laser. The task of Chap. 5 is to show how to describe the motion of a two-level atom driven by such a field.

iii) *Atomic Recoil is Neglected*: The atom is considered to be so heavy that it may be treated as being fixed in space. This means that we do not include any kinetic term in the atomic Hamiltonian.

These restrictions will be lifted as we proceed further into the book; thus in Chap. 6 we will consider three-level systems, which play a very important part in quantum devices, and in Chap. 7 we will include atomic motion, and therefore allow the electromagnetic field to transfer momentum to the atom. Then in Part III and in all of the remainder of the book we will introduce the effects of quantization of the electromagnetic field.

4. The Two-Level System Approximation

The model of quantum devices presented in Chap. 2 relies on the availability of two-level systems and of techniques to manipulate them and thus produce quantum gates. The basic ideas have already been presented in *Bk. I: Sect. 12.3.* The aim of this chapter is to show how it is possible to use real atoms, and nevertheless restrict our description to only two levels. Such a restriction is appropriate when the atom interacts with a classical driving field which is almost resonant with a particular transition. In this chapter we will consider more carefully how the two-level description arises from a full multi-level atom, and what allowance has to be made for the non-resonant levels. The work of this chapter draws on the pioneering monograph on the two-level atom and its application in quantum optics by *Allen* and *Eberly* [4.1].

4.1 An Atom Driven by a Classical Driving Field

This chapter will build on the more elementary development given in *Bk. I: Ch. 12,* in which we presented the basic ideas of the interaction between an atom and a quantized light field, and established the notation we shall use. Here we want to consider the effect of a *classical driving field*, which can be achieved by taking the quantized driving field to be in a coherent state—experimentally, this corresponds to the use of a highly coherent laser input. The correct way of describing this was first introduced by *Mollow* [4.2], and in Sect. 9.5 we give our own treatment, based on the quantum stochastic Schrödinger equation. There, we show that the coherent state driving field is equivalent to a using a quantized vacuum driving field, and modifying the Hamiltonian by including a classical driving field. If the classical driving field is sufficiently strong, the quantized vacuum driving field can be neglected.

Let us consider the interaction of a classical optical field with an atom, with centre of mass considered to be located at the point $\boldsymbol{X} = 0$. The electric field generated by a laser will have optical frequency ω and polarization vector $\boldsymbol{\epsilon}$, modulated by a very much slower time-dependence. More precisely, at the position $\boldsymbol{X} = 0$ of the atom, this field has the form

$$\boldsymbol{E}_{\rm cl}(0,t) = \boldsymbol{\epsilon}\mathcal{E}(t)e^{-i\omega t} + \boldsymbol{\epsilon}^*\mathcal{E}^*(t)e^{i\omega t} \equiv \boldsymbol{E}_{\rm cl}^{(+)}(0,t) + \boldsymbol{E}_{\rm cl}^{(-)}(0,t). \tag{4.1.1}$$

Here $\mathcal{E}(t)$ represents the modulation, slowly varying on the optical time scale, so that $|\dot{\mathcal{E}}(t)| \ll \omega|\mathcal{E}(t)|$.

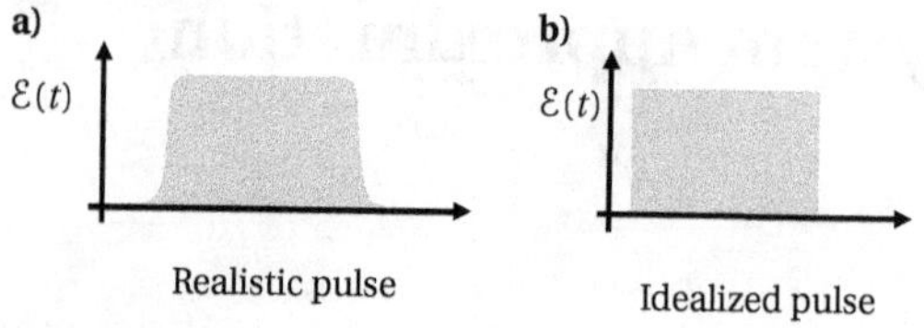

Fig. 4.1. Comparison of realistic and idealized pulses for manipulation the state of a two-level system.

The most important special case of a pulse shape is the *rectangular pulse*

$$\mathcal{E}(t) = \begin{cases} \mathcal{E}, & \text{for } 0 \leqslant t \leqslant T, \\ 0, & \text{otherwise.} \end{cases} \tag{4.1.2}$$

Here the step function must be understood as being "slow" on the optical time scale that is, the step function is an idealization of the turn on and turn off of an optical field, both of which take place over many optical cycles.

4.1.1 Schrödinger Equation for a Multilevel Atom

The relevant Hamiltonian we choose will assume the electric dipole approximation, and is adapted from that given in *Bk. I: (12.1.17)–(12.1.21)*. However, since the electromagnetic field is to be treated as an imposed classical field, we can omit the field Hamiltonian term $H_{\rm EM}$, so that if the Hamiltonian of the atom is $H_{\rm Atom}$, and $\boldsymbol{d} \equiv e\boldsymbol{x}$ is the operator for the electric dipole, the Schrödinger equation takes the form

$$i\hbar \frac{d}{dt}|\psi(t)\rangle = (H_{\rm Atom} - \boldsymbol{d}\cdot\boldsymbol{E}_{\rm cl}(0,t))\,|\psi(t)\rangle. \tag{4.1.3}$$

We introduce the eigenstates and eigenvectors of the atomic Hamiltonian, which satisfy

$$H_{\rm Atom}|n\rangle = \hbar\omega_n|n\rangle, \tag{4.1.4}$$

and we can decompose the interaction in terms of these as

$$\boldsymbol{d}\cdot\boldsymbol{E}_{\rm cl}(0,t) = \sum_{n,k} \boldsymbol{d}_{nk}\cdot\boldsymbol{E}_{\rm cl}^{(+)}(0,t)|n\rangle\langle k| + \sum_{n,k} \boldsymbol{d}_{nk}\cdot\boldsymbol{E}_{\rm cl}^{(-)}(0,t)|n\rangle\langle k|. \tag{4.1.5}$$

where $\boldsymbol{d}_{nk} = \langle n|\boldsymbol{d}|k\rangle = e\langle n|\boldsymbol{x}|k\rangle$ are the matrix elements of the electric dipole operator.

This interaction induces mixing between all pairs of states $|k\rangle$, $|n\rangle$, and these can be interpreted as *transitions* such as $|k\rangle \to |n\rangle$ when frequency ω of the driving field matches the transition frequency $\omega_{n,k} \equiv \omega_n - \omega_k$; that is:

i) *Absorption*: When $\omega_{n,k} = \omega$ the term $\boldsymbol{d}_{nk}\cdot\boldsymbol{E}_{\rm cl}^{(+)}(0,t)|n\rangle\langle k|$ in the interaction Hamiltonian is associated with the absorption of a unit of energy $\hbar\omega$ from the field;

ii) *Emission*: Similarly, when $\omega_{n,k} = -\omega$ the term $\boldsymbol{d}_{nk}\cdot\boldsymbol{E}_{\rm cl}^{(-)}(0,t)|n\rangle\langle k|$ in the interaction Hamiltonian is associated with the emission of a unit of energy ω into the field.

4.1.2 Non-Resonant Driving and the AC Stark Shift

When no transition is on resonance, energy is neither absorbed from nor emitted into the field, and the net effect is a shift in the energy eigenvalues of the states $|n\rangle$, known as the *dynamic Stark shift*, or the *AC Stark shift*. This is similar to the *static Stark shift*, which occurs when an atom experiences a strong static electric field, and is easily treated by perturbation theory. An energy level shift proportional to the square of the electric field is found[1].

a) The Schrödinger Equation: To show the AC Stark shift arises, we expand the wavefunction in the atomic energy eigenstates

$$|\psi(t)\rangle = \sum_n a_n(t)|n\rangle, \tag{4.1.6}$$

and we choose the initial condition that the system is in a particular energy eigenstate $|g\rangle$, so that

$$a_n(0) = \delta_{ng}. \tag{4.1.7}$$

The notation $|g\rangle$ implies that $|g\rangle$ is usually the ground state, but this is not necessary for the derivation of the AC Stark shift formula, which is valid for any state.

Using the Hamiltonian (4.1.5), this leads to the Schrödinger equation for the coefficients $a_n(t)$

$$\frac{d}{dt}a_n(t) = -\mathrm{i}\omega_n a_n(t) + \frac{\mathrm{i}}{\hbar}\sum_k \boldsymbol{d}_{nk}\cdot\boldsymbol{E}_{cl}(0,t)a_k(t). \tag{4.1.8}$$

b) Perturbative Solution in the Interaction Picture: We introduce an interaction picture by writing

$$a_n(t) = \tilde{a}_n(t)\mathrm{e}^{-\mathrm{i}\omega_n t}, \tag{4.1.9}$$

and in this interaction picture the Schrödinger equation takes the form

$$\frac{d}{dt}\tilde{a}_g(t) = \frac{\mathrm{i}}{\hbar}\sum_k \boldsymbol{d}_{gk}\cdot\left(\mathcal{E}(t)\boldsymbol{\epsilon}\,\mathrm{e}^{-\mathrm{i}\omega t} + \mathcal{E}^*(t)\boldsymbol{\epsilon}^*\,\mathrm{e}^{\mathrm{i}\omega t}\right)\tilde{a}_k(t)\mathrm{e}^{\mathrm{i}(\omega_g-\omega_k)t}. \tag{4.1.10}$$

This equation can be treated perturbatively in exactly the same way as we did for only one excited level in *Bk. I: Sect.12.3.4*:

i) We solve for the coefficients a_k for $k \neq g$, and neglect the initial condition terms.

ii) *Adiabatic Approximation*: We then made an adiabatic approximation; that is, we assume that the interaction strength is weak, and the resulting condition that the rate of change of the coefficients is very much slower than the optical time scales leads to a perturbation theory with the solution

$$\tilde{a}_k(t) \approx \left(-\mathrm{e}^{\mathrm{i}(\omega_k-\omega_g-\omega)t}\frac{\boldsymbol{d}_{kg}\cdot\boldsymbol{\epsilon}\,\mathcal{E}(t)}{\hbar(\omega_g+\omega-\omega_k)} - \mathrm{e}^{\mathrm{i}(\omega_k-\omega_g+\omega)t}\frac{\boldsymbol{d}_{kg}\cdot\boldsymbol{\epsilon}^*\,\mathcal{E}^*(t)}{\hbar(\omega_g-\omega-\omega_k)}\right)\tilde{a}_g(t). \tag{4.1.11}$$

[1] The linear Stark shift only occurs when the system does not have inversion symmetry.

iii) We then substitute back into the Schrödinger equation (4.1.10), getting the modified equation

$$i\hbar\frac{d}{dt}\tilde{a}_g(t) = \delta E_g(t)\tilde{a}_g(t). \tag{4.1.12}$$

iv) *AC Stark Shift*: Here $\delta E_g(t) \equiv \hbar\delta\omega_g(t)$ is a *quadratic* AC Stark shift given by

$$\delta E_g(t) \equiv \hbar\delta\omega_g(t) = \sum_k \frac{2\omega_{gk}|\boldsymbol{d}_{kg}\cdot\boldsymbol{\epsilon}\,\mathcal{E}(t)|^2}{\omega_{gk}^2-\omega^2}. \tag{4.1.13}$$

This is a shift in the energy level of the state $|g\rangle$, and it can be either positive or negative.

c) Expression in Terms of Polarizability: We can write $\delta E_g = -\frac{1}{2}\alpha(\omega)|E|^2$, where $\alpha(\omega)$ is a frequency dependent polarizability, so that the shift is proportional to the intensity of the electromagnetic field—thus the AC Stark shift is a quadratic effect.

Exercise 4.1 Derivation of the AC Stark Shift: Work out the explicit details in the derivation of the formula (4.1.13) for the Stark shift $\delta\omega_g(t)$.

Exercise 4.2 Optical Potential and FORT: If $\omega \approx \omega_{eg}$, so that the contribution from a particular excited state $n = e$ dominates, show that

$$\delta\omega_g = \frac{1}{4}\frac{\Omega^2}{\Delta}, \qquad \Delta = \omega - \omega_{eg}, \qquad \Omega = \frac{|2\boldsymbol{d}_{gk}\cdot\boldsymbol{\epsilon}\mathcal{E}|}{\hbar}. \tag{4.1.14}$$

This is the basis of the *far off resonance optical trap,* conventionally referred to by the acronym *FORT*, in which the optical field (and hence the Rabi frequency) is made spatially dependent. The result is in agreement with that we found for the two-level system in the rotating wave approximation in *Bk. I: Sect.12.3.4.*

It should be noted that in practice fluctuations in the laser field are very important. These can fluctuations in either the intensity or the direction of the laser field, and give rise to heating of the atoms contained in the trap. For a treatment of these effects see [4.3, 4.4].

Exercise 4.3 Convergence of the Summation over Levels in the Stark Shift Formula: Note that $\left|\omega_{gk}/(\omega_{gk}^2-\omega^2)\right|$ is bounded as a function of k. Hence use the completeness of the states $|k\rangle$ to show that the summation in (4.1.13) converges.

d) Effective Hamiltonian and Application to Quantum State Engineering: An atom in a non-resonant driving field is thus described by the effective Hamiltonian

$$H_{\text{eff}} = \hbar\left(\omega_g + \delta\omega_g(t)\right)|g\rangle\langle g|. \tag{4.1.15}$$

Here the AC Stark shift $\delta\omega_g$ is given by (4.1.13), and is the frequency shift of the state $|g\rangle$ which results from the application of the optical field. The solution of this Schrödinger equation is

$$a_g(t) = a_g(0)e^{-i\omega_g t - i\int_0^t dt'\,\delta\omega_g(t')}, \tag{4.1.16}$$

so that, as a result, the amplitude acquires a phase shift which depends on the intensity of the electromagnetic field applied.

Hence, if we want to phase-shift an atomic amplitude, we can use the AC Stark shift to do this by applying an appropriate *non-resonant* laser pulse. For maximum effect, however, it is normal to choose the laser frequency reasonably close to a transition as in the case of the FORT in Ex. 4.2, since this enhances one of the denominators in (4.1.13).

4.1.3 Two-Level Atoms and the AC Stark Shift

We now consider the case of an atom driven by laser light of frequency $\omega \approx \omega_{eg}$, that is, light almost resonant with the transition between two levels, $|g\rangle \longrightarrow |e\rangle$, as in Fig. 4.2. At the same time, we will also consider the effect of the other non-resonant transitions, of which there will be a significant number. The strategy for solving the problem separates the consideration of the near resonant transitions from that of the off-resonant transitions, which we treat in a similar way to that used in the previous section.

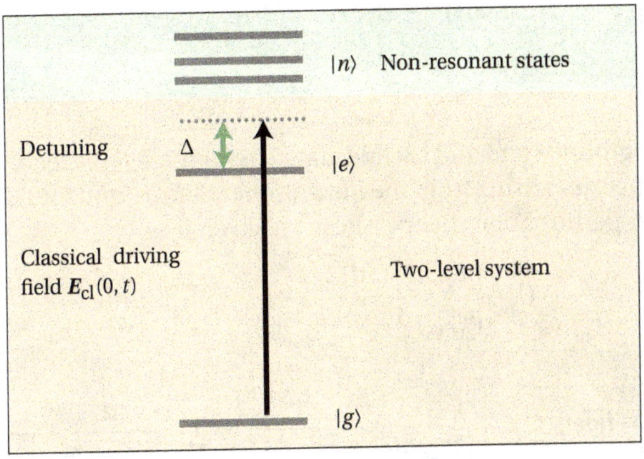

Fig. 4.2. Near-resonant excitation of a two-level system. The position of the dotted line is not drawn to scale—the deviation Δ from exact resonance with $|e\rangle$ is very small, so that Δ is typically in the microwave or radio frequency regime.

a) Hamiltonian: We group the Hamiltonian into two parts; the first represents the two-level system in the rotating-wave approximation, and the second represents the non-resonant terms (including the anti-resonant terms that were explicitly dropped when using the rotating-wave approximation). The Hamiltonian thus takes the form

$$H = H_{\text{RWA}} + H_{\text{Stark}}, \tag{4.1.17}$$

$$H_{\text{RWA}} \equiv \hbar\omega_{eg}|e\rangle\langle e| - \boldsymbol{d}_{eg}\cdot\boldsymbol{\epsilon}\,\mathcal{E}(t)\mathrm{e}^{-\mathrm{i}\omega t}|e\rangle\langle g| - \boldsymbol{d}_{ge}\cdot\boldsymbol{\epsilon}^*\,\mathcal{E}^*(t)\mathrm{e}^{\mathrm{i}\omega t}|g\rangle\langle e|, \tag{4.1.18}$$

$$H_{\text{Stark}} \equiv \sum_{n\neq e,g}\hbar\omega_n|n\rangle\langle n| - \sum_{\substack{n\neq e\\ k\neq g}}\left(\boldsymbol{d}_{nk}\cdot\boldsymbol{\epsilon}\,\mathcal{E}(t)\mathrm{e}^{-\mathrm{i}\omega t}|n\rangle\langle k| - \boldsymbol{d}_{kn}\cdot\boldsymbol{\epsilon}^*\,\mathcal{E}^*(t)\mathrm{e}^{\mathrm{i}\omega t}|k\rangle\langle n|\right). \tag{4.1.19}$$

i) The term H_{RWA} represents the two-level system in the rotating-wave approximation, which we have solved exactly in *Bk. I: Sect.12.3.*

ii) The term H_{Stark} includes the non-resonant interactions, which are treated using perturbation theory, in the same way as in the previous section.

b) The Effective Hamiltonian: Implementing the perturbation theory for this system, we can write

$$H_{\text{eff}} = \hbar\big(\omega_g + \delta\omega_g(t)\big)\,|g\rangle\langle g| \;+\; \hbar\big(\omega_e + \delta\omega_e(t)\big)\,|e\rangle\langle e| \\ -\,\boldsymbol{d}_{eg}\cdot\boldsymbol{\epsilon}\,\mathcal{E}(t)\mathrm{e}^{-\mathrm{i}\omega t}|e\rangle\langle g| \;-\; \boldsymbol{d}_{eg}\cdot\boldsymbol{\epsilon}^*\,\mathcal{E}^*(t)\mathrm{e}^{\mathrm{i}\omega t}|g\rangle\langle e|. \tag{4.1.20}$$

Notice that:

i) The bare ground state and excited state energies are replaced by the AC Stark shifted energies arising from the non-resonant terms, and found by using second order perturbation theory. These are given by

$$\delta E_g(t) \equiv \hbar\delta\omega_g(t) = \left\{\sum_{k\neq e,g}\frac{|\boldsymbol{d}_{kg}\cdot\boldsymbol{\epsilon}|^2}{\hbar(\omega_g+\omega-\omega_k)} + \sum_{k\neq e,g}\frac{|\boldsymbol{d}_{kg}\cdot\boldsymbol{\epsilon}|^2}{\hbar(\omega_g-\omega-\omega_k)}\right\}|\mathcal{E}(t)|^2, \tag{4.1.21}$$

$$\delta E_e(t) \equiv \hbar\delta\omega_e(t) = \left\{\sum_{k\neq e,g}\frac{|\boldsymbol{d}_{ke}\cdot\boldsymbol{\epsilon}|^2}{\hbar(\omega_e+\omega-\omega_k)} + \sum_{k\neq e,g}\frac{|\boldsymbol{d}_{ke}\cdot\boldsymbol{\epsilon}|^2}{\hbar(\omega_e-\omega-\omega_k)}\right\}|\mathcal{E}(t)|^2. \tag{4.1.22}$$

It is important to note that the near resonant contributions from g and e are excluded from the sum.

ii) There is a also third order perturbation correction to the Rabi frequency, but this is normally not important.

Exercise 4.4 Derivation of AC Stark Shift Formulae: Adapt the arguments used in Sect. 4.1.2 to derive the formulae (4.1.21, 4.1.22).

5. Coherent Manipulation of Two-Level Systems

The manipulation of the quantum state of a two-level atom is the key tool in the construction of quantum devices. We have already given a simplified treatment of the driven two-level atom in *Bk. I: Sect.14.2*, and in this chapter we will provide a more elaborate description, adapted to the use of coherent optical pulses to transfer a two-level atom from one quantum state to another. The key parameters are the amplitude, the pulse length, and the detuning of the laser frequency from resonance with the atomic transition. The available motions in phase space can be parametrized by the *Bloch vector*, a three dimensional vector, whose motion takes place on the *Bloch sphere*—a picture introduced by *Bloch* [5.1] to describe nuclear magnetic resonance. This provides a visualization of the quantum state of the two-level atom which is almost universally used when any two-level system is being considered.

For an optical pulse of constant amplitude, one can transform to a "rotating frame", in which the Hamiltonian becomes time-independent, and is known as the *Rabi Hamiltonian*. The eigenstates of this Hamiltonian, known as the *dressed states* of the two-level atom, provide a very useful way of viewing the behaviour of the system as a function of the detuning Δ from resonance. By sweeping the detuning of the laser field slowly, one can effect a transfer of the quantum state of an atom from the ground state to the excited state.

5.1 The Ideal Two-Level Atom Driven by a Classical Field

As we did in Sect. 4.1, will idealize the electromagnetic field as a classical driving field, neglecting all fluctuations and the quantized nature of light. Since the techniques used are so fundamental, we feel it is best to describe them in the simplest form first, in order to establish their nature clearly. Consequently, at this stage we will not justify this "classical" representation of a electromagnetic field, leaving a full quantum description to Part III of this book.

5.1.1 General Pulse Shape

We will formulate the equations of motion for the case of an applied field pulse $\mathcal{E}(t)$ with an arbitrary time dependence. In practice one tries to achieve the ideal square pulse shape as in Fig. 4.1, and this will form the main part of the discussion in this chapter.

a) Equations of Motion in the Schrödinger Picture: We solve the time dependent Schrödinger equation for the two-level system in the rotating-wave approximation arising from the Hamiltonian H_{RWA}, as in (4.1.18). We write the wavefunction in terms of the ground and excited states as

$$|\psi, t\rangle = a_e(t)|e\rangle + a_g(t)|g\rangle, \tag{5.1.1}$$

and the Schrödinger equation is equivalent to equations of motion for the coefficients in the form

$$i\hbar\dot{a}_e(t) = \hbar\omega_e a_e(t) - \boldsymbol{d}_{eg}\cdot\boldsymbol{\epsilon}\,\mathcal{E}(t)e^{-i\omega t}a_g(t), \tag{5.1.2}$$
$$i\hbar\dot{a}_g(t) = \hbar\omega_g a_g(t) - \boldsymbol{d}_{ge}\cdot\boldsymbol{\epsilon}^*\,\mathcal{E}^*(t)e^{i\omega t}a_e(t). \tag{5.1.3}$$

b) Transformation of the Two-Level System to a Rotating Frame: The transformation to the "rotating frame" is intended both to remove the explicit fast time dependence $\exp(\pm i\omega t)$, and to refer all energies to the ground state energy $\hbar\omega_g$. It is achieved by setting

$$a_e(t) = \tilde{a}_e(t)e^{-i(\omega+\omega_g)t}, \tag{5.1.4}$$
$$a_g(t) = \tilde{a}_g(t)e^{-i\omega_g t}. \tag{5.1.5}$$

After inserting these transformations, we can write the equations of motion in terms of the *Rabi Hamiltonian*

$$H_{\text{Rabi}}(t) \equiv \hbar\begin{pmatrix} -\Delta & -\boldsymbol{d}_{eg}\cdot\boldsymbol{\epsilon}\mathcal{E}(t)/\hbar \\ -\boldsymbol{d}_{ge}\cdot\boldsymbol{\epsilon}^*\,\mathcal{E}(t)^*/\hbar & 0 \end{pmatrix}, \tag{5.1.6}$$

in which $\Delta \equiv \omega - \omega_{eg}, \quad \omega_{eg} \equiv \omega_e - \omega_g.$ (5.1.7)

The equations of motion the take the usual form giving the equations

$$i\hbar\frac{d}{dt}\begin{pmatrix}\tilde{a}_e \\ \tilde{a}_g\end{pmatrix} = H_{\text{Rabi}}(t)\begin{pmatrix}\tilde{a}_e \\ \tilde{a}_g\end{pmatrix}. \tag{5.1.8}$$

The optical frequencies have now been transformed away, and for an ideal square pulse the Hamiltonian is time independent during the pulse.

c) Characteristic Parameters of the Rabi Hamiltonian: The characteristic parameters are:

i) *The Rabi frequency*: This is normally defined to be positive, and given by

$$\Omega_R(t) \equiv \left|\frac{2\mathcal{E}(t)\boldsymbol{d}_{eg}\cdot\boldsymbol{\epsilon}}{\hbar}\right|. \tag{5.1.9}$$

(However, it is sometimes convenient to omit the modulus, and choose the Rabi frequency to be complex number $2\mathcal{E}(t)\boldsymbol{d}_{eg}\cdot\boldsymbol{\epsilon}/\hbar$.)

ii) *The detuning*: defined by

$$\Delta \equiv \omega - \omega_{eg}. \tag{5.1.10}$$

iii) The duration T of the pulse of electromagnetic radiation.

The definition of the Rabi frequency is as in *Bk. I: Sect.12.3.2*, but there $\mathcal{E}$ is defined to include the electric dipole moment which is not convenient in this book, where we must normally consider many transitions, each with its own dipole moment.

5.1.2 Validity of the Rotating Wave Approximation

As already noted in *Bk. I: Sect.12.3*, the condition for the off-resonant terms to be negligible requires that all time dependences retained and of interest must be on a slow time scale. This means that

$$\Omega_R \ll \omega \approx \omega_{eg}, \tag{5.1.11}$$

$$|\Delta| \ll \omega \approx \omega_{eg}, \tag{5.1.12}$$

and as well that $\mathcal{E}(t)$ be slowly varying on the optical time scale, which requires

$$\left|\frac{1}{\mathcal{E}(t)}\frac{d\mathcal{E}(t)}{dt}\right| \ll \omega. \tag{5.1.13}$$

This means that a "instantaneous" turn on of a square pulse as in (4.1.2) has to be interpreted as being slow on the optical time scale, but fast on the time scales given by Ω_R and Δ.

Under these conditions, the coefficients $a_e(t)$ and $a_g(t)$ are also slowly varying on the optical time scale

5.2 The Rabi Problem for a Square Pulse

We want to consider the simplest situation, in which the driving field $\mathcal{E}(t)$ has the idealized square form depicted in Fig. 4.1 b; essentially, this means we want to study the evolution of the quantum state of the system when:

i) The driving field has a constant time-independent value $\mathcal{E}$,

ii) The system evolves during a time interval $0 \leqslant t \leqslant T$,

iii) There is a given initial condition for the quantum state at the beginning of the time interval.

5.2.1 The Pseudospin Formalism

All two-level systems can be described using the Pauli matrices as a basis for their operators, and the resulting formalism is called the *pseudospin formulation* in which $\frac{1}{2}\hbar\boldsymbol{\sigma}$ is called a pseudospin, since it obeys all the algebraic properties of a spin, but is not directly related to angular momentum.

a) Hamiltonian: Following the Pauli matrix properties given in *Bk. I: Sect.10.7*, the Rabi Hamiltonian (5.1.6) can be written in terms of the Pauli matrices as

$$H_{\text{Rabi}} = -\tfrac{1}{2}\hbar\left(\Delta + \Omega_{\text{eff}}\,\boldsymbol{n}\cdot\boldsymbol{\sigma}\right), \tag{5.2.1}$$

where $\boldsymbol{n}$ is a unit vector, and use of the formula *Bk. I: (10.7.9)* shows that

$$\Omega_{\text{eff}}\,\boldsymbol{n} = \left(\text{Re}\left(\boldsymbol{d}_{eg}\cdot\boldsymbol{\epsilon}\mathcal{E}\right),\ \text{Im}\left(\boldsymbol{d}_{eg}\cdot\boldsymbol{\epsilon}\mathcal{E}\right),\ \tfrac{1}{2}\Delta\right). \tag{5.2.2}$$

b) The Density Matrix and the Bloch Vector: Even when we are dealing with a pure state, it is often simpler to use the density matrix to describe the state of the two-level system, and we will use the pseudospin formalism to write the density matrix, in a notation similar to that in *Bk. I: Sect.14.2*, as

$$\rho(t) = \tfrac{1}{2}\left(1 + \boldsymbol{s}(t)\cdot\boldsymbol{\sigma}\right). \tag{5.2.3}$$

The vector $\boldsymbol{s}(t)$ is known as the *Bloch vector*, and completely specifies the quantum state defined by the density operator. In the case of a pure state, as here, $|\boldsymbol{s}(t)| = 1$, so that we can specify $\boldsymbol{s}(t)$ by a pair of polar angles thus

$$\boldsymbol{s}(t) = \left(\sin\Theta(t)\cos\Phi(t),\ \sin\Theta(t)\sin\Phi(t),\ \cos\Theta(t)\right). \tag{5.2.4}$$

We will work in the rotating frame, so that corresponding to the notation in (5.1.8), the density matrix can be written

$$\rho(t) = \begin{pmatrix} |\tilde{a}_e(t)|^2 & \tilde{a}_g(t)^*\,\tilde{a}_e(t) \\ \tilde{a}_g(t)\tilde{a}_e(t)^* & |\tilde{a}_g(t)|^2 \end{pmatrix}. \tag{5.2.5}$$

The representation in terms of polar angles (5.2.4) translates into the expression for the amplitudes

$$\begin{pmatrix} \tilde{a}_e(t) \\ \tilde{a}_g(t) \end{pmatrix} = \begin{pmatrix} \cos\frac{\Theta(t)}{2}\,\mathrm{e}^{-\frac{\mathrm{i}\Phi(t)}{2}} \\ \sin\frac{\Theta(t)}{2}\,\mathrm{e}^{\frac{\mathrm{i}\Phi(t)}{2}} \end{pmatrix}. \tag{5.2.6}$$

This can be checked by direct substitution into (5.2.5). Of course, as with all wavefunctions, it is possible to multiply by an arbitrary phase, and indeed we will show that time evolution introduces such a phase.

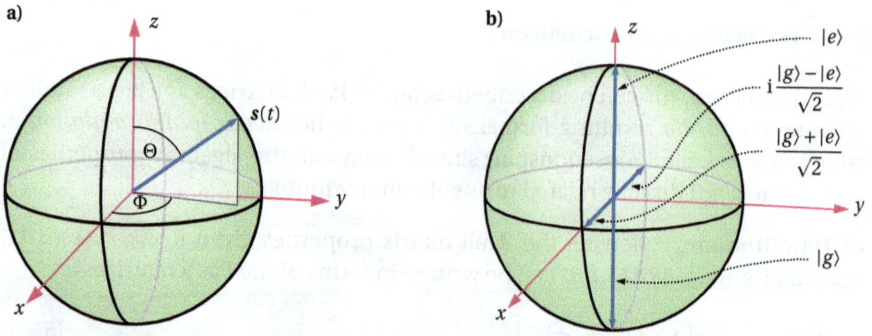

Fig. 5.1. The Bloch Sphere. **a)** Definition of the angle Θ and Φ for the Bloch vector **s**; **b)** Bloch vectors for the ground state, the excited state and two superposition states, as defined in (5.2.7).

5.2.2 The Bloch Sphere

The Bloch vector can be plotted on a sphere, as illustrated in Fig. 5.1 a. The motion of the Bloch vector on the Bloch sphere provides a very intuitive understanding of the evolution of the quantum state of a two-level system. For, example as illustrated in Fig. 5.1 b, particular states of interest are:

$$\left.\begin{array}{llll} \text{Ground state} & |g\rangle & \Theta = \pi, & \Phi = 0, \\ \text{Excited state} & |e\rangle & \Theta = 0, & \Phi = 0, \\ \text{Even superposition} & \dfrac{|g\rangle + |e\rangle}{\sqrt{2}} & \Theta = \tfrac{1}{2}\pi, & \Phi = 0, \\ \text{Odd superposition} & \pm i\dfrac{|g\rangle - |e\rangle}{\sqrt{2}} & \Theta = \tfrac{1}{2}\pi, & \Phi = \pm\pi. \end{array}\right\} \tag{5.2.7}$$

a) Polar Notation for the Rabi Hamiltonian: For general detuning we will solve the Rabi problem for a square pulse, that is, a pulse for which

$$\mathcal{E}(t) = \mathcal{E}_0 = |\mathcal{E}_0|\, e^{-i\varphi}, \tag{5.2.8}$$

during the period $0 \leqslant t \leqslant T$, and is zero otherwise. The Rabi Hamiltonian (5.1.6) can be written in the form

$$H = \hbar \begin{pmatrix} -\Delta & \frac{1}{2}\Omega_R e^{-i\varphi} \\ \frac{1}{2}\Omega_R e^{i\varphi} & 0 \end{pmatrix}, \tag{5.2.9}$$

or in terms of Pauli matrices as

$$H_{\text{Rabi}} = -\tfrac{1}{2}\hbar\left(\Delta + \Omega_{\text{eff}}\,\boldsymbol{n}\cdot\boldsymbol{\sigma}\right), \tag{5.2.10}$$

$$\boldsymbol{n} \equiv -\left(\sin\theta\cos\varphi,\ \sin\theta\sin\varphi,\ \cos\theta\right), \tag{5.2.11}$$

$$\Omega_{\text{eff}} \equiv \sqrt{\Omega_R^2 + \Delta^2}\,. \tag{5.2.12}$$

The angle φ is the phase of the pulse as in (5.2.8), while the angle θ is defined by

$$\tan\theta = -\frac{\Omega_R}{\Delta}, \quad \sin\theta = \frac{\Omega_R}{\sqrt{\Delta^2+\Omega_R^2}} \geqslant 0, \quad \cos\theta = -\frac{\Delta}{\sqrt{\Delta^2+\Omega_R^2}}\,. \tag{5.2.13}$$

b) Ranges of θ and Δ: From these definitions, the range of θ is given by

$$0 \leqslant \theta \leqslant \pi, \tag{5.2.14}$$

and the ranges of Δ and θ are related by

$$\left.\begin{aligned} -\infty < \Delta < 0 \quad &\Longleftrightarrow \quad 0 \leqslant \theta \leqslant \tfrac{1}{2}\pi\,, \\ 0 < \Delta < \infty \quad &\Longleftrightarrow \quad \tfrac{1}{2}\pi \leqslant \theta \leqslant \pi\,. \end{aligned}\right\} \tag{5.2.15}$$

The relationship is illustrated in Fig. 5.2.

5.2.3 The Bloch Equation

Using von Neumann's equation for the density matrix, and the Pauli matrix property *Bk. I: (10.7.6)*, the equation of motion for the Bloch vector is

$$\frac{d\boldsymbol{s}(t)}{dt} = -\Omega_{\text{eff}}\boldsymbol{n}\times\boldsymbol{s}(t). \tag{5.2.16}$$

This is the famous *Bloch equation*, introduced by *Bloch* [5.1] to describe the motion of a nuclear spin induced by the interaction of a radio frequency field with its magnetic moment. Here, we deal with a two-level system interacting via an electric dipole moment with an optical field. This is not a genuine spin system, and is often called a *pseudospin system.*

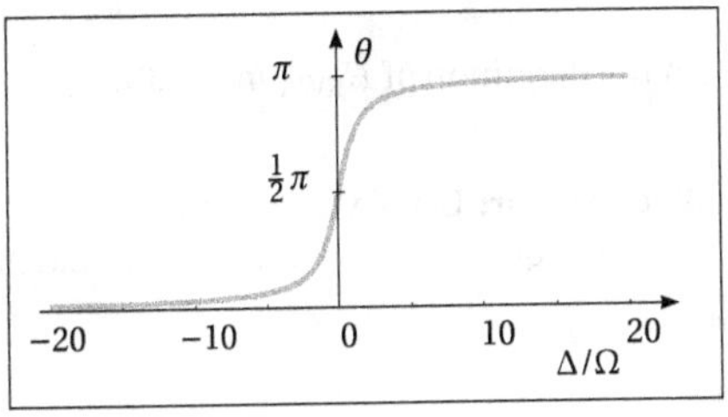

Fig. 5.2. Variation of θ as defined by (5.2.13, 5.2.15) as a function of Δ/Ω.

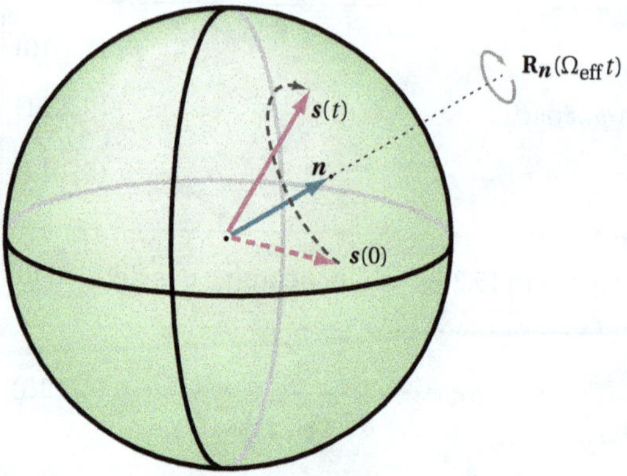

Fig. 5.3. Motion of the Bloch vector on the Bloch sphere as induced by the Rabi Hamiltonian H_{Rabi}.

a) Time Evolution as a Rotation of the Bloch Vector in Pseudospin Space: The time evolution operator during the pulse can then be written as a rotation operator in the pseudospin space

$$U_t = \exp(-\mathrm{i}H_{\text{Rabi}}\,t/\hbar) = \mathrm{e}^{\mathrm{i}\Delta t/2}\exp\left(\tfrac{1}{2}\mathrm{i}\boldsymbol{n}\cdot\boldsymbol{\sigma}\Omega_{\text{eff}}\,t\right), \tag{5.2.17}$$

$$= \mathrm{e}^{\mathrm{i}\Delta t/2}\left\{\cos\left(\tfrac{1}{2}\Omega_{\text{eff}}\,t\right) + \mathrm{i}\boldsymbol{n}\cdot\boldsymbol{\sigma}\sin\left(\tfrac{1}{2}\Omega_{\text{eff}}\,t\right)\right\}. \tag{5.2.18}$$

This represents a rotation of angle $\Omega_{\text{eff}}t$ about the axis $\boldsymbol{n}$, as illustrated in Fig. 5.3. When applied to the density operator, we find

$$\rho(t) = U_t\rho(0)U_t^\dagger = \tfrac{1}{2}\left(1 + \boldsymbol{s}(t)\cdot\boldsymbol{\sigma}\right), \tag{5.2.19}$$

where $\boldsymbol{s}(t)$ is the solution of the Bloch equation (5.2.16).

b) Time Evolution Operator: By explicitly writing out the matrix elements as given in (5.2.18), this can be written in terms of the *effective Rabi frequency* (5.2.12) in the form

$$U_t = \mathrm{e}^{\mathrm{i}\Delta t/2}\begin{pmatrix} \cos\frac{1}{2}\Omega_{\text{eff}}t - \mathrm{i}\cos\theta\sin\frac{1}{2}\Omega_{\text{eff}}t & -\mathrm{i}\mathrm{e}^{-\mathrm{i}\varphi}\sin\theta\sin\frac{1}{2}\Omega_{\text{eff}}t \\ -\mathrm{i}\mathrm{e}^{\mathrm{i}\varphi}\sin\theta\sin\frac{1}{2}\Omega_{\text{eff}}t & \cos\frac{1}{2}\Omega_{\text{eff}}t + \mathrm{i}\cos\theta\sin\frac{1}{2}\Omega_{\text{eff}}t \end{pmatrix}. \tag{5.2.20}$$

Exercise 5.1 Bloch Sphere Polar Co-ordinates: By using the result *Bk. I: (10.7.9)*, note that (5.2.10) implies that

$$\mathrm{Tr}\left\{H_{\text{RWA}}\right\} = -\hbar\Delta, \tag{5.2.21}$$

$$\mathrm{Tr}\left\{\boldsymbol{\sigma}H_{\text{RWA}}\right\} = -\hbar\Omega_{\text{eff}}(t)\,\boldsymbol{n}. \tag{5.2.22}$$

Hence show that the results (5.2.13) follow from the definition of H_{RWA} in (5.1.6).

Exercise 5.2 Equation of Motion of the Bloch Vector: Using von Neumann's equation for the density matrix, explicitly show that the equation of motion for the Bloch vector is given by (5.2.16).

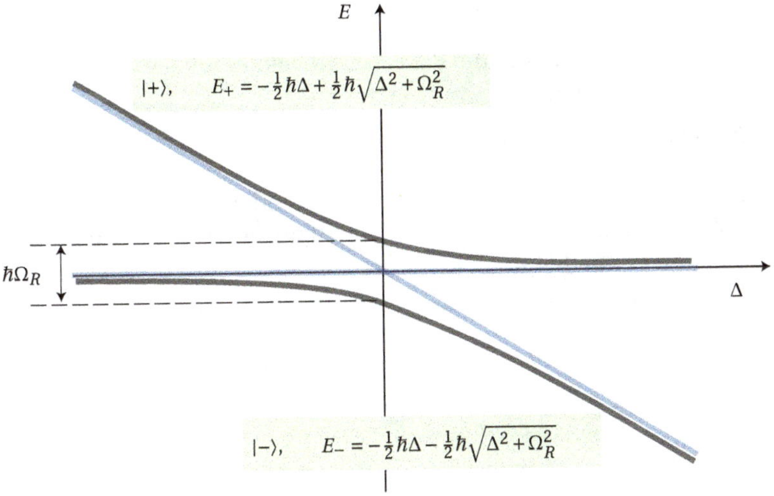

Fig. 5.4. Dressed eigenvalues of a two-level system as a function of detuning Δ. The light blue lines indicate the eigenvalues in the case that $\Omega_R = 0$, in which case the two curves become two straight lines crossing at the origin. Inclusion of a non-zero value of the Rabi frequency creates an avoided crossing at the origin, giving a smooth transition from one asymptote to the other.

5.3 Dressed States of the Rabi Hamiltonian

The eigenvectors of the Rabi Hamiltonian—using the form (5.2.1)—correspond to those values of the Bloch vector $\boldsymbol{s}$ which are unchanged by the time evolution operator, and these arise when $\boldsymbol{s}$ and the axis of rotation $\boldsymbol{n}$ are parallel or antiparallel. Using the notation θ, φ as in (5.2.8, 5.2.13), this means that $\boldsymbol{s}$ is characterized by

$$\Theta, \Phi = \begin{cases} \theta, \varphi, \\ \pi - \theta, \varphi + \pi. \end{cases} \tag{5.3.1}$$

5.3.1 Eigenvectors—the Dressed States

Substituting these values into the definition (5.2.6), the corresponding eigenvectors are given by

$$|+\rangle = \cos\tfrac{1}{2}\theta e^{-i\varphi/2}|e\rangle + \sin\tfrac{1}{2}\theta e^{i\varphi/2}|g\rangle, \tag{5.3.2}$$

$$|-\rangle = -\sin\tfrac{1}{2}\theta e^{-i\varphi/2}|e\rangle + \cos\tfrac{1}{2}\theta e^{i\varphi/2}|g\rangle. \tag{5.3.3}$$

These eigenvectors are known as the *dressed states.*

5.3.2 Eigenvalues—the Dressed Energy Levels

The corresponding eigenvalues of the Rabi Hamiltonian (5.1.6) are known as the *dressed energy levels*, given by

$$E_{\pm} = -\tfrac{1}{2}\hbar\Delta \pm \tfrac{1}{2}\hbar\sqrt{\Delta^2 + \Omega_R^2}\,. \tag{5.3.4}$$

a) Limiting Cases: The qualitative behaviours at the end points and centre of the Δ range are illustrated in Fig. 5.4 and Fig. 5.2, and these are summarized in Table 5.1. In brief these correspond to:

i) *Red Detuning* corresponds to $\Delta < 0$. The limiting behaviours of the dressed states are

- $|+\rangle \approx |g\rangle$, Stark shifted up from $E = 0$.
- $|-\rangle \approx |e\rangle$, Stark shifted down from $E = -\Delta$.
- Perturbation theory is applicable.

ii) *On Resonance*

- The dressed states are composed of equal proportions of the bare states $|e\rangle$ and $|g\rangle$.
- The energy levels experience AC Stark splitting of value $\pm\frac{1}{2}\Omega_R$.
- Perturbation theory cannot be applied.

iii) *Blue Detuning* corresponds to $\Delta < 0$. The limiting behaviours of the dressed states are

- $|+\rangle \approx |e\rangle$, Stark shifted up from $E = -\Delta$.
- $|-\rangle \approx |g\rangle$, Stark shifted down from $E = 0$.
- Perturbation theory is applicable.

5.3.3 Quantum State Engineering

Use of a timed pulse to transform the state of a two-level system is a central technique in quantum optics, using which it is possible to engineer any particular state of the system. The form (5.2.20) of the time evolution operator enables one to select values of Δ/Ω and Ωt so that any unitary transformation in two dimensions can be realized. This means that a pulse can always be chosen to transform an arbitrary initial state into any chosen final state.

a) Particular Cases: The particular cases when $\Omega_{\text{eff}} t = 2\pi$, π or $\frac{1}{2}\pi$ yield the following simple but important special cases.

i) *The 2π-pulse*: Here $\Omega_{\text{eff}} t = 2\pi$ so that

$$U_{2\pi/\Omega_{\text{eff}}} = e^{i\pi\Delta/\Omega_{\text{eff}}} \begin{pmatrix} -1 & 0 \\ 0 & -1 \end{pmatrix}. \tag{5.3.5}$$

Limiting Forms Related to θ The following follow from the definitions:

i) Large *red* detuning $\quad \Delta \to -\infty \quad \begin{cases} \theta \to -\frac{\Omega_R}{\Delta} \\ \cos\frac{1}{2}\theta \to 1 \\ \sin\frac{1}{2}\theta \to -\frac{\Omega_R}{2\Delta} \end{cases}$

ii) At resonance, approached from below $\quad \Delta \to 0^- \quad \begin{cases} \theta \to \frac{1}{2}\pi^- \\ \cos\frac{1}{2}\theta \to \frac{1}{\sqrt{2}} \\ \sin\frac{1}{2}\theta \to \frac{1}{\sqrt{2}} \end{cases}$

iii) At resonance, approached from above $\quad \Delta \to 0^+ \quad \begin{cases} \theta \to \frac{1}{2}\pi^+ \\ \cos\frac{1}{2}\theta \to \frac{1}{\sqrt{2}} \\ \sin\frac{1}{2}\theta \to \frac{1}{\sqrt{2}} \end{cases}$

iv) Large *blue* detuning $\quad \Delta \to +\infty \quad \begin{cases} \theta \to \pi - \frac{\Omega_R}{\Delta} \\ \cos\frac{1}{2}\theta \to \frac{\Omega_R}{2\Delta} \\ \sin\frac{1}{2}\theta \to 1 \end{cases}$

Limiting Forms for the Dressed Eigenvalues and Eigenstates

v) Large *red* detuning $\quad \Delta \to -\infty$

$$E_+ = -\hbar\Delta - \hbar\frac{\Omega_R^2}{2\Delta} \qquad |+\rangle \sim |e\rangle - \frac{\Omega_R}{2\Delta}\mathrm{e}^{\mathrm{i}\varphi}|g\rangle \longrightarrow |e\rangle$$

$$E_- = -\hbar\frac{\Omega_R^2}{2\Delta} \qquad |-\rangle \sim |g\rangle + \frac{\Omega_R}{2\Delta}\mathrm{e}^{-\mathrm{i}\varphi}|e\rangle \longrightarrow |g\rangle$$

vi) On resonance $\quad \Delta = 0$

$$E_+ = +\frac{1}{2}\hbar\Omega_R \qquad |+\rangle = \frac{1}{\sqrt{2}}\left(\mathrm{e}^{-\mathrm{i}\varphi/2}|e\rangle + \mathrm{e}^{\mathrm{i}\varphi/2}|g\rangle\right)$$

$$E_- = -\frac{1}{2}\hbar\Omega_R \qquad |-\rangle = \frac{1}{\sqrt{2}}\left(-\mathrm{e}^{-\mathrm{i}\varphi/2}|e\rangle + \mathrm{e}^{\mathrm{i}\varphi/2}|g\rangle\right)$$

vii) Large *blue* detuning $\quad \Delta \to +\infty$

$$E_+ = +\hbar\frac{\Omega_R^2}{2\Delta} \qquad |+\rangle \sim |g\rangle + \frac{\Omega_R}{2\Delta}e^{-i\varphi}|e\rangle \longrightarrow |g\rangle$$

$$E_- = -\hbar\Delta - \hbar\frac{\Omega_R^2}{2\Delta} \qquad |-\rangle \sim |e\rangle - \frac{\Omega_R}{2\Delta}e^{i\varphi}|g\rangle \longrightarrow |e\rangle$$

Table 5.1. Limiting forms of formulae for a two-level system driven by coherent light.

ii) *The π-pulse*: Here $\Omega_{\text{eff}}t = \pi$ so that

$$U_{\pi/\Omega_{\text{eff}}} = \mathrm{i}e^{\mathrm{i}\pi\Delta/2\Omega_{\text{eff}}}\boldsymbol{n}\cdot\boldsymbol{\sigma} = -\mathrm{i}e^{\mathrm{i}\pi\Delta/2\Omega_{\text{eff}}}\begin{pmatrix} \cos\theta & e^{-\mathrm{i}\varphi}\sin\theta \\ e^{\mathrm{i}\varphi}\sin\theta & -\cos\theta \end{pmatrix}. \tag{5.3.6}$$

iii) *The $\frac{1}{2}\pi$-pulse*: Here $\Omega_{\text{eff}}t = \frac{1}{2}\pi$ so that

$$U_{\pi/2\Omega_{\text{eff}}} = e^{\mathrm{i}\pi\Delta/4\Omega_{\text{eff}}}\frac{1+\mathrm{i}\boldsymbol{n}\cdot\boldsymbol{\sigma}}{\sqrt{2}} = \frac{e^{\mathrm{i}\pi\Delta/4\Omega_{\text{eff}}}}{\sqrt{2}}\begin{pmatrix} 1-\mathrm{i}\cos\theta & -\mathrm{i}e^{-\mathrm{i}\varphi}\sin\theta \\ -\mathrm{i}e^{\mathrm{i}\varphi}\sin\theta & 1+\mathrm{i}\cos\theta \end{pmatrix}. \tag{5.3.7}$$

b) Transfer Probabilities from the Ground State: Starting from the ground state, the state after a time t is given by

$$|\psi, t\rangle = U_t|g\rangle = \tilde{a}_g(t)|g\rangle + \tilde{a}_e(t)|e\rangle, \tag{5.3.8}$$

in which

$$\tilde{a}_g(t) = e^{\mathrm{i}\Delta t/2}\left(\cos\tfrac{1}{2}\Omega_{\text{eff}}t + \mathrm{i}\cos\theta\sin\tfrac{1}{2}\Omega_{\text{eff}}t\right), \tag{5.3.9}$$

$$\tilde{a}_e(t) = -\mathrm{i}e^{\mathrm{i}(\Delta t/2-\varphi)}\sin\theta\sin\tfrac{1}{2}\Omega_{\text{eff}}t, \tag{5.3.10}$$

in which the angles θ, φ, are as defined in (5.2.13). From these amplitudes we find:

i) The corresponding transition probability to the excited state is

$$|\tilde{a}_e(t)|^2 \equiv = \frac{\Omega_R^2}{\Omega_{\text{eff}}^2}\sin^2\tfrac{1}{2}\Omega_{\text{eff}}t = \frac{1}{2}\frac{\Omega_R^2}{\Delta^2+\Omega_R^2}(1-\cos\Omega_{\text{eff}}t). \tag{5.3.11}$$

The behaviour can be seen in Fig. 5.5 a for different values of Δ.

ii) Viewed as a function of Δ, this probability displays a Lorentzian resonance shape, with width Ω_R. The width is thus determined by the power of the incident light field, and this phenomenon is known as *power broadening*, and is illustrated in Fig. 5.5 b.

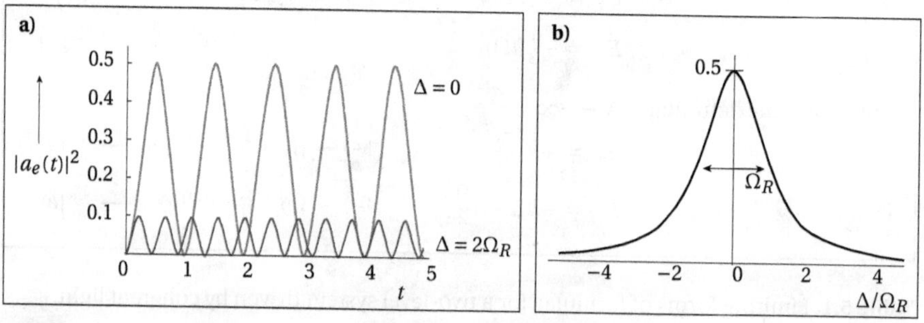

Fig. 5.5. Values of $|a_e(t)|^2$ as a function of time t and the detuning Δ, as given by (5.3.11): **a)** $|a_e(t)|^2$ as a function of time for two different detuning values; **b)** Peak value of $|a_e(t)|^2$ as a function of Δ for a fixed value of Ω_R, showing the power broadened spectral response.

iii) The survival probability in the ground state is

$$|\tilde{a}_g(t)|^2 \equiv \cos^2 \tfrac{1}{2}\Omega_{\text{eff}} t + \frac{\Delta^2}{\Omega_{\text{eff}}^2} \sin^2 \tfrac{1}{2}\Omega_{\text{eff}} t = \tfrac{1}{2}\left(1 + \frac{\Delta^2}{\Omega_{\text{eff}}^2}\right) + \tfrac{1}{2}\frac{\Omega^2}{\Omega_{\text{eff}}^2} \cos \Omega_{\text{eff}} t. \tag{5.3.12}$$

iv) Both of these probabilities display oscillations at the effective Rabi frequency Ω_{eff}.

c) **The On-Resonance Case:** Here we set $\Delta = 0$, corresponding to

$$\theta \longrightarrow \tfrac{1}{2}\pi, \qquad \Omega_{\text{eff}} \longrightarrow \Omega_R, \tag{5.3.13}$$

and the formulae of the previous sections simplify as follows.

i) The occupation probabilities become

$$|\tilde{a}_e(t)|^2 = \tfrac{1}{2}(1 - \cos \Omega_R t), \tag{5.3.14}$$

$$|\tilde{a}_g(t)|^2 = \tfrac{1}{2}(1 + \cos \Omega_R t). \tag{5.3.15}$$

ii) The dressed states defined in (5.3.2, 5.3.3) become

$$|+\rangle = \frac{1}{\sqrt{2}}\left(e^{-i\varphi/2}|e\rangle + e^{i\varphi/2}|g\rangle\right), \tag{5.3.16}$$

$$|-\rangle = \frac{1}{\sqrt{2}}\left(-e^{-i\varphi/2}|e\rangle + e^{i\varphi/2}|g\rangle\right). \tag{5.3.17}$$

iii) The time evolution can be written in terms of the dressed states as

$$U_t = e^{-iE_+ t/\hbar}|+\rangle\langle+| \; + \; e^{-iE_- t/\hbar}|-\rangle\langle-|. \tag{5.3.18}$$

iv) The time evolution operator takes the explicit form

$$U_t = \begin{pmatrix} \cos \tfrac{1}{2}\Omega_R t & -ie^{-i\varphi} \sin \tfrac{1}{2}\Omega_R t \\ -ie^{i\varphi} \sin \tfrac{1}{2}\Omega_R t & \cos \tfrac{1}{2}\Omega_R t \end{pmatrix}. \tag{5.3.19}$$

v) The 2π, π and $\frac{1}{2}\pi$ pulses correspond to the unitary operators:

$$U_{2\pi/\Omega_R} = \begin{pmatrix} -1 & 0 \\ 0 & -1 \end{pmatrix}, \tag{5.3.20}$$

$$U_{\pi/\Omega_R} = \begin{pmatrix} 0 & -ie^{-i\varphi} \\ -ie^{i\varphi} & 0 \end{pmatrix}, \tag{5.3.21}$$

$$U_{\pi/2\Omega_R} = \frac{1}{\sqrt{2}}\begin{pmatrix} 1 & -i \\ -i & 1 \end{pmatrix}. \tag{5.3.22}$$

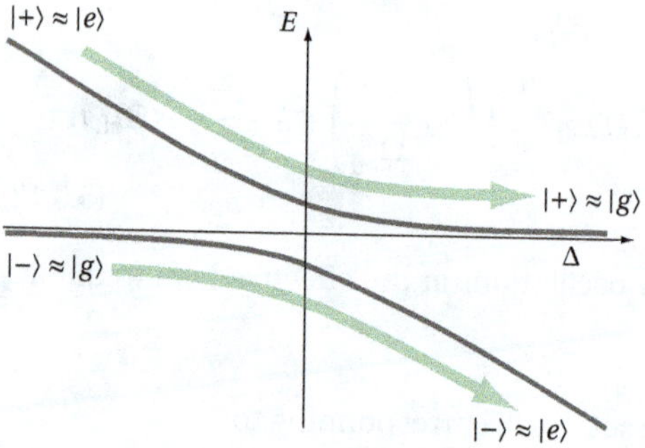

Fig. 5.6. Adiabatic passage along the dressed states—see also Fig. 5.4. Sweeping Δ adiabatically from $\Delta \ll -\Omega_R$ to $\Delta \gg \Omega_R$ transfers $|e\rangle \to |g\rangle$ along the dressed state $|+\rangle$, and transfers $|g\rangle \to |e\rangle$ along the dressed state $|-\rangle$.

5.4 Adiabatic Passage along Dressed States

The behaviour as Δ moves over the range $(-\infty,\infty)$ can be applied to transfer population between the states $|g\rangle$ and $|e\rangle$. For example the state $|+\rangle$ for $\Delta \ll -\Omega_R$ is essentially $|g\rangle$, and as Δ passes through zero and eventually to $\Delta \gg \Omega_R$, the state $|+\rangle$ gains an increasing proportion of $|e\rangle$, eventually becoming almost the same as $|e\rangle$.

Thus, if we sweep the detuning Δ slowly through the resonance (so that the adiabatic theorem applies) we can transform the ground state $|g\rangle$ to the excited state $|e\rangle$. The behaviour is illustrated in Fig. 5.6.

5.4.1 The Adiabatic Theorem and Berry's Phase

The idea of adiabatic passage comes from the adiabatic theorem, which considers a Hamiltonian of the form $H(t)$ for which the time dependence is very weak, so that an eigenstate of the Hamiltonian can adjust to the gradually changing Hamiltonian.

a) Derivation: To derive the adiabatic theorem we proceed as follows:

i) We define a set of eigenstates for each time, so that

$$H(t)|\varphi_n, t\rangle = E_n(t)|\varphi_n, t\rangle. \tag{5.4.1}$$

We now take the time derivative of this equation, and project onto $\langle\varphi_k, t|$, which gives

$$\langle\varphi_k, t|\left\{\frac{dH(t)}{dt}|\varphi_n, t\rangle + H(t)\frac{d}{dt}|\varphi_n, t\rangle\right\} = \langle\varphi_k, t|\left\{\frac{dE_n(t)}{dt}|\varphi_n, t\rangle + E_n(t)\frac{d}{dt}|\varphi_n, t\rangle\right\}, \tag{5.4.2}$$

and therefore, rearranging, we get

$$\langle\varphi_k, t|\left(\frac{d}{dt}|\varphi_n, t\rangle\right) = \frac{\langle\varphi_k, t|\frac{dH(t)}{dt}|\varphi_n, t\rangle}{E_n(t) - E_k(t)}, \qquad (k \neq n). \tag{5.4.3}$$

ii) Notice also that since $\langle\varphi_n, t|\varphi_n, t\rangle = 1$, we can say that

$$\text{Re}\left\{\left(\frac{d}{dt}\langle\varphi_n, t|\right)|\varphi_n, t\rangle\right\} = 0. \tag{5.4.4}$$

The imaginary part does not vanish, and can be written in terms of an *adiabatic following energy* $E_n^{\text{Adia}}(t)$ as

$$\langle\varphi_n, t|\left(\frac{d}{dt}|\varphi_n, t\rangle\right) \equiv -\frac{\mathrm{i}}{\hbar}E_n^{\text{Adia}}(t). \tag{5.4.5}$$

iii) If we now expand the wavefunction in terms of the adiabatic eigenstates as

$$|\psi, t\rangle = \sum_n c_n(t)|\varphi_n, t\rangle, \tag{5.4.6}$$

and then compute

$$\langle\varphi_k, t|H(t)|\psi, t\rangle = \mathrm{i}\hbar\langle\varphi_k, t|\left(\frac{d}{dt}|\psi, t\rangle\right), \tag{5.4.7}$$

$$= \mathrm{i}\hbar\sum_n\left\{\frac{dc_n(t)}{dt}\langle\varphi_k, t|\varphi_n, t\rangle + \mathrm{i}\hbar c_n(t)\langle\varphi_k, t|\frac{d|\varphi_n, t\rangle}{dt}\right\}. \tag{5.4.8}$$

iv) We now substitute:

a) Setting $\langle\varphi_k, t|H(t) = E(t)\langle\varphi_k, t|$ in the left hand side of (5.4.7);

b) Using (5.4.5) in the second term of (5.4.8) for $n = k$;

c) Using (5.4.3) in the second term of (5.4.8) for $n \neq k$;

to get

$$\mathrm{i}\hbar\frac{dc_k(t)}{dt} = \left(E_k(t) - E_k^{\text{Adia}}(t)\right)c_k(t) - \sum_{n\neq k}\frac{\langle\varphi_k, t|\mathrm{i}\hbar\left(\frac{d}{dt}H(t)\right)|\varphi_n, t\rangle}{E_n(t) - E_k(t)}c_n(t). \tag{5.4.9}$$

v) *The adiabatic approximation* assumes the last terms of this equation are negligible, which will be valid for sufficiently slow rate of change of $H(t)$, or for sufficiently large energy differences in the denominator, leaving the approximate equation

$$\mathrm{i}\hbar\frac{dc_k(t)}{dt} \approx \left(E_k(t) - E_k^{\text{Adia}}(t)\right)c_k(t). \tag{5.4.10}$$

vi) Since this equation does not couple the different coefficients $c_k(t)$ to each other, during the time evolution the magnitudes of the $c_k(t)$ are not changed. In particular, if the system starts in an eigenstate $|\varphi_k, t_0\rangle$ at time t_0, it will progress to the eigenstate $|\varphi_k, t_1\rangle$ at time t_1, but with an additional phase arising from the term $E_k^{\text{Adia}}(t)$.

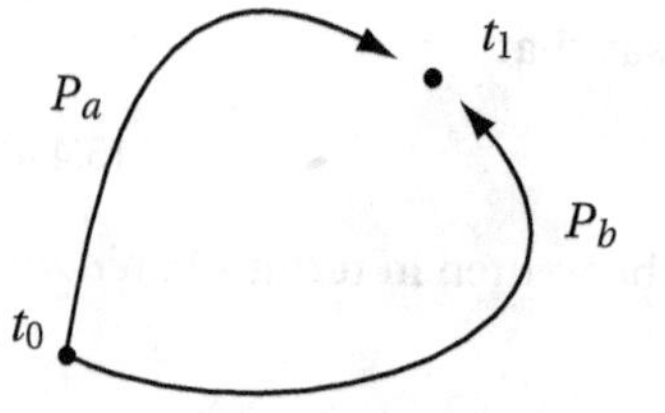

Fig. 5.7. A Berry phase results when the parameters determining the Hamiltonian (the Berry vector) are moved from an initial value at time t_0 to a final value at time t_1. The final phase of an adiabatically moved eigenstate depends on the path (such as P_a or P_b) along which we choose to move the Berry vector $\boldsymbol{B}(t)$.

b) The Berry Phase: The additional phase arising from $E_k^{\text{Adia}}(t)$ is known as a *Berry phase*. The problem posed and solved by *Berry* [5.2] was as follows.

The time-dependence of the Hamiltonian $H(t)$ normally arises because it is a function of certain parameters, in the case of the two-level system there will be a vector of two possibly time-dependent parameters, which we shall call the Berry vector, $\boldsymbol{B}(t) \equiv \{\Delta(t), \Omega_R(t)\}$. In the time period (t_0, t_1) under consideration, the Berry vector moves from an initial value to a final value,

$$\boldsymbol{B}(t_0) \quad \underset{P}{\longrightarrow} \quad \boldsymbol{B}(t_0). \tag{5.4.11}$$

Here P represents the path in the two-dimensional parameter space that the Berry vector takes as it proceeds from its initial value to its final value, and this path is not unique—see Fig. 5.7. The question that arises is: Does the final state depend in any way on the path chosen?

We have shown in (5.4.10) that the magnitudes of the coefficients are unchanged, but Berry showed that, in general, the phases of the coefficients $c_k(t_1)$ depend on the path followed by the Berry vector. Thus if the initial state under consideration is an eigenvector of $H(t_0)$, only one of the $c_k(t_0)$ is nonzero, and we reach a final state which is an eigenstate of $H(t_1)$, with only the phase depending on the path. The physical final state is unique.

But if the initial state is a superposition, the final phase of each coefficient depends on $E_k^{\text{Adia}}(t)$, which is in general different for each k and for each path. Thus the final state will be a different superposition for each path chosen, and therefore a completely different state.

The implications of the Berry phase are wide ranging, and have generated many publications. Even so, the initial work of Berry [5.2] is probably the clearest exposition on the subject.

c) Geometric Phase: Berry's phase is formulated within the framework of the adiabatic theorem—the change of the Hamiltonian is considered to be very slow, and the object of interest is an eigenstate of this Hamiltonian. However, the adiabatic formulation is not fundamental. The same kind of phase change can occur during the ordinary, and quite rapid, dynamical evolution of a quantum system governed by a time-dependent Hamiltonian. If there are two different dynamical paths joining two quantum states, then path-dependent phase difference can arise. We discuss this further in Sect. 24.4 in the context of a geometric phase quantum gate.

6. Coherent Manipulation of Multilevel Systems

While the basic idea of a two-level system is simple and elegant, in practice the direct use of two levels of any particular atom can present difficulties; for example, if there is sufficiently strong coupling between the ground and excited states, spontaneous emission from the excited state will also be present, destroying coherence. More flexibility can be obtained by using at least three levels and two coherent driving fields. This chapter is largely concerned with the techniques for using three-level atomic systems in conjunction with two optical driving fields to produce effective two-level systems. Such optical fields can be distinguished from each other by either frequency or polarization, and are normally chosen to address distinct transitions in the multilevel system under consideration.

As in the previous chapter, we shall treat the driving fields as being purely classical—we cover the fully quantized treatment in Chap. 13.

The driven three-level system plays a very prominent role in the theory of quantum devices, and this is largely because a classical driving field on one transition can be used to manipulate the properties of the other transitions in the system. In effect, one is able to create a two-level atom with parameters which can be manipulating using the driving field on another transition. A major application is given in Sect. 19.4, where we show how quantum jumps on a weak transition in a three-level system can be monitored with high efficiency using the fluorescence on a strong transition. This enables efficient measurement in the ion trap quantum quantum computer, which is mandatory for its operation.

Three-level systems will also play a prominent part in Chap. 27, Chap. 28 and Chap. 29, where we show how quantum information can be transferred and stored using devices based on ensembles of three-level atoms

6.1 Three-Level Systems

We will formulate the equations of motion for a three-level system driven by two laser fields, and then treat the various situations in which these can be used. There is a number of different ways in which the levels can be arranged, as shown in Fig. 6.1, but for definiteness, we will formulate the problem for the situation illustrated in Fig. 6.1 a.

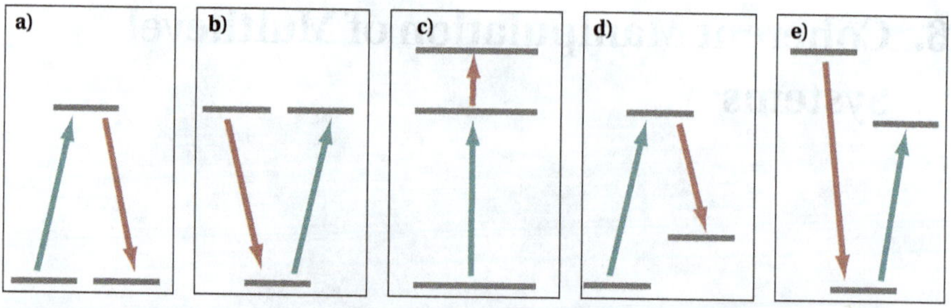

Fig. 6.1. The possible arrangements of three levels and transitions: **a)** The Λ-configuration; **b)** The V-configuration; **c)** The "ladder" configuration; **d)** The Λ-configuration with unequal lower energy levels, corresponding to a Raman process; **e)** The V-configuration with unequal upper energy levels, also corresponding to a Raman process.

6.1.1 The Λ-Configuration

For this system, in which the transitions driven by the laser fields have the appearance of the Greek letter Λ, we can write a Hamiltonian

$$\begin{aligned} H_{3\,\text{level}} &= \hbar\omega_e|e\rangle\langle e| + \hbar\omega_a|a\rangle\langle a| + \hbar\omega_b|b\rangle\langle b| \\ &\quad + \tfrac{1}{2}\hbar\left(\Omega_a e^{-i\nu_a t}|e\rangle\langle a| + \Omega_a^* e^{i\nu_a t}|a\rangle\langle e|\right) \\ &\quad + \tfrac{1}{2}\hbar\left(\Omega_b e^{-i\nu_b t}|e\rangle\langle b| + \Omega_b^* e^{i\nu_b t}|b\rangle\langle e|\right). \end{aligned} \tag{6.1.1}$$

The solution to the corresponding Schrödinger equation can be conveniently written

$$|\Psi_{3\,\text{level}}, t\rangle = A_e(t)|e\rangle + A_a(t)e^{i\nu_a t}|a\rangle + A_b(t)e^{i\nu_b t}|b\rangle, \tag{6.1.2}$$

which yields the resulting equation of motion

$$i\frac{d}{dt}\begin{pmatrix} A_a(t) \\ A_e(t) \\ A_b(t) \end{pmatrix} = \begin{pmatrix} \omega_a + \nu_a & \frac{1}{2}\Omega_a & 0 \\ \frac{1}{2}\Omega_a^* & \omega_e & \frac{1}{2}\Omega_b \\ 0 & \frac{1}{2}\Omega_b^* & \omega_b + \nu_b \end{pmatrix} \begin{pmatrix} A_a(t) \\ A_e(t) \\ A_b(t) \end{pmatrix}. \tag{6.1.3}$$

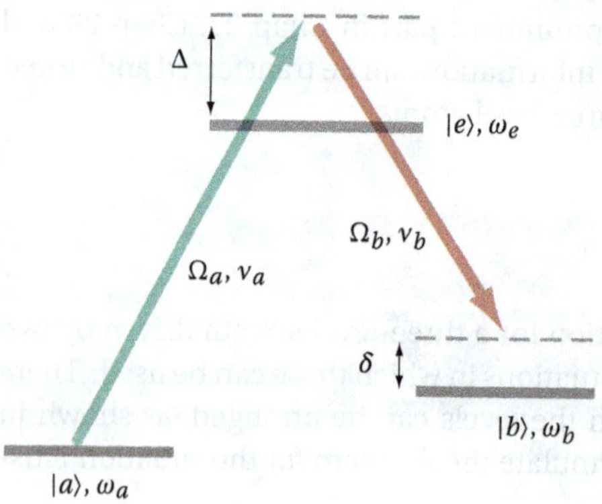

Fig. 6.2. A three-level atom being driven by two laser fields in a Λ-configuration, with angular frequencies ν_a, ν_b, and amplitudes given by the Rabi frequencies Ω_a, Ω_b. In the case drawn, the driving frequencies are both detuned from the corresponding resonant frequency, so that $\nu_a - (\omega_e - \omega_a) = \Delta$, $\nu_b - (\omega_e - \omega_b) = \Delta - \delta$.

a) Detunings: These equations are formulated for arbitrary driving frequencies, but the case of most interest is that in which:

i) The difference between the driving frequencies is close to the transition frequency between the two lower levels, as illustrated in Fig. 6.2. In this case the net effect of the two driving fields is to transfer population between state a and state b, since the difference frequency is almost on resonance with the frequency for the transition $a \leftrightarrow b$. This leads us to define a *small* detuning δ such that

$$\omega_a + \nu_a = \omega_b + \nu_b + \delta. \tag{6.1.4}$$

The detuning δ is often chosen to be zero.

ii) There is also another detuning, Δ, the detuning of ν_a from the transition frequency from $|a\rangle$ to the excited level $|e\rangle$, as illustrated in Fig. 6.2. It follows from (6.1.4) that each driving frequency is detuned from its corresponding transition frequency; that is

$$\nu_a = \omega_e - \omega_a + \Delta, \tag{6.1.5}$$

$$\nu_b = \omega_e - \omega_b + \Delta - \delta. \tag{6.1.6}$$

In practice Δ is chosen to be significantly larger than the the linewidth of the excited state $|e\rangle$, that is, the lasers are *far detuned* from the transition frequencies. This means that no significant population of the excited state arises as a result of the driving fields, and therefore spontaneous emission from $|e\rangle$ is minimized.

b) Standard Form of the Rabi Equations of Motion: In this case, we define

$$a_r(t) \equiv e^{i(\Delta+\omega_e)t} A_r(t), \qquad r = e,\ a,\ b, \tag{6.1.7}$$

and the equations can be written in the accepted standard form for the Raman configuration, in which only the detunings and the Rabi frequencies occur:

$$i\frac{d}{dt}\begin{pmatrix} a_a(t) \\ a_e(t) \\ a_b(t) \end{pmatrix} = \begin{pmatrix} 0 & \frac{1}{2}\Omega_a & 0 \\ \frac{1}{2}\Omega_a^* & -\Delta & \frac{1}{2}\Omega_b \\ 0 & \frac{1}{2}\Omega_b^* & -\delta \end{pmatrix}\begin{pmatrix} a_a(t) \\ a_e(t) \\ a_b(t) \end{pmatrix} \equiv \mathbf{\Omega}_{\text{Raman}}\begin{pmatrix} a_a(t) \\ a_e(t) \\ a_b(t) \end{pmatrix}. \tag{6.1.8}$$

Taking account of the approximate resonance conditions (6.1.4–6.1.6), the quantum state (6.1.2) can be written in terms of the time dependent states defined by

$$|a,t\rangle \equiv e^{-i\omega_a t}|a\rangle, \tag{6.1.9}$$

$$|e,t\rangle \equiv e^{-i(\Delta+\omega_e)t}|e\rangle, \tag{6.1.10}$$

$$|b,t\rangle \equiv e^{-i(\omega_b+\delta)t}|b\rangle, \tag{6.1.11}$$

in the straightforward form

$$|\Psi_{3\,\text{level}}, t\rangle = a_a(t)|a,t\rangle + a_e(t)|e,t\rangle + a_b(t)|b,t\rangle. \tag{6.1.12}$$

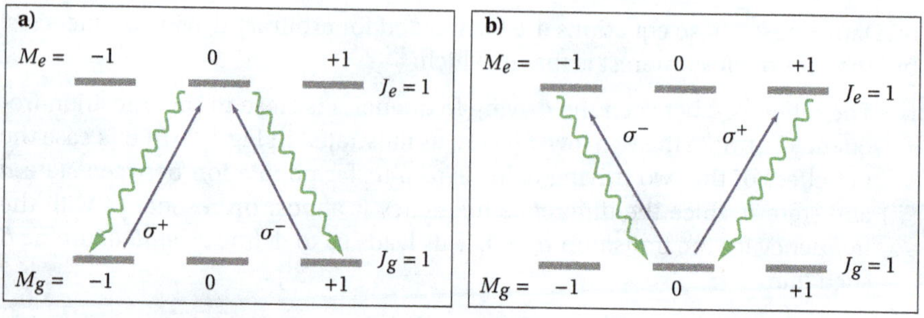

Fig. 6.3. a) Practical implementation of the Λ-configuration using counterpropagating circularly polarized laser beams. The vertical transitions $M_e \longleftrightarrow M_g$ with $M_e = M_g$ are forbidden by angular momentum conservation. **b)** A corresponding V-configuration uses the same laser beams, but is decoupled from the Λ-configuration.

6.1.2 Experimental Realization of the Λ-Configuration

An important experimental realization of a Λ-configuration uses a transition between levels e and g with $J_g = 1$ and $J_e = 1$ with two driving fields, as illustrated in Fig. 6.3. The optical electric field is a superposition of $\sigma_\pm$ circularly polarized waves

$$\boldsymbol{E}(\boldsymbol{x}, t) = \big(\hat{\boldsymbol{e}}^{(-1)}\mathcal{E}_{-1}(\boldsymbol{x}) + \hat{\boldsymbol{e}}^{(+1)}\mathcal{E}_{+1}(\boldsymbol{x})\big)\mathrm{e}^{-\mathrm{i}\omega t}. \tag{6.1.13}$$

Here $\mathcal{E}_\pm(\boldsymbol{x})$ can correspond either to two counterpropagating circularly polarized travelling waves, or to two standing waves, one with σ_+ polarization, and the other with σ_- polarization. The polarization vectors $\hat{\boldsymbol{e}}^{\pm 1}$ are as defined in (8.5.13).

This system has two sets of three levels, but the angular momentum conservation rules resulting from the photon polarizations allow transitions only within the two sets

$$\text{Λ-Configuration:} \quad M_g = -1 \quad \longleftrightarrow \quad M_e = 0 \quad \longleftrightarrow \quad M_g = 1, \tag{6.1.14}$$

$$\text{V-Configuration:} \quad M_e = -1 \quad \longleftrightarrow \quad M_g = 0 \quad \longleftrightarrow \quad M_e = 1. \tag{6.1.15}$$

Spontaneous emission with $M_e = M_g$ is also forbidden, so the Λ and V systems are quite independent. This means that we can ignore the dynamics of the V system as long as we can ensure that the initial state involves only those states which are part of the Λ-configuration.

6.2 Far-Detuned Raman Processes

Let us now consider the usual experimental configuration for a *far-detuned Raman process* in which

$$\Delta \gg \text{all of } \Omega_a, \Omega_b, \delta. \tag{6.2.1}$$

We will consider only the case where there is initially no population in the excited state $|e\rangle$, so that we can set $a_e(0) = 0$, and solve the equation for $a_e(t)$ in terms of the other amplitudes in the form

$$a_e(t) = -\mathrm{i}\int_0^t dt'\, \mathrm{e}^{\mathrm{i}\Delta(t-t')}\big(\Omega_a^* a_a(t') + \Omega_b^* a_b(t')\big). \tag{6.2.2}$$

The far-detuned condition means that the variation of $a_a(t')$ and $a_b(t')$ is much slower than that of $\mathrm{e}^{\mathrm{i}\Delta(t-t')}$, so we can approximate this equation as

$$a_e(t) \approx -\mathrm{i}\big(\Omega_a^* a_a(t) + \Omega_b^* a_b(t)\big)\int_0^t dt'\, \mathrm{e}^{\mathrm{i}\Delta(t-t')}, \tag{6.2.3}$$

$$= \left(\frac{1-\mathrm{e}^{\mathrm{i}\Delta t}}{\Delta}\right)\big(\Omega_a^* a_a(t) + \Omega_b^* a_b(t)\big). \tag{6.2.4}$$

We can now substitute for $a_e(t)$ in the equations for $a_a(t)$ and $a_b(t)$, and neglect the rapidly varying $\mathrm{e}^{\mathrm{i}\Delta t}$, to arrive at the approximate equations for these two levels

$$\mathrm{i}\frac{d}{dt}\begin{pmatrix} a_a(t) \\ a_b(t) \end{pmatrix} = \frac{1}{4\Delta}\begin{pmatrix} |\Omega_a|^2 & \Omega_a\Omega_b \\ \Omega_a^*\Omega_b^* & |\Omega_b|^2 - 4\Delta\delta \end{pmatrix}\begin{pmatrix} a_a(t) \\ a_b(t) \end{pmatrix}. \tag{6.2.5}$$

a) Equivalent Two-Level System: The resulting equations correspond to a two-level system, as illustrated in Fig. 6.4, with

i) Energy levels $|\Omega_a|^2/4\Delta$, $|\Omega_b|^2/4\Delta$;

ii) A driving field with Rabi frequency $\Omega_a\Omega_b/4\Delta$;

iii) A detuning of δ.

The approximation (6.2.5) is not essential, since the full equation (6.1.8) is easily solved numerically. The picture provided by the approximate two-level system is however valuable because of its simplicity, and because it is a very good description experimental situation in which the technique is normally used.

b) Practical Application: Raman processes are very frequently used to simulate two-level systems, because of their great flexibility. Using a Raman transition, it is possible to access transitions for which appropriate lasers are unavailable, or would require microwaves or even radio-frequencies if addressed directly. The two laser frequencies ν_a and ν_b must be locked to each other by some means, since the difference frequency can be very sensitive to relative drift. For example, by using appropriate nonlinear optical media, very small frequency differences can be induced by acousto-optic or electro-optic modulation of light arising from a single laser. Any drift of laser frequency is transferred to the light frequencies produced, and cancels in the difference frequency used in the Raman process.

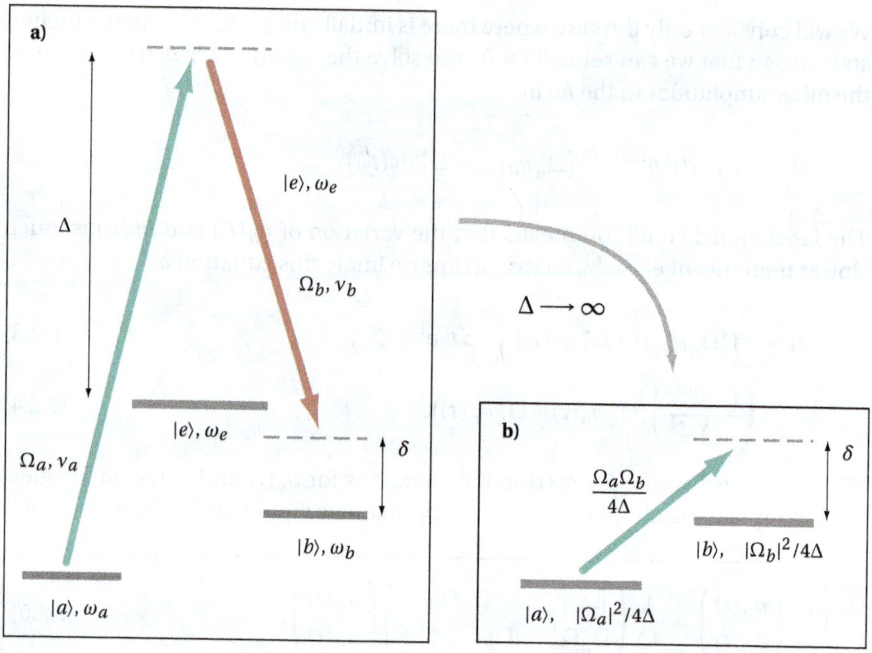

Fig. 6.4. a) A three level system driven by two fields, both far-detuned form their corresponding transitions $a \leftrightarrow e$ and $b \leftrightarrow e$. **b)** In the limit that $\Delta \to \infty$, the dynamics becomes effectively that of a two-level system, with parameters as shown.

6.2.1 The Dark State Configuration

The case when the detuning $\delta = 0$ is of very great importance, since it provides a nontrivial *dark state*, that is, a state which does not interact with the driving fields. This is clear from the approximate driving equation (6.2.5), which becomes when $\delta \to 0$

$$i\frac{d}{dt}\begin{pmatrix} a_a(t) \\ a_b(t) \end{pmatrix} = \frac{1}{4\Delta}\begin{pmatrix} \Omega_a \\ \Omega_b^* \end{pmatrix}\begin{pmatrix} \Omega_a^* & \Omega_b \end{pmatrix}\begin{pmatrix} a_a(t) \\ a_b(t) \end{pmatrix}. \qquad (6.2.6)$$

From this we see that the eigenvalues and eigenvectors are

$$\lambda_{\text{dark}} = 0, \qquad \boldsymbol{v}_{\text{dark}} = \begin{pmatrix} -\Omega_b \\ \Omega_a^* \end{pmatrix}, \qquad (6.2.7)$$

$$\lambda_{\text{bright}} = \frac{|\Omega_a|^2 + |\Omega_b|^2}{4\Delta}, \qquad \boldsymbol{v}_{\text{bright}} = \begin{pmatrix} \Omega_a \\ \Omega_b^* \end{pmatrix}. \qquad (6.2.8)$$

The first of these, the dark state of course does not evolve in time. The bright state has a rather slow time evolution in practice since the Rabi frequencies are not very large compared to the detuning Δ.

a) Exact Eigenvalues and Eigenvectors for the Three-Level System: In the case that $\delta = 0$, the matrix $\boldsymbol{\Omega}_{\text{Raman}}$ in (6.1.8) is easily diagonalized analytically, and we can study the behaviour exactly. The eigenvalues of $\boldsymbol{\Omega}_{\text{Raman}}$ are

$$\Omega_0 = 0, \qquad \Omega_\pm = = -\tfrac{1}{2}\Delta \pm \tfrac{1}{2}\sqrt{\Omega_{\text{eff}}^2 + \Delta^2}\,, \tag{6.2.9}$$

where we have defined the *effective Rabi frequency*

$$\Omega_{\text{eff}} \equiv \sqrt{|\Omega_a|^2 + |\Omega_b|^2}\,. \tag{6.2.10}$$

The corresponding normalized eigenvectors are

$$\boldsymbol{u}_0 = \frac{1}{\Omega_{\text{eff}}}\begin{pmatrix} -\Omega_b \\ 0 \\ \Omega_a^* \end{pmatrix}, \qquad \boldsymbol{u}_\pm = \frac{1}{\sqrt{4\Omega_\pm^2 + \Omega_{\text{eff}}^2}}\begin{pmatrix} \Omega_a \\ 2\Omega_\pm \\ \Omega_b^* \end{pmatrix}. \tag{6.2.11}$$

b) The Dark State: The eigenvector for *dark state* $\boldsymbol{u}_0$ corresponds exactly to its approximate form in (6.2.7). Most importantly, it *involves no component of the excited state* $|e\rangle$—an exact result. The situation is particularly interesting when the two lower levels are degenerate ground states, so that $\omega_a = \omega_b$. In this case the excited state $|e\rangle$ can decay by spontaneous emission to either of the ground states, but the dark state, which has no component of $|e\rangle$, cannot decay at all.

c) Bright States: The two states $\boldsymbol{u}_\pm$ are called *bright states*, and are analogous to the dressed states we have discussed for two level systems in Sect. 5.3. Because they have an excited state admixture, they couple to the incident light fields, and they also can decay by spontaneous emission.

6.2.2 Approximations in the Far-Detuned Case

Because of the importance of the far-detuned configuration, it is worthwhile to put down the approximate forms of eigenvalues, eigenvectors etc. for future reference, and to assist in the qualitative understanding of the behaviour of the three-level system.

a) Approximate Eigenvalues: When the far-detuned conditions (6.2.1) are satisfied, the eigenvalues (6.2.9) can be approximated by

$$\Omega_0 = 0, \tag{6.2.12}$$

$$\Omega_+ \approx \frac{\Omega_{\text{eff}}^2}{4\Delta}, \tag{6.2.13}$$

$$\Omega_- \approx -\Delta - \frac{\Omega_{\text{eff}}^2}{4\Delta} \approx -\left(\frac{\Delta}{\Omega_{\text{eff}}}\right)^2 \Omega_+\,. \tag{6.2.14}$$

b) Relevant Time Scales: Thus it is clear that in the far-detuned case, the time scale represented by Ω_- is very much faster than that represented by Ω_+, and in fact we can also say that

$$\Omega_+ \ll \Omega_{\text{eff}} \ll |\Omega_-|\,. \tag{6.2.15}$$

The time scales of the system are thus very different from the Rabi frequencies Ω_a and Ω_b; the fast time scale is that of the detuning Δ, while the slow time scale is of the same kind as that found in an optical FORT, which we have already met in *Bk. I: Sect.12.3.4* and Ex. 4.2.

c) Approximate Eigenvectors: The eigenvectors $\boldsymbol{u}_\pm$ now take the forms, in the lowest order of approximation,

$$\boldsymbol{u}_+ \approx \frac{1}{\sqrt{(1+\eta)}}\begin{pmatrix}\Omega_a/\Omega_{\text{eff}}\\ 2\sqrt{\eta}\\ \Omega_b^*/\Omega_{\text{eff}}\end{pmatrix} \xrightarrow{\eta\to 0} \frac{1}{\Omega_{\text{eff}}}\begin{pmatrix}\Omega_a\\ 0\\ \Omega_b^*\end{pmatrix} \tag{6.2.16}$$

$$\boldsymbol{u}_- \approx \frac{1}{\sqrt{1+3\eta}}\begin{pmatrix}\Omega_a/2\Delta\\ -1-\eta\\ \Omega_b^*/2\Delta\end{pmatrix} \xrightarrow{\eta\to 0} \begin{pmatrix}0\\ -1\\ 0\end{pmatrix}, \tag{6.2.17}$$

$$\text{where } \eta \equiv \frac{\Omega_{\text{eff}}^2}{4\Delta^2} \ll 1. \tag{6.2.18}$$

6.2.3 Effect of Spontaneous Emission

When using the Raman configuration, we have to take account of the possible decay of the excited state $|e\rangle$ by spontaneous emission. We will postpone an explicit consideration of spontaneous emission in three-level systems until Chap. 28, but at this stage we do need to make sure that the decay constant Γ is not so large as to cause difficulties. To accomplish this, it is most advantageous to set the parameters so that the detuning Δ corresponds to a far-detuned system, as in (6.2.1). Under these conditions we can show that the population of the excited state $|e\rangle$ is always small for processes of interest, and thus provided Γ is not very large, we can probably neglect spontaneous emission. Quantitatively, we would require

$$\Gamma \ll \Omega_a, \qquad \Gamma \ll \Omega_b. \tag{6.2.19}$$

The basic effect can be accounted for by setting $\Delta \to \Delta + i\Gamma$ in the equation of motion (6.1.8). In the case that $\Gamma \ll \Omega_{\text{eff}} \ll \Delta$, the eigenvalue $\Omega_0 = 0$ is not changed, while the other eigenvalues become approximately (to the lowest non-zero order)

$$\Omega_+ \approx \frac{\Omega_{\text{eff}}^2}{4\Delta}\left(1-\frac{i\Gamma}{4\Delta}\right), \tag{6.2.20}$$

$$\Omega_- \approx -\Delta - \frac{\Omega_{\text{eff}}^2}{4\Delta} - i\Gamma. \tag{6.2.21}$$

The decay rate relative to the oscillation frequency is therefore of order of magnitude Γ/Δ in both cases.

6.3 Quantum State Engineering in the Raman Configuration

The equivalence of the Raman configuration to a driven two-level system makes it possible to manipulate the two ground state levels $|a\rangle$ and $|b\rangle$ in exactly the same way as described in Chap. 5, provided the effects of spontaneous emission of the excited state $|e\rangle$ are not too serious.

The existence of the dark state provides another possibility for manipulating these levels which has many advantages. The dark state amplitude $\boldsymbol{u}_0$ as defined in (6.2.11) depends on the values of the Rabi frequencies Ω_a and Ω_b, so that the dark state itself can be written

$$|\Psi_{\text{dark}}, \Omega_a, \Omega_b\rangle = \frac{\Omega_b|a\rangle - \Omega_a|b\rangle}{\Omega_{\text{eff}}} \longrightarrow \begin{cases} |a\rangle & \text{as} \quad \Omega_a \to 0, \\ -|b\rangle & \text{as} \quad \Omega_b \to 0. \end{cases} \tag{6.3.1}$$

The fact that the dark state depends parametrically on the Rabi frequencies means that if we can vary these Rabi frequencies adiabatically as in (6.3.1), we can transfer from $|a, t\rangle$ to $|b, t\rangle$ without ever involving the excited state $|e, t\rangle$, and thus avoiding spontaneous emission from the excited state during the transfer.

6.3.1 Adiabatic Population Transfer along a Dark State—STIRAP

This parametric dependence on the Rabi frequencies is used in the technique of *stimulated Raman adiabatic passage*, nowadays most commonly referred to by its acronym—STIRAP [6.1–6.3]. It is a method of using partially overlapping pulses, one from each laser, to produce complete population transfer between two quantum states of an atom or molecule.

The method is illustrated in Fig. 6.5, using an adiabatic pulse sequence as shown in Fig. 6.5 b, and proceeds as follows:

i) The atom is first prepared in the state $|a\rangle$.

ii) We first turn on Ω_b, leaving $\Omega_a = 0$. The state $|a\rangle$ is then by (6.3.1), in the dark state $|\Psi_{\text{dark}}, 0, \Omega_b\rangle$.

iii) The frequencies Ω_a and Ω_b are then varied smoothly, so that $\Omega_b \to 0$ at the same time as Ω_a acquires a non-zero value, and at no time are both of Ω_a and Ω_b equal to zero. This leaves the atom in the state $|\Psi_{\text{dark}}, \Omega_a, 0\rangle = -|b\rangle$.

   This can be done by using a smooth pulse shape which is the same for both Ω_a and Ω_b, but with a time delay T between the two pulses.

iv) This provides a *coherent* transfer between the two ground states, $|a\rangle \to -|b\rangle$. Unlike the case of a π-pulse on a two-level system (as treated in Sect. 5.3.3) the transfer does not require a precise time or precise Rabi frequencies.

   As long as the time variation is sufficiently slow, the process can be described purely qualitatively. Both of Ω_a and Ω_b can both be zero initially, and then Ω_b should increase, followed by an increase in Ω_a. After that both should go to

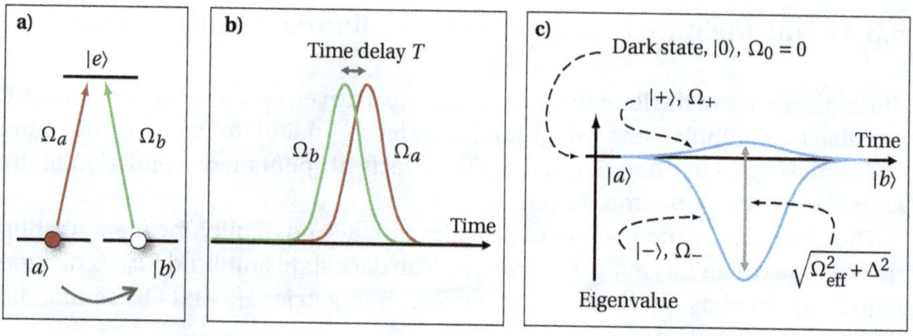

Fig. 6.5. Adiabatic transfer along a dark state. **a)** Energy levels of the bare states; **b)** The pulse sequence used, in which the field producing the $e \leftrightarrow b$ transition is applied first, and then that for the $e \leftrightarrow a$ is applied after a time delay T; **c)** Plot of the eigenvalues of $\boldsymbol{\Omega}_{\text{Raman}}$, as defined in (6.1.8) corresponding to the dark state $|0\rangle$ and the two bright states $|\pm\rangle$.

zero in the reverse order. Independently of the particular values of Ω_a and Ω_b during this process, the transfer $|a\rangle \to -|b\rangle$ will occur.

v) This procedure is also different from a π-pulse of (5.3.21), where the *same pulse* (if we choose $e^{i\phi} = -i$) transfers $-|a\rangle \to |b\rangle$ and $|b\rangle \to |a\rangle$—it thus effects a *swap* of the two states.

For the adiabatic passage procedure, if we start in $|b\rangle$ and use the *same* adiabatic pulse sequence, this does not yield $|a\rangle$. Instead we will have admixtures of the other dressed states, which are bright states.

Hence this procedure does *not* swap the two ground states, or represent a swap on two qubit states.

vi) For an early experimental implementation, see [6.1].

6.3.2 Conditions for the Validity of the Adiabatic Transfer Procedure

There will always be a small population of the excited state, and the effect of this can be estimated using the procedures introduced for the adiabatic theorem in Sect. 5.4.1. We start from the equations of motion in the form (6.1.8), from which we see that the time derivative of the Hamiltonian required for the adiabatic theorem is

$$\frac{dH(t)}{dt} = \hbar\dot{\boldsymbol{\Omega}}_{\text{Raman}} = \tfrac{1}{2}\hbar\begin{pmatrix} 0 & \dot{\Omega}_a & 0 \\ \dot{\Omega}_a^* & 0 & \dot{\Omega}_b \\ 0 & \dot{\Omega}_b^* & 0 \end{pmatrix}. \tag{6.3.2}$$

The time dependent eigenvectors are simply those given by (6.2.11), so we can write the state of the system in terms of them as vector

$$\boldsymbol{w}(t) = c_0(t)\boldsymbol{u}_0(t) + c_+(t)\boldsymbol{u}_+(t) + c_-(t)\boldsymbol{u}_-(t). \tag{6.3.3}$$

We will consider the case when $c_\pm(t)$ are both very small, so that we can use (5.4.9) to get

$$\mathrm{i}\frac{dc_\pm(t)}{dt} \approx \left(\Omega_\pm(t) - \Omega_\pm^{\mathrm{Adia}}(t)\right)c_\pm(t) - \frac{\boldsymbol{u}_\pm^\dagger(t)\dot{\boldsymbol{\Omega}}_{\mathrm{Raman}}\boldsymbol{u}_0(t)}{\Omega_\pm(t) - \Omega_0(t)}\,c_0(t), \tag{6.3.4}$$

where we have omitted the very small terms proportional to $c_\pm(t)$ in the second term of the right-hand side of this equation. We can now also use the approximate eigenvectors (6.2.16–6.2.18) to get

$$\frac{\boldsymbol{u}_+^\dagger\dot{\boldsymbol{\Omega}}_{\mathrm{Raman}}\boldsymbol{u}_0}{\Omega_+ - \Omega_0} \approx \frac{2}{\sqrt{\eta}}\left(\frac{\dot{\Omega}_b\Omega_a^* - \dot{\Omega}_a^*\Omega_b}{\Omega_{\mathrm{eff}}\Delta}\right) = 4\left(\frac{\dot{\Omega}_b\Omega_a^* - \dot{\Omega}_a^*\Omega_b}{\Omega_{\mathrm{eff}}^2}\right), \tag{6.3.5}$$

$$\frac{\boldsymbol{u}_-^\dagger\dot{\boldsymbol{\Omega}}_{\mathrm{Raman}}\boldsymbol{u}_0}{\Omega_- - \Omega_0} \approx -2\left(\frac{\dot{\Omega}_b\Omega_a^* - \dot{\Omega}_a^*\Omega_b}{\Omega_{\mathrm{eff}}\Delta}\right). \tag{6.3.6}$$

a) Source and Measure of Non-Adiabaticity: Let us assume that Ω_a and Ω_b are of the same order of magnitude, and their rates of change as a function of time are also of the same order of magnitude. Then in this case, (6.3.5) shows the rate of transfer of amplitude from the dark state $\boldsymbol{u}_0$ to the bright state $\boldsymbol{u}_+$ is of the same order of magnitude as the typical logarithmic rate of change, such as $\dot{\Omega}_a/\Omega_a$.

For the rate of transfer to the other bright state, $\boldsymbol{u}_-$, there is an additional factor $\sqrt{\eta} = \Omega_{\mathrm{eff}}/2\Delta$, which is very small in the far-detuned case we are considering. Thus population loss as a result of the finite rate of sweep of the Rabi frequencies is largely to $|+\rangle$, the upper bright state.

b) Source and Measure of Spontaneous Emission: However, the approximate forms for the eigenvectors (6.2.16–6.2.18) show that the component of the bare excited state $|e,t\rangle$ arising from $\boldsymbol{u}_+$ is approximately $2\sqrt{\eta}$, while the component of the bare excited state in $\boldsymbol{u}_-$ is approximately -1. This means that the bare excited state contributes in approximately equal amounts to both terms in (6.3.5, 6.3.6), and this in turn means that the amount of spontaneous emission associated with both terms will be approximately the same.

7. The Two-Level System Including Atomic Motion

When it interacts with an atom, an optical field exerts a force which we have so far neglected, because this force is indeed very small. However, if the optical field is sufficiently intense, this force can easily be detected. In this chapter we will consider the mechanical effects which arise when an atom moves in a very strong coherent laser field, and set up the quantum mechanical formalism needed to describe such processes. This will be an extension of the basic formalism of the previous two chapters, in which we omitted the term in the Hamiltonian corresponding to the kinetic energy of the atom—that is, of the centre of mass motion. This approximation is equivalent to allowing the atomic mass to be infinite, and thus to the neglect of recoil when the atom emits or absorbs a photon. The remedy is very simple, we add an appropriate kinetic energy term.

An atom under the influence of an optical field displays a variety of different kinds of behaviour, depending on the nature of any other forces the atom experiences in the particular experimental arrangement. For example:

i) The quantized motion of electromagnetically trapped ions interacting with classical light fields is the basis for the ion trap quantum computer. This forms the subject matter of Chap. 23 and of Chap. 24, where the detailed description of the ion trap quantum computer is formulated.
ii) In the case of laser cooling, the subject matter of Part IV, the transfer of population $e \leftrightarrow g$ induced by the coherent optical field is combined with the incoherent transfer $e \to g$ arising from spontaneous emission. With an appropriate choice of detuning between the transition frequency and the frequency of the optical field, a net loss of kinetic energy is induced—thus cooling arises. In addition, incoherent effects arise from the thermalized state of the atoms.

This chapter, however, is concerned only with the interactions between *freely moving* two-level atoms and *coherent* light—of which the most important case is an optical standing wave. Even when the electromagnetic field is treated classically, this gives rise to genuinely quantum phenomena, such as the Bragg scattering of a matter field by a standing light wave. Such effects were predicted by *Kapitsa* and *Dirac* in 1933 [7.1], but only became experimentally accessible with the availability of strong laser fields. Thus, it was only in the 1985 that the diffraction of atoms by light was experimentally observed by *Gould, Ruff, Moskowitz* and *Pritchard* [7.2, 7.3].

7.1 Hamiltonian Including Kinetic and Electronic Terms

We will proceed by explicitly putting a kinetic energy term into the atomic Hamiltonian, and include the spatial dependence of the electromagnetic field, which we shall not at this stage quantize. Thus, we take the Hamiltonian for the two-level system in the rotating-wave approximation, as in (4.1.18), and modify it so that it becomes

$$H = \frac{\boldsymbol{P}^2}{2M} + \hbar\omega_{eg}|e\rangle\langle e| - \boldsymbol{d}_{eg}\cdot\boldsymbol{E}^{(+)}_{\rm cl}(\boldsymbol{X},t)|e\rangle\langle g| - \boldsymbol{d}_{ge}\cdot\boldsymbol{E}^{(-)}_{\rm cl}(\boldsymbol{X},t)|g\rangle\langle e|. \tag{7.1.1}$$

Here $\boldsymbol{X}$ and $\boldsymbol{P}$ are the position and momentum operators for the atom.

The argument of the field operator is now $\boldsymbol{X}$, the position operator of atom. This means that the electromagnetic field, even when treated classically, becomes an operator in the translational Hilbert space as a result of its position dependence. We will develop a fully quantized treatment of the mechanical effects of light in Chap. 15 in the context of laser cooling, and in Chap. 23 in the context of ion trapping quantum processing. In these treatments the electromagnetic field operator has this kind of dependence on the position operator of the atom in addition to its intrinsic quantization in terms of photon creation and destruction operators.

7.1.1 Interaction with a Travelling Wave

Let us consider the transition $e \to g$ being induced, as in Fig. 7.1, by an incident travelling wave laser beam with wavevector $\boldsymbol{k}$, which can be written as

$$\boldsymbol{E}^{(+)}_{\rm cl}(\boldsymbol{x},t) = \boldsymbol{\epsilon}\mathcal{E}(t)e^{i\boldsymbol{k}\cdot\boldsymbol{x}-i\omega t}. \tag{7.1.2}$$

We will also introduce the *Rabi frequency*, in much the same way as in (5.1.9), in this case defined as by

$$\Omega_R(t) \equiv \frac{2\boldsymbol{d}_{eg}\cdot\boldsymbol{\epsilon}\mathcal{E}(t)}{\hbar}. \tag{7.1.3}$$

7.1.2 Basis States

The basis states for the Hilbert space are now

$$|g,\boldsymbol{p}\rangle \equiv |g\rangle\otimes|\boldsymbol{p}\rangle, \qquad |e,\boldsymbol{p}\rangle \equiv |e\rangle\otimes|\boldsymbol{p}\rangle, \tag{7.1.4}$$

where $|\boldsymbol{p}\rangle$ is of course the eigenstate of the momentum operator $\boldsymbol{P}$.

The classical field is now evaluated at the operator position of the atom, so that it will be proportional to $\exp(i\boldsymbol{k}\cdot\boldsymbol{X})$. Noting that

$$e^{i\boldsymbol{k}\cdot\boldsymbol{X}}|\boldsymbol{p}\rangle = |\boldsymbol{p}+\hbar\boldsymbol{k}\rangle, \tag{7.1.5}$$

it follows that the interaction couples these states according to

$$\left(-\boldsymbol{d}_{eg}\cdot\boldsymbol{E}^{(+)}_{\rm cl}(\boldsymbol{X},t)|e\rangle\langle g|\right)|g,\boldsymbol{p}\rangle = -\tfrac{1}{2}\hbar\Omega_R(t)|e,\boldsymbol{p}+\hbar\boldsymbol{k}\rangle, \tag{7.1.6}$$

This means that absorption of a photon $g \to e$ is associated with a recoil momentum $\hbar\boldsymbol{k}$. In a similar way, emission of a photon corresponds to

$$\left(-\boldsymbol{d}_{ge}\cdot\boldsymbol{E}^{(-)}_{\rm cl}(\boldsymbol{X},t)|g\rangle\langle e|\right)|e,\boldsymbol{p}+\hbar\boldsymbol{k}\rangle = -\tfrac{1}{2}\hbar\Omega_R(t)|g,\boldsymbol{p}\rangle. \tag{7.1.7}$$

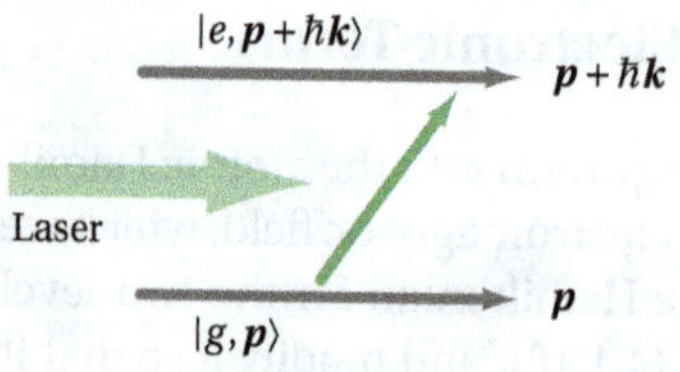

Fig. 7.1. An incident travelling wave laser beam induces a momentum change of $\hbar \boldsymbol{k}$ when a photon absorbed induces the transition $e \to g$.

7.1.3 Equations of Motion

The set of states

$$\{|g, \boldsymbol{p}\rangle, |e, \boldsymbol{p}+\hbar\boldsymbol{k}\rangle\}, \tag{7.1.8}$$

forms a closed family under absorption and emission from a travelling wave. This means that we can write a wavefunction in terms of this set of basis states, and we will choose to write it in the form

$$|\psi, t\rangle = a_g(t)|g, \boldsymbol{p}\rangle + a_e(t)\mathrm{e}^{-\mathrm{i}\omega t}|e, \boldsymbol{p}+\hbar\boldsymbol{k}\rangle. \tag{7.1.9}$$

From the Hamiltonian (7.1.1), the Schrödinger equation becomes a pair of equations for the coefficients:

$$\mathrm{i}\dot{a}_e(t) = \left(\frac{(\boldsymbol{p}+\hbar\boldsymbol{k})^2}{2M\hbar} - \Delta\right) a_e(t) - \tfrac{1}{2}\Omega_R(t) a_g(t), \tag{7.1.10}$$

$$\mathrm{i}\dot{a}_g(t) = \frac{\boldsymbol{p}^2}{2M\hbar} a_g(t) - \tfrac{1}{2}\Omega_R(t) a_e(t), \tag{7.1.11}$$

$$\text{with} \qquad \Delta \equiv \omega - \omega_{eg}. \tag{7.1.12}$$

a) Rabi Equations of Motion and the Doppler Shift: If we explicitly include the phase of free particle motion (in the ground state) by writing

$$A_g(t) = a_g(t)\mathrm{e}^{-\mathrm{i}\boldsymbol{p}^2 t/2M\hbar}, \qquad A_e(t) = a_e(t)\mathrm{e}^{-\mathrm{i}\boldsymbol{p}^2 t/2M\hbar}, \tag{7.1.13}$$

the equations of motion become

$$\mathrm{i}\frac{d}{dt}\begin{pmatrix} A_e(t) \\ A_g(t) \end{pmatrix} = \begin{pmatrix} -\Delta + \boldsymbol{k}\cdot\boldsymbol{v} + \omega_{\text{Recoil}} & -\tfrac{1}{2}\Omega_R(t) \\ -\tfrac{1}{2}\Omega_R(t) & 0 \end{pmatrix} \begin{pmatrix} A_e(t) \\ A_g(t) \end{pmatrix}. \tag{7.1.14}$$

In these equations we note the following points:

i) *Recoil Energy:* The energy

$$E_R \equiv \hbar\omega_{\text{Recoil}} = \frac{\hbar^2 k^2}{2M}, \tag{7.1.15}$$

is known as the *recoil energy*, that is, the kinetic energy of an atom with momentum exactly $\hbar\boldsymbol{k}$, the momentum of a single photon.

The term "recoil energy" reflects the fact that momentum conservation requires that, when an atom at rest absorbs a photon through a transition $e \to g$, some of the photon's energy must supply the kinetic energy of the excited atom; in fact an amount equal to the recoil energy.

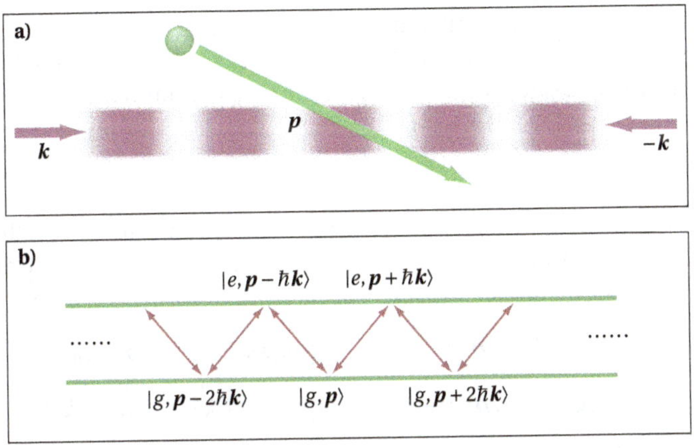

Fig. 7.2. a) Schematic diagram of a two-level atom moving through a standing wave field, formed from counterpropagating laser beams with wavenumbers $\pm\boldsymbol{k}$. In practice the optical wavelength is very much larger—by a factor of about 1000—than the diameter of the atom. **b)** Schematic of possible transitions between the excited and ground states; odd multiples of $\hbar k$ are added to the excited state, while even multiples of $\hbar k$ are added in the ground state.

ii) *Doppler Shift*: If the atom is moving with velocity $\boldsymbol{v} = \boldsymbol{p}/M$, the energy required is different from the recoil energy, and is in fact $\hbar(\boldsymbol{k}\cdot\boldsymbol{v} + \omega_{\text{Recoil}})$. Here the additional term $\boldsymbol{k}\cdot\boldsymbol{v}$ represents the Doppler shift of the electromagnetic field as seen by the atom moving with velocity $\boldsymbol{v}$.

iii) The equations in this form (7.1.14) are of exactly the same form as the Rabi equations given by (5.1.8, 5.1.6), with the replacement $\Delta \to \Delta - \boldsymbol{k}\cdot\boldsymbol{v} + \omega_{\text{Recoil}}$; hence all of the results for the driven two-level atom can be taken over the system in which recoil is included.

b) Applications: This process forms the most elementary example of the manipulation of an atom's motion using light. In principle, it can be seen as a kind of beam splitter for atomic interferometry. For further information on optics and interferometry with atoms and molecules see the review of by *Cronin, Schmiedmayer* and *Pritchard* [7.4].

7.2 Motion of an Atom in a Standing Light Wave

When an atom moves in a standing light wave, the description in terms of a two-level system and a Rabi Hamiltonian, as in (7.1.14), is no longer possible. In this section we will formulate a description of an atom moving in a standing wave of the form

$$\boldsymbol{E}^{(+)}_{\text{cl}}(\boldsymbol{x}, t) = \boldsymbol{\epsilon}\mathcal{E}(t)\mathrm{e}^{-\mathrm{i}\omega t}\cos\boldsymbol{k}\cdot\boldsymbol{x}, \tag{7.2.1}$$

as illustrated in Fig. 7.2. We can show that the standing wave can act as a diffraction grating for the matter wave field of the atom, an effect predicted by *Kapitsa* and *Dirac* in 1933 [7.1] for the motion of electrons in a standing wave light field.

7.2.1 Basis States

Because the standing wave consists of two counterpropagating travelling waves, absorption and emission of photons with momenta $\pm\hbar\boldsymbol{k}$ takes place. Correspondingly, the interaction couples the states according to

$$\left(-\boldsymbol{d}_{eg}\boldsymbol{E}_{\text{cl}}^{(+)}(\boldsymbol{X},t)|e\rangle\langle g|\right)|g,\boldsymbol{p}\rangle = -\tfrac{1}{2}\hbar\Omega_R(t)\Big(|e,\boldsymbol{p}+\hbar\boldsymbol{k}\rangle + |e,\boldsymbol{p}-\hbar\boldsymbol{k}\rangle\Big), \tag{7.2.2}$$

$$\left(-\boldsymbol{d}_{ge}\boldsymbol{E}_{\text{cl}}^{(-)}(\boldsymbol{X},t)|g\rangle\langle e|\right)|e,\boldsymbol{p}\rangle = -\tfrac{1}{2}\hbar\Omega_R(t)\Big(|g,\boldsymbol{p}+\hbar\boldsymbol{k}\rangle + |g,\boldsymbol{p}-\hbar\boldsymbol{k}\rangle\Big). \tag{7.2.3}$$

In this case all of the states $|g,\boldsymbol{p}+n\hbar\boldsymbol{k}\rangle$ and $|e,\boldsymbol{p}+m\hbar\boldsymbol{k}\rangle$ with either n even and m odd, or n odd and m even, are coupled to each other.

7.2.2 Equations of Motion

Let us set $\boldsymbol{q}\equiv\hbar\boldsymbol{k}$ and, corresponding to (7.1.9) and (7.1.14), write the wavefunction in the form

$$|\psi,t\rangle = \mathrm{e}^{\mathrm{i}\boldsymbol{p}^2 t/2M\hbar}\sum_n\Big(B_g(n,t)|g,\boldsymbol{p}+n\boldsymbol{q}\rangle + \mathrm{e}^{-\mathrm{i}\omega t}B_e(n,t)|e,\boldsymbol{p}+n\boldsymbol{q}\rangle\Big). \tag{7.2.4}$$

The resulting equations of motion can then be written

$$\left.\begin{aligned}\mathrm{i}\dot{B}_e(n,t) &= (\omega_n-\Delta)B_e(n,t) - \tfrac{1}{2}\Omega_R\Big(B_g(n+1,t)+B_g(n-1,t)\Big),\\ \mathrm{i}\dot{B}_g(n,t) &= \quad\omega_n B_g(n,t) \quad - \tfrac{1}{2}\Omega_R\Big(B_e(n+1,t)+B_e(n-1,t)\Big),\end{aligned}\right\} \tag{7.2.5}$$

$$\omega_n \equiv n\boldsymbol{k}\cdot\boldsymbol{v} + n^2\omega_{\text{Recoil}}. \tag{7.2.6}$$

The dependence on the velocity $\boldsymbol{v}\equiv\boldsymbol{p}/M$ is only through the quantity ω_n, so that the component of $\boldsymbol{v}$ perpendicular to $\boldsymbol{k}$ is unimportant. The original experiments [7.2, 7.3, 7.5] on this subject used a supersonic jet of sodium atoms with speeds of of up to 1000 m/s, while experiments with ultra-cold atoms [7.6] use velocities of the order 1 mm/s. The theoretical description is nevertheless unchanged.

a) Parameter Regimes: If we measure time in units of Ω_R^{-1}, there are three dimensionless parameters determining the behaviour of solutions of these equations, namely

$$\frac{\boldsymbol{k}\cdot\boldsymbol{v}}{\Omega_R} = \frac{\boldsymbol{k}\cdot\boldsymbol{p}}{M\Omega_R}, \tag{7.2.7}$$

$$w \equiv \frac{\omega_{\text{Recoil}}}{\Omega_R} = \frac{\hbar k^2}{2M\Omega_R} \tag{7.2.8}$$

$$\delta \equiv \frac{\Delta}{\Omega_R}. \tag{7.2.9}$$

However, the first parameter is in practice equivalent to w, since the component of $\boldsymbol{p}$ parallel to $\boldsymbol{k}$ is almost always a small integral multiple of $\boldsymbol{k}$. The parameter regimes of interest are then as follows:

i) *Far-Detuned Regime*: Here we set $\Delta \gg \Omega$, and $\Delta \gg \omega_{\text{recoil}}$. We can make an adiabatic solution for $B_e(n,t)$ of the first equation and substitute this into the second equation, getting equations of motion for the ground state amplitudes:

$$\mathrm{i}\dot{B}_g(n,t) = \omega_n B_g(n,t) - \frac{\Omega_R^2}{4\Delta}\Big(B_g(n+2,t) + 2B_g(n,t) + B_g(n-2,t)\Big). \quad (7.2.10)$$

The far-detuned regime is very commonly used, and is the same as that which produces an optical potential, as discussed in Ex. 4.2. Here we get an optical lattice, which we shall discuss further in Sect. 7.2.3.

ii) *Raman–Nath Regime*: This is defined by $w \ll 1$. The Raman–Nath *approximation* sets $w = 0$, and amounts to assuming $M \to \infty$. It also applies, however, to the situation which pertains when the optical field is applied for a very short time, during which any spatial recoil of the atom is very small

iii) *Bragg Regime*: Here w is finite, and the optical pulse is applied for a time T such that $\omega_{\text{Recoil}} T$ is of the order of magnitude of 1, or greater.

Let us set the initial condition to be

$$B_g(n,0) = \delta_{n,0}\,. \quad (7.2.11)$$

Bragg scattering occurs when there is a non-zero integral value $2N$ such that $\omega_{2N} = \omega_0 = 0$. The reasoning is most easily explained in the far-detuned case using (7.2.10), which couples $B_g(n)$ for all even n to each other. The states corresponding to $n = 0$ and $n = N$ are in resonance, whereas all others are non-resonant.

This resonance occurs when $\omega_{2N} = \omega_0 = 0$, and this corresponds to the *Bragg condition*; namely

$$2N = \frac{2\boldsymbol{k}\cdot\boldsymbol{p}}{\hbar \boldsymbol{k}^2}. \quad (7.2.12)$$

This can be rewritten by setting

$$d \equiv \frac{\pi}{|\boldsymbol{k}|}, \qquad \lambda_{\text{Matter}} \equiv \frac{h}{|\boldsymbol{p}|}, \qquad \boldsymbol{k}\cdot\boldsymbol{p} \equiv |\boldsymbol{k}||\boldsymbol{p}|\sin\theta, \quad (7.2.13)$$

and then takes the familiar form

$$N\lambda_{\text{Matter}} = 2d\sin\theta. \quad (7.2.14)$$

Here, d is half of the optical wavelength, that is, the distance between planes formed by the maxima of $\cos^2 \boldsymbol{k}\cdot\boldsymbol{x}$, which forms the effective optical potential.

b) Behaviour of Solutions: For appropriate values of the parameter, the solutions of the equations of motion (7.2.5) can display the Raman–Nath regime at

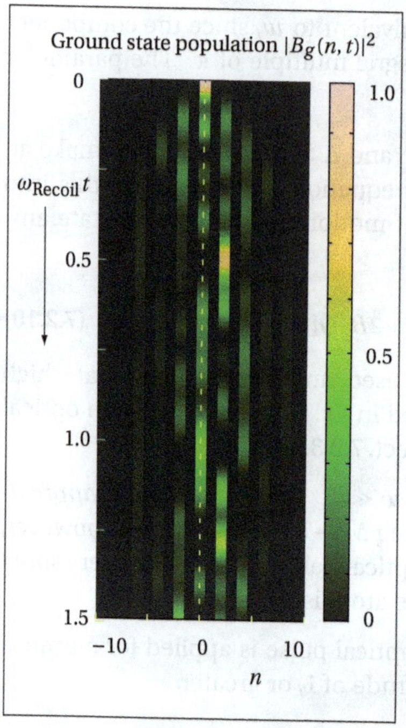

Fig. 7.3. The figure illustrates both the Raman–Nath regime and the Bragg regime. The equations (7.2.5) are simulated by setting $\hbar = 1$, and with the parameter values $M = 0.3$, $\Omega = 2$, $\hbar|\boldsymbol{p}| = 10^6$, $|\hbar|\boldsymbol{k}| = 0.3$, corresponding to the resonance condition for first order Bragg scattering. With these parameters, the recoil frequency is $\omega_{\text{Recoil}} = 0.15$. Initially, when $\omega_{\text{Recoil}} t \ll 1$, the process corresponds to Raman–Nath scattering, with the occupation being transferred to successively higher values of n, and with successively lower occupation for each n. By $\omega_{\text{Recoil}} t \approx 0.2$, the population is spread over a large number of values of n, with no clear preference for any particular value. However, for longer times the Bragg resonance condition starts to take effect, and the first order Bragg peak occurs at $\omega_{\text{Recoil}} t \approx 0.5$, when most of the population has been transferred to $n = 2$, corresponding to first order scattering. This population then is eventually transferred (less completely) back to $n = 0$, passing first through another episode of Raman–Nath scattering.

short times,and the Bragg regime at longer times, and we give an example of this in Fig. 7.3. The choice of the parameter values is designed to make the effects visible easily in the same plot, and is probably unrealistic. However, this plot demonstrates the two main regimes. Raman–Nath scattering amounts to a modulation of the amplitudes which happens purely locally. In the far-detuned regime this amounts to a spatial phase modulation of the amplitude, corresponding to the production of the sidebands visible in the illustration. After a relatively short time, all but the resonant values of $n = 0, 2$ have dephased, and we see Bragg scattering appear.

c) Example—Bragg Scattering of Sodium Atoms: The transition chosen is usually the yellow sodium D line, and typical parameters are:

Sodium D line wavelength	589 nm
Recoil angular frequency ω_{Recoil}	$2\pi \times 0.0000245$ GHz
Δ is typically about	-10 GHz
Rabi frequency Ω_R	Order of magnitude $\Delta/100$

7.2.3 Off-Resonant Excitation—Optical Lattice Potentials

In the position representation we can write

$$\langle \boldsymbol{x}|\psi, t\rangle = \psi_g(\boldsymbol{x}, t)|g\rangle + \psi_e(\boldsymbol{x}, t)|e\rangle e^{-i\omega t}. \tag{7.2.15}$$

The equations of motion then become

$$i\hbar\frac{\partial\psi_e(\boldsymbol{x}, t)}{\partial t} = \left(-\frac{\hbar^2\nabla^2}{2M} - \hbar\Delta\right)\psi_e(\boldsymbol{x}, t) - \tfrac{1}{2}\hbar\Omega_R(\boldsymbol{x})\psi_g(\boldsymbol{x}, t), \tag{7.2.16}$$

$$i\hbar\frac{\partial\psi_g(\boldsymbol{x}, t)}{\partial t} = -\frac{\hbar^2\nabla^2}{2M}\psi_g(\boldsymbol{x}, t) - \tfrac{1}{2}\hbar\Omega_R(\boldsymbol{x})\psi_e(\boldsymbol{x}, t), \tag{7.2.17}$$

where $\Omega_R(\boldsymbol{x}) \equiv \Omega_0 \cos \boldsymbol{k}\cdot\boldsymbol{x}$.

Assuming a large detuning from the excited state, $\Delta \gg \Omega_R$ (weak excitation) and $\Delta \gg p^2/2M\hbar$ (detuning larger than the relevant kinetic energies), we can eliminate the excited state in perturbation theory, as in the derivation of the AC Stark shift in Sect. 4.1.2. Thus, we get

$$i\hbar\frac{\partial}{\partial t}\psi_g(\boldsymbol{x}, t) = \left(-\frac{\hbar^2\nabla^2}{2M} + V_{\rm opt}(\boldsymbol{x})\right)\psi_g(\boldsymbol{x}, t), \tag{7.2.18}$$

corresponding to the atom moving in the periodic optical potential

$$V_{\rm opt}(\boldsymbol{x}) \equiv \frac{\hbar\Omega_0^2}{4\Delta}\cos^2 \boldsymbol{k}\cdot\boldsymbol{x}. \tag{7.2.19}$$

This periodic potential is the simplest form of an *optical lattice potential.*

a) Effect of Other Energy levels: The derivation here assumes a two-level system and the rotating-wave approximation. However, an optical potential is much more general phenomenon, and occurs when there are several excited states. In such a case, one can perform a calculation similar to that in Sect. 4.1.2, keeping all the excited states, and the optical potential then reflects the complete AC Stark shift of the ground state.

b) Bragg and Raman–Nath Regimes: A detailed treatment of the two regimes, their use, validity and application is given in [7.4] and [7.7].

7.3 Momentum Transfer by Adiabatic Passage

The adiabatic passage technique of Sect. 6.3.1 can be used to transfer momentum between atomic states, a technique proposed [7.8] by *Marte, Zoller* and *Hall* in 1991. Atoms are scattered by two counterpropagating σ^+ and σ^- light beams, with frequency ω, and wave vectors directed in opposite directions along the z-axis.

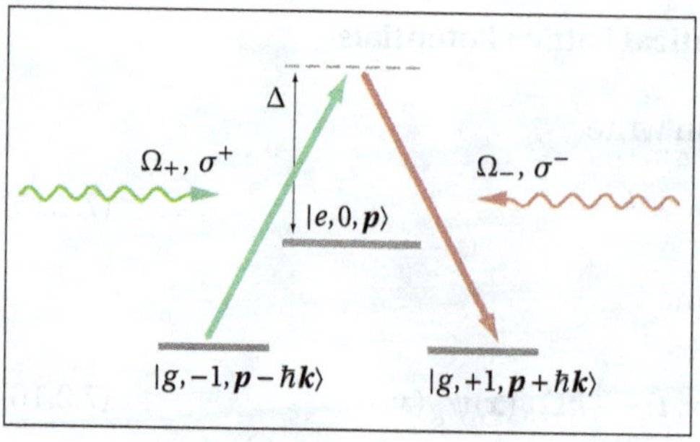

Fig. 7.4. Three-level system with Zeeman ground states $|g,\pm1,\boldsymbol{p}\pm\hbar\boldsymbol{k}\rangle$, coupled by $\sigma^\pm$ polarized light to an upper state $|e,\boldsymbol{p},0\rangle$.

This involves the generalization of the system described by the Hamiltonian (7.1.1) to the Λ configuration shown in Fig. 7.4, which is similar to that illustrated in Fig. 6.3. Thus the two ground states are Zeeman states characterized by $M_g = \pm1$, momenta $\boldsymbol{p}\pm\hbar\boldsymbol{k}$, and their electronic energy is taken as zero. The excited state has $M_e = 0$, momentum $\boldsymbol{p}$, and electronic energy $\hbar\omega_{eg}$.

The Hamiltonian will couple only the states

$$|e,0,\boldsymbol{p}\rangle \qquad |g,+1,\boldsymbol{p}+\hbar\boldsymbol{k}\rangle, \qquad |g,-1,\boldsymbol{p}-\hbar\boldsymbol{k}\rangle, \tag{7.3.1}$$

so that we can write the wavefunction in the form analogous to that used in (7.1.9–7.1.14), that is, as

$$|\Psi,t\rangle = \mathrm{e}^{-\mathrm{i}p^2t/2M}\Big(A_0(t)\mathrm{e}^{-\mathrm{i}\omega_{eg}t}|e,0,\boldsymbol{p}\rangle + A_{+1}(t)|g,+1,\boldsymbol{p}+\hbar\boldsymbol{k}\rangle + A_{-1}(t)|g,-1,\boldsymbol{p}-\hbar\boldsymbol{k}\rangle\Big). \tag{7.3.2}$$

The equation of motion can then be written as

$$\mathrm{i}\frac{d}{dt}\begin{pmatrix}A_0(t)\\ A_{+1}(t)\\ A_{-1}(t)\end{pmatrix} = \boldsymbol{\Omega}_{\text{Transfer}}(t)\begin{pmatrix}A_0(t)\\ A_{+1}(t)\\ A_{-1}(t)\end{pmatrix}, \tag{7.3.3}$$

in which

$$\boldsymbol{\Omega}_{\text{Transfer}}(t) \equiv \begin{pmatrix}-\Delta-\frac{1}{2}\mathrm{i}\kappa & -\frac{1}{2}\Omega_-(t) & -\frac{1}{2}\Omega_+(t)\\ -\frac{1}{2}\Omega_-(t)^* & \boldsymbol{k}\cdot\boldsymbol{v}+\omega_{\text{Recoil}} & 0\\ -\frac{1}{2}\Omega_+(t)^* & 0 & -\boldsymbol{k}\cdot\boldsymbol{v}+\omega_{\text{Recoil}}\end{pmatrix}. \tag{7.3.4}$$

i) This is simply the appropriate modification of (7.1.14) for the case of two ground state levels, forming a Λ configuration.

ii) Note that in (7.1.14), the excited state is written $|e,\boldsymbol{p}+\hbar\boldsymbol{k}\rangle$, whereas here the excited state is written $|e,0,\boldsymbol{p}\rangle$, thus changing the locations of the recoil terms involving $\pm\boldsymbol{k}\cdot\boldsymbol{v}+\omega_{\text{Recoil}}$ in the matrix.

iii) Without using the full quantum stochastic description, we have followed [7.8] by including a decay term in the equation of motion for the excited state by adding $-\frac{1}{2}\mathrm{i}\kappa$ to the top left element of $\boldsymbol{\Omega}_{\text{Transfer}}(t)$.

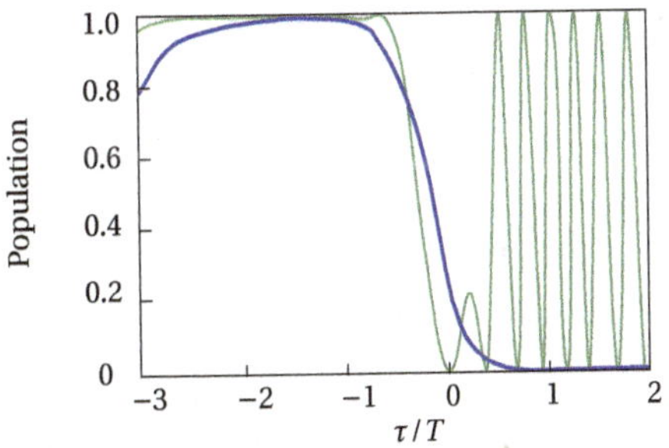

Fig. 7.5. Solution for the population of the final $|g,+1\rangle$ state as a function of the time delay τ between the pulses. The pulses are Gaussian, and T is their full width at half maximum. The Rabi frequencies are $\Omega_{\pm}T = 50$. The blue curve corresponds to a spontaneous decay rate $\kappa T = 5$, while the green curve is for $\kappa T = 0$.

7.3.1 Solutions of the Equations of Motion

The population transfer predicted by these equations is shown in Fig. 7.5, where the population of the final state $|g,+1,\boldsymbol{p}+\hbar\boldsymbol{k}\rangle$ is plotted as a function of the time delay τ between the Ω_- pulse and the Ω_+ pulse. The pulses are both Gaussian with equal lengths (full width at half maximum $T_1 = T_2 = T = 1$) and amplitude $\Omega_+ = \Omega_- = 50$. The curves are plotted both without damping ($\kappa = 0$), and with damping ($\kappa T = 5$).

The behaviour during the pulses is very similar in both cases; as expected, the transfer takes place almost completely for a range of negative values of τ, centred on $\tau = -1.57$. For positive τ there is a significant difference between the damped and undamped cases; there are strong Rabi oscillations in the undamped case, which are destroyed by the presence of damping.

Exercise 7.1 Quantum Stochastic Equations: Using the techniques of Chap. 9, formulate appropriate quantum stochastic equations of motion corresponding to (7.3.3, 7.3.4). What is the correct quantum stochastic interpretation of the equations of this section?

8. Modelling Real Atoms

Real atoms exist because of the electron, the atomic nucleus and the electromagnetic interaction that holds a collection of electrons in a tightly organized shell structure around the nucleus. The interaction with the electromagnetic field is very highly structured because of quantum mechanics, which forces the electrons into precisely defined energy levels, and enforces conservation laws which severely restrict the kinds of absorptions and emissions of photons allowed. Most importantly, the atomic energy levels are very stable, leading to sharp spectral lines—this means that it is necessary to tune the frequency of light very precisely to induce any significant interaction with the atom.

In Chap. 4 we showed how to isolate a two-level atom from a real atom with many energy levels, in a way which did not depend significantly on the detailed structure of the atom. However, some knowledge of atomic structure is necessary for a full understanding of atomic manipulation techniques, particularly since laboratory experiments are done with real atoms, whose connection with the idealized two-level atom is not always straightforward.

In this chapter we will outline the basic ideas of atomic structure, and those techniques necessary for their use in practical situations. The theory of atomic spectra of real atoms is a very complex subject, especially when the effects of nuclear spin and magnetic moments are taken into account. The subject is, however, very mature and well understood, even though some calculations necessary for the study of ultra-cold atoms are quite demanding.

This chapter is not essential for the understanding of most of this book, rather, it is presented as a brief reference to the details of atomic structure, since it is inevitable that at some stage every person working in quantum optics will need to know some of the details we present here. For a very full description of the theory of atomic spectra and radiative transitions we recommend the books of *Sobelman* [8.1,8.2].

8.1 The Alkali Atoms

Let us therefore look at the structure of the simplest real atoms. The alkali atoms have a single valence electron in the state (n, s) outside a closed shell. In the case of sodium, for example, the ground state has the configuration $1s^2\,2s^2\,2p^6\,3s$.

The spectroscopy of the alkali atoms relevant to this book is described by the possible excited states of this single valence electron—the inner core comprising the closed shell and the nucleus is very tightly bound, and its state does not change during processes of interest to us. The antisymmetrization of the wavefunction of the valence electron with respect to those in the core can be ignored because the core wave function and valence electron wavefunction do not overlap significantly

8.1.1 Separation of Valence Electron and Core Electrons

We write the wavefunction of the atom in the centre of mass system as

$$\Psi = \Phi_{\text{core}} \frac{u_{nl}(r)}{r} Y_{lm}(\theta, \phi), \tag{8.1.1}$$

where the core wavefunction corresponds to the electrons in the filled shell, and $u(r)$ is the radial wavefunction of the valence electron. The resulting situation can then be summarized as:

i) The electron core is frozen; that is, for the processes we consider, there is not sufficient energy to remove an electron from the core.

ii) The frozen electron core provides an effective potential for the valence electron.

iii) Thus, the description of the valence electron is that of a single electron moving in the effective potential of the core.

The interactions which determine the precise structure and energy levels are, in order of importance, the *electrostatic interaction* between the charges of the particles; *fine structure* arising from the magnetic effects of the spin and orbital angular momenta of the electrons; and finally *hyperfine structure*, arising from the magnetic moment of the nucleus.

8.1.2 Electrostatic Interactions

The electrostatic interactions experienced by the valence electron can be reduced to that arising from the nuclear charge, and that arising from the core electrons. These are not necessarily perfectly spherically symmetric charge distributions,

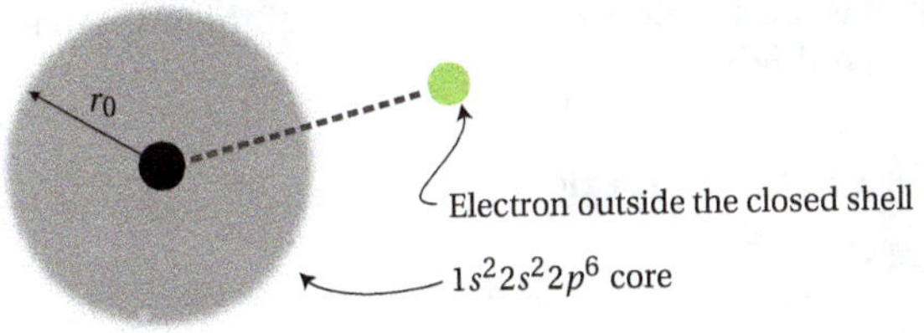

Fig. 8.1. Structure of a typical alkali atom.

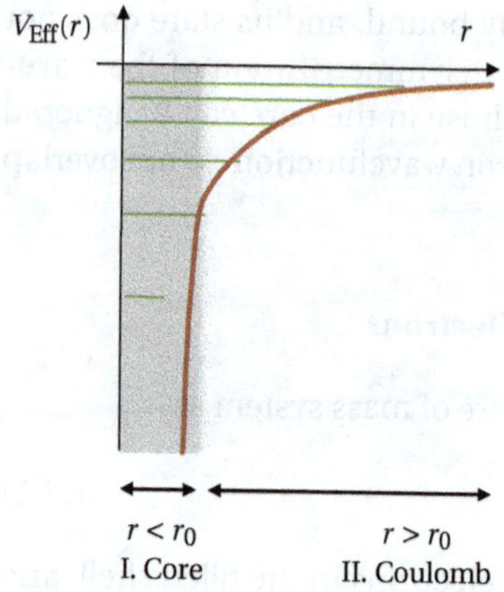

Fig. 8.2. Potential and energy level structure of an alkali atom showing the inner core and the outer Coulomb potential. The grey shaded area represents the core, where the potential becomes increasingly attractive. Quantum defect theory replaces the effect of this core with an appropriate boundary condition at $r = r_0$.

but deviations from spherical symmetry are very small. The nucleus may have an electric quadrupole moment, and the core charge distribution may be perturbed by this quadrupole moment.

The spherically symmetric part of electrostatic interaction is overwhelmingly stronger than all other interactions, and provides spherically symmetric effective potential $V_{\rm Eff}(r)$ whose eigenfunctions describe the atom, with all other effects being treated by perturbation theory.

8.1.3 The Radial Wavefunction of the Valence Electron

Taking into account the spherical symmetry of the effective potential, this obeys the radial Schrödinger equation

$$\left[-\frac{\hbar^2}{2m}\left(\frac{d^2}{dr^2}-\frac{l(l+1)}{r^2}\right)+V_{\rm Eff}(r)\right]u_{nl}(r)=\epsilon_{nl}u_{nl}(r). \tag{8.1.2}$$

Using the notation m_e for the electron mass, and M for the total of the mass M_N of the nucleus and that of the $Z-1$ core electrons, the *effective mass* which appears in this equation is

$$m \equiv \frac{m_e M}{m_e + M} \approx \frac{m_e M_N}{m_e + M_N} \quad \text{to 1st order in } m_e/M_N. \tag{8.1.3}$$

The effective potential $V_{\rm Eff}(r)$ has a transition at r_0, the radius of the frozen core, with the limiting behaviours on either side

$$V_{\rm Eff}(r) \longrightarrow \begin{cases} -\dfrac{Ze^2}{4\pi\varepsilon_0 r}, & \text{for } r \ll r_0, \quad \text{(Region I)} \\[2ex] -\dfrac{e^2}{4\pi\varepsilon_0 r}, & \text{for } r \gg r_0. \quad \text{(Region II)} \end{cases} \tag{8.1.4}$$

The situation is illustrated in Fig. 8.2.

a) Quantum Defect Theory: The eigenfunctions in this kind of potential can be very well approximated using the methods of *quantum defect theory* [8.3]. This is an approximation method, in which the wavefunction is considered separately in the two regions.

i) For the region I, $r < r_0$, the potential is very much stronger than in region II, and is not of the Coulomb form. The corresponding wavefunction, $F_{\rm I}(r)$, has a boundary condition at $r = 0$, which sets it to zero there.

ii) For the region II, $r > r_0$, the potential is the Coulomb potential, and the corresponding wavefunction, $F_{\rm II}(r)$, is a Coulomb wavefunction, with a boundary condition at infinity, where it vanishes.

iii) The two functions must be matched at $r = r_0$, leading to the three conditions

$$F_{\rm I}(r) = 0, \qquad r = 0, \tag{8.1.5}$$

$$F_{\rm II}(r) = 0, \qquad r = \infty, \tag{8.1.6}$$

$$\frac{d\log F_{\rm I}(r)}{dr} = \frac{d\log F_{\rm II}(r)}{dr}, \qquad r = r_0. \tag{8.1.7}$$

iv) The power of the method comes from the very different sizes of the potentials in the two regions. The magnitude of the potential in the core is very much greater than that in the Coulomb region, and also therefore very much greater than the magnitude of the energy eigenvalue ϵ_{nl}. A result of this is that, for a given value of l, the wavefunction $F_{\rm I}(r)$ is relatively insensitive to the value of the eigenvalue ϵ_{nl}. Thus, in enforcing the boundary condition (8.1.6), the logarithmic derivative $d\log F(r)/dr|_{r=r_0}$ is almost independent of n, and this gives a boundary condition on the Coulomb wavefunction $F_{\rm II}(r)$ which is the same for all n.

b) The Energy Spectrum: The result of imposing the boundary condition leads to a spectrum given by the *quantum defect formula*

$$\epsilon_{nl} = -\frac{R_M}{(n-\mu_l)^2} \equiv -\frac{R_M}{\nu_{nl}^2}. \tag{8.1.8}$$

Here:

i) R_M is the Rydberg constant for core mass M (essentially the nuclear mass M_N, as discussed in (8.1.3) above),

$$R_M = R_\infty \frac{m}{m_e}, \qquad R_\infty = \frac{m_e e^4}{8\varepsilon_0 h^3 c} = 1.0974 \times 10^7\,\mathrm{m}^{-1}. \tag{8.1.9}$$

ii) n is the principal quantum number.

iii) μ_l is known as the the *quantum defect*, and is approximately energy independent.

iv) $\nu_{nl} = n - \mu_l$ is known as the *effective quantum number.*

The spectra given by the quantum defect formula (8.1.8) are quantitatively quite different from the hydrogen spectrum, even though the qualitative description in terms of the quantum numbers n and l is very similar. In hydrogen the energy levels given by the Rydberg formula depend only on the principal quantum number n, a degeneracy which is removed by presence in the quantum defect formula of the l-dependent quantum defect μ_l. This dependence arises because the radial wavefunctions for higher l are very small in the core region, compared to those for small and zero values of l. Consequently the energy eigenvalues for smaller l are reduced, because the electron has a higher probability to be in the core region where the potential is strongly negative.

c) The Sodium Atom: The reduction of energy eigenvalues for smaller l is a very significant effect. If we consider the sodium atom, for example, the ground state of the valence election is $3s$, and the first excited state is $3p$. These levels are degenerate in hydrogen, but the energy difference between them in sodium is 2.105eV, leading to a 589 nm transition between them, the well known Na D line.

8.2 Atoms with More than One Valence Electron

The model of the valence electrons moving in the field of a frozen core is still very useful when there is more that one valence electron, but it can become very complex. The most significant complication is the necessity to couple the angular momentum states of the individual valence electrons into a wavefunction of definite total angular momentum, which can be done in many different ways, as described in detail in [8.1, 8.2].

8.2.1 Spectroscopic Notation

In the central field approximation, in which valence electrons move *without mutual interaction* in the central field $V_{\rm Eff}(r)$ of the frozen core—see (8.1.4)—a multi-electron state can be described by the quantum numbers n_1, l_1, n_2, l_2, n_3, l_3, etc., which determine the energy levels, and the corresponding orbital and spin angular momentum quantum numbers m_1^l, m_1^s, m_2^l, m_2^s, m_3^l, m_3^s, etc., which do not affect the energy levels. This gives $2(2l_1+1)\times 2(2l_2+1)\times 2(2l_3+1)\times\ldots$ different quantum states.

The spectroscopic notation relevant for our purposes does not describe the core electrons, since transitions of interest involve only the valence electrons. Hence we can write, for example

Li:	$2s$,	Na:	$3s$,	K:	$4s$,	Rb:	$5s$,	Cs:	$6s$,
Be:	$2s^2$,	Mg:	$3s^2$,	Ca:	$4s^2$,	Sr:	$5s^2$,	Ba:	$6s^2$,
B:	$2s^2 2p$,	Al:	$3s^2 3p$,	Sc:	$3d4s^2$,	Y:	$4d5s^2$,	La:	$5d6s^2 p$.

This table describes the observed configurations of the *ground states* of the each of the atomic species when only the central field is taken into account.

8.2.2 *L*-*S* Coupling of Angular Momenta

In the central field approximation without interactions, the individual spin and orbital angular momenta of the electrons $\boldsymbol{l}_i$, $\boldsymbol{s}_i$, as well as $\boldsymbol{j}_i \equiv \boldsymbol{l}_i + \boldsymbol{s}_i$, are conserved. When interactions between the electrons are taken into account, the most appropriate description is in terms of the $\boldsymbol{L}$-$\boldsymbol{S}$ coupling scheme

$$\boldsymbol{L} \equiv \sum_i \boldsymbol{l}_i, \qquad \boldsymbol{S} \equiv \sum_i \boldsymbol{s}_i, \qquad \boldsymbol{J} \equiv \boldsymbol{L} + \boldsymbol{S}, \tag{8.2.1}$$

$$\text{with eigenstates} \qquad |L, M_L\rangle, \quad |S, M_S\rangle, \quad |J, M_J\rangle, \tag{8.2.2}$$

$$\text{and where} \qquad L + S \geqslant J \geqslant |L - S|. \tag{8.2.3}$$

a) Electrostatic Interaction and Hund's Rule: The electrostatic interaction between the valence electrons (when there are two or more of them) results in a splitting of the level corresponding to a given electronic configuration into a number of levels, and these levels correspond to the different values of the total orbital and spin quantum numbers L and S. The electrostatic interaction between electrons is repulsive, so this leads to an increase in the energy eigenvalues above the central field result. The lowest energy eigenvalue, that is, the ground state, normally follows *Hund's rule*, according to which, for a given configuration, the level with the greatest possible value of S, and the greatest possible value of L allowable for this value of S, has the lowest energy. For a detailed explanation the reader should consult the books by Sobel'man [8.1,8.2].

b) Spectral Terms: Energy levels corresponding to definite values of L and S are called *spectral terms*, or more briefly simply *terms*. In a traditional scheme whose origin lies in the early days of spectroscopy, it is customary to describe a given spectral term by the quantum numbers S, L, J, using a notation ${}^{2S+1}L_J$, in which L takes on integral values coded by capital letters thus:

$$L = \begin{cases} 0 \;\; 1 \;\; 2 \;\; 3 \;\; 4 \;\; 5 \;\; 6 \;\; 7 \;\; 8 \;\; 9 \;\; 10\,, \\ S \;\; P \;\; D \;\; F \;\; G \;\; H \;\; I \;\; K \;\; L \;\; M \;\; N\,. \end{cases} \tag{8.2.4}$$

According to the $\boldsymbol{L}$-$\boldsymbol{S}$ coupling scheme, the term ${}^{2S+1}L_J$ corresponds to the set of quantum states coupled using Clebsch–Gordan coefficients

$$|L, S, J, M_J\rangle = \sum_{M_S, M_L} \langle SM_S, LM_L | JM_J\rangle \, |S, M_S\rangle \, |L, M_L\rangle. \tag{8.2.5}$$

i) *Multiplicity*: For given values of L and S the number of values of J available is, according to the rule (8.2.3), the minimum of $2L+1$ and $2S+1$. Nevertheless, the value $2S+1$ is normally called the *multiplicity* of the term.

ii) *Parity*: The parity of a state is also necessary for a full description, since this is not determined by the L value except in the case of the alkali atoms, where there is only one valence electron. The notation is a superscript "o" when the parity is odd, otherwise there is not superscript; for example in calcium the first excited state is ${}^3P_0^{\mathrm{o}}$, meaning that $S = 1$, $L = 1$, $J = 0$, and the parity is negative.

Configuration	Term	J	Energy (eV)
$4s^2$	$^1S_0$	0	−6.11316
$4s4p$	$^3P^o_0$	0	−4.23382
	$^3P^o_1$	1	−4.22735
	$^3P^o_2$	2	−4.21422
$3d4s$	$^3D_1$	1	−3.59190
	$^3D_3$	3	−3.58748
$3d4s$	$^1D_2$	2	−3.40415
$4s4p$	$^1P^o_1$	1	−3.18065
$4s5s$	$^3S_1$	1	−2.20276
$4s5s$	$^1S_0$	0	−1.98234
$3d4p$	$^3F^o_2$	2	−1.68315
	$^3F^o_3$	3	−1.67220
	$^3F^o_4$	4	−1.66251
$3d4p$	$^1D^o_2$	2	−1.67013
$4s5p$	$^3P^o_0$	0	−1.58182
	$^3P^o_1$	1	−1.58095
	$^3P^o_2$	2	−1.57842
$4s5p$	$^1P^o_1$	1	−1.55902
$4s4d$	$^1D_2$	2	−1.48876
$4s4d$	$^3D_1$	1	−1.43298
	$^3D_2$	2	−1.43252
	$^3D_3$	3	−1.43183
$3d4p$	$^3D^o_1$	1	−1.37791
	$^3D^o_2$	2	−1.37459
	$^3D^o_3$	3	−1.36963
$4p^2$	$^3P_0$	0	−1.34999
	$^3P_1$	1	−1.34413
	$^3P_2$	2	−1.33337

Table 8.1. Configurations, terms and energy levels for calcium.

c) Notation for Spectral Configuration and Term: For a given configuration and term, the two notations are joined to each other; thus calcium has the ground state configuration $4s^2$, and the spectral term of the ground state is ${}^1S_0$, leading to the notation $(4s^2)\,{}^1S_0$.

Some of the excited states are given in Table 8.1, showing excited states which come from the excitation of one or both of the two valence electrons from the $4s$ level into the $3d$, $4p$, $4d$, $5s$, $5p$ levels.

8.2.3 Spectral Terms for Alkali Atoms

The same notation is used for alkali atoms, even though there is only one valence electron, so that the quantum numbers L, S are the same as l, s, the quantum numbers describing a single electron. The quantum states in the L-S coupling scheme are written

$$|n(ls)j,m\rangle = \sum_{m_s,m_l} \langle sm_s, lm_l|jm_j\rangle\,|s,m_s\rangle\,|n,l,m_l\rangle, \tag{8.2.6}$$

corresponding to the terms and configurations in Table 8.2.

8.3 Fine and Hyperfine Structure

The interactions between the magnetic field produced by the motion of the electronic charge and the intrinsic magnetic moments of the electron and the nucleus remove most of the degeneracies within the spectral terms. In addition to magnetic interactions, if the charge distribution of the nucleus is not spherically symmetrical, there will be a further contribution, which in practice is of the same order of magnitude as the effect of the nuclear magnetic moment.

8.3.1 Magnetic Moments

Let us now consider the three sources of magnetic moments in detail.

i) *Electronic Magnetic Effects*: The magnetic moment arising from the motion of the electron in the atom is related to its orbital angular momentum $\boldsymbol{l}$ by

$$\boldsymbol{\mu}_l = \frac{e}{2m_e}\boldsymbol{l}, \qquad \text{where } e = -|e| \text{ is the charge on the } \textit{electron}. \tag{8.3.1}$$

The magnetic moment of the electron as given by the Dirac equation is related to the electron spin $\boldsymbol{s}$ by

$$\boldsymbol{\mu}_e = \frac{e}{m_e}\boldsymbol{s}. \tag{8.3.2}$$

The interactions related to $\boldsymbol{\mu}_L$ and $\boldsymbol{\mu}_e$ are of the same order of magnitude as each other, and give rise to small changes in the energy levels, which are called *fine structure.*

Configuration	Term	J	Energy (eV)
$3s$	$^2S_{\frac{1}{2}}$	$\frac{1}{2}$	−5.13908
$3p$	$^2P^o_{\frac{1}{2}}$	$\frac{1}{2}$	−3.03678
	$^2P^o_{\frac{3}{2}}$	$\frac{3}{2}$	−3.03465
$4s$	$^2S_{\frac{1}{2}}$	$\frac{1}{2}$	−1.94772
$3d$	$^2D_{\frac{5}{2}}$	$\frac{5}{2}$	−1.52211
	$^2D_{\frac{3}{2}}$	$\frac{3}{2}$	−1.52210
$4p$	$^2P^o_{\frac{1}{2}}$	$\frac{1}{2}$	−1.38645
	$^2P^o_{\frac{3}{2}}$	$\frac{3}{2}$	−1.38576
$4f$	$^2F^o_{\frac{5}{2}}$	$\frac{5}{2}$	−0.85085
	$^2F^o_{\frac{7}{2}}$	$\frac{7}{2}$	−0.85085
$5p$	$^2P^o_{\frac{1}{2}}$	$\frac{1}{2}$	−0.79462
$5p$	$^2P^o_{\frac{3}{2}}$	$\frac{3}{2}$	−0.79432

Table 8.2. Configurations, terms and energy levels for sodium.

ii) *Nuclear Magnetic Effects*: The magnetic moment of the nucleus can be written in terms of the nuclear angular momentum $\boldsymbol{I}$ as

$$\boldsymbol{\mu}_N = \mu_N \frac{|e|}{m_p}\boldsymbol{I}. \tag{8.3.3}$$

Here m_p is the mass of the proton, and μ_N is a typically a number between −3 and 10, dependent on the particular nucleus. Hence, the magnitude of a nuclear magnetic moment is of order of magnitude m_e/m_p smaller than the electron's magnetic moment, and interactions involving the nuclear magnetic moment are consequently very much smaller than those involving $\boldsymbol{\mu}_L$ and $\boldsymbol{\mu}_e$. The effect of these interactions is one of the two principal sources of what is know as the *hyperfine structure.*

iii) *The g-Factor—Expression of Magnetic Moments in Terms of Bohr Magnetons*: It is convenient to express all magnetic moments in terms of the Bohr magneton

$$\mu_B \equiv \frac{|e|\hbar}{2m_e} = \begin{cases} 9.27400968 \times 10^{-24}\,\mathrm{J\,T^{-1}}, \\ 5.78838181 \times 10^{-5}\,\mathrm{eV\,T^{-1}}. \end{cases} \tag{8.3.4}$$

In these units we write

$$\boldsymbol{\mu}_l = -g_l \mu_B \boldsymbol{l}, \qquad g_l = 1, \tag{8.3.5}$$

$$\boldsymbol{\mu}_s = -g_s \mu_B \boldsymbol{s}, \qquad g_s = 2(1+a) \approx 2, \tag{8.3.6}$$

$$\boldsymbol{\mu}_I = -g_I \mu_B \boldsymbol{I}, \qquad g_I = \frac{\mu_N m_e}{m_p}. \tag{8.3.7}$$

Note that:

a) In the first line, to correct for the finite mass m_N of the nucleus, we should more accurately write

$$g_l = \frac{m_N}{m_e + m_N}. \tag{8.3.8}$$

b) In the second line, the quantity a is the QED correction given by

$$a \equiv 0.5\left(\frac{\alpha}{\pi}\right) - 0.32848\left(\frac{\alpha}{\pi}\right)^2 + 1.49\left(\frac{\alpha}{\pi}\right)^3 + \cdots \tag{8.3.9}$$

Here α is the fine-structure constant, defined by

$$\alpha \equiv \frac{e^2}{4\pi\epsilon_0 \hbar c} \approx \frac{1}{137}. \tag{8.3.10}$$

8.3.2 Electric Quadrupole Moment of the Nucleus

The most significant effect of a non-spherically symmetric nuclear charge distribution comes from the nuclear electric quadrupole moment, and this provides the other principal contribution to the hyperfine structure.

8.3.3 Spin Orbit Interaction and Fine Structure

The fine structure arises from the the coupling of the spin angular momentum $\boldsymbol{s}$ and the orbital angular momentum $\boldsymbol{l}$ of the electron.

a) Alkali Atoms: In this case the Hamiltonian takes the form

$$H_{\text{SO}} = \frac{1}{2m^2c^2 r}\frac{\partial V_{\text{Eff}}(r)}{\partial r}\boldsymbol{s}\cdot\boldsymbol{l}. \tag{8.3.11}$$

This can be derived from the low energy approximation to the Dirac equation for an electron in the potential $V_{\text{Eff}}(r)$. In first-order perturbation theory, the change in the energy level of the state $|n(sl)jm_j\rangle$ can be evaluated using the wavefunction (8.1.1) and by writing $\boldsymbol{l}\cdot\boldsymbol{s} = \left(\boldsymbol{j}^2 - \boldsymbol{l}^2 - \boldsymbol{s}^2\right)/2$, which yields

$$\begin{aligned}\langle n(sl)jm_j|H_{\text{SO}}|n(sl)jm_j\rangle &= \big(j(j+1) - l(l+1) - s(s+1)\big)\\ &\times\frac{\hbar^2}{2m^2c^2}\int dr\,|u_{nl}(r)|^2\frac{1}{r}\frac{\partial V_{\text{Eff}}(r)}{\partial r}.\end{aligned} \tag{8.3.12}$$

Notice that

$$\left.\begin{aligned} u_{nl}(r) &\sim r^{l+1}, \\ \frac{1}{r}\frac{\partial V_{\text{Eff}}(r)}{\partial r} &\sim r^{-3}, \end{aligned}\right\} \quad \text{as} \quad r \to 0. \tag{8.3.13}$$

If $l = 0$, then $j = s$, and the first factor in (8.3.12) vanishes, so the correction is zero. Otherwise, when $l \geqslant 0$ the integrand approaches zero like r^{2l-1}, so the integral converges. At larger r the factor $|u_{nl}(r)|^2$ is bounded, while the potential term becomes very small. The net effect is that the correction to the energy becomes very much smaller as l increases, and this can be seen in the data in Table 8.2. In particular, notice that in the $4f$ configuration (for which $l = 3$) there is no discernible splitting between the energy levels for the two different values of j.

b) The Sodium Spectrum: As shown in Table 8.2, the ground state of sodium is $(3s)\ ^2S_{\frac{1}{2}}$ and the first two excited states are $(3p)\ ^2P_{\frac{1}{2}}$ and $(3p)\ ^2P_{\frac{3}{2}}$. The energies of these two states are very close, and transitions between them and the ground state give a fine structure doublet, corresponding to the well known Na D doublet at 589 nm, which is split by 515.5 GHz, which is in the microwave regime, with a wavelength of about 0.58 cm.

8.3.4 Nuclear Spin and Hyperfine Structure

Denoting the the nuclear spin by $\boldsymbol{I}$ as in (8.3.4), the hyperfine Hamiltonian can be written as [8.4]

$$\mathcal{H}_{\text{hfs}} = \sum_{k>0} \boldsymbol{T}^{(k)} \cdot \boldsymbol{M}^{(k)}, \tag{8.3.14}$$

where $\boldsymbol{T}^{(k)}$ and $\boldsymbol{M}^{(k)}$ are spherical tensor operators of rank k which represent respectively the electronic and the nuclear parts of the interactions. The magnetic terms have odd k and the electric terms have even k.

a) Magnetic Dipole Interaction: The nuclear term is the magnetic moment of the nucleus, which is from (8.3.7),

$$\boldsymbol{M}^{(1)} = \boldsymbol{\mu}_I = -g_I \boldsymbol{\mu}_B \boldsymbol{I}. \tag{8.3.15}$$

The electronic term represents the magnetic field produced by the electron at the position of the nucleus, and takes the form

$$\boldsymbol{T}^{(1)} = \frac{\mu_0}{4\pi} 2\mu_B \left\{ \frac{\boldsymbol{l}}{r^3} - \frac{1}{r^3}\left[\boldsymbol{s} - 3\frac{\boldsymbol{s}\cdot\boldsymbol{r}}{r^2}\boldsymbol{r} \right] + \frac{2}{3}\frac{\delta(r)}{r^2}\boldsymbol{s} \right\}, \tag{8.3.16}$$

where μ_0 is the vacuum permeability.

i) The term proportional to $\boldsymbol{l}$ is that part of the magnetic field at the location of the nucleus which arises from the motion of the electronic charge.

ii) The term proportional to $\boldsymbol{s}$ takes the form of a magnetic field at point $\boldsymbol{r}$ appropriate to a classical magnetic dipole. The term proportional to $\delta(r)$ can be thought of as the magnetic field *inside* an infinitesimal solenoid which represents the magnetic dipole.

This term only gives a non-zero contribution for s-electrons, for which the radial wavefunction does not vanish at the origin.

b) Electric Quadrupole Interaction: The second order term in the hyperfine interaction is the electric quadrupole part, and is expressed as

$$\boldsymbol{M}^{(2)} \cdot \boldsymbol{T}^{(2)} = -\frac{Q\sqrt{6}}{2I(I-1)} \frac{e^2 (\boldsymbol{I}:\boldsymbol{I})^{(2)} \cdot \boldsymbol{C}^{(2)}(\theta,\phi)}{4\pi\varepsilon_0 r^3}. \tag{8.3.17}$$

The notation $(\boldsymbol{I}:\boldsymbol{I})^{(2)}$ represents the second-rank spherical tensor formed from the products of the components of $\boldsymbol{I}$, as explained in Appendix 8.A, and $\boldsymbol{C}^{(2)}(\theta,\phi)$ is the spherical harmonic in the notation of (8.A.17–8.A.19). This expression represents the potential energy of an electron in the quadrupole electric field produced by the nucleus at the position $\boldsymbol{r} \equiv (r,\theta,\phi)$.

c) Angular Momentum Coupling: In order to evaluate the hyperfine structure effects, an appropriate angular momentum coupling of $\boldsymbol{j}$, $\boldsymbol{l}$, $\boldsymbol{I}$ is necessary. To do this, the nuclear angular momentum $\boldsymbol{I}$ is coupled to $\boldsymbol{j} = \boldsymbol{l} + \boldsymbol{s}$ to give a total angular momentum operator

$$\boldsymbol{F} = \boldsymbol{j} + \boldsymbol{I}, \tag{8.3.18}$$

which corresponds to the coupling of the states

$$|n(I(sl)j)FM_F\rangle = \sum_{M_I,m_j} |IM_I\rangle |n(sl)jm_j\rangle \langle IM_I, jm_j|FM_F\rangle, \tag{8.3.19}$$

$$F = |I-j|, \ldots I+j. \tag{8.3.20}$$

d) Evaluation of the Hyperfine Structure: The evaluation involves the use of angular momentum algebra, including the Wigner–Eckart theorem, and is outlined in [8.4]. The mean hyperfine energy in these states becomes

$$W_F = \tfrac{1}{2}\hbar AK + \hbar B \frac{\frac{3}{2}K(K+1) - 2I(I+1)j(j+1)}{2I(2I-1)2j(2j-1)}, \tag{8.3.21}$$

$$K = F(F+1) - I(I+1) - j(j+1). \tag{8.3.22}$$

The parameters A and B relate to the magnetic and electric terms respectively, as follows:

$$A = \begin{cases} -g_I \mu_B^2 \dfrac{\mu_0}{4\pi\hbar} \dfrac{2l(l+1)}{j(j+1)} \langle r^{-3} \rangle_{nl}, & l \neq 0, \\[2ex] -g_I \mu_B^2 \dfrac{\mu_0}{4\pi\hbar} \dfrac{16\pi}{3} |\psi_s(0)|^2, & l = 0 \end{cases} \tag{8.3.23}$$

$$B = Qe^2 \frac{1}{4\pi\varepsilon_0\hbar} \frac{2j-1}{2j+2} \langle r^{-3} \rangle_{nl}. \tag{8.3.24}$$

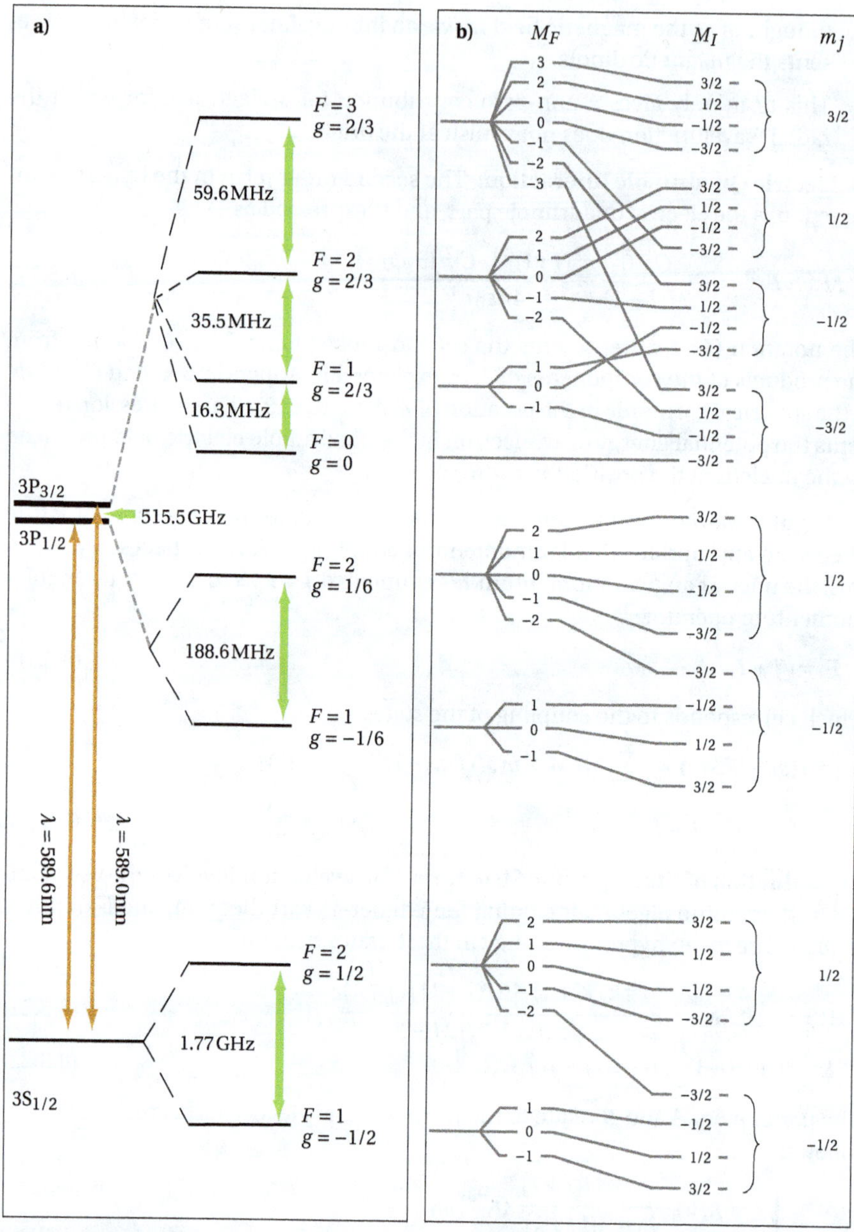

Fig. 8.3. The 3S and 3P energy levels of ^{23}Na. **a)** The fine structure and hyperfine structure; **b)** Zeeman splitting of the hyperfine levels, showing the effect for weak and for strong magnetic fields.

e) Hyperfine Levels of Sodium: The nuclear angular momentum for ^{23}Na is $3/2$, and the resulting fine structure and hyperfine structure is shown in Fig. 8.3. The splittings correspond to the radio frequency regime, ranging from 16.3 MHz to 1.77 GHz.

The terms of most interest—$3s$ and $3p$—are split into hyperfine levels as follows:

i) The hyperfine structure of the excited state $(3p)\,{}^2P_{\frac{3}{2}}$, which has $j = 3/2$, giving levels with $F = 0, 1, 2, 3$.

ii) The hyperfine structure of the excited state $(3p)\,{}^2P_{\frac{1}{2}}$ has $j = 1/2$, giving levels with $F = 1$ and $F = 2$.

iii) The hyperfine structure of the ground state $(3s)\,{}^2S_{\frac{1}{2}}$, has $j = 1/2$, giving levels with $F = 1$ and $F = 2$.

8.3.5 The Zeeman Effect

Applying a magnetic field splits the hyperfine levels into their M_F components, provided the field is not so strong as to overwhelm the hyperfine Hamiltonian. The Zeeman Hamiltonian for an applied field $\boldsymbol{B}$ is

$$\mathcal{H}_{\text{Zeeman}} = \boldsymbol{B}\cdot(\boldsymbol{\mu}_l + \boldsymbol{\mu}_s + \boldsymbol{\mu}_I) = -\mu_B \boldsymbol{B}\cdot(g_l \boldsymbol{l} + g_s \boldsymbol{s} + g_I \boldsymbol{I}), \tag{8.3.25}$$

where we have used the notation of (8.3.1–8.3.7).

As we have already seen in Sect. 8.3.1, the nuclear magnetic moment is very much smaller than the orbital and spin magnetic moments. This means that we can in fact neglect the term proportional to $\boldsymbol{I}$ in the Zeeman Hamiltonian (8.3.25). The effect of the nuclear magnetic moment for the Zeeman effect comes from its contribution to the hyperfine Hamiltonian as given in Sect. 8.3.4. The treatment of the effect then depends on the size of the applied magnetic field $\boldsymbol{B}$ compared to the magnetic field produced at the nucleus by the spin and orbital magnetic moments. We will first consider the case of relatively large field, in which we neglect the nuclear magnetic moment entirely, and then consider hyperfine effects.

a) Zeeman Effect without Hyperfine Effects: We need to evaluate the mean of $\mathcal{H}_{\text{Zeeman}}$ in a state of definite j and m_j,

$$\Delta E_{\text{Zeeman}} \equiv \langle n(sl)\,j m_j | \mathcal{H}_{\text{Zeeman}} | n(sl)\,j m_j\rangle \tag{8.3.26}$$

$$\approx -\mu_B \boldsymbol{B}\cdot\langle n(sl)\,j m_j | \boldsymbol{l} + 2\boldsymbol{s} | n(sl)\,j m_j\rangle. \tag{8.3.27}$$

We can now use the vector model result (8.A.54) to make the replacement (where we will omit the labels $n(sl)$ for compactness)

$$\langle j m_j | \boldsymbol{l} + 2\boldsymbol{s} | j m_j\rangle \longrightarrow \langle j m_j | \boldsymbol{j} | j m_j\rangle \frac{\langle j m_j | (\boldsymbol{l} + 2\boldsymbol{s})\cdot\boldsymbol{j} | j m_j\rangle}{\hbar^2 j(j+1)}. \tag{8.3.28}$$

We now evaluate this expression by writing

$$(\boldsymbol{l} + 2\boldsymbol{s})\cdot\boldsymbol{j} = (\boldsymbol{j} + \boldsymbol{s})\cdot\boldsymbol{j} = \boldsymbol{j}^2 + \tfrac{1}{2}(\boldsymbol{j}^2 + \boldsymbol{s}^2 - \boldsymbol{l}^2), \tag{8.3.29}$$

and take $\boldsymbol{B}$ parallel to the z-axis, and with magnitude B, so that

$$\Delta E_{\text{Zeeman}} = -g_j(l,s)\,\mu_B B m_j, \tag{8.3.30}$$

$$\text{where}\quad g_j(l,s) \equiv 1 + \frac{j(j+1)+s(s+1)-l(l+1)}{j(j+1)}. \tag{8.3.31}$$

b) The Weak-Field Case where Hyperfine Effects are Significant: In this case we need to evaluate the mean value of $\mathcal{H}_{\text{Zeeman}}$ in the eigenstates $|n(I(sl)j)FM_F\rangle$ of (8.3.19). In this set of states each of j, l and s has a single well-defined value, valid for the whole set. This means that we can use the results of (8.3.28–8.3.31) in any calculation involving the mean value of $\boldsymbol{l}+2\boldsymbol{s}$. This amounts to simply making the substitution

$$\boldsymbol{l}+2\boldsymbol{s} \to g_j(l,s)\,\boldsymbol{j}. \tag{8.3.32}$$

In the following, we will omit the terms $n(I(sl)j)$ for simplicity, and thus do the calculation

$$\langle FM_F|\mathcal{H}_{\text{Zeeman}}|FM_F\rangle \tag{8.3.33}$$

$$\approx -\mu_B \boldsymbol{B}\cdot\langle FM_F|\boldsymbol{l}+2\boldsymbol{s}|FM_F\rangle, \tag{8.3.34}$$

$$= -g_j(l,s)\mu_B \boldsymbol{B}\cdot\frac{\langle FM_F|\boldsymbol{j}\cdot\boldsymbol{F}\,\boldsymbol{F}|FM_F\rangle}{\hbar^2 F(F+1)}, \tag{8.3.35}$$

$$= -g_j(l,s)\mu_B B M_F\frac{\langle FM_F|\frac{1}{2}(\boldsymbol{F}^2+\boldsymbol{j}^2-\boldsymbol{I}^2)|FM_F\rangle}{\hbar^2 F(F+1)}, \tag{8.3.36}$$

$$= -g_j(l,s)\frac{F(F+1)+j(j+1)-I(I+1)}{2F(F+1)}\mu_B B M_F. \tag{8.3.37}$$

8.4 Interaction with Electromagnetic Radiation

The Hamiltonian which we will use to describe the system composed of an electron of mass m_1 interacting with the frozen core of mass m_2 made up of the remaining electrons and the nucleus is

$$H_{\text{Eff}} = \frac{\boldsymbol{p}_1{}^2}{2m_1} + \frac{\boldsymbol{p}_2{}^2}{2m_2} + V_{\text{Eff}}\big(|\boldsymbol{x}_1-\boldsymbol{x}_2|\big). \tag{8.4.1}$$

The interaction with the radiation field in the electric dipole approximation is given by the appropriate terms for the electron, and the frozen core:

$$H_{\text{EM}} = Ze\boldsymbol{x}_1\cdot\boldsymbol{E}(\boldsymbol{x}_1) - e\boldsymbol{x}_2\cdot\boldsymbol{E}(\boldsymbol{x}_2). \tag{8.4.2}$$

Here $-e$ is the charge of the electron, and the integer Z is included as the net charge of the core, which in the case of an ion (such as Ca^+, as is used in ion trapping) is different from one. It is assumed that the wavelength of the electromagnetic radiation is sufficiently large that the interaction with the core can be characterized by using only the net charge.

8.4.1 Centre of Mass Hamiltonian

We now convert to centre of mass coordinates, in which

$$\boldsymbol{x} = \boldsymbol{x}_2 - \boldsymbol{x}_1, \qquad \boldsymbol{X} = \frac{m_1\boldsymbol{x}_1 + m_2\boldsymbol{x}_2}{m_1 + m_2}, \tag{8.4.3}$$

$$m = \frac{m_1 m_2}{m_1 + m_2}, \qquad M = m_1 + m_2. \tag{8.4.4}$$

In practice it is normal to make the approximation that the core mass m_1 is very much larger than the electron mass m_2, and this allows approximations based on this disparity to be made, so that the Hamiltonians become

$$H_{\text{Eff}} = \frac{\boldsymbol{P}^2}{2M} + \frac{\boldsymbol{p}^2}{2m} + V_{\text{Eff}}(|\boldsymbol{x}|), \tag{8.4.5}$$

$$H_{\text{EM}} \approx (Z-1)e\boldsymbol{X}\cdot\boldsymbol{E}(\boldsymbol{X}) + e\boldsymbol{x}\cdot\boldsymbol{E}(\boldsymbol{X}). \tag{8.4.6}$$

8.5 Electric Dipole Radiation for a Multilevel Transition

In this section we will consider an atom, in the definite quantum state, emitting electromagnetic radiation as it makes a transition to a lower energy state. The basic formalism is as presented for a simple two-level atom in *Bk. I: Sect.12.2*, and the main modification here will be to include the consequences of the initial and final angular momentum states, and their different couplings to the electromagnetic field. In practical terms, these are all very important.

8.5.1 Electric Dipole Matrix Elements and Rabi Frequency

Let us consider an atomic transition in which a ground state, with Zeeman levels $|\gamma_g J_g M_g\rangle$, is coupled to an excited state, with Zeeman levels $|\gamma_e J_e M_e\rangle$, by an electric dipole transition. Here γ_g and γ_e represent all other quantum numbers relevant to the situation, but independent of angular momentum.

a) Dipole Interaction Hamiltonian in a Spherical Basis: The interaction Hamiltonian in the dipole approximation and rotating-wave approximation can be written as

$$H_{\text{Int}} = \boldsymbol{d}^{(-)}\cdot\boldsymbol{E}^{(+)} + \boldsymbol{d}^{(+)}\cdot\boldsymbol{E}^{(-)}. \tag{8.5.1}$$

i) To transform to the spherical basis we proceed as in Appendix 8.A i, and denote the electric dipole operator in a spherical basis by d_q, as an abbreviation for the full notation $d_{1,q}$.

ii) *Projection Operators*: To separate the different possible transitions we write

$$\boldsymbol{d}^{(-)}\cdot\boldsymbol{E}^{(+)} = (\mathcal{P}_e\boldsymbol{d}\,\mathcal{P}_g)\cdot\boldsymbol{E}^{(+)} = \sum_{q=0,\pm1}(\mathcal{P}_e d_q\mathcal{P}_g)\left(\boldsymbol{e}_q^*\cdot\boldsymbol{E}^{(+)}\right), \tag{8.5.2}$$

where $\mathcal{P}_e$ and $\mathcal{P}_g$ denote the projection operators on the excited and ground state manifolds, so that

$$\mathcal{P}_e d_q \mathcal{P}_g = \sum_{M_e,M_g} |\gamma_e J_e M_e\rangle\langle\gamma_e J_e M_e|d_q|\gamma_g J_g M_g\rangle\langle\gamma_g J_g M_g| \,. \tag{8.5.3}$$

iii) *Application of the Wigner–Eckart Theorem*: Using the 3j-symbol form (8.A.48) we can write the dipole transition operator in the form

$$\langle\gamma_e J_e M_e|d_q|\gamma_g J_g M_g\rangle = (-)^{J_e-M_e}(\gamma_e J_e\|\boldsymbol{d}\|\gamma_g J_g)\begin{pmatrix} J_e & 1 & J_g \\ -M_e & q & M_g \end{pmatrix}. \tag{8.5.4}$$

Combining these expressions we get

$$\boldsymbol{d}^{(-)}\cdot\boldsymbol{E}^{(+)} = \frac{(\gamma_e J_e\|\boldsymbol{d}\|\gamma_g J_g)}{\sqrt{2J_e+1}}\sum_q A_q^\dagger\left(\boldsymbol{e}_q^*\cdot\boldsymbol{E}^{(+)}\right), \tag{8.5.5}$$

where we have defined a *raising operator* in the atomic space by

$$A_q^\dagger \equiv \sum_{M_e,M_g}\langle J_g,1,M_g,q|J_e,M_e\rangle\left(|\gamma_e J_e M_e\rangle\langle\gamma_g J_g M_g|\right), \tag{8.5.6}$$

$$\equiv (-)^{-J_g+1-q}\sqrt{2J_e+1}\sum_{M_e,M_g}\begin{pmatrix} J_g & 1 & J_e \\ M_g & q & M_e \end{pmatrix}\left(|J_e M_e\rangle\langle J_g M_g|\right). \tag{8.5.7}$$

The operator $A_q^\dagger$ induces a transition from the ground state to the excited state with $M_e = M_g + q$.

b) Definition of the Rabi Frequency: For a given laser field $\boldsymbol{E}^{(+)} = \mathcal{E}\boldsymbol{\epsilon}\,\mathrm{e}^{-\mathrm{i}\omega t}$, the Rabi frequency is conventionally defined as the interaction strength on a transition for which the Clebsch-Gordan coefficient is equal to one, which means that it is expressed in terms of the reduced matrix element as

$$\hbar\Omega_R \equiv 2\mathcal{E}\frac{(\gamma_e J_e\|\boldsymbol{d}\|\gamma_g J_g)}{\sqrt{2J_e+1}} = 2\mathcal{E}\langle\gamma_e J_e\|\boldsymbol{d}\|\gamma_g J_g\rangle\,. \tag{8.5.8}$$

8.5.2 Spontaneous Emission Rate for a $J_g \to J_e$ Transition

The differential spontaneous decay rate for emission of a photon by the transition $\gamma_e J_e M_e \to \gamma_g J_g M_g$ in the solid angle element $d\Omega$ is

$$dW_\sigma(\gamma_e J_e M_e \to \gamma_g J_g M_g) = \frac{1}{8\pi^2\varepsilon_0\hbar}\frac{\omega_{eg}^3}{c^3}|\hat{\boldsymbol{e}}_{\boldsymbol{k}}^{(\sigma)}\cdot\langle\gamma_e J_e M_e|\boldsymbol{d}|\gamma_g J_g M_g\rangle|^2\,d\Omega. \tag{8.5.9}$$

Here $\hat{\boldsymbol{e}}_{\boldsymbol{k}}^{(\sigma)}$ is the polarization vector, and $\boldsymbol{k}$ is the direction of the emitted photon into the solid angle $d\Omega$.

a) Polarization State of the Photon Emitted: The coefficient of $A_q^\dagger$ in (8.5.5) defines the components of the electric field which couple to the transitions for which $\Delta M \equiv M_e - M_g = q$. A particular transition $J_g M_g \to J_e M_e$ is defined with respect to the co-ordinate system in which the angular momentum eigenstates are defined. This may be completely arbitrary, but may also relate to an imposed magnetic field.

We now introduce a notation for the spherical components of a polarization vector $\hat{\boldsymbol{e}}_{\boldsymbol{k}}^{(\sigma)}$, defined by

$$E_q(\sigma, \boldsymbol{k}) \equiv \boldsymbol{e}_q^* \cdot \hat{\boldsymbol{e}}_{\boldsymbol{k}}^{(\sigma)}. \tag{8.5.10}$$

For a given q, (8.5.5) shows that the coefficient of a plane wave component with wavevector $\boldsymbol{k}$ will involve the linear combination of destruction operators for the two photon polarizations

$$B_q \equiv \sum_\sigma E_q(\sigma, \boldsymbol{k})\, a_{\sigma \boldsymbol{k}}\,, \tag{8.5.11}$$

and this then selects the particular kind of polarization of a photon emitted in this transition with wavevector $\boldsymbol{k}$.

b) Evaluation of Polarizations: If we write the wavevector as $\boldsymbol{k} \equiv k\hat{\boldsymbol{k}}$, this provides an orthonormal set comprising $\hat{\boldsymbol{k}}$ and:

i) Two plane polarization vectors in the form

$$\hat{\boldsymbol{k}} = \begin{pmatrix} \sin\theta\cos\varphi \\ \sin\theta\sin\varphi \\ \cos\theta \end{pmatrix}, \qquad \hat{\boldsymbol{e}}_{\boldsymbol{k}}^{(1)} = \begin{pmatrix} -\cos\theta\cos\varphi \\ -\cos\theta\sin\varphi \\ \sin\theta \end{pmatrix}, \qquad \hat{\boldsymbol{e}}_{\boldsymbol{k}}^{(2)} = \begin{pmatrix} \sin\varphi \\ -\cos\varphi \\ 0 \end{pmatrix}. \tag{8.5.12}$$

ii) Two circular polarization vectors with photon spin ± 1

$$\hat{\boldsymbol{e}}_{\boldsymbol{k}}^{(\pm 1)} = \frac{\hat{\boldsymbol{e}}_{\boldsymbol{k}}^{(1)} \pm \mathrm{i}\hat{\boldsymbol{e}}_{\boldsymbol{k}}^{(2)}}{\sqrt{2}}. \tag{8.5.13}$$

The operator $B_q^\dagger$ creates a linear combination of the states with polarizations $\hat{\boldsymbol{e}}^{(1)}$ and $\hat{\boldsymbol{e}}^{(2)}$ with coefficients $E_q(1,\boldsymbol{k})^*$ and $E_q(2,\boldsymbol{k})^*$, so the polarization created is

$$\boldsymbol{u}_q \propto E_q(1,\boldsymbol{k})^*\, \hat{\boldsymbol{e}}_{\boldsymbol{k}}^{(1)} + E_q(2,\boldsymbol{k})^*\, \hat{\boldsymbol{e}}_{\boldsymbol{k}}^{(2)} = \boldsymbol{e}_q - \hat{\boldsymbol{k}}(\boldsymbol{e}_q \cdot \hat{\boldsymbol{k}}), \tag{8.5.14}$$

that is, the polarization is the projection of $\boldsymbol{e}_q$ on the plane orthogonal to $\boldsymbol{k}$. This means that:

i) If $\Delta M = q = 0$, the photon is plane polarized, since all components of $\boldsymbol{u}_0$ are real.

ii) If $\Delta M = q = \pm 1$, the photon is elliptically or circularly polarized, since $\boldsymbol{u}_{\pm 1}$ has complex components.

8.5.3 Selection Rules, Polarization and Angular Distribution

We can use the spherical basis to write the scalar product $\hat{\boldsymbol{e}}_{\boldsymbol{k}}^{(\sigma)} \cdot \boldsymbol{d}$, which appears in (8.5.8), in terms of $E_q(\sigma, \boldsymbol{k})$, and the spherical components d_q of the dipole operator, which are given by

$$d_q \equiv \boldsymbol{e}_d^* \cdot \boldsymbol{d}. \tag{8.5.15}$$

The expression for the scalar product takes the form

$$\hat{\boldsymbol{e}}_{\boldsymbol{k}}^{(\sigma)} \cdot \boldsymbol{d} = \sum_q E_q(\sigma, \boldsymbol{k})\, d_q. \tag{8.5.16}$$

The quantities $E_q(\sigma, \boldsymbol{k})$, being the spherical components of the vector $\hat{\boldsymbol{e}}_{\boldsymbol{k}}^{(\sigma)}$, can be expressed in terms of spherical harmonics, as shown in Appendix 8.A i, j. To do this we need the polar co-ordinates θ_e, φ_e of $\hat{\boldsymbol{e}}_{\boldsymbol{k}}^{(\sigma)}$, which depend on the particular polarization chosen. For example, with the choices (8.5.12), we find the correspondences

$$\left.\begin{array}{lll} \hat{\boldsymbol{e}}_{\boldsymbol{k}}^{(1)}: & \theta_e = \theta - \tfrac{1}{2}\pi, & \phi_e = \phi, \\ \hat{\boldsymbol{e}}_{\boldsymbol{k}}^{(2)}: & \theta_e = \tfrac{1}{2}\pi, & \phi_e = \phi - \tfrac{1}{2}\pi. \end{array}\right\} \tag{8.5.17}$$

Using this notation, the transition amplitude becomes

$$\hat{\boldsymbol{e}}_{\boldsymbol{k}}^{(\sigma)} \cdot \langle \gamma J M | \boldsymbol{d} | \gamma' J' M' \rangle = \sum_q C_{1q}^*(\theta_\epsilon, \phi_\epsilon) \langle \gamma_e J_e M_e | d_q | \gamma_g J_g M_g \rangle, \tag{8.5.18}$$

$$\equiv \mathcal{M}_{eg}(\theta_\epsilon, \varphi_\epsilon). \tag{8.5.19}$$

We now apply the Wigner–Eckart theorem to the dipole operator

$$\langle \gamma_e J_e M_e | d_q | \gamma_g J_g M_g \rangle = (-1)^{J_e - M_e} (\gamma_e J_e \| \boldsymbol{d} \| \gamma_g J_g) \begin{pmatrix} J_e & 1 & J_g \\ -M_e & q & M_j \end{pmatrix}. \tag{8.5.20}$$

a) Selection rules: From the properties of the 3j-symbol we have the selection rules:

i) *Angular momentum conservation*:

$$J_g + J_e \geqslant 1, \tag{8.5.21}$$

$$\Delta J = J_e - J_g = 0,\ \pm 1, \tag{8.5.22}$$

$$\Delta M = M_e - M_g = q = 0,\ \pm 1. \tag{8.5.23}$$

ii) *Parity*: In addition we have the selection rule that the parity must change.

b) Transitions:

i) $\Delta M = 0$ or $M_e = M_g$ and $q = 0$ (linear polarization)

$$\mathcal{M}_{eg}(\theta_\epsilon, \varphi_\epsilon) = e_0 \langle \gamma_e J_e M_e | d_0 | \gamma_g J_g M_g \rangle. \tag{8.5.24}$$

ii) $\Delta M = +1$ or $M_e = M_g + 1$ and $q = 1$ (right hand elliptically polarized)

$$\mathcal{M}_{eg}(\theta_\epsilon, \varphi_\epsilon) = e_{+1}^* \langle \gamma_e J_e M_e | d_{+1} | \gamma_g J_g M_g \rangle. \tag{8.5.25}$$

iii) $\Delta M = -1$ or $M_e = M_g - 1$ and $q = -1$ (left hand elliptically polarized)

$$\mathcal{M}_{eg}(\theta_\epsilon, \varphi_\epsilon) = e^*_{-1} \langle \gamma_e J_e M_e | d_{+1} | \gamma_g J_g M_g \rangle. \tag{8.5.26}$$

c) Angular Distributions: The angular distribution for the $\Delta M = M_e - M_g = q$ transition is determined by $|C_{1q}(\theta_\epsilon, \phi_\epsilon)|^2$.

8.5.4 Total Transition Rate

In order to find the lifetime of the atom, we need to sum over all polarizations and directions of emission $\hat{\boldsymbol{k}}$ of the photon. Thus, the total rate of transitions $g \to e$ is

$$W(\gamma_e J_e M_e \to \gamma_g J_g M_g) = \frac{1}{8\pi^2 \varepsilon_0 \hbar} \frac{\omega_{eg}^3}{c^3} \sum_{\sigma=1,2} \int \left| \hat{\boldsymbol{e}}^{(\sigma)}_{\boldsymbol{k}} \cdot \langle \gamma_e J_e M_e | \boldsymbol{d} | \gamma_g J_g M_g \rangle \right|^2 d\Omega. \tag{8.5.27}$$

a) Sum over Polarizations and Integration over Solid Angle: If we set the matrix element $\langle \gamma_e J_e M_e | \boldsymbol{d} | \gamma_g J_g M_g \rangle \to \boldsymbol{D}$, and choose a co-ordinate system in which $\boldsymbol{D}$ is parallel to the z-axis, the summation and integration in (8.5.27) has the form

$$\sum_{\sigma=1,2} \int \left| \hat{\boldsymbol{e}}^{(\sigma)}_{\boldsymbol{k}} \cdot \boldsymbol{D} \right|^2 d\Omega = |\boldsymbol{D}|^2 \int \left[1 - \cos^2\theta\right] \sin\theta \, d\theta \, d\varphi = \frac{8\pi}{3} |\boldsymbol{D}|^2. \tag{8.5.28}$$

b) Total Transition Rate for $M_e \to M_g$: The transition rate for the transition from a definite initial state M_e of the atom to a definite final state M_g is therefore

$$W(\gamma_e J_e M_e \to \gamma_g J_g M_g) = \frac{1}{8\pi^2 \varepsilon_0 \hbar} \frac{\omega_{eg}^3}{c^3} \frac{8\pi}{3} \left| \langle \gamma_e J_e M_e | \boldsymbol{d} | \gamma_g J_g M_g \rangle \right|^2. \tag{8.5.29}$$

c) Total Decay Rate: The total transition rate to *any* final state of the atom in the $\gamma_g J_g$ multiplet is found by summing over the possible final states to obtain the *total decay rate*

$$\sum_{M_g} W(\gamma_e J_e M_e \to \gamma_g J_g M_g) = \frac{1}{8\pi^2 \varepsilon_0 \hbar} \frac{\omega_{eg}^3}{c^3} \sum_{M_g} \frac{8\pi}{3} \left| \langle \gamma_e J_e M_e | \boldsymbol{d} | \gamma_g J_g M_g \rangle \right|^2, \tag{8.5.30}$$

$$= \frac{1}{3\pi \varepsilon_0 \hbar} \frac{\omega_{eg}^3}{c^3} \frac{1}{2J_e + 1} \left| (\gamma_e J_e \| \boldsymbol{d} \| \gamma_g J_g) \right|^2, \tag{8.5.31}$$

which is the same for all M_e (as expected from rotational invariance).

To do this calculation, we have used (8.A.15) and

$$\sum_{M_g} \left| \langle \gamma_e J_e M_e | \boldsymbol{d} | \gamma_g J_g M_g \rangle \right|^2 = \sum_{M_g} \sum_q \left| \langle \gamma_e J_e M_e | d_q | \gamma_g J_g M_g \rangle \right|^2, \tag{8.5.32}$$

$$= \left| (\gamma_e J_e \| \boldsymbol{d} \| \gamma_g J_g) \right|^2 \sum_{M_g} \sum_q \begin{pmatrix} J_e & 1 & J_g \\ -M_e & q & M_g \end{pmatrix}^2, \tag{8.5.33}$$

$$= \left| (\gamma_e J_e \| \boldsymbol{d} \| \gamma_g J_g) \right|^2 \frac{1}{2J_e + 1}. \tag{8.5.34}$$

8.5.5 Example

We will choose a transition with $q = \Delta M = 0$, so that the angular distribution is given by

$$|C_{10}(\theta_e,\phi_e)|^2 = \cos^2\theta_e. \tag{8.5.35}$$

Using the polarizations and notations in (8.5.12), the two choices of angles are

$$\cos\theta_{e1} = \sin\theta, \quad \cos\theta_{e2} = 0, \tag{8.5.36}$$

so that

$$dW_1(\gamma_e J_e M_e \to \gamma_g J_g M_g) = \frac{1}{8\pi^2\varepsilon_0\hbar}\frac{\omega_{eg}^3}{c^3}\left|\langle\gamma_e J_e M_e|d_0|\gamma_g J_g M_g\rangle\right|^2 \sin^2\theta\, d\Omega, \tag{8.5.37}$$

$$dW_2(\gamma_e J_e M_e \to \gamma_g J_g M_g) = 0. \tag{8.5.38}$$

Summing and integrating over the solid angle we obtain

$$W(\gamma_e J_e M_e \to \gamma_g J_g M_g) = \left(\frac{8\pi}{3}\right)\frac{1}{8\pi^2\varepsilon_0\hbar}\frac{\omega_{eg}^3}{c^3}\left|\langle\gamma_e J_e M_e|d_0|\gamma_g J_g M_g\rangle\right|^2. \tag{8.5.39}$$

8.5.6 Fine Structure and Hyperfine Structure

We consider a single electron outside a closed shell, as in an alkali atom, with quantum numbers nlm_l, and wavefunction

$$\psi_{nlm_l} = R_{nl}(r)Y_{lm_l}(\theta,\varphi). \tag{8.5.40}$$

The dipole matrix element for nlm_l coupling is

$$\langle n_e l_e m_{l_e}|d_q|n_g l_g m_{l_g}\rangle = \int_0^\infty dr\, R_{n_e}(r)\, r\, R_{n_g}(r)\langle l_e m_{l_e}|C_{1q}|l_g m_{l_g}\rangle, \tag{8.5.41}$$

$$= R_{n_e l_e n_g l_g}(l_e\,\|C_1\|\,l_g)\frac{\langle l_g 1 m_{l_g} q|l_e m_{l_e}\rangle}{\sqrt{2l_e+1}}. \tag{8.5.42}$$

Sobelman [8.1] gives a formula for the reduced matrix element in his equation (14.32). In our notation, this result can be written in the form

$$(l_e\,\|C_1\|\,l_g) = \begin{cases} (-1)^{l_e - l_{\max}}\sqrt{l_{\max}}, & \text{for } l_e = l_g \pm 1, \\ 0, & \text{otherwise.} \end{cases} \tag{8.5.43}$$

Here $l_{\max} \equiv \max(l_e, l_g)$. (8.5.44)

8.A Appendix: Spherical Tensors and the Wigner–Eckart Theorem

This appendix is a brief summary of relevant aspects of the quantum theory of angular momentum, in a notation consistent with [8.5], in which a full presentation of angular momentum theory is given.

We will use the operators $J_0, J_\pm$, defined through

$$J_z \equiv \hbar J_0, \qquad J_x \pm \mathrm{i}J_y \equiv \hbar J_\pm, \tag{8.A.1}$$

which have the commutation relations

$$[J_0, J_\pm] = \pm J_\pm, \qquad [J_+, J_-] = 2J_0. \tag{8.A.2}$$

Notice that we make a distinction between $\boldsymbol{J}$, which has the dimensions of angular momentum, and these operators, which are more appropriate for angular momentum algebra. The eigenstates $|j,m\rangle$ satisfy the algebraic relations

$$J_0|j,m\rangle = m|j,m\rangle, \tag{8.A.3}$$

$$J_\pm|j,m\rangle = \sqrt{(j\mp m)(j\pm m+1)}\,|j,m\pm 1\rangle. \tag{8.A.4}$$

These relationships are equivalent to the property of the operators $\boldsymbol{J}$ as infinitesimal generators of quantum mechanical rotation operators, that is, a rotation of magnitude α about an axis $\boldsymbol{n}$ is given by $U = \exp\left(-\mathrm{i}\alpha \boldsymbol{J}\cdot\boldsymbol{n}/\hbar\right)$.

a) Addition of Angular Momentum: In the case of a particle with spin, or in many body systems, one often finds two (or more) sets of angular momentum operators $\hat{\boldsymbol{J}}_1, \hat{\boldsymbol{J}}_2$, and the quantum states are then characterized as $|j_1, j_2, m_1, m_2\rangle$.

However, in general only total angular momentum is conserved, so a set including the conserved operator $\hat{\boldsymbol{J}} \equiv \hat{\boldsymbol{J}}_1 + \hat{\boldsymbol{J}}_2$ is more convenient. Since both $\hat{\boldsymbol{J}}_1^2$ and $\hat{\boldsymbol{J}}_2^2$ are scalar operators, they commute with $\hat{\boldsymbol{J}}$, so we can choose as a complete set of commuting observables the set $\hat{\boldsymbol{J}}_1^2, \hat{\boldsymbol{J}}_2^2, \hat{\boldsymbol{J}}^2, \hat{J}_z$, with eigenvalues $\hbar^2 j_1(j_1+1), \hbar^2 j_2(j_2+1), \hbar^2 j(j+1), \hbar m$, and with eigenstates written as $|j_1, j_2, j, m\rangle$.

For a given j_1, j_2 the range of values of j satisfies the condition

$$|j_1 - j_2| \leqslant j \leqslant j_1 + j_2. \tag{8.A.5}$$

Any written proof of this is tedious, but available in any good textbook. It is algebraically rather simple to show.

b) Clebsch–Gordan Coefficients: The states of either kind can be written in terms of the other kind using the completeness of either form:

$$|j_1, j_2, m_1, m_2\rangle = \sum_{jm} \langle j_1, j_2, j, m | j_1, j_2, m_1, m_2\rangle |j_1, j_2, j, m\rangle, \tag{8.A.6}$$

$$|j_1, j_2, j, m\rangle = \sum_{m_1 m_2} \langle j_1, j_2, m_1, m_2 | j_1, j_2, j, m\rangle |j_1, j_2, m_1, m_2\rangle. \tag{8.A.7}$$

Furthermore, the coefficients $\langle j_1, j_2, m_1, m_2 | j_1, j_2, j, m\rangle$ are real with the usual choice of phase convention for the angular momentum states (defined by 8.A.4), so that the coefficients for both equations are the same.

These coefficients are called the *Clebsch–Gordan coefficients*, are conventionally written in an abbreviated notation $\langle j_1, j_2, m_1, m_2 | j, m\rangle$, the repetition of the j_1, j_2 in the first part being unnecessary.

c) Properties of Clebsch–Gordan Coefficients:

$$\langle j_1, j_2, m_1, m_2 | j, m\rangle = 0, \qquad \text{unless } m = m_1 + m_2, \tag{8.A.8}$$

$$\sum_{jm} \langle j_1, j_2, m_1, m_2 | j, m\rangle \langle j_1, j_2, m_1', m_2' | j, m\rangle = \delta_{m_1 m_1'} \delta_{m_2 m_2'}, \tag{8.A.9}$$

$$\sum_{m_1 m_2} \langle j_1, j_2, m_1, m_2 | j, m\rangle \langle j_1, j_2, m_1, m_2 | j', m'\rangle = \delta_{mm'} \delta_{jj'}. \tag{8.A.10}$$

The rotation matrices $\mathcal{D}^j_{m'm}(R)$, as defined in [8.5], satisfy the identity

$$\mathcal{D}^j_{m'm}(R) = \sum_{m_1 m_1' m_2 m_2'} \langle j_1, j_2, m_1, m_2 | j, m\rangle \langle j_1, j_2, m_1', m_2' | j, m\rangle \mathcal{D}^{j_1}_{m_1' m_1}(R) \mathcal{D}^{j_2}_{m_2' m_2}(R). \tag{8.A.11}$$

d) 3j-Symbols: The Clebsch–Gordan coefficients have many symmetries, and these are displayed most clearly by the 3j-symbols, which are defined by

$$\langle j_1, j_2, m_1, m_2 | j, m\rangle = (-)^{-j_1 + j_2 - m} \sqrt{2j+1} \begin{pmatrix} j_1 & j_2 & j \\ m_1 & m_2 & -m \end{pmatrix} \tag{8.A.12}$$

$$\begin{pmatrix} j_1 & j_2 & j \\ m_1 & m_2 & m \end{pmatrix} = \frac{(-)^{j_1 - j_2 - m}}{\sqrt{2j+1}} \langle j_1, j_2, m_1, m_2 | j, -m\rangle. \tag{8.A.13}$$

e) Properties of 3j-Symbols:

i) *Permutation Symmetry:* Under even permutations of the columns, the 3j-symbol does not change, under odd permutations of the columns the 3j-symbol is multiplied by $(-)^{j_1+j_2+j}$.

ii) *Orthogonality:* The orthogonality conditions (8.A.9, 8.A.10) correspond to the properties

$$\sum_{jm} (2j+1) \begin{pmatrix} j_1 & j_2 & j \\ m_1 & m_2 & m \end{pmatrix} \begin{pmatrix} j_1 & j_2 & j \\ m_1' & m_2' & m \end{pmatrix} = \delta_{m_1, m_1'} \delta_{m_2, m_2'}, \tag{8.A.14}$$

$$\sum_{m_1 m_2} \begin{pmatrix} j_1 & j_2 & j \\ m_1 & m_2 & m \end{pmatrix} \begin{pmatrix} j_1 & j_2 & j' \\ m_1 & m_2 & m' \end{pmatrix} = \frac{1}{2j+1} \delta_{jj'} \delta_{mm'}. \tag{8.A.15}$$

iii) *Conservation of Angular Momentum:*

$$\begin{pmatrix} j_1 & j_2 & j \\ m_1 & m_2 & m \end{pmatrix} = 0, \qquad \text{unless } m_1 + m_2 + m = 0. \tag{8.A.16}$$

f) Spherical Harmonics: In atomic physics it is customary to use a spherical harmonic definition more directly related to the Legendre functions $P_{l,|m|}(\cos\theta)$, and therefore with a normalization different from that of the usual orthonormal spherical harmonics $Y_{l,m}(\theta, \phi)$. Explicitly, one defines

$$C_{l,m}(\theta,\phi) \equiv \sqrt{\frac{4\pi}{2l+1}}\, Y_{l,m}(\theta,\phi), \tag{8.A.17}$$

so that

$$\int d\Omega\, C^*_{l',m'}(\theta,\phi) C_{l,m}(\theta,\phi) = \frac{4\pi}{2l+1}\delta_{ll'}\,\delta_{mm'}, \tag{8.A.18}$$

and

$$C_{l,0}(\theta,\phi) = P_l(\cos\theta). \tag{8.A.19}$$

g) Definition of a Spherical Tensor: A spherical tensor $T_{k,q}$ is a set of operators in which $q=-k,k-1,\dots,k$, and whose transformation $U\,T_{k,q}\,U^\dagger$ in exactly the same way as an angular momentum eigenstate $|k,,q\rangle$. This leads to algebraic relations similar to (8.A.3), but using commutators thus:

$$[J_0, T_{k,q}] = q\,T_{k,q}, \tag{8.A.20}$$

$$[J_\pm, T_{k,q}] = \sqrt{(k\mp q)(k\pm q+1)}\,T_{k,q\pm 1}. \tag{8.A.21}$$

h) Vector Operator in a Spherical Basis: The most important kind of spherical tensor is formed from a vector operator $\boldsymbol{A}$, whose defining property is

$$[J_i, A_j] = \mathrm{i}\hbar\epsilon_{ijk}A_k, \tag{8.A.22}$$

where i,j,k are different from each other, and take on the values x,y,z. These commutation relations are equivalent to those for a rank 1 spherical tensor $A_{1,q}$ defined by

$$A_{1,0} = A_z, \qquad A_{1,\pm1} = \mp\frac{1}{\sqrt{2}}(A_x \pm \mathrm{i}A_y). \tag{8.A.23}$$

i) The Spherical Basis: Let us now write the vector $\boldsymbol{A}$ in terms of unit vectors as

$$\boldsymbol{A} = A_x\boldsymbol{e}_x + A_y\boldsymbol{e}_y + A_z\boldsymbol{e}_z, \tag{8.A.24}$$

with Cartesian components A_x, A_y, A_z, so that

$$A_x = \boldsymbol{e}_x\cdot\boldsymbol{A}, \quad A_y = \boldsymbol{e}_y\cdot\boldsymbol{A}, \quad A_z = \boldsymbol{e}_z\cdot\boldsymbol{A}. \tag{8.A.25}$$

We can define a corresponding basis set of *spherical unit vectors* by

$$\boldsymbol{e}_1 = -\frac{1}{\sqrt{2}}(\boldsymbol{e}_x + \mathrm{i}\boldsymbol{e}_y), \quad \boldsymbol{e}_0 = \boldsymbol{e}_z, \quad \boldsymbol{e}_{-1} = \frac{1}{\sqrt{2}}(\boldsymbol{e}_x - \mathrm{i}\boldsymbol{e}_y), \tag{8.A.26}$$

which have the properties

$$\boldsymbol{e}^*_q = (-1)^q\boldsymbol{e}_{-q}, \quad \boldsymbol{e}^*_q\cdot\boldsymbol{e}_{q'} = \delta_{qq'}. \tag{8.A.27}$$

The expansion in a spherical basis corresponding to the Cartesian basis expansion (8.A.24) takes the form

$$\boldsymbol{A} = \sum_{q=0,\pm1} A_{1,q}\boldsymbol{e}^*_q = \sum_{q=0,\pm1} A^*_{1,q}\boldsymbol{e}_q = \sum_{q=0,\pm1} (-1)^q A_{1,-q}\boldsymbol{e}_q, \tag{8.A.28}$$

where the spherical components are defined by (8.A.23), or equivalently by

$$A_{1,q} \equiv \boldsymbol{e}^*_q\cdot\boldsymbol{A}. \tag{8.A.29}$$

j) Expression of a Vector in Spherical Harmonics: The spherical components of a vector $\boldsymbol{A}$ whose components commute with each other (such as position $\boldsymbol{x}$ or wavevector $\boldsymbol{k}$) can be written explicitly in terms of spherical harmonics as

$$A_0 \equiv A_z = |\boldsymbol{A}|\cos\theta = |\boldsymbol{A}|\, C^1_0(\theta,\phi), \tag{8.A.30}$$

$$A_{+1} \equiv -\frac{1}{\sqrt{2}}(A_x + \mathrm{i}A_y) = -|\boldsymbol{A}|\frac{\sin\theta\,\mathrm{e}^{\mathrm{i}\varphi}}{\sqrt{2}} = |\boldsymbol{A}|\, C^1_{+1}(\theta,\phi), \tag{8.A.31}$$

$$A_{-1} \equiv \frac{1}{\sqrt{2}}(A_x + \mathrm{i}A_y) = |\boldsymbol{A}|\frac{\sin\theta\,\mathrm{e}^{-\mathrm{i}\varphi}}{\sqrt{2}} = |\boldsymbol{A}|\, C^1_{-1}(\theta,\phi). \tag{8.A.32}$$

k) Second Rank Tensors in a Spherical Basis: In a Cartesian basis, a rank-2 tensor is written T_{ij}, and and satisfies the commutation relations

$$[J_i, T_{lm}] = \mathrm{i}\hbar\left(\epsilon_{ilk}T_{km} + \epsilon_{imk}T_{lk}\right). \tag{8.A.33}$$

The Cartesian tensor can be separated into three irreducible tensors, namely

i) A scalar, $\boldsymbol{T}^{(0)} = \sum_i T_{ii}$,

ii) A vector, $\boldsymbol{T}^{(1)} = \sum_{jk}\epsilon_{ijk}T_{jk}$,

iii) A traceless symmetric rank-2 tensor, $\boldsymbol{T}^{(2)} = T_{ij} + T_{ji} - \frac{2}{3}\delta_{ij}\sum_i T_{ii}$.

In a spherical basis we write

$$T_{0,0} = T^{(0)}_0, \tag{8.A.34}$$

$$T_{1,0} = T^{(1)}_z, \qquad T_{1,\pm1} = \mp\tfrac{1}{\sqrt{2}}\left(T^{(1)}_x \pm \mathrm{i}T^{(1)}_y\right), \tag{8.A.35}$$

$$T_{2,0} = T^{(2)}_{zz}, \qquad T_{2,\pm1} = \pm\sqrt{\tfrac{2}{3}}\left(T^{(2)}_{zx} \pm T^{(2)}_{yz}\right), \qquad T_{2,\pm2} = \sqrt{\tfrac{1}{6}}\left(T^{(2)}_{xx} - T^{(2)}_{yy} \pm 2\mathrm{i}\,T^{(2)}_{xy}\right). \tag{8.A.36}$$

All of these results can be demonstrated by showing they are consistent with the commutation relations both in the spherical form (8.A.20, 8.A.21), and in the Cartesian form (8.A.33).

l) The Product of two Vectors as a Rank-2 Tensor: The products A_iB_j of the components of two vector operators obviously form the components of a rank-2 Cartesian tensor, which we will call $\boldsymbol{A}:\boldsymbol{B}$. This can be expressed as three spherical tensors, as above, thus:

$$(\boldsymbol{A}:\boldsymbol{B})^{(0)} = \boldsymbol{A}\cdot\boldsymbol{B}, \tag{8.A.37}$$

$$(\boldsymbol{A}:\boldsymbol{B})^{(1)} = \boldsymbol{A}\times\boldsymbol{B}, \tag{8.A.38}$$

$$(\boldsymbol{A}:\boldsymbol{B})^{(2)}_{ij} = A_iB_j + A_jB_i - \tfrac{2}{3}\delta_{ij}\boldsymbol{A}\cdot\boldsymbol{B}. \tag{8.A.39}$$

Two special cases are worth noting.

i) If $\boldsymbol{A} = \boldsymbol{B} = \boldsymbol{x}$, then the components of $\boldsymbol{x}$ commute with each other, so that

$$(\boldsymbol{x}:\boldsymbol{x})^{(0)} = \boldsymbol{x}^2, \tag{8.A.40}$$

$$(\boldsymbol{x}:\boldsymbol{x})^{(1)} = 0, \tag{8.A.41}$$

$$(\boldsymbol{x}:\boldsymbol{x})^{(2)} = 2\left(x_ix_j - \tfrac{1}{3}\delta_{ij}\boldsymbol{x}^2\right). \tag{8.A.42}$$

ii) On the other hand, that if $\boldsymbol{A} = \boldsymbol{B} = \boldsymbol{J}$, then the angular momentum commutation relations mean that $\boldsymbol{J} \times \boldsymbol{J} = i\hbar \boldsymbol{J}$, so that

$$(\boldsymbol{J} : \boldsymbol{J})^{(0)} = \boldsymbol{J}^2, \tag{8.A.43}$$

$$(\boldsymbol{J} : \boldsymbol{J})^{(1)} = i\hbar \boldsymbol{J}, \tag{8.A.44}$$

$$(\boldsymbol{J} : \boldsymbol{J})^{(2)} = \left(J_i J_j + J_j J_i - \tfrac{2}{3}\delta_{ij} \boldsymbol{J}^2\right). \tag{8.A.45}$$

m) Scalar product of Two Spherical Tensors: We use the notation

$$\boldsymbol{T}^{(k)} \cdot \boldsymbol{R}^{(k)} \equiv \sum_q (-)^q T_{k,-q} R_{k,q}, \tag{8.A.46}$$

to define the *scalar product* of two rank k spherical tensors.

n) The Wigner–Eckart Theorem: In atomic, molecular and nuclear physics one often needs to compute the matrix elements of a tensor operator between initial and final states of well-defined angular momentum.

If we write the particular states of well-defined angular momentum as $|\gamma j m\rangle$, where γ represents all quantum numbers not related to the angular momentum quantum numbers j, m, then the Wigner–Eckart theorem states that for an arbitrary spherical tensor operator $T_{k,q}$

$$\langle \gamma j m | T_{k,q} | \gamma' j' m' \rangle = \langle k, j', q, m' | j m \rangle \, \langle \gamma j \| T_k \| \gamma' j' \rangle. \tag{8.A.47}$$

The first factor on the right is a Clebsch–Gordan coefficient, and the second is called the *reduced matrix element*, and does not depend on m, q or m'. This provides a great simplification, since all matrix elements of this kind are related to the one reduced matrix element. To compute the reduced matrix element, one chooses values if m', m which make for the easiest calculation, and then all others are given by (8.A.47).

The proof of the theorem is really rather obvious—it relies on the fact that the product $T_{k,q}|\gamma j m\rangle$ transforms under rotations in exactly the same way as a state of two different angular momenta k, q and j, m, i.e. a state like $|\beta k, j, q, m\rangle$, and one can use the orthogonality relations (8.A.10, 8.A.11) to complete the proof.

o) Wigner–Eckart Theorem Using 3j-Symbols: In atomic physics the 3j-symbol is much preferred, and the Wigner–Eckart theorem is then rewritten as

$$\langle \gamma j m | T_{k,q} | \gamma' j' m' \rangle = (-)^{j-m} \begin{pmatrix} j & k & j' \\ -m & q & m' \end{pmatrix} (\gamma j \| T_k \| \gamma' j'). \tag{8.A.48}$$

The quantity $(\gamma j \| T_k \| \gamma' j')$ is called the reduced matrix element, as is the corresponding quantity in (8.A.47), even though they are not equal; in fact the relationship between them is

$$(\gamma j \| T_k \| \gamma' j') = \sqrt{2j+1}\, \langle \gamma j \| T_k \| \gamma' j' \rangle. \tag{8.A.49}$$

p) The Wigner–Eckart Theorem for a Vector Operator: There are some simple results for vector operators of particular use in atomic spectroscopy.

i) If we take an arbitrary vector operator $\boldsymbol{A}$, we can show that, for any values of m and other quantum numbers γ, γ',

$$\langle \gamma j m | \boldsymbol{J}\cdot\boldsymbol{A} | \gamma' j m\rangle = \langle j \| \boldsymbol{J} \| j\rangle \, \langle \gamma j \| \boldsymbol{A} \| \gamma' j\rangle. \tag{8.A.50}$$

Notice that the value of m is arbitrary, since $\boldsymbol{J}\cdot\boldsymbol{A}$ is a scalar operator, which therefore commutes with $\boldsymbol{J}$.

ii) Setting $\boldsymbol{A} \to \boldsymbol{J}$ leads to

$$\langle j \| \boldsymbol{J} \| j\rangle = \hbar\sqrt{j(j+1)}\,. \tag{8.A.51}$$

The sign of the square root is not given by this argument, but must be positive to be consistent with the positive sign of the square root in (8.A.4).

iii) From these results, it follows that

$$\langle \gamma j \| \boldsymbol{A} \| \gamma' j\rangle = \frac{\langle \gamma j m | \boldsymbol{J}\cdot\boldsymbol{A} | \gamma' j m\rangle}{\hbar\sqrt{j(j+1)}}\,, \tag{8.A.52}$$

$$(\gamma j \| \boldsymbol{A} \| \gamma' j) = \frac{\langle \gamma j m | \boldsymbol{J}\cdot\boldsymbol{A} | \gamma' j m\rangle}{\hbar}\sqrt{\frac{2j+1}{j(j+1)}}\,. \tag{8.A.53}$$

Exercise 8.1 Proof of (8.A.50): Write $\boldsymbol{J}\cdot\boldsymbol{A} = \sum_q (-)^q J_q A_q$, and insert a complete set of states between the J_q and the A_q operators. Use the the Wigner–Eckart theorem on both operators, and the orthogonality properties of the Clebsch-Gordan coefficients or 3j-symbols given in Appendix 8.A a to prove (8.A.50).

Exercise 8.2 The Vector Model: Show that the relationship (8.A.53) means that for any vector operator

$$\langle \gamma j m | \boldsymbol{A} | \gamma' j m'\rangle = \langle \gamma j m | \boldsymbol{J} | \gamma' j m'\rangle \frac{\langle \gamma j m | \boldsymbol{J}\cdot\boldsymbol{A} | \gamma' j m\rangle}{\hbar^2 j(j+1)}\,. \tag{8.A.54}$$

As in (8.A.50), note that value of m in the matrix element of $\boldsymbol{J}\cdot\boldsymbol{A}$ is arbitrary.

This result is usually interpreted to mean that, within a subspace of fixed j, a vector $\boldsymbol{A}$ may be replaced by its projection along the direction of the vector $\boldsymbol{J}$, whose length within this subspace is $\hbar\sqrt{j(j+1)}$.

III ATOMS AND QUANTIZED OPTICAL FIELDS

Part III: Atoms and Quantized Optical Fields

In contrast to the point of view of Part II, in which coherent optical fields were used to manipulate the quantum state of the atom in a wide variety of ways, the aim of this part of the book is to develop the techniques necessary to describe in detail the interaction of atoms with the fully quantized electromagnetic field.

The Quantum Optical System-Environment Model: An atom interacting with a classical electromagnetic field necessarily also interacts with the quantized electromagnetic field. The interaction of an atom with the quantized electromagnetic field has three main aspects:

i) *Dissipation and Decoherence*: Excited states of atoms decay, emitting photons in the process, and quantum superpositions lose their coherence very rapidly. These aspects have been explained in *Bk. I: Ch.12–Ch.14* as well as in *Bk. I: Ch.18–Ch.20*, where the electromagnetic field was seen to behave largely as an *environment* or *heat bath* that is, a system in which an enormous number of modes interact with the atom. The effect on the atom is significant, while the effect on the very much larger environment is small.

ii) *Quantum State Preparation*: The effect of the environment on the atom can, in fact, be beneficial. For example, when an atom decays, the result is nevertheless a pure state. It is possible to engineer the coupling of atomic systems to the electromagnetic field (by the simultaneous application of appropriate coherent laser fields) so that the state to which the system decays is a non-trivial pure state. For example, the dark states of the three-level atoms we discussed in Sect. 6.2.1 can be prepared dissipatively, as we will show in Sect. 13.3.3, by a process which is essentially that of atomic decay, but under the influence of applied coherent optical fields. This process can also be regarded as *optical pumping*, in which the combination of driving fields and spontaneous emission leads the atomic system to move into a chosen pure state.

iii) *Quantum Information Transfer to and from the Quantized Electromagnetic Field*: The processes of damping, dissipation and decoherence are all ways in which quantum information is transferred from the atom to the electromagnetic field. However, this information is not necessarily lost. When an atom decays, it creates a photon, and during the decay process the atom-photon system is in an entangled state. This is not a very useful entangled state, but it does carry information about the atom. Using more complex atomic systems, one can engineer very effective ways both of transferring quantum information from the atomic system to the electromagnetic field, and of performing the reverse process—such methods will be discussed in Part IX.

By regarding the electromagnetic field as merely an environment, we lose the ability to perform measurements on it and manipulate its quantum state. The signals produced from atoms as they interact with the atom are always contained in a well-defined set of field modes. While this set of modes covers a sufficiently broad range of frequencies for the effect on the atom to be dissipative, the polarization and spatial modes are normally quite precisely defined, which makes effective detection or processing of the output field generated by the atom quite feasible. If we choose never to measure the field radiated by an atom, all we see is decoherence and decay; on the other hand, when we do measure the radiated field, we see information about the atom.

The Quantum Stochastic Framework: The theoretical framework needed to describe the effects of the quantized electromagnetic field and an atom on each other provides an extensive and essentially complete description of the way in which the quantum properties of atoms and those of the quantized optical field are connected to each. *This enables the full process of quantum information transfer between atoms and fields* to be described. The structure we will present has the following essential features:

i) *The Quantum Stochastic Schrödinger Equation*: Here we show how to set up a quantum stochastic formalism in which the wavefunction of the atomic system is driven by a quantum noise field arising from the vacuum modes of the electromagnetic field, in addition to any coherent fields involved.
   The result is called the *quantum stochastic Schrödinger equation*, from which all of the methodologies which we list below can be derived. It describes, in principle, all information about the atom, the radiated optical field, and the relationship between them.
ii) *The Master Equation*: The master equation, which we have already introduced and derived in *Bk. I: Ch.13*, can also be straightforwardly derived from the quantum stochastic Schrödinger equation, by tracing over all electromagnetic field variables. As we saw in *Book I*, it provides a description of the dissipation and decoherence processes induced by the quantized electromagnetic field, seen strictly from the point of view of the atom. The information transferred to the electromagnetic field is not observed, and the resulting equation of motion for the density operator is irreversible and dissipative.
iii) *The Input-Output Formalism*: Here we give the procedure by which the quantum properties of the optical field radiated by a system are connected to the properties of the radiating system.
iv) *Cascaded Quantum Systems*: Quantum information transfer from light to an atom requires a way to describe an atom driven by non-classical light. This is achieved by the cascaded quantum system theory, itself a consequence of the quantum stochastic Schrödinger equation and the input-output formalism.

Together with the procedures given in Part II, the quantum stochastic formalism provides all of the tools needed to track quantum information as it proceeds through a system of atoms, cavities, optical fields and measurements. The concepts of *continuous measurement* and *Quantum trajectories*—the subject matter of Part V—augment this formalism by providing a comprehensive picture of the relationship between measurements of the optical field and the quantum state of the atomic system, as well as a practical set of algorithms for simulation of atomic systems as they emit and absorb photons.

9. The Quantum Stochastic Schrödinger Equation

An atom, when interacting with the quantized electromagnetic field, is influenced by the incoming modes, and exerts an influence on the outgoing modes. As used in quantum optics, the atom is a device for receiving information from the electromagnetic field, processing this information, and transmitting the information into the electromagnetic field. As we have seen in *Book I*, the essential behaviour of an atom in the electromagnetic field can be simplified considerably, yielding in *Bk. I: Ch.13* the quantum optical master equation, and the more general idea of a *quantum Markov process.* However, that theory represents only the behaviour of the atom—the optical field remains undescribed.

In this chapter we will give a description of atoms and light which is complete. We want to be able to describe the quantum state of the optical field, that of the atom, and indeed also the joint quantum state of the system comprising the field and the atom. Most importantly, we want to do this in a way which preserves all of the simplicity of a quantum Markov description.

The central part of the methodology we shall use, in this chapter, in the other chapters in this part of the book and in most of the remainder of this book, is the *quantum stochastic Schrödinger equation.* This is an abstraction of the Schrödinger equation for the full system of atom and electromagnetic field, valid under the conditions we normally find in quantum optics. The electromagnetic field itself, in the Markov approximation, becomes a kind of operator valued white noise, with its own *quantum Ito calculus.* Using this Ito calculus, the Schrödinger equation becomes a linear quantum stochastic differential equation, for the full wavefunction describing the combined atom-field system.

The most important cases we consider occur when the optical field is initially in the vacuum state, or in a coherent state of a few modes. The solution of the quantum stochastic Schrödinger equation is a wavefunction for the full quantum-mechanical state of the atom-light system. It is normally characterized by entanglement between the atomic and optical states.

From such a solution, any measurable quantity can be calculated. The quantum optical approximations made mean that all results are approximate, but in quantum optical situations, these are very accurate approximations.

The quantum stochastic Schrödinger equation can be seen as the fundamental formalism from which it is then possible to derive further formalisms. In particular:

i) *The Master Equation*: The density operator which results from tracing over the noise degrees of freedom obeys a quantum mechanical master equation; we show how this is done in Chap. 10.

ii) *Quantum Langevin Equations and the Input-Output Formalism*: If we transform the description to the Heisenberg picture, and this yields an operator quantum Langevin equation description, in which input and output fields in driven systems naturally appear—this is the *input-output formalism* of *Collett* and *Gardiner* [9.1, 9.2], which provides the logical description of the interaction of quantized systems with non-classical light. We show how this is done in Chap. 11.

iii) *Photon Counting, Continuous Measurements and Quantum Trajectories*: One can derive a very powerful numerical and conceptual formalism, in which *quantum jumps* in the atomic wavefunction arise naturally as photons are counted in the optical field emitted by an atom. It is possible both to simulate these jumps, and to measure them experimentally. The simulation method provides a very powerful computational tool.

iv) *Cascaded Quantum Systems*: The final tool needed is a way of calculating the behaviour of a quantum system being driven by an arbitrary optical field. Central to such an ambition is the need to characterize the field of interest, and this presents considerable difficulties. Any optical field described by a quantum Gaussian state can be characterized by a few parameters, but this is not in general possible for non-Gaussian states. Fortunately, there is a way of making a characterization which covers almost every experimental situation. If we have the correct quantum stochastic differential equation to describe the system which produces the optical field, then this characterizes the field completely, and under these conditions, the cascaded system approach shows how to describe a system driven by this field.

We should finally note that the quantum stochastic Schrödinger equation can often be solved directly, since it is a linear equation for the wavefunction, and we will present some solutions in this book, in particular in Ex. 9.3, Ex. 9.4, Ex. 9.7 and Sect. 27.1.3.

9.1 The Quantum Optical System-Environment Model

The method we used in *Book I* to describe the interaction between atoms and light can be written more abstractly, so that the total Hamiltonian includes only the interaction with electromagnetic field modes in a rather narrow range around a central resonance frequency ω_0—in *Bk. I: Sect.12.1.4* we called this "stripped down" quantum electrodynamics, because only the minimum number of modes necessary to produce the desired formulation is included in the description.

As we noted in *Bk. I: Sect.13.1*, the situation we find in quantum optics is in fact only one implementation of the *system-environment model*, in which a *system* with relatively few degrees of freedom interacts with an *environment*, which has essentially an infinite number of degrees of freedom. In this case, the environment is the quantized electromagnetic field, and the system is an atom or some other small system. In *Bk. I: Sect.13.1* we also used another terminology, namely the *heat bath*, for the environment. This terminology has as its origin the idea of a thermalized system, which is so large that its interaction with the small system only results in minimal energy transfer, causing therefore only a minimal change to its temperature.

These terminologies are valid for the electromagnetic field, but the most useful case we will consider is that when the temperature of the environment is absolute zero. Most obviously, this corresponds to the vacuum state of the electromagnetic field. However, a *coherent optical field* also has zero temperature, and represents the situation of an atom interacting with an incoming laser field.

9.1.1 The Hamiltonian

In a given atom there are of course many possible transitions, each with its own frequency, and for each such transition we introduce an interaction. However, as we have seen in Sect. 4.1.3, when considering a single transition, we can represent all of the effects of the other transitions in terms of Stark shifts of the relevant levels, leading to a small adjustment to the transition frequency.

Hence we will write down the Hamiltonian in the case that only one transition is relevant, and we explicitly specify the modes to be included as those modes with frequencies in a narrow range ϑ around the transition frequency ω_0, which we shall call the *coupling bandwidth*. Within this range, the coupling between the electromagnetic field and the atomic system varies smoothly, as schematically represented in Fig. 9.1. Thus, the Hamiltonian has the form

$$H_{\text{tot}} = H_{\text{sys}} + H_{\text{EM}} + H_{\text{int}}, \tag{9.1.1}$$

with the simplified electromagnetic field Hamiltonian

$$H_{\text{EM}} = \sum_{|\omega_k - \omega_0| < \vartheta} \hbar\omega_k\, b_k^\dagger b_k. \tag{9.1.2}$$

Here, b_k is the annihilation operator

$$\left[b_k, b_{k'}^\dagger\right] = \delta_{k,k'}, \tag{9.1.3}$$

and the index k is interpreted as in *Bk. I: Sect.12.1.4*, that is, as an abbreviation for $\mathbf{k}, \lambda$, the wavevector and polarization of an individual mode. Since we want to develop formalism in which system is not just a single two-level atom, but something quite general, the system Hamiltonian H_{sys} is left unspecified.

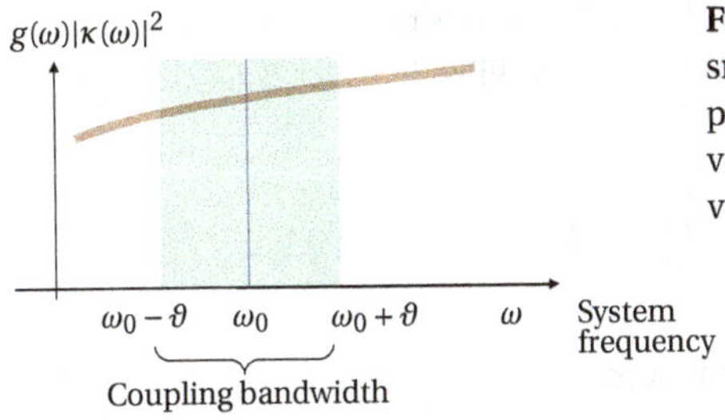

Fig. 9.1. Schematic representation of the smoothly varying behaviour of the coupling coefficient $g(\omega)|\kappa(\omega)|^2$ over the interval $(\omega_0-\vartheta, \omega_0+\vartheta)$, which is necessary for the validity of (9.2.14).

9.1.2 The Coupling Operators

The interaction Hamiltonian is treated in the rotating wave approximation, and involves an operator c which couples to the electromagnetic field. It then takes the form

$$H_{\rm int}^{(\vartheta)}(t) = \mathrm{i}\hbar \sum_{|\omega_k-\omega_0|<\vartheta} \left(\kappa_k^* b_k^\dagger c - \kappa_k b_k c^\dagger\right). \tag{9.1.4}$$

In order to express the fact that this represents a transition at a frequency ω_0, the coupling operator c must satisfy the commutation relation

$$[H_{\rm sys}, c] = -\hbar\omega_0 c. \tag{9.1.5}$$

This is an abstraction of the commutation relation between the atomic Hamiltonian *Bk. I: (12.1.22)* and the operator σ^- which we used when developing our description of atomic decay in *Bk. I: Sect.12.1*.

9.1.3 Examples of Hamiltonians and Coupling Operators

This operator c is intentionally written in a notation which is independent of the various physical realizable possibilities, which are quite varied. Some common situations are:

i) *Two-Level System*: In this case there is a ground state $|g\rangle$ and an excited state $|e\rangle$, and the mapping is

$$c \to |g\rangle\langle e| = \sigma^-, \qquad c^\dagger \to |e\rangle\langle g| = \sigma^+, \tag{9.1.6}$$

$$H_{\rm sys} \to \hbar\big(\omega_e|e\rangle\langle e| + \omega_g|g\rangle\langle g|\big), \qquad \omega_e - \omega_g = \omega_0. \tag{9.1.7}$$

ii) *Multi-Level Atom*: If there is only one resonant transition, this is the the same as for the two-level atom; however if there are several transitions $g_i \leftrightarrow e_i$, coupling to the same field mode, then we would write

$$c \to \sum_i r_i |g_i\rangle\langle e_i|, \qquad c^\dagger \to \sum_i r_i^* |e_i\rangle\langle g_i|, \tag{9.1.8}$$

$$H_{\rm sys} \to \hbar\sum_i \big(\omega_{e_i}|e_i\rangle\langle e_i| + \omega_{g_i}|g_i\rangle\langle g_i|\big), \qquad \omega_{e_i} - \omega_{g_i} = \omega_0 \text{ for all } i. \tag{9.1.9}$$

In these expressions the quantities r_i represent the relative strengths of the couplings.

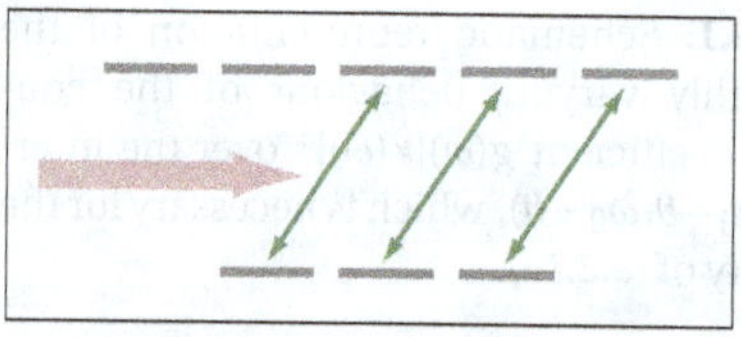

Fig. 9.2. Transitions between Zeeman sublevels, all of which couple to the same mode of the optical field.

An example of this kind of situation would be the transitions between Zeeman sublevels of two atomic level, as shown in Fig. 9.2.

iii) *Harmonic Oscillator*: When the creation and destruction operators are $a^\dagger$, a, the mapping is

$$c \to a, \qquad c^\dagger \to a^\dagger. \tag{9.1.10}$$

In the case of multiple harmonic oscillators (for example different modes in an optical cavity coupled to each other by a non-linear optical medium) with operators $a_i^\dagger$, a_i, corresponding to frequencies ω_i, transfer of quanta from mode i to mode j will occur if $\omega_0 = \omega_i - \omega_j$, and in this case the mapping would be

$$c \to a_j^\dagger a_i, \qquad c^\dagger \to a_i^\dagger a_j. \tag{9.1.11}$$

9.2 Physics of the Quantum Stochastic Schrödinger Equation

In this section we will show how the Schrödinger equation for a system described by the system-environment model we formulated in the previous section can be transformed to yield the quantum stochastic Schrödinger equation, provided appropriate physically reasonable conditions are satisfied. There are two conditions which must be satisfied:

i) The coupling to the electromagnetic field varies smoothly with frequency over the coupling bandwidth $(\omega_0 - \vartheta, \omega_0 + \vartheta)$.

ii) The electromagnetic coupling is sufficiently weak to ensure that a perturbation theory is valid for times very much less than $1/\vartheta$.

When both of these conditions are satisfied, the approximation that results is known as the *Born–Markov approximation.* For typical quantum-optical situations it is a very accurate approximation, whose use enables the description of quantum optical systems in a simple and powerful form, that of the *quantum stochastic Schrödinger equation.*

9.2.1 The Schrödinger Equation in the Interaction Picture

We write the state vector of the composite system of atom plus electromagnetic field as $|\Psi, t\rangle$, and this is a vector in the direct product Hilbert space $\mathcal{H}_{\text{sys}} \otimes \mathcal{H}_{\text{EM}}$.

This state vector obeys the Schrödinger equation

$$\frac{d}{dt}|\Psi,t\rangle = -\frac{\mathrm{i}}{\hbar}H_{\mathrm{tot}}|\Psi,t\rangle. \tag{9.2.1}$$

We will also use the time evolution operator $U(t)$, such that $|\Psi,t\rangle = U(t)|\Psi,0\rangle$, which obeys the similar equation

$$\frac{d}{dt}U(t) = -\frac{\mathrm{i}}{\hbar}H_{\mathrm{tot}}U(t), \tag{9.2.2}$$

with the initial condition $U(0) = 1$.

a) The Interaction Picture: In the interaction picture the operators obey the free dynamics, while the state vector changes according to the interaction Hamiltonian. This picture is defined through the transformation

$$U_I(t) = \exp\left(-\frac{\mathrm{i}}{\hbar}(H_{\mathrm{EM}}+H_{\mathrm{sys}})t\right), \tag{9.2.3}$$

$$O_I(t) = U_I^\dagger(t)O_S U_I(t), \tag{9.2.4}$$

$$|\Psi_I,t\rangle = U_I^\dagger(t)|\Psi_S,t\rangle. \tag{9.2.5}$$

b) Application of the Interaction Picture: The interaction Hamiltonian in the interaction picture requires knowledge of the evolution of the operators c under the Hamiltonian H_{sys}, and this follows from the transition frequency requirement (9.1.5):

$$c_I(t) = c\mathrm{e}^{-\mathrm{i}\omega_0 t}, \qquad c_I^\dagger(t) = c^\dagger\mathrm{e}^{\mathrm{i}\omega_0 t}. \tag{9.2.6}$$

The interaction picture field operators follow from the form of H_{EM}

$$b_{I,k}(t) = b_k\mathrm{e}^{-\mathrm{i}\omega_k t}, \qquad b_{I,k}^\dagger(t) = b_k^\dagger\mathrm{e}^{\mathrm{i}\omega_k t}. \tag{9.2.7}$$

Thus, both sets of operators can be written in terms of the time-independent Schrödinger picture operators. Using these, the interaction Hamiltonian in the interaction picture becomes

$$H_{\mathrm{int},I}^{(\vartheta)}(t) = \mathrm{i}\hbar\sum_{|\omega_k-\omega_0|<\vartheta}\left(\kappa_k^* b_k^\dagger\mathrm{e}^{\mathrm{i}(\omega_k-\omega_0)t}c - \kappa_k b_k\mathrm{e}^{-\mathrm{i}(\omega_k-\omega_0)t}c^\dagger\right). \tag{9.2.8}$$

9.2.2 The Noise Operator

We can define a time-dependent noise operator in terms of the field operators in the form

$$f(t) \equiv \sum_{|\omega_k-\omega_0|<\vartheta}\kappa_k b_k\mathrm{e}^{-\mathrm{i}(\omega_k-\omega_0)t}, \tag{9.2.9}$$

and write the interaction Hamiltonian as

$$H_{\mathrm{int},I}^{(\vartheta)}(t) = \mathrm{i}\hbar\left(f^\dagger(t)c - c^\dagger f(t)\right). \tag{9.2.10}$$

a) Commutation Relations: The noise operator has the commutation relation

$$[f(t), f^\dagger(t')] = \gamma(t-t'), \tag{9.2.11}$$

with

$$\gamma(\tau) = \sum_{|\omega_k-\omega_0|<\vartheta} |\kappa_k|^2 \mathrm{e}^{-\mathrm{i}(\omega_k-\omega_0)\tau}, \tag{9.2.12}$$

$$\to \int_{\omega_0-\vartheta}^{\omega_0+\vartheta} d\omega\, g(\omega)|\kappa(\omega)|^2 \mathrm{e}^{-\mathrm{i}(\omega-\omega_0)\tau}. \tag{9.2.13}$$

Here we have made the transition to a description in terms of a continuum of modes and a density of states, as in *Bk. I: Sect.12.2.* When dealing with the electromagnetic field, the notation k represents both the propagation vector and the polarization, so $k \equiv \{\boldsymbol{k}, \lambda\}$. There can therefore be a number of different k all corresponding to the same frequency ω_k, and therefore the notation $|\kappa(\omega)|^2$ is to be interpreted as the *average value* of $|\kappa_k|^2$ for all k for which $\omega_k = \omega$.

b) Properties of $\gamma(t-t')$: If $g(\omega)|\kappa(\omega)|^2$ is a smooth function of ω over the relevant bandwidth, as in Fig. 9.1, then

$$\gamma(\tau) \to 0, \qquad \text{when} \quad |\tau| \gg 1/\vartheta. \tag{9.2.14}$$

This means that $f(t)$ and $f^\dagger(t+\tau)$ commute for τ satisfying this condition.

9.2.3 The Born–Markov Approximation

We can transform the interaction picture Schrödinger equation, using a procedure which depends on the *Born–Markov* approximation, as follows.

a) Solving the Schrödinger Equation in Small Time Steps: We want to solve the interaction picture Schrödinger equation

$$\frac{d}{dt}|\Psi, t\rangle = -\frac{\mathrm{i}}{\hbar} H_{\mathrm{int},I}^{(\vartheta)}(t)|\Psi, t\rangle, \tag{9.2.15}$$

$$= \left(f^\dagger(t)c - c^\dagger f(t)\right)|\Psi, t\rangle. \tag{9.2.16}$$

By iteration we can write

$$|\Psi, t\rangle = \left(1 + \int_0^t dt_1 \left(f^\dagger(t_1)c - c^\dagger f(t_1)\right)\right)|\Psi, 0\rangle$$
$$+ \int_0^t dt_1 \left(f^\dagger(t_1)c - c^\dagger f(t_1)\right) \int_0^{t_1} dt_2 \left(f^\dagger(t_2)c - c^\dagger f(t_2)\right)|\Psi, t_2\rangle. \tag{9.2.17}$$

This is an exact equation, to which we will apply approximation procedures to eventually derive the quantum stochastic Schrödinger equation.

b) Initial Conditions: In this section we will only consider the case that the electromagnetic field is in the vacuum state, and therefore will take an initial condition

$$|\Psi, 0\rangle = |\psi\rangle_{\mathrm{sys}} \otimes |0\rangle_{\mathrm{EM}}. \tag{9.2.18}$$

We shall treat the case of an initial coherent electromagnetic field in Sect. 9.5, and the case of an initial thermal electromagnetic field in Sect. 9.6. However, in all cases, the initial state will be taken as a direct product of "system" and "field" states.

c) Born Approximation: The Born approximation requires the interaction to be sufficiently weak, so that we can choose a time interval Δt such that for $0 < t_2 < \Delta t$ we can replace $|\Psi, t_2\rangle \to |\Psi, 0\rangle$ in the second line. In this case, because the initial state is the vacuum, $f(t_1)$ and $f(t_2)$ acting on $|\Psi,0\rangle$ give zero, so that the term on the second line reduces to

$$\int_0^{\Delta t} dt_1 \int_0^{t_1} dt_2 \left(-c^\dagger c f(t_1) f^\dagger(t_2) + cc f^\dagger(t_1) f^\dagger(t_2)\right) |\Psi,0\rangle. \tag{9.2.19}$$

We can use the commutator (9.2.11) to write

$$f(t_1) f^\dagger(t_2) |\Psi,0\rangle = \left(f^\dagger(t_2) f(t_1) + \gamma(t_1 - t_2)\right) |\Psi,0\rangle = \gamma(t_1 - t_2) |\Psi,0\rangle, \tag{9.2.20}$$

and this part becomes a zero photon state, while the other term, which involves $ccf^\dagger f^\dagger$, is a two-photon state. The expression (9.2.17) contains terms with zero, one and two photons, and for $t \to \Delta t$ its non-zero terms become

$$\begin{aligned} |\Psi, \Delta t\rangle = & \left(1 - c^\dagger c \int_0^{\Delta t} dt_1 \int_0^{t_1} dt_2\, \gamma(t_1 - t_2)\right) |\Psi,0\rangle && \text{(no photons)} \\ & + c \int_0^{\Delta t} dt_1\, f^\dagger(t_1) |\Psi,0\rangle && \text{(one photon)} \\ & + cc \int_0^{\Delta t} dt_1 \int_0^{t_1} dt_2\, f^\dagger(t_1) f^\dagger(t_2) |\Psi,0\rangle. && \text{(two photons)} \end{aligned} \tag{9.2.21}$$

d) Markov Approximation: From (9.2.14) we know the time scale Δt must be chosen so that $\Delta t \gg 1/\vartheta$. Taking this into account, one can calculate that

$$\begin{aligned} & \int_0^{\Delta t} dt_1 \int_0^{t_1} dt_2\, \gamma(t_1 - t_2) \\ & = \int_{\omega_0 - \vartheta}^{\omega_0 + \vartheta} d\omega\, g(\omega) |\kappa(\omega)|^2 \\ & \quad \times \left\{ \frac{2 \sin^2\left((\omega - \omega_0)\Delta t/2\right)}{(\omega - \omega_0)^2} - \mathrm{i} \left(\frac{\Delta t}{\omega - \omega_0} - \frac{\sin\left((\omega - \omega_0)\Delta t\right)}{(\omega - \omega_0)^2} \right) \right\}, \end{aligned} \tag{9.2.22}$$

$$\longrightarrow \left(\tfrac{1}{2}\Gamma + \mathrm{i}\delta\omega\right)\Delta t, \quad \text{for } \Delta t \gg 1/\vartheta, \tag{9.2.23}$$

where Γ, the damping constant, and $\delta\omega$, the line-shift, are given by

$$\Gamma = 2\pi g(\omega_0) |\kappa(\omega_0)|^2, \tag{9.2.24}$$

$$\delta\omega = -\mathrm{P} \int_{\omega_0 - \vartheta}^{\omega_0 + \vartheta} d\omega \frac{g(\omega) |\kappa(\omega)|^2}{\omega - \omega_0}. \tag{9.2.25}$$

These expressions are identical with the damping constant and line-shift as given by *Bk. I: (12.2.23)* and *Bk. I: (12.2.24)*, where they were derived using the method of Weisskopf and Wigner.

This result can be regarded as a Markov approximation, since its validity is equivalent to setting

$$\gamma(t_1 - t_2) \to 2\left(\tfrac{1}{2}\Gamma + \mathrm{i}\delta\omega\right)\delta(t_1 - t_2), \tag{9.2.26}$$

provided we are only concerned with a time scale slower than $1/\vartheta$. This is the same kind of approximation we made when introducing the Langevin equation in *Bk. I: Sect.3.1.*

9.2.4 Line-Shifts and Renormalization

The line-shift $\delta\omega$ depends on the cutoff ϑ, and this means that a renormalization procedure must be used. The value of ϑ is chosen sufficiently small to make sure that the coefficient $g(\omega)|\kappa(\omega)|^2$ does not vary too much over the range $(\omega_0 - \vartheta, \omega_0 + \vartheta)$. However, it must also be sufficiently large to allow us to choose the time step size Δt to satisfy the requirement $\Delta t \gg 1/\vartheta$ and at the same time permit Δt to be small on the evolution time scale of interest.

Otherwise ϑ is arbitrary. The renormalization procedure is logically as follows. We have chosen a model theory, valid within the bandwidth $(\omega_0 - \vartheta, \omega_0 + \vartheta)$, and the line-shifts, even though very small, depend on ϑ. This means that we must adjust the transition frequency ω_0 which parametrizes $H_{\rm sys}$ so that the physically measurable frequency matches that given by including the line-shift. The net effect is that wherever $\delta\omega$ arises it is cancelled by this preliminary adjustment—in practice, this means we can omit the frequency shifts under most circumstances.

However, should we wish to compare two physical situations which differ only by having different values of $g(\omega)|\kappa(\omega)|^2$, the line-shifts predicted will differ, and we will not be able to adjust for both shifts simultaneously. This means that we will see a measurable frequency difference arising from the difference in line-shifts between the two situations. Two examples are:

i) We could adjust the density of states $g(\omega)$ by the use of an optical cavity.

ii) When two atoms are close together, each feels the electromagnetic field radiated from the other. This means the the line-shift for each atom depends on the distance to the other atom, giving rise to a detectable force between the atoms. This is treated in Chap. 14, where it is shown that, although this force arises from the line-shift, it does not depend on the cutoff ϑ.

9.3 Formulation of Quantum Stochastic Calculus

The central results of the previous section are the single time step evolution algorithm (9.2.21), and the Markovian ansatz (9.2.26). In this section we will use these results to generate a quantum stochastic calculus, analogous to the Ito calculus of *Bk. I: Ch.4*, but involving non-commuting operator noise increments. At the same time we will derive the corresponding quantum stochastic differential equation for the state vector $|\Psi, t\rangle$. This kind of equation contains all of the physics—both reversible and irreversible—of the quantum optical system-environment model. And remarkably, the only physical parameters needed to describe the effects of the coupling between the system and the environment are the decay constant Γ, and the line-shift $\delta\omega$.

9.3.1 Coarse Graining in Time and the Quantum Ito Increments

We can now use this formalism to study the distribution of the photons produced during the time evolution, from which we will be able to show that only no-photon and one-photon terms in (9.2.21) need be retained. Only the one photon term involves a noise operator, and its form leads to the definition of the *quantum Ito increment* by

$$\Delta B(t) \equiv \frac{1}{\sqrt{\Gamma}} \int_t^{t+\Delta t} f(t')\,dt'. \tag{9.3.1}$$

a) Probabilities of 0, 1 and 2 Photons: The probabilities of the terms in (9.2.21) are computed as follows:

i) The probability of zero photons is the norm of the first line in (9.2.21), and using the results (9.2.22–9.2.24), we get

$$P_0(\Delta t) = \langle\Psi,0|\Big(1 - cc^\dagger\big(\tfrac{1}{2}\Gamma - i\delta\omega\big)\Delta t\Big)\Big(1 - cc^\dagger\big(\tfrac{1}{2}\Gamma + i\delta\omega\big)\Delta t\Big)|\Psi,0\rangle, \tag{9.3.2}$$

$$\to 1 - \langle cc^\dagger\rangle_0 \Gamma\,\Delta t, \qquad \text{to order } \Delta t. \tag{9.3.3}$$

Here the notation $\langle\,\rangle_0$ means the average over the initial system state vector.

ii) Similarly, the probability of one photon is the norm of the second line in (9.2.21)

$$P_1(\Delta t) = \langle cc^\dagger\rangle_0 \Gamma\,\Delta t. \tag{9.3.4}$$

iii) Finally, the probability of two photons can be calculated as the norm of the third line in (9.2.21), and its value is

$$P_2(\Delta t) = \tfrac{1}{2}\Gamma^2\langle ccc^\dagger c^\dagger\rangle_0 \Delta t^2 \longrightarrow 0, \qquad \text{to order } \Delta t. \tag{9.3.5}$$

iv) Note that $P_0(\Delta t) + P_1(\Delta t) = 1$ to order Δt, consistently with $P_2(\Delta t)$ being of order Δt^2.

b) Coarse Graining in Time: Thus, if Δt is sufficiently large compared to $1/\vartheta$, but $\Gamma\Delta t$ and $\delta\omega\,\Delta t$ are nevertheless sufficiently small, then we can neglect the two photon terms and write

$$\Delta|\Psi,0\rangle \equiv |\Psi,\Delta t\rangle - |\Psi,0\rangle, \tag{9.3.6}$$

$$= \left(-\left(\tfrac{1}{2}\Gamma + i\delta\omega\right)c^\dagger c\,\Delta t + \sqrt{\Gamma}\,c\,\Delta B^\dagger(0)\right)|\Psi,0\rangle, \tag{9.3.7}$$

This difference equation gives the time evolution for the first time step. To proceed further we need to analyse the nature of the Hilbert space in which the noise operators act.

9.3.2 The Quantum Noise Hilbert Space

Provided $\Delta t \gg 1/\vartheta$, we can use the properties of $\gamma(\tau)$ to show that the quantum Ito increments satisfy

$$[\Delta B(t), \Delta B^\dagger(t)] = \Delta t, \tag{9.3.8}$$

$$[\Delta B(t), \Delta B^\dagger(s)] = 0, \tag{9.3.9}$$

for non-overlapping intervals $(t, t+\Delta t)$, $(s, s+\Delta s)$—see Ex. 9.1.

a) Number Operators: We will define

$$T \equiv t_n - t_0, \tag{9.3.10}$$

and divide the interval (t_0, t_n) into n equal non-overlapping subintervals of length

$$\Delta t \equiv \frac{T}{n}. \tag{9.3.11}$$

In each interval the operators

$$A_i \equiv \Delta B(t_i)/\sqrt{\Delta t}\,, \tag{9.3.12}$$

provide a set of independent destruction operators such that $[A_i, A_j^\dagger] = \delta_{i,j}$. Corresponding to each of these is a number operator

$$N(t_i) \equiv A_i^\dagger A_i = \frac{\Delta B^\dagger(t_i)\Delta B(t_i)}{\Delta t}. \tag{9.3.13}$$

The number operators all commute with each other, so for $i \neq j$ the measurement of $N(t_i)$ does not affect the measurement of $N(t_j)$.

In the limit of sufficiently small Δt, the eigenvalues which can occur with significant probability as a result of the time evolution from the initial vacuum state are 0 and 1; that is, in any interval of length Δt there is either a photon present or not. Measuring each of the set of $N(t_i)$ in the full interval (t_0, t_n) amounts to measuring the photon statistics of the electromagnetic field.

b) Resolution into Hilbert Spaces at Each Time Interval: The description expressed in terms of the operators $A_i^\dagger$, A_i forms a Hilbert space $\mathcal{H}_i$ on the i-th time interval, and a Hilbert space for the whole time interval composed of the direct product of each of these individual Hilbert spaces, each representing the number of photons created during that subinterval. This is illustrated in Fig. 9.3. This is

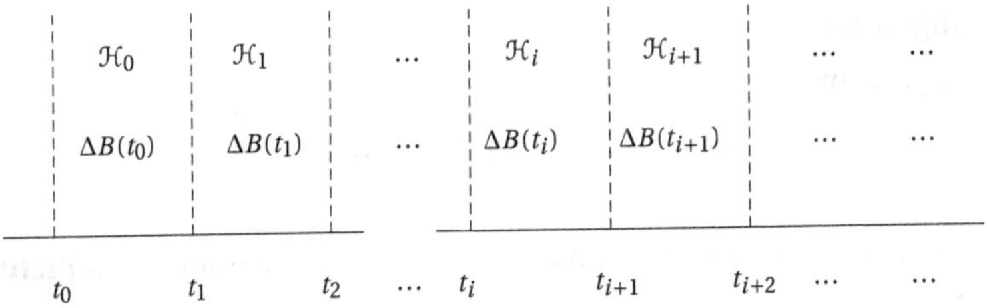

Fig. 9.3. The resolution of the Hilbert space into the direct product of Hilbert spaces at each time interval, each with an appropriate set of destruction (and creation) operators. During the evolution of the quantum stochastic Schrödinger equation, photons are successively created in the time intervals Δt_i, so that the quantum state is always the vacuum for later times.

in contrast to the description in terms of creation and destruction operators b_k, each of which corresponds to to a frequency ω_k, giving a description in terms of a direct product of Hilbert spaces $\mathcal{H}(\omega_k)$.

As a result, this gives a very important simplification to the description of the evolution process, since evolution up to a time t does not affect the Hilbert spaces for times later than t, meaning that all of the Hilbert spaces $\mathcal{H}_i$ for $t_i > t$ remain in the vacuum state.

Exercise 9.1 Proof of the Ito Commutation Relations: Show that

$$[\Delta B(t), \Delta B^\dagger(t)] = \frac{1}{\Gamma}\left\{\int_t^{t+\Delta t} dt' \int_t^{t'} dt''\,\gamma(t'-t'') + \int_t^{t+\Delta t} dt' \int_t^{t'} dt''\,\gamma(t''-t')\right\}, \tag{9.3.14}$$

and hence, using (9.2.22–9.2.24), prove the Ito commutation relation (9.3.8).

The Ito commutation relation (9.3.9) is obvious when $\Delta t \gg 1/\vartheta$, as a result of the property (9.2.14) of $\gamma(\tau)$.

9.3.3 Solution of the Schrödinger Equation for a Finite Time Interval

The derivation of the coarse grained first time step equation (9.3.7) required that $|\Psi,0\rangle$ be the vacuum for the operators $\Delta B(0)$, $\Delta B^\dagger(0)$. Looking at $|\Psi,\Delta t\rangle$, we see that the Ito increments on the time subinterval $(\Delta t, 2\Delta t)$ commute with those at the previous time step, so for these increments, $|\Psi,\Delta t\rangle$ is a vacuum state.

Therefore we can do another iteration

$$\Delta|\Psi,\Delta t\rangle \equiv |\Psi,2\Delta t\rangle - |\Psi,\Delta t\rangle, \tag{9.3.15}$$

$$= \left(-\left(\tfrac{1}{2}\Gamma + i\delta\omega\right)c^\dagger c\,\Delta t + \sqrt{\Gamma}\,c\,\Delta B^\dagger(\Delta t)\right)|\Psi,\Delta t\rangle, \tag{9.3.16}$$

and we can repeat this indefinitely, since for the Ito increments for the $(n+1)$th time subinterval, the state $|\Psi,n\Delta t\rangle$ is a vacuum state. The result is that we can write a *difference equation*

$$\Delta|\Psi, n\Delta t\rangle \equiv |\Psi, (n+1)\Delta t\rangle - |\Psi, n\Delta t\rangle, \tag{9.3.17}$$

$$= \left(-\left(\tfrac{1}{2}\Gamma + \mathrm{i}\delta\omega\right)c^\dagger c\,\Delta t + \sqrt{\Gamma}\,c\,\Delta B^\dagger(n\Delta t)\right)|\Psi, n\Delta t\rangle. \tag{9.3.18}$$

9.3.4 Quantum Stochastic Schrödinger Equation in the Interaction Picture

We cannot take the limit $\Delta t \to 0$, but we can take a limit in which the strength of the interaction, characterized by Γ and $\delta\omega$, becomes very weak, so that the change in each time step becomes very small. In that limit, we can rewrite the difference equation as a quantum stochastic differential equation to which we give the name *the quantum stochastic Schrödinger equation*:

$$d|\Psi, t\rangle = \left(-\left(\tfrac{1}{2}\Gamma + \mathrm{i}\delta\omega\right)c^\dagger c\,dt + \sqrt{\Gamma}\,c\,dB^\dagger(t)\right)|\Psi, t\rangle. \tag{9.3.19}$$

Note that in this form, the quantum stochastic Schrödinger equation is an interaction picture equation.

In Chap. 18 we will discuss, in the context of continuous measurement theory, the nature and interpretation of solutions of the quantum stochastic Schrödinger equation.

a) Relationship of $dB(t)$ to $\Delta B(t)$: This is similar to that which occurs in classical stochastic differential equations, as described in *Bk. I: Ch.4.* This means that an equation of this kind is to be regarded as a shorthand for an integral equation, and

$$\sum F(t_n)\,\Delta B(t_n) \to \int F(t)\,dB(t). \tag{9.3.20}$$

The increments $\Delta B(t_n)$ represent the field in the future of t_n, and are thus both statistically and quantum-mechanically independent of $F(t_n)$. Thus the integral is of the *Ito* kind.

b) Zero Temperature Quantum Ito Algebra: The quantities $dB(t)$, $dB^\dagger(t)$—the infinitesimal quantum Ito increments—have the same property as the classical stochastic Ito increments $dW(t)$ in that they are quantities of order $\sqrt{dt}$. The algebra of these increments in the case we are considering, that we apply any finite number of increments to the vacuum state, is the *zero temperature quantum Ito algebra*

$$\begin{aligned} &dB(t)\,dB^\dagger(t) = dt, \\ &dB^\dagger(t)\,dB(t) = dB(t)^2 = dB^\dagger(t)^2 = 0, \\ &dB(t)\,dt = dt\,dB(t) = dB^\dagger(t)\,dt = dt\,dB^\dagger(t) = dt^2 = 0. \end{aligned} \tag{9.3.21}$$

These can be modified to include the case of non-zero temperature; see Sect. 9.6.

Exercise 9.2 Unitary Evolution: Demonstrate that the quantum stochastic Schrödinger equation corresponds to unitary evolution by showing $d(\langle\Psi, t|\Psi, t\rangle) = 0$, as a result of the quantum stochastic Schrödinger equation (9.3.19) and the quantum Ito algebra.

9.4 Transformation back to the Schrödinger Picture

Transforming the equations of motion into the quantum stochastic Schrödinger equation gives the correction $-\left(\frac{1}{2}\Gamma - i\delta\omega\right)c^\dagger c$ to the system motion. The full system motion is then described partially by the time dependence of the interaction picture operators as described by the Hamiltonian H_{sys}, and partially as the time dependence of the interaction picture quantum states as given by this correction.

It is more convenient to have all of the system motion described by the time evolution of the quantum states, while preserving the quantum Ito increments $dB(t)$ and $dB^\dagger(t)$, effectively as quantized driving fields. This is then a Schrödinger picture for an atomic system driven by these fields, which are regarded as known time-dependent operators.

The appropriate transformation is implemented by means of the unitary operator

$$U_{sys}(t) = \exp\left(\frac{i}{\hbar} H_{sys}\, t\right). \tag{9.4.1}$$

We then make the transformations

$$O(t) \equiv U^\dagger_{sys}(t) O_I(t) U_{sys}(t), \tag{9.4.2}$$

$$|\Psi, t\rangle \equiv U^\dagger_{sys}(t)|\Psi_I, t\rangle. \tag{9.4.3}$$

These are now operators in this Schrödinger picture.

9.4.1 Ito Stochastic Increments in the Schrödinger Picture

If we define the Ito increments in the Schrödinger picture by

$$dB_{Sch}(t) = e^{i\omega_0 t} dB(t), \quad dB^\dagger_{Sch}(t) = e^{-i\omega_0 t} dB^\dagger(t), \tag{9.4.4}$$

then these Ito increments obey exactly the same algebra (9.3.21) as the original increments. Therefore it will be unnecessary to use the explicit notation $dB_{Sch}(t)$, instead of which we shall always write $dB(t)$, etc.

9.4.2 Quantum Stochastic Schrödinger Equation in the Schrödinger Picture

The quantum stochastic Schrödinger equation takes the form

$$d|\Psi,t\rangle = \left(-\frac{i}{\hbar}H_{\text{eff}}\,dt + \sqrt{\Gamma}\,c\,dB^\dagger(t)\right)|\Psi,t\rangle. \tag{9.4.5}$$

a) Definitions of Terms: Here H_{eff} is called the *effective Hamiltonian*, which is non-Hermitian, and is defined by the three equations:

$$H_{\text{eff}} \equiv H_{\text{sys}} + \hbar\delta\omega\, c^\dagger c - \tfrac{1}{2}i\hbar\Gamma\, c^\dagger c, \tag{9.4.6}$$

$$\Gamma = 2\pi g(\omega_0)|\kappa(\omega_0)|^2, \tag{9.4.7}$$

$$\delta\omega = -\mathrm{P}\int_{\omega_0-\vartheta}^{\omega_0+\vartheta} d\omega\,\frac{g(\omega)|\kappa(\omega)|^2}{\omega-\omega_0}. \tag{9.4.8}$$

It is in this form that we shall mostly use the quantum stochastic Schrödinger equation.

b) Compact Notation for the Quantum Stochastic Schrödinger Equation: In further developments we shall often use the compact notation

$$d|\Psi,t\rangle = \mathsf{K}|\Psi,t\rangle\,dt + \mathsf{J}|\Psi,t\rangle\,dB^\dagger(t), \tag{9.4.9}$$

$$\mathsf{K} \equiv -\frac{i}{\hbar}H_{\text{eff}} = -\frac{i}{\hbar}\left(H_{\text{sys}} + \hbar\delta\omega\, c^\dagger c\right) - \tfrac{1}{2}\Gamma\, c^\dagger c, \tag{9.4.10}$$

$$\mathsf{J} \equiv \sqrt{\Gamma}\,c. \tag{9.4.11}$$

9.4.3 The Central Role of the Quantum Stochastic Schrödinger Equation

The quantum stochastic Schrödinger equation forms the basis from which three other important formalisms can be derived, namely:

i) The quantum optical master equation, which we covered in *Bk. I: Ch.13.* It is also the subject of Chap. 10, where it is derived directly from the quantum stochastic Schrödinger equation.

ii) The quantum Langevin equation, and corresponding quantum stochastic differential equations, which form the most straightforward way of relating light emitted by a system to the light absorbed by the system and the dynamics of the system itself. This is the subject of Chap. 11.

iii) The theory of continuous measurement, by which the detections of photons are related directly to jumps in the quantum state of the system generating the photons. This is the subject of Chap. 18 and Chap. 19.

9.4.4 Solving the Quantum Stochastic Schrödinger Equation

Because the quantum stochastic Schrödinger equation is a linear equation in the wavefunction, it is often possible to solve it explicitly, using straightforward methods like those in the following exercises.

Exercise 9.3 Decay of a Single-Photon Cavity Mode: An optical cavity with frequency ω_{cav} can be modelled by a harmonic oscillator, for which we can write in the effective Hamiltonian terms of creation and destruction operators, so that $c \to a$ and $c^\dagger \to a^\dagger$. Consider the case in which a single photon is initially in the cavity, and the electromagnetic field corresponds to the vacuum.

i) Explicitly write the quantum stochastic Schrödinger equation for this system, omitting the line-shift terms, which are absorbed into the transition frequency according to the procedure of Sect. 9.2.4.

ii) Write the wavefunction in the form

$$|\Psi, t\rangle = |\mathrm{cav}, 1\rangle \otimes |\Phi_0, t\rangle + |\mathrm{cav}, 0\rangle \otimes |\Phi_1, t\rangle. \tag{9.4.12}$$

Here $|\mathrm{cav}, n\rangle$ is an n-photon cavity state, while $|\Phi_0, t\rangle$ and $|\Phi_1, t\rangle$ are respectively vacuum and one photon states of the radiation field, normalized so that

$$\langle\Phi_0, t|\Phi_0, t\rangle = \text{probability of 0 photons in the radiation field}, \tag{9.4.13}$$

$$\langle\Phi_1, t|\Phi_1, t\rangle = \text{probability of 1 photon in the radiation field}. \tag{9.4.14}$$

iii) Set the initial condition at $t = 0$ to be $|\Psi, 0\rangle = |\mathrm{cav}, 1\rangle \otimes |0\rangle$, where $|0\rangle$ is the *normalized* vacuum state of the radiation field.

iv) Show that the quantum stochastic Schrödinger equation is equivalent to the equations

$$d|\Phi_0, t\rangle = -\left(i\omega_{\mathrm{cav}} + \tfrac{1}{2}\Gamma\right)|\Phi_0, t\rangle, \tag{9.4.15}$$

$$d|\Phi_1, t\rangle = \sqrt{\Gamma}\, dB^\dagger(t)|\Phi_0, t\rangle. \tag{9.4.16}$$

v) Show that the solutions of these equations are

$$|\Phi_0, t\rangle = \mathrm{e}^{-i\omega_{\mathrm{cav}} t}\mathrm{e}^{-\Gamma t/2}|0\rangle, \tag{9.4.17}$$

$$|\Phi_1, t\rangle = \sqrt{\Gamma}\int_0^t \mathrm{e}^{-i\omega_{\mathrm{cav}} t'}\mathrm{e}^{-\Gamma t'/2} dB^\dagger(t')|0\rangle. \tag{9.4.18}$$

vi) Define the operator $f_{\mathrm{rad}}(t)$ by

$$f_{\mathrm{rad}}(t)\, dt \equiv \sqrt{\frac{\Gamma}{1-\mathrm{e}^{-\Gamma t}}}\int_0^t \mathrm{e}^{-i\omega_{\mathrm{cav}} t'}\mathrm{e}^{-\Gamma t'/2} dB^\dagger(t'). \tag{9.4.19}$$

Show that this operator is the destruction operator for a photon emitted into the spatial mode function $F_{\mathrm{rad}}(x, t)$ at time t, defined by

$$F_{\mathrm{rad}}(x, t) = \begin{cases} \sqrt{\frac{\Gamma}{1-\mathrm{e}^{-\Gamma t}}}\,\mathrm{e}^{-i\omega_{\mathrm{cav}}(t-x/c)}\mathrm{e}^{-\Gamma(t-x/c)/2}, & 0 < x < ct, \\ 0, & \text{otherwise}. \end{cases} \tag{9.4.20}$$

The function $F_{\mathrm{rad}}(x, t)$ is schematically illustrated in Fig. 9.4.

vii) Formulate and solve the problem for the case of an initial two-photon cavity state.

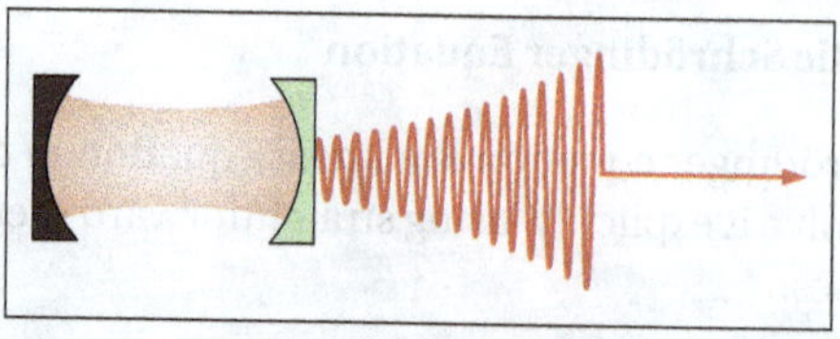

Fig. 9.4. Illustration of the mode function $F_{\rm rad}(x,t)$ defined in (9.4.20) produced during the loss of a single photon from an optical cavity.

Exercise 9.4 Solving the Quantum Stochastic Schrödinger Equation for Atomic Decay: Consider now the very similar case of a two-level atom with excited state $|e\rangle$ and ground state $|g\rangle$, so that $c \to |g\rangle\langle e|$ and $c^\dagger \to |e\rangle\langle g|$, and $H_{\rm sys} \to \hbar\omega_{eg}|e\rangle\langle e|$.

i) Explicitly write the quantum stochastic Schrödinger equation for this system, omitting the line-shift terms.

ii) Write the wavefunction in the form

$$|\Psi,t\rangle = |e\rangle \otimes |\Phi_0,t\rangle + |g\rangle \otimes |\Phi_1,t\rangle, \tag{9.4.21}$$

in which $|\Phi_0,t\rangle$ and $|\Phi_1,t\rangle$ are respectively vacuum and one photon states of the radiation field, normalized so that

$$\langle\Phi_0,t|\Phi_0,t\rangle = \text{probability of 0 photons in the radiation field}, \tag{9.4.22}$$

$$\langle\Phi_1,t|\Phi_1,t\rangle = \text{probability of 1 photon in the radiation field}. \tag{9.4.23}$$

iii) We set the initial condition at $t=0$ to be $|\Psi,0\rangle = |e\rangle \otimes |0\rangle$, where $|0\rangle$ is the *normalized* vacuum state of the radiation field.

iv) Show that the quantum stochastic Schrödinger equation is equivalent to the equations

$$d|\Phi_0,t\rangle = -\left(i\omega_{eg} + \tfrac{1}{2}\Gamma\right)|\Phi_0,t\rangle, \tag{9.4.24}$$

$$d|\Phi_1,t\rangle = \sqrt{\Gamma}\,dB^\dagger(t)|\Phi_0,t\rangle. \tag{9.4.25}$$

v) Show that the solutions of these equations are

$$|\Phi_0,t\rangle = \mathrm{e}^{-i\omega_{eg}t}\mathrm{e}^{-\Gamma t/2}|0\rangle, \tag{9.4.26}$$

$$|\Phi_1,t\rangle = \sqrt{\Gamma}\int_0^t \mathrm{e}^{-i\omega_{eg}t'}\mathrm{e}^{-\Gamma t'/2}\,dB^\dagger(t')|0\rangle. \tag{9.4.27}$$

vi) Compare this solution with the Wigner–Weisskopf solution *Bk. I: (12.2.28)* for the same problem, as given in *Bk. I: Sect.12.2.2*, and show they are the same.

vii) The harmonic oscillator solutions (9.4.17, 9.4.18) are very similar to the two-level atom solutions (9.4.26, 9.4.27). Explain why this is so, and under what conditions the solutions would differ from each other.

In Sect. 27.1.3 we will use similar methods for the three-level atom.

9.4.5 The Evolution Operator

The evolution of an arbitrary state can be defined in terms of the evolution operator $\mathcal{V}(t, t_0)$ which obeys the equations

$$\mathcal{V}(t_0, t_0) = 1, \tag{9.4.28}$$

$$d\mathcal{V}(t, t_0) = \left(-\frac{\mathrm{i}}{\hbar} H_{\text{eff}}\, dt + \sqrt{\Gamma}\, c\, dB^\dagger(t)\right)\mathcal{V}(t, t_0). \tag{9.4.29}$$

This is of course an operator in the rotating frame.

9.4.6 Coupling to Several Independent Light Fields

We can easily generalize to the situation in which the system has a number of transition operators c_i, each with its own transition frequency ω_i such that

$$[H_{\text{sys}}, c_i] = -\hbar\omega_i c_i. \tag{9.4.30}$$

Each operator couples to the part of the electromagnetic field with frequency in the range $(\omega - \vartheta_i, \omega + \vartheta_i)$ and these ranges should not overlap. This will mean that these fields are described by independent operators, which commute with each other.

This will then yield a set of Ito increments (in the rotating frame) which have the Ito algebra

$$\left.\begin{aligned} &dB_i(t)\, dB_j^\dagger(t) = \delta_{i,j}\, dt, \\ &dB_i^\dagger(t)\, dB_j(t) = dB_i(t)\, dB_j(t) = dB_i^\dagger(t)\, dB_j^\dagger(t) = 0, \\ &dB_i(t)\, dt = dt\, dB_i(t) = dB_i^\dagger(t)\, dt = dt\, dB_i^\dagger(t) = dt^2 = 0. \end{aligned}\right\} \tag{9.4.31}$$

The quantum stochastic Schrödinger equation then takes the form

$$d|\Psi, t\rangle = \left(-\frac{\mathrm{i}}{\hbar} H_{\text{eff}}\, dt + \sum_i \sqrt{\Gamma_i}\, c_i\, dB_i^\dagger(t)\right)|\Psi, t\rangle. \tag{9.4.32}$$

In this equation we have used the effective Hamiltonian

$$H_{\text{eff}} \equiv H_{\text{sys}} + \hbar \sum_i \delta\omega_i\, c_i^\dagger c_i - \tfrac{1}{2}\mathrm{i}\hbar \sum_i \Gamma_i\, c_i^\dagger c_i, \tag{9.4.33}$$

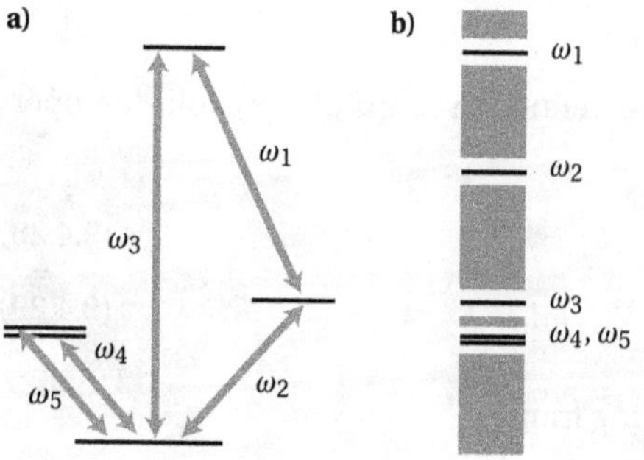

Fig. 9.5. Coupling the optical field to several transitions.
a) Energy levels with well separated transition frequencies, apart from ω_4 and ω_5;
b) The electromagnetic spectrum, with the bandwidths $(\omega_i - \vartheta, \omega_i + \vartheta)$ indicated by white bands, with the relevant transitions in the centres. Each band has its own noise, which in the case of ω_4 and ω_5 covers two closely separated transitions.

Exercise 9.5 Nearly Degenerate Energy Levels: In Fig. 9.5 the frequencies ω_4 and ω_5 both fall within the same energy band. How does this modify the formulation of the quantum stochastic Schrödinger equation (9.4.32)?

9.5 Inclusion of a Coherent Input Field

Let us suppose the modes of the electromagnetic field are initially coherently occupied. By this we mean that each mode of the electromagnetic field is in a coherent state, which can be written $|k, \beta_k\rangle$, so that

$$b_k|k,\beta_k\rangle = \beta_k|k,\beta_k\rangle, \tag{9.5.1}$$

where the symbol β_k represents the coherent amplitude in the mode k. We will represent the set of all mode amplitudes using the vector notation

$$\boldsymbol{\beta} \equiv \{\beta_1, \beta_2, \beta_3, \ldots, \beta_k, \ldots\}, \tag{9.5.2}$$

and write the state of the field as a whole as

$$|\boldsymbol{\beta}\rangle \equiv |1,\beta_1\rangle \otimes |2,\beta_2\rangle \otimes \cdots |k,\beta_k\rangle \otimes \cdots \tag{9.5.3}$$

From this definition we find that the action of the noise operator $f(t)$, defined by (9.2.9), is

$$f(t)|\boldsymbol{\beta}\rangle = \mathcal{F}(t)|\boldsymbol{\beta}\rangle, \tag{9.5.4}$$

where $\mathcal{F}(t)$ is a c-number function defined by

$$\mathcal{F}(t) = \sum_{|\omega_k - \omega_0| < \vartheta} \kappa_k \beta_k e^{-i(\omega_k - \omega_0)t}. \tag{9.5.5}$$

The function $\mathcal{F}(t)$ represents a time-dependent coherent field, which we will show is equivalent to a classical driving field.

This method of including a classical driving field was introduced by *Mollow* [9.3], and the transformation is often called the *Mollow transformation*.

9.5.1 Quantum Stochastic Schrödinger Equation

We now return the discretized description we use in Sect. 9.3.1–Sect. 9.3.3, and instead of the definition (9.3.1) of the quantum Ito increment, we first subtract the coherent component from the field operator; that is, we define the quantum Ito increment as

$$\Delta B(t) \equiv \frac{1}{\sqrt{\Gamma}} \int_t^{t+\Delta t} \left(f(t') - \mathcal{F}(t')\right) dt'. \tag{9.5.6}$$

Then the same algebra yields, instead of (9.4.5), a quantum stochastic Schrödinger equation in the form

$$d|\Psi, t\rangle = \left\{ -\frac{\mathrm{i}}{\hbar} H_{\text{eff}}\, dt + \sqrt{\Gamma}\left(\mathcal{F}^*(t) c - \mathcal{F}(t) c^\dagger\right) dt + \sqrt{\Gamma}\, c\, dB^\dagger(t) \right\} |\Psi, t\rangle. \tag{9.5.7}$$

This equation of motion is exactly that which corresponds to the addition of a classical driving field $\mathcal{F}(t)$ to the original system in addition to the vacuum state of the quantized electromagnetic field.

Exercise 9.6 Quantum Stochastic Differential Equation for the Driven Detuned Two-Level System: This corresponds to the situations treated in Chap. 5, specifically to the Rabi Hamiltonian (5.2.9). Show that the quantum stochastic Schrödinger equation arising from (9.5.7) involves the mapping (in the rotating frame)

$$c \to |g\rangle\langle e|, \qquad c^\dagger \to |e\rangle\langle g| \tag{9.5.8}$$

$$H_{\text{eff}} \to -\hbar\left(\Delta + \tfrac{1}{2}\mathrm{i}\Gamma\right)|e\rangle\langle e| + \tfrac{1}{2}\hbar\Omega_R\left(\mathrm{e}^{-\mathrm{i}\varphi}|e\rangle\langle g| + \mathrm{e}^{\mathrm{i}\varphi}|g\rangle\langle e|\right), \qquad \omega_0 = 0. \tag{9.5.9}$$

Exercise 9.7 The Mollow Solution:
The quantum stochastic Schrödinger equation for the driven detuned two-level system can be written using the compact notation of Sect. 9.4.2b

$$d|\Psi, t\rangle = \left(\mathbf{K}\, dt + \mathbf{J}\, dB^\dagger(t)\right)|\Psi, t\rangle, \tag{9.5.10}$$

in which

$$\mathbf{K} \equiv -\frac{\mathrm{i}}{\hbar} H_{\text{eff}}, \qquad \mathbf{J} \equiv \sqrt{\Gamma}\, |g\rangle\langle e|. \tag{9.5.11}$$

i) Take the initial condition as $|\Psi, 0\rangle = |r\rangle \equiv \alpha|g\rangle + \beta|e\rangle$, where α and β are some appropriately normalized coefficients.

ii) Show that the solution of this quantum stochastic Schrödinger equation can be written

$$|\Psi, t\rangle = \sum_0^\infty |\psi_n, t\rangle, \tag{9.5.12}$$

$$|\psi_0, t\rangle = \mathrm{e}^{\mathbf{K}t}|r\rangle, \tag{9.5.13}$$

$$|\psi_n, t\rangle = \int_0^t \mathrm{e}^{\mathbf{K}(t-t')}\mathbf{J}\, dB^\dagger(t)|\psi_{n-1}, t'\rangle. \tag{9.5.14}$$

This is equivalent to the solution of *Mollow* [9.3] for the problem of fluorescence from a driven two-level atom. The vectors $|\psi_n, t\rangle$ can be interpreted as those components of the wavefunction in which n photons have been emitted.

We will return to this kind of interpretation in Chap. 18.

9.6 Interaction with a Finite Temperature Light Field

The quantum stochastic Schrödinger equation is explicitly an equation of motion for a pure state. However, in the case that the initial field has a non-zero temperature, its quantum state would normally be described by a thermal density operator ρ_T, whose use would appear to prevent the formulation of a quantum stochastic Schrödinger equation for a time-dependent pure state $|\Psi, t\rangle$. Nevertheless, there is a solution to this difficulty, and indeed we can formulate an appropriate quantum stochastic Schrödinger equation for finite temperature situations.

9.6.1 Finite Temperature Ito Increments

The thermal statistics of the Ito increments are determined by:

i) The thermal statistics of the Bose operators b_k, $b_k^\dagger$.

ii) The definition (9.3.1) of the increment in terms of the noise operators, themselves defined in (9.2.9).

In a thermal state, we know from *Bk. I: Ch.10*, that the density operator is

$$\rho_T = \frac{1}{Z_T}\exp\left(-\frac{H_{\mathrm{EM}}}{k_{\mathrm{B}}T}\right), \tag{9.6.1}$$

for which we can say

$$\langle b_k^\dagger b_{k'}\rangle = \delta_{k,k'}\bar{N}(T,\omega_k). \tag{9.6.2}$$

From these it follows that

$$\langle \Delta B^\dagger(t)\,\Delta B(t)\rangle = \bar{N}(T,\omega_0)\,\Delta t, \qquad \langle \Delta B(t)\,\Delta B^\dagger(t)\rangle = \big(\bar{N}(T,\omega_0)+1\big)\,\Delta t. \tag{9.6.3}$$

The correlation functions involving two different times are zero, so that the noise increments for different times are independent of each other. The density operator at time t is therefore a quantum Gaussian with the second order correlations (9.6.3), which is

$$\left(1-\mathrm{e}^{-\lambda}\right)\exp\left(-\frac{\lambda\,\Delta B^\dagger(t)\,\Delta B(t)}{\Delta t}\right), \tag{9.6.4}$$

in which

$$\lambda = \mathrm{e}^{-\hbar\omega_0/k_{\mathrm{B}}T}, \qquad \bar{N}(T,\omega_0) = \frac{\lambda}{1-\lambda} = \frac{1}{\mathrm{e}^{\hbar\omega_0/k_{\mathrm{B}}T}-1}. \tag{9.6.5}$$

Exercise 9.8 Statistics of the Thermal Ito Increments: Using the approximations used for deriving the zero temperature quantum Ito algebra (namely, that the bandwidth ϑ is small, and that $\Delta t \gg 1/\vartheta$) and the methods of *Bk. I: Ch.10*, derive the results (9.6.3–9.6.5).

9.6.2 Ensemble Formulation for Finite Temperature Fields

The density operator for any system is defined in terms of a statistical ensemble of quantum states $|a\rangle$ which occur with probabilities $p(a)$, by the formula $\rho \equiv \sum_a p(a)|a\rangle\langle a|$. The measurable physics is determined by the density operator, so that two ensembles which give the same density operator are equivalent.

If we use the discretized description of the Hilbert space given in Sect. 9.3.1, in the finite temperature case there will be a thermal density operator $\rho_i(T)$ at temperature T for each of the Hilbert spaces $\mathcal{H}_i$, and the density operator in the full Hilbert space is

$$\rho(T) \equiv \rho_1(T)\otimes\rho_2(T)\otimes\cdots\rho_i(T)\otimes\rho_{i+1}(T)\otimes\rho_{i+2}(T)\ldots \tag{9.6.6}$$

a) Finite Temperature Equation of Motion: This will be an equation of motion for $|\Psi,t\rangle$, in which the initial state of the bath is a member of an ensemble which gives rise to the density operator $\rho(T)$. This means that

i) The wavefunction $|\Psi,t\rangle$ is interpreted as a member of an ensemble of wavefunctions.

ii) Each member of the ensemble is characterized by a state of the electromagnetic field, which itself is a member of the ensemble described by $\rho(T)$.

b) The Thermal Ensemble: To describe our choice of ensemble, let us use the number operators we introduced in Sect. 9.3.2a, and introduce their eigenstates defined by

$$A_i^\dagger A_i|r,i\rangle = r|r,i\rangle. \tag{9.6.7}$$

The density operator at time step i is given by

$$\rho_i(T) = \left(1-e^{-\lambda}\right)\sum_r e^{-\lambda r}|r,i\rangle\langle r,i|, \tag{9.6.8}$$

where λ is as defined in (9.6.5). This density operator is simply a rewriting of that given by (9.6.4).

The natural choice of ensemble is given by the quantum states

$$|r_1,1\rangle\otimes|r_2,2\rangle\otimes\cdots|r_i,i\rangle\otimes|r_{i+1},i+1\rangle\otimes|r_{i+2},i+2\rangle\otimes\cdots \tag{9.6.9}$$

with the ensemble probabilities

$$p_T(r_1,r_2,\ldots,r_{i+1},r_{i+2},\ldots) \equiv \prod_i\left\{\left(1-e^{-\lambda}\right)e^{-\lambda r_i}\right\}. \tag{9.6.10}$$

c) Interpretation of a Member of the Ensemble: The quantum state (9.6.9) is defined on a set of time intervals Δt_i, and in each of these there is a definite number r_i of photons. The ensemble probabilities (9.6.10) describe independent but identical thermal statistics at each time, so this state can be seen as a field which streams in with time, and in each time interval Δt_i the number of photons is r_i. Time-averaged quantities will be identical with averages of the same quantities over the ensemble.

d) Coherent State Thermal Ensemble: It is also possible to represent a thermal density operator by a coherent state ensemble, whose probabilities are given by the thermal P-function, as given in *Bk. I: (15.2.16)*. The individual quantum states in this ensemble are now

$$|\beta_1,1\rangle\otimes|\beta_2,2\rangle\otimes\cdots|\beta_i,i\rangle\otimes|\beta_{i+1},i+1\rangle\otimes|\beta_{i+2},i+2\rangle\otimes\cdots \tag{9.6.11}$$

with ensemble probability densities

$$p_T(\beta_1,\beta_2,\dots,\beta_{i+1},\beta_{i+2},\dots)\equiv\prod_i\left\{\left(\frac{1}{\pi\bar{N}(T,\omega_0)}\right)\exp\left(-\frac{|\beta_i|^2}{\bar{N}(T,\omega_0)}\right)\right\}, \tag{9.6.12}$$

and of course, in this ensemble,

$$\langle\beta_i^*\beta_j\rangle=\delta_{ij}\bar{N}(T,\omega_0),\qquad \langle\beta_i\beta_j\rangle=\langle\beta_i^*\beta_j^*\rangle=0. \tag{9.6.13}$$

9.6.3 Derivation of the Quantum Stochastic Differential Equation

The derivation of the appropriate quantum stochastic differential equation follows the same basic formulation as given in Sect. 9.2.1–Sect. 9.2.3, with a more general strategy, which does not depend on the field being in the vacuum state. We proceed from (9.2.17), making the Born approximation as in Sect. 9.2.3c.

From that point, however, we have a different strategy; for compactness, let us define the Hermitian operator

$$\mathcal{C}(t)\equiv-\mathrm{i}\Big(f^\dagger(t)c-c^\dagger f(t)\Big). \tag{9.6.14}$$

which leads to

$$|\Psi,\Delta t\rangle=\left(1+\mathrm{i}\int_0^{\Delta t}dt_1\,\mathcal{C}(t_1)-\int_0^{\Delta t}dt_1\int_0^{t_1}dt_2\,\mathcal{C}(t_1)\mathcal{C}(t_2)\right)|\Psi,0\rangle. \tag{9.6.15}$$

The essence of the derivation relies on the following points:

i) We want to evaluate $|\Psi,\Delta t\rangle$ to order Δt, which will mean in particular that the norm $\langle\Psi,\Delta t|\Psi,\Delta t\rangle=1$ is preserved to this order, that is, any deviation from 1 is of order Δt^2 or higher.

ii) The thermal mean of the the second order term is of order Δt. We therefore rewrite the algorithm as

$$\begin{aligned}|\Psi,\Delta t\rangle&=\left(1+\mathrm{i}\int_0^{\Delta t}dt_1\,\mathcal{C}(t_1)-\int_0^{\Delta t}dt_1\int_0^{t_1}dt_2\,\langle\mathcal{C}(t_1)\mathcal{C}(t_2)\rangle_T\right)|\Psi,0\rangle\\&\quad+\int_0^{\Delta t}dt_1\int_0^{t_1}dt_2\Big(\langle\mathcal{C}(t_1)\mathcal{C}(t_2)\rangle_T-\mathcal{C}(t_1)\mathcal{C}(t_2)\Big)|\Psi,0\rangle.\end{aligned} \tag{9.6.16}$$

iii) We now compute the thermal average of the norm, which we denote by the symbol

$$N \equiv \big\langle \langle \Psi, \Delta t | \Psi, \Delta t \rangle \big\rangle_T . \tag{9.6.17}$$

Since $f(t)$ is defined as a linear combination of the operators b_k, thermal averages of an odd number of $\mathcal{C}(t)$ operators are zero. Bearing this in mind, we find

a) The first line of algorithm is norm-preserving to order Δt.

b) Cross terms arising from the two lines are zero.

c) We will shown in Sect. 9.6.4c that the term arising from the norm of the second line is of order of magnitude Δt^2, so that this term does not affect the norm of the wavefunction to order Δt. We can interpret this term as representing fluctuations of order of magnitude Δt, which are negligible compared to those coming from the second term in the first line, which are of order of magnitude $\Delta t^{1/2}$.

We conclude that we can omit the second line of the algorithm (9.6.16), and that the first line of (9.6.16) is the algorithm we need.

9.6.4 Evaluation of Thermal Averages

In our discussion we have not explicitly stated how the orders of magnitude asserted are derived. To do this we need explicit formulae for the thermal correlation functions of the noise, and these are also needed to transform the algorithm into a quantum stochastic differential equation.

a) Thermal Correlation Functions: The non-zero thermal averages we will need are the *thermal correlation functions*

$$\langle f^\dagger(t) f(t') \rangle = \gamma_T^{(-)}(t-t'), \qquad \langle f(t) f^\dagger(t') \rangle = \gamma_T^{(+)}(t-t'). \tag{9.6.18}$$

The thermal correlation functions have a form similar to that of $\gamma(t-t')$ in (9.2.12, 9.2.13), namely

$$\gamma_T^{(-)}(\tau) = \int_{\omega_0-\vartheta}^{\omega_0+\vartheta} d\omega\, g(\omega) |\kappa(\omega)|^2 \bar{N}(T,\omega_0) e^{-i(\omega-\omega_0)\tau}, \tag{9.6.19}$$

$$\gamma_T^{(+)}(\tau) = \int_{\omega_0-\vartheta}^{\omega_0+\vartheta} d\omega\, g(\omega) |\kappa(\omega)|^2 \big(\bar{N}(T,\omega_0)+1\big) e^{-i(\omega-\omega_0)\tau}. \tag{9.6.20}$$

b) Evaluation of Terms in the Algorithm: In the algorithm (9.6.16) itself only two-time averages arise, and these can be evaluated using the methods presented in Sect. 9.2.3d, and in the appropriate limit as given in Sect. 9.3.1b, namely Δt should be sufficiently large compared to $1/\vartheta$, but $\Gamma\, \Delta t$ and $\delta\omega\, \Delta t$ should be nevertheless sufficiently small.

c) Orders of Magnitude in the Derivation: The evaluation of the mean square of the fluctuation term in the second line of (9.6.16) involves four point averages in expressions like

$$\int_0^{\Delta t} dt_1 \int_0^{t_1} dt_2 \int_0^{\Delta t} ds_1 \int_0^{t_1} ds_2 \langle f(s_2) f^\dagger(s_1) f(t_2) f^\dagger(t_1) \rangle_T . \tag{9.6.21}$$

Since the operators are quantum Gaussian, we can use the Hartree–Fock factorization formulae in *Bk. I: Sect.10.4.1*, so that the four-time average becomes

$$\begin{aligned} &\langle f(s_2) f^\dagger(s_1)\rangle_T \langle f(t_2) f^\dagger(t_1)\rangle_T + \langle f(s_2) f^\dagger(t_1)\rangle_T \langle f(t_2) f^\dagger(t_1)\rangle_T \\ &\quad = \gamma_T^{(+)}(s_2 - s_1)\gamma_T^{(+)}(t_2 - t_1) + \gamma_T^{(+)}(s_2 - t_1)\gamma_T^{(+)}(t_2 - s_1). \end{aligned} \tag{9.6.22}$$

The evaluation of the resulting expressions like

$$\int_0^{\Delta t} dt_1 \int_0^{t_1} dt_2 \int_0^{\Delta t} ds_1 \int_0^{t_1} ds_2 \gamma_T^{(+)}(s_2 - t_1)\gamma_T^{(+)}(t_2 - s_1), \tag{9.6.23}$$

is then done using the method of (9.2.22–9.2.24), and the result is always proportional to Δt^2.

9.6.5 Finite Temperature Evolution Operator

The end result is a quantum stochastic differential equation which we shall present as an equation for the *finite temperature evolution operator* $\mathcal{V}_T(t, t_0)$, a generalization of that defined in Sect. 9.4.5. This operator is unitary, and satisfies the quantum stochastic differential equation

$$\mathcal{V}_T(t_0, t_0) = 1, \tag{9.6.24}$$

$$d\mathcal{V}_T(t, t_0) = \left(-\frac{i}{\hbar} H_{\text{eff}}\, dt + \sqrt{\Gamma}\, c\, dB^\dagger(t) - \sqrt{\Gamma}\, c^\dagger\, dB(t)\right) \mathcal{V}_T(t, t_0), \tag{9.6.25}$$

$$\begin{aligned} H_{\text{eff}} &\equiv H + \hbar\left(\delta\omega^- c^\dagger c - \delta\omega^+ c c^\dagger\right) \\ &\quad - \tfrac{1}{2} i\hbar\Gamma\left(\left(\bar{N}(T,\omega_0) + 1\right) c^\dagger c + \bar{N}(T,\omega_0)\, c c^\dagger\right), \end{aligned} \tag{9.6.26}$$

$$\Gamma = 2\pi g(\omega_0) |\kappa(\omega_0)|^2, \tag{9.6.27}$$

$$\delta\omega^- = -\mathrm{P} \int_{\omega_0 - \vartheta}^{\omega_0 + \vartheta} \frac{g(\omega)|\kappa(\omega)|^2 \left(\bar{N}(T,\omega) + 1\right) d\omega}{\omega}, \tag{9.6.28}$$

$$\delta\omega^+ = -\mathrm{P} \int_{\omega_0 - \vartheta}^{\omega_0 + \vartheta} \frac{g(\omega)|\kappa(\omega)|^2 \bar{N}(T,\omega)\, d\omega}{\omega}. \tag{9.6.29}$$

The quantum state $|\Psi, t\rangle$ also satisfies (9.6.25), and is given by

$$|\Psi, t_0\rangle = \mathcal{V}_T(t, t_0) |\Psi, t\rangle. \tag{9.6.30}$$

For finite temperature the system can be described by a density operator $\boldsymbol{\rho}(t)$, whose evolution is given by

$$\boldsymbol{\rho}(t) = \mathcal{V}_T(t,t_0)\boldsymbol{\rho}(t_0)\mathcal{V}_T^\dagger(t,t_0). \tag{9.6.31}$$

a) Frequency Shifts: For non-zero temperature there are two kinds of frequency shift, with a similar structure as for the zero temperature case. The formula (9.6.5) for the thermal occupation $\bar{N}(T,\omega)$ shows that this vanishes rapidly at high frequency, and that the principal value integrals involving this converge for any value of the cutoff ϑ. However, the situation is really much the same as for the zero temperature case, since in practice ϑ is chosen too small for the convergence issue the be relevant.

b) Finite Temperature Ito Algebra: This procedure leads to the Ito rules for finite temperatures and several noise fields in the form

$$\begin{aligned}
&dB(t)\,dB^\dagger(t) = \big(\bar{N}(T,\omega_0)+1\big)\,dt,\\
&dB^\dagger(t)\,dB(t) = \bar{N}(T,\omega_0)\,dt,\\
&dB(t)^2 = dB^\dagger(t)^2 = 0,\\
&dB(t)\,dt = dt\,dB(t) = dB^\dagger(t)\,dt = dt\,dB^\dagger(t) = dt^2 = 0.
\end{aligned} \tag{9.6.32}$$

c) Equivalence to Added Classical Noise: We can introduce the classical complex Ito increments, similar to those we defined in *Bk. I: Sect.7.3.2*, which have the Ito algebra

$$dw^*(t)\,dw(t) = \bar{N}(T,\omega_0)\,dt, \qquad dw(t)^2 = dw^*(t)^2 = 0. \tag{9.6.33}$$

The finite temperature Ito algebra is then equivalent to adding these classical increments to the zero-temperature quantum Ito increments thus:

$$dB(t) \to dB(t) + dw(t), \qquad dB^\dagger(t) \to dB^\dagger(t) + dw^*(t), \tag{9.6.34}$$

and including the additional rules

$$dB(t)\,dw(t) = dB(t)^\dagger\,dw(t) = dB(t)^\dagger\,dw^*(t) = dB(t)\,dw^*(t) = 0. \tag{9.6.35}$$

10. The Master Equation

In *Bk. I: Ch.13* we presented the quantum-mechanical master equation, and there we derived it directly from the full Hamiltonian of a system interacting with a light field, or more generally in *Bk. I: Sect.13.3,* when the system interacted with an almost arbitrary heat bath. This chapter is brief, and is intended mainly to show how the master equation, as formulated and developed in *Book I,* fits into the quantum stochastic framework developed here. For the benefit of readers looking for material on methods and applications of the master equation, we give at the end of this chapter, in Sect. 10.3, a guide to the numerous examples and applications that we have presented in *Book I.*

The master equation describes the evolution of the system when the observations of the light field (or heat bath) are not recorded, and for this it uses the *reduced density operator*. Being an operator in the system space only, the reduced density operator is often a matrix of quite small dimension, whose matrix elements can be written down explicitly, and usually can be measured experimentally. The importance of the master equation arises from the basic nature of quantum optical systems, which is *modular,* characterized by a structure of building blocks made of matter linked together by light beams. The elementary structure is always similar to that we have considered in the previous two chapters, a single atom driven by a light field and then emitting light. It is the light that is measured, but these optical measurements are directly related to the matrix elements of the the reduced density operator.

In contrast, the quantum stochastic differential equation description we have given in the previous chapter involves a description of both the system and the light field—it necessarily involves quite complicated and infinite dimensional structures. These structures can be very illuminating, but it is out of the question to measure everything within them which we can consider theoretically. Furthermore, the solutions of the master equation provide a great deal of information about the optical field. In fact, as we shall see in Sect. 11.3, the correlation functions of the optical field produced by the system are directly related to those of the system variables, and these can be determined by solving the master equation.

The aim of this chapter is to derive the master equation, and the associated results, such as the quantum regression theorem covered in *Bk. I: Sect.13.4.3,* directly from the quantum stochastic Schrödinger equation. In this way, we find a situation is analogous to that presented in *Bk. I: Sect.4.2* for the classical stochastic differential equations, where, starting from the stochastic differential equation, we can derive the Fokker–Planck equation, an equation of motion for the probability density which does not involve the classical Ito increments.

10.1 The Master Equation within the Quantum Stochastic Framework

The method by which we can derive the master equation is in fact very simple, once the quantum stochastic differential equation formalism is accepted. The reduced density operator corresponding to the solution of the quantum stochastic differential equation is defined as

$$\bar{\rho}(t) \equiv \mathrm{Tr}_{\mathrm{EM}}\left\{|\Psi, t\rangle\langle\Psi, t|\right\}, \tag{10.1.1}$$

and we implement the trace over the electromagnetic field modes as the quantum stochastic average over the increments $dB(t)$, $dB^\dagger(t)$, and in doing so we use the full power of the quantum Ito calculus.

10.1.1 Derivation of the Master Equation

Let us consider the case of finite temperature, so that the evolution of the full density operator is given by adapting the equations (9.6.24–9.6.29) to the case where

$$|\Psi, t\rangle = \mathcal{V}_T(t, t_0), |\Psi, t_0\rangle, \tag{10.1.2}$$

$$\boldsymbol{\rho}(t_0) = |\Psi, t_0\rangle\langle\Psi, t_0|, \tag{10.1.3}$$

so that

$$d\bar{\rho}(t) = -\frac{\mathrm{i}}{\hbar}\left(H_{\mathrm{eff}}\bar{\rho}(t) - \bar{\rho}(t)H_{\mathrm{eff}}^\dagger\right)dt$$
$$-\mathrm{Tr}_{\mathrm{EM}}\left\{\Gamma\left(c\,dB^\dagger(t) - c^\dagger\,dB(t)\right)\boldsymbol{\rho}(t)\left(c\,dB^\dagger(t) - c^\dagger\,dB(t)\right)\right\}. \tag{10.1.4}$$

We can now use the cyclic property of the trace over the electromagnetic field to move the increments inside the trace to the left-hand side, and then use the Ito rules (9.6.32). For example,

$$\mathrm{Tr}_{\mathrm{EM}}\left\{c\,dB^\dagger(t)\boldsymbol{\rho}(t)c^\dagger\,dB(t)\right\} = \mathrm{Tr}_{\mathrm{EM}}\left\{dB(t)dB^\dagger(t)c\boldsymbol{\rho}(t)c^\dagger\right\}, \tag{10.1.5}$$

$$= \tfrac{1}{2}\Gamma\left(\bar{N}(T,\omega_0) + 1\right)c\bar{\rho}(t)c^\dagger\,dt. \tag{10.1.6}$$

Applying this methodology, we arrive at the *master equation* in the form

$$\frac{d\bar{\rho}}{dt} = -\frac{\mathrm{i}}{\hbar}\left(H_{\mathrm{eff}}\bar{\rho} - \bar{\rho}H_{\mathrm{eff}}^\dagger\right) + \Gamma\left(\bar{N}(T,\omega_0)+1\right)c\bar{\rho}c^\dagger + \Gamma\bar{N}(T,\omega_0)c^\dagger\bar{\rho}c, \tag{10.1.7}$$

$$= -\frac{\mathrm{i}}{\hbar}\left[H_{\mathrm{sys}}, \bar{\rho}\right] - \mathrm{i}\left[\delta\omega^- c^\dagger c - \delta\omega^+ cc^\dagger, \bar{\rho}\right]$$
$$+\tfrac{1}{2}\Gamma\left(\bar{N}(T,\omega_0)+1\right)\left(2c\bar{\rho}c^\dagger - c^\dagger c\bar{\rho} - \bar{\rho}c^\dagger c\right)$$
$$+\tfrac{1}{2}\Gamma\bar{N}(T,\omega_0)\left(2c^\dagger\bar{\rho}c - cc^\dagger\bar{\rho} - \bar{\rho}cc^\dagger\right). \tag{10.1.8}$$

where

$$\Gamma = 2\pi g(\omega_0)|\kappa(\omega_0)|^2, \tag{10.1.9}$$

$$\delta\omega^- = \mathrm{P}\int_{\omega_0-\vartheta}^{\omega_0+\vartheta} \frac{g(\omega)|\kappa(\omega)|^2\big(\bar{N}(T,\omega)+1\big)\,d\omega}{\omega_0-\omega}, \tag{10.1.10}$$

$$\delta\omega^+ = \mathrm{P}\int_{\omega_0-\vartheta}^{\omega_0+\vartheta} \frac{g(\omega)|\kappa(\omega)|^2\bar{N}(T,\omega)\,d\omega}{\omega_0-\omega}. \tag{10.1.11}$$

a) Generality: The form (10.1.8) assumes that the operators $c^\dagger$, c have the commutation relations arising from (9.1.5), that is, they are *eigenoperators* of the commutator $[H_{\text{sys}},\ ]$. Other than that they are completely general; however in practice we choose operators of the form $|f\rangle\langle e|$, where the states $|e\rangle$, $|f\rangle$ are eigenstates of H_{sys} such that $E_f - E_e = \hbar\omega_0$. These operators form a basis in terms of which any eigenoperator of H_{sys} can be expressed.

b) The Lindblad Form: The form of the master equation above, that is, a combination of Hamiltonian terms, like that in the first line, and dissipative terms as in the next two lines, is known as the *Lindblad form*, which we have discussed in some detail in *Bk. I: Sect.13.1.3*.

In brief, a master equation of the Lindblad form is one which can be written in the form

$$\frac{\partial\bar{\rho}}{\partial t} = L\bar{\rho} \equiv -\frac{i}{\hbar}[H,\bar{\rho}] + \sum_i \tfrac{1}{2}\gamma_i\bar{\mathcal{M}}_i\left\{2c_i\bar{\rho}c_i^\dagger - \bar{\rho}c_i^\dagger c_i - c_i^\dagger c_i\bar{\rho}\right\}$$
$$+ \sum_i \tfrac{1}{2}\gamma_i\bar{\mathcal{N}}_i\left\{2c_i^\dagger\bar{\rho}c_i - \bar{\rho}c_i c_i^\dagger - c_i c_i^\dagger\bar{\rho}\right\}, \tag{10.1.12}$$

where H is Hermitian, $\bar{\mathcal{M}}_i, \bar{\mathcal{N}}_i$ are real dimensionless non-negative c-numbers (either of which can be the larger), and the c_i are *arbitrary* dimensionless operators. The coefficient $\frac{1}{2}\gamma_i$ is a positive damping rate, and could in principle be absorbed into the other two coefficients, which would then acquire its dimension of inverse time.

If a master equation is of the Lindblad form, the time evolution is completely consistent with quantum mechanics, and in particular, this means that

i) The solution for the density operator is always a positive definite operator—that is, no negative probabilities occur.

ii) The trace of $\bar{\rho}$ is time independent, so that probability is conserved.

c) Time Dependence of the Master Equation Operator: It is possible, and quite common, to have a time-dependent master equation operator $L(t)$, such as would arise from a time-dependent coherent classical field. This does not change any of the results of this chapter.

10.1.2 Superoperator Notations for Master Equation Operators

It is often convenient to use a notation in terms of *superoperators* $\mathcal{J}$ and $\mathcal{K}$. These are two-sided linear operators which act on the density operator, and for the master equation (10.1.12) defined by

i) *Recycling Operators*:

$$\mathcal{J}_i^{(-)}\rho \equiv \gamma_i\bar{\mathcal{M}}_i\, c_i\rho c_i^\dagger, \tag{10.1.13}$$

$$\mathcal{J}_i^{(+)}\rho \equiv \gamma_i\bar{\mathcal{N}}_i\, c_i^\dagger\rho c_i\,. \tag{10.1.14}$$

ii) *Non-Hermitian Effective Hamiltonian*: Here we include the Hamiltonian term, and get the single operator,

$$\mathcal{K}\rho \equiv -\frac{\mathrm{i}}{\hbar}\left[H_{\text{sys}},\rho\right]-\sum_i \tfrac{1}{2}\gamma_i\bar{\mathcal{M}}_i\left[c_i^\dagger c_i,\rho\right]_+ -\sum_i \tfrac{1}{2}\gamma_i\bar{\mathcal{N}}_i\left[c_i c_i^\dagger,\rho\right]_+, \tag{10.1.15}$$

$$\equiv -\frac{\mathrm{i}}{\hbar}H_{\text{eff}}\rho+\frac{\mathrm{i}}{\hbar}\rho H_{\text{eff}}^\dagger\,. \tag{10.1.16}$$

iii) *Liouvillian*: The full evolution can be written

$$\frac{\partial\rho}{\partial t} = L\rho, \tag{10.1.17}$$

where L is known as the *Liovillian operator*, defined by the sum of these terms

$$L \equiv \mathcal{K}+\sum_i\left\{\mathcal{J}_i^{(-)}+\mathcal{J}_i^{(+)}\right\}. \tag{10.1.18}$$

Exercise 10.1 The Meaning of "Recycling": Show that non-Hermitian evolution operator $\mathcal{K}$ gives rise to a loss of probability, that is $\mathrm{Tr}\left\{\mathcal{K}\rho\right\}\leqslant 0$ for all ρ. The recycling terms balance this loss, giving rise to the concept of their "recycling" the probability lost by the non-Hermitian evolution.

10.1.3 Non-Lindblad Master Equations

Master equations which are not of the Lindblad form can arise, for example if the rotating wave approximation is not made. An example is the quantum Brownian motion master equation, treated in *Bk. I: Sect.16.3.5*,which is generally regarded as a useful tool.

Such equations do not have the property that every positive definite initial condition yields a positive definite solution. However, this does not mean that they cannot be used in an approximate description of physical processes. The deviations from positivity are often small and transient, and usually arise from quite extreme initial conditions. Solutions with reasonable initial conditions can be both positive definite and quite accurate [10.1].

10.2 Time Correlation Functions

In this section we will use the quantum stochastic Schrödinger equation formalism to derive the correlation function formulae derived in *Bk. I: Sect.13.4.* We do this by relating the evolution operator for the master equation to the evolution operator for the quantum stochastic Schrödinger equation, which we call the *full evolution operator.*

10.2.1 The Evolution Operator for the Master Equation

The solutions to a master equation

$$\frac{\partial \bar{\rho}(t)}{\partial t} = L\bar{\rho}(t), \tag{10.2.1}$$

where $L\bar{\rho}$ is the linear operation such as in the right-hand side of (10.1.12), can be formally written

$$\bar{\rho}(t) = V(t,t_0)\bar{\rho}(t_0), \tag{10.2.2}$$

where $V(t,t_0)$ is the *evolution operator* which satisfies the equation of motion

$$\frac{d}{dt}V(t,t_0) = LV(t,t_0). \tag{10.2.3}$$

Clearly, $V(t,t_0)$ obeys the *semigroup property*

$$V(t,t_1)V(t_1,t_0) = V(t,t_0), \qquad \text{where } t \geqslant t_1 \geqslant t_0, \tag{10.2.4}$$

and $V(t_0,t_0) = 1$.

a) Two-Sided Operators: Like the superoperators of Sect. 10.1.2, both of the operators L and $V(t,t_0)$ are *two-sided* operators—that is, if $\bar{\rho}$ can be represented as an $n\times n$ matrix then L and $V(t,t_0)$ are to be understood as matrices of size $n^2\times n^2$, acting on an n^2 dimensional column vector composed of all the elements of $\bar{\rho}$ written in a chosen order.

10.2.2 Relation to the Full Evolution Operator

The full evolution operator $\mathcal{V}(t,t_0)$, defined either at zero temperature through its equation of motion (9.4.28, 9.4.29) or at finite temperature by (9.6.24–9.6.29), depends on the quantum stochastic increments only within the the time interval (t_0,t). The independence of the increments at different times means that averages computed with respect to the noise increments have a factorization property which can be regarded as the quantum version of the Markov property of classical stochastic processes. To make this clear, take the evolution equation (9.6.31) of the density operator and write the evolution equation for the reduced density operator as

$$\bar{\rho}(t) = \mathrm{Tr}_{\mathrm{EM}}\left\{\mathcal{V}(t,t_0)\boldsymbol{\rho}(t_0)\mathcal{V}^\dagger(t,t_0)\right\}, \tag{10.2.5}$$

$$\equiv V(t,t_0)\bar{\rho}(t_0). \tag{10.2.6}$$

Now consider a point t_1 inside the interval (t_0, t), so that $\mathcal{V}(t, t_0) = \mathcal{V}(t, t_1)\mathcal{V}(t_1, t_0)$. Then from this equation we can use the independence of the increments at different times to write the trace over the electromagnetic field in terms of the separate traces over the intervals (t_0, t_1) and (t_1, t)

$$\bar{\rho}(t) = \mathrm{Tr}_{\mathrm{EM}}\left\{\mathcal{V}(t, t_1)\mathcal{V}(t_1, t_0)\boldsymbol{\rho}\mathcal{V}^\dagger(t_1, t_0)\mathcal{V}^\dagger(t, t_1)\right\}, \tag{10.2.7}$$

$$= \mathrm{Tr}_{(t_1,t)}\left\{\mathcal{V}(t, t_1)\,\mathrm{Tr}_{(t_0,t_1)}\left\{\mathcal{V}(t_1, t_0)\boldsymbol{\rho}\mathcal{V}^\dagger(t_1, t_0)\right\}\mathcal{V}^\dagger(t, t_1)\right\}, \tag{10.2.8}$$

$$= V(t, t_1)V(t_1, t_0)\bar{\rho}(t_0). \tag{10.2.9}$$

Thus the semigroup property, as formulated in (10.2.4), also follows from the ability to break the trace into independent traces over the two time intervals.

10.2.3 Multitime Averages

If F and G are two system operators, then the correlation function $\langle F(t+\tau)G(t)\rangle$ can be got by using the Heisenberg picture; it is

$$\langle F(t+\tau)G(t)\rangle = \mathrm{Tr}_{\mathrm{sys}}\mathrm{Tr}_{\mathrm{EM}}\{F(t+\tau)G(t)\,\rho_{\mathrm{sys}}\otimes\rho_{\mathrm{EM}}\}. \tag{10.2.10}$$

Here ρ_{sys} and ρ_{EM} are the Heisenberg picture density operators for the system and electromagnetic field respectively. The Heisenberg operators can be written in terms of the Schrödinger picture operators as

$$G(t) = \mathcal{V}^\dagger(t, t_0)G\mathcal{V}(t, t_0), \qquad F(t+\tau) = \mathcal{V}^\dagger(t+\tau, t_0)F\mathcal{V}(t+\tau, t_0), \tag{10.2.11}$$

so that we can write

$$\langle F(t+\tau)G(t)\rangle = \mathrm{Tr}_{\mathrm{sys}}\mathrm{Tr}_{\mathrm{EM}}\left\{\mathcal{V}^\dagger(t+\tau, t_0)F\mathcal{V}(t+\tau, t_0)\mathcal{V}^\dagger(t, t_0)G\mathcal{V}(t, t_0)\rho_{\mathrm{sys}}\otimes\rho_{\mathrm{EM}}\right\}, \tag{10.2.12}$$

$$= \mathrm{Tr}_{\mathrm{sys}}\mathrm{Tr}_{\mathrm{EM}}\left\{F\mathcal{V}(t+\tau, t)G\mathcal{V}(t, t_0)\rho_{\mathrm{sys}}\otimes\rho_{\mathrm{EM}}\mathcal{V}^\dagger(t+\tau, t_0)\right\}. \tag{10.2.13}$$

We now recognize the independence of the noises at different times, so that we can write

$$\mathrm{Tr}_{\mathrm{EM}} = \mathrm{Tr}_{(t,t+\tau)}\mathrm{Tr}_{(t_0,t)}, \tag{10.2.14}$$

$$\rho_{\mathrm{EM}} = \rho_{(t,t+\tau)}\otimes\rho_{(t_0,t)}, \tag{10.2.15}$$

which enables us to rewrite (10.2.13) as

$$\langle F(t+\tau)G(t)\rangle = \mathrm{Tr}_{\mathrm{sys}}\Bigg\{F\,\mathrm{Tr}_{(t,t+\tau)}\Bigg(\mathcal{V}(t+\tau, t)G\,\mathrm{Tr}_{(t_0,t)}$$
$$\times\left(\mathcal{V}(t, t_0)\rho_{\mathrm{sys}}\otimes\rho_{(t_0,t)}\mathcal{V}^\dagger(t, t_0)\right)\otimes\rho_{(t,t+\tau)}\mathcal{V}^\dagger(t+\tau, t)\Bigg)\Bigg\},$$
$$= \mathrm{Tr}_{\mathrm{sys}}\left\{FV(t+\tau, t)GV(t, t_0)\bar{\rho}(t_0)\right\}. \tag{10.2.16}$$

By similar reasoning we can work out the correlation function with reversed time ordering, giving us the two formulae

$$\langle F(t+\tau)G(t)\rangle = \text{Tr}_{\text{sys}}\left\{FV(t+\tau,t)\big(G\bar{\rho}(t)\big)\right\}, \tag{10.2.17}$$

$$\langle F(t)G(t+\tau)\rangle = \text{Tr}_{\text{sys}}\left\{GV(t+\tau,t)\big(\bar{\rho}(t)F\big)\right\}. \tag{10.2.18}$$

It is not difficult to see that this process can be extended to yield general time ordered correlation functions, in the sense that we can evaluate any correlation function of the form

$$\langle F_0(s_0)F_1(s_1)\ldots F_m(s_m)G_n(t_n)G_{n-1}(t_{n-1})\ldots G_0(t_0)\rangle, \tag{10.2.19}$$

where the terms are ordered

$$t_n \geqslant t_{n-1} \geqslant \ldots t_0, \qquad s_m \geqslant s_{m-1} \geqslant \ldots s_0. \tag{10.2.20}$$

A fuller formulation is given in *Quantum Noise.*

10.2.4 Quantum Regression Theorem

The reader is referred to *Bk. I: Sect.13.4.3* for a derivation and discussion of the quantum regression theorem. This relies only on the use of the master equation evolution operator, and therefore is not affected by our choice of the quantum stochastic Schrödinger equation formalism in this chapter.

10.3 Applications of the Master Equation

In practice the master equation produces relatively simple and compact sets of equations of motion for the density matrix elements, and in atomic systems there are normally only a relatively small number of matrix elements. It is widely used in the quantum optical literature, where numerous techniques have been developed for its application to specific systems.

As noted in the introduction to this chapter, we have treated the application of the master equation in some depth in *Book I*, and have also given explicit treatments of the simplest examples of master equations. In summary, we covered such material as follows:

i) *The Excitation and Decay of a Two-Level System*: We cover this in some detail in *Bk. I: Sect.14.1*

ii) *A Two-Level System Driven by a Coherent Laser Field*: The basic equations and solutions are presented in *Bk. I: Sect.14.2.* In Chap. 13 we shall also give a more detailed treatment of two-level and three-level systems, while in Part IV the equations are extended to include mechanical effects of light, and laser cooling.

iii) *Systems of Harmonic Oscillators*: In *Bk. I: Sect.14.3* we have presented the basic theory of damping, gain and decoherence. Harmonic oscillator master equations are often very elegantly treated using phase space methods, which we have formulated in *Bk. I: Ch.15–Ch.17.* These methods give ways of transforming—often exactly but also approximately—the master equation into classical Fokker–Planck equations. We have given an extensive range of examples using the *Wigner function* in *Bk. I: Ch.16,* and using the *P-function* in *Bk. I: Ch.17.*

11. Inputs, Outputs and Quantum Langevin Equations

In the laboratory, we manipulate atoms using light beams, and deduce the behaviour of the atoms by measurements on light they emit. There is a clear separation between the three basic components; the input field; the atomic system; and the output field, that is the field we measure. Although to the experimenter, the input field and the output field are clearly distinct, the description we have used in the previous two chapters has not made any clear distinction between them, preferring to talk about the electromagnetic field as a whole.

The input-output formalism was introduced by *Collett* and *Gardiner* [11.1, 11.2], and was based on the formalism of quantum network theory, devised by *Yurke* and *Denker* [11.3] for use in superconducting microwave systems. The aim was to provide a method by which one could:

i) Specify the input field—that is, specify the quantum state of the incoming electromagnetic field.

ii) Write down a master equation for the density operator of the atomic system being studied, and solve it for the quantities of interest.

iii) From the solutions of the master equation, specify the output field emitted from the atomic system.

While this appears to be a perfectly obvious way of proceeding, it broke new ground by insisting that, quantum mechanically, even in the vacuum state there is always an input field. In essence, it noted that vacuum fluctuations were always impinging on the atom, and that they behaved very much like quantum mechanical versions of the Langevin noise sources.

The input-output formalism is most naturally formulated directly in the Heisenberg picture, as was originally done by Collett and Gardiner. However, for the full quantum stochastic development of the formalism it is simplest to transfer the results of Chap. 9 into the Heisenberg picture. In this way, the direct connection between the two ways of looking at the problem becomes very clear. We will show that the quantum Ito increments can be interpreted in terms of the input field operator as

$$dB(t) = b_{\rm in}(t)\,dt, \qquad dB^\dagger(t) = b_{\rm in}^\dagger(t)\,dt.$$

Here $b_{\rm in}(t)$ is essentially the incoming electromagnetic field evaluated at the position of the atom.

To obtain the output field $b_{\rm out}(t)$ proves to be very simple, there is a boundary condition at the position of the atom, which can be written

$$b_{\rm out}(t) - b_{\rm in}(t) = \sqrt{\Gamma}\, c(t),$$

where Γ represents the atomic damping rate, and $c(t)$ is the operator to which the field couples, essentially the dipole moment operator.

11.1 The Quantum Langevin Equation

The strategy for the derivation of the quantum Langevin equation employs all of the approximation procedures used in the derivation of the quantum stochastic Schrödinger equation. We will therefore keep our notation very much the same as we used in Chap. 9.

a) Heisenberg Equations of Motion: From the Hamiltonian (9.1.1–9.1.4) we derive the Heisenberg equations of motion for b_k, and for an arbitrary system operator a. They are

$$\dot{b}_k(t) = -\mathrm{i}\omega_k b_k(t) + \kappa_k^* c(t), \tag{11.1.1}$$

$$\dot{a}(t) = -\frac{\mathrm{i}}{\hbar}\left[a(t), H_{\rm sys}\right] + \sum_{|\omega_k-\omega_0|<\vartheta}\Big(\kappa_k^* b_k^\dagger(t)\,[a(t), c(t)] - [a(t), c^\dagger(t)]\kappa_k b_k(t)\Big). \tag{11.1.2}$$

b) Solution for Field Operators: We now solve (11.1.1) to obtain

$$b_k(t) = \mathrm{e}^{-\mathrm{i}\omega_k(t-t_0)}\, b_k(t_0) + \kappa_k^* \int_{t_0}^{t} \mathrm{e}^{-\mathrm{i}\omega_k(t-t')}\, c(t')\, dt'. \tag{11.1.3}$$

c) Expression in Terms of Noise Operators: We substitute for $b_k(t)$ in (11.1.2) to obtain

$$\begin{aligned}\dot{a} &= -\frac{\mathrm{i}}{\hbar}\left[a, H_{\rm sys}\right] + \int_{t_0}^{t} dt' \Big(\bar{\gamma}^*(t-t')c^\dagger(t')[a,c] - \bar{\gamma}(t-t')[a,c^\dagger]c(t')\Big)\\ &\qquad + \bar{f}^\dagger(t)[a,c] - [a,c^\dagger]\bar{f}(t),\end{aligned} \tag{11.1.4}$$

$$\bar{f}(t) \equiv \sum_{|\omega_k-\omega_0|<\vartheta} \mathrm{e}^{-\mathrm{i}\omega_k t}\kappa_k b_k(t_0) = \mathrm{e}^{-\mathrm{i}\omega_0 t} f(t), \tag{11.1.5}$$

$$\bar{\gamma}(\tau) = \sum_{|\omega_k-\omega_0|<\vartheta} \mathrm{e}^{-\mathrm{i}\omega_k \tau}|\kappa_k|^2 = \mathrm{e}^{-\mathrm{i}\omega_0\tau}\gamma(\tau). \tag{11.1.6}$$

In these equations:

i) For notational convenience in (11.1.4) we omit the time argument on the system operators when it is t but write it explicitly otherwise. (Thus $a \equiv a(t)$.)

ii) The electromagnetic field operators are essentially the same as the operators $b(\omega_k)$ as defined in (9.2.7); the precise relationship is

$$b(\omega_k)_{(9.2.7)} = e^{-i\omega_k t_0} b_k(t_0). \tag{11.1.7}$$

iii) Thus $f(\tau)$ is a noise operator, and is as defined in (9.2.9). Similarly, $\gamma(\tau)$ represents the same damping kernel as defined in (9.2.12).

d) Markov Approximations: The equations are exact so far, but we can now make the same kinds of approximations as were made in deriving the quantum stochastic Schrödinger equation in Sect. 9.2.1. These amount to:

i) We make the approximation that $\gamma(\tau)$ is sharply peaked at $\tau = 0$. Since the integrals involving $\gamma(\tau)$ are only for $\tau > 0$, we formalize this by writing, for any function $F(\tau)$,

$$\int_0^\infty \gamma(\tau')F(\tau')\,d\tau' \to F(0)\int_0^\infty \gamma(\tau')\,d\tau' = \left(\tfrac{1}{2}\Gamma + i\delta\omega\right)F(0). \tag{11.1.8}$$

ii) We define an "in" field by

$$b_{in}(t) = \sum_{|\omega_k-\omega_0|<\vartheta} e^{-i(\omega_k-\omega_0)(t-t_0)} b_k(t_0), \tag{11.1.9}$$

which satisfies the commutation relation

$$[b_{in}(t), b_{in}^\dagger(t')] = \delta(t-t'), \tag{11.1.10}$$

for time scales much slower than$1/\vartheta$.

iii) The commutator $[f(t), f(t')] = \gamma(t-t')$, as shown in (9.2.11), and here $t-t'$ may have either sign. Therefore the approximation that $\gamma(\tau)$ is sharply peaked at $\tau = 0$ requires

$$\gamma(\tau) \to \delta(\tau)\int_{-\infty}^\infty \gamma(\tau')\,d\tau' = \Gamma\delta(\tau). \tag{11.1.11}$$

Consistent with these we therefore set

$$f(t) \to \sqrt{\Gamma}\, b_{in}(t). \tag{11.1.12}$$

11.1.1 The Quantum Langevin Equation

Using (11.1.8–11.1.12) we can transform the equation of motion (11.1.4) into a *quantum Langevin equation*

$$\begin{aligned}\dot{a} = &-\frac{i}{\hbar}[a, H_{sys}] + \frac{i}{\hbar}[a, \delta\omega^+ c^\dagger c - \delta\omega^- c c^\dagger] \\ &- [a, c^\dagger]\left(\tfrac{1}{2}\Gamma c + \sqrt{\Gamma} b_{in}(t)\right) + \left(\tfrac{1}{2}\Gamma c^\dagger + \sqrt{\Gamma} b_{in}^\dagger(t)\right)[a, c].\end{aligned} \tag{11.1.13}$$

The quantized noise inputs $b_{\rm in}(t)$ and $b_{\rm in}^\dagger(t)$ have the commutation relations

$$[b_{\rm in}(t)\,,\, b_{\rm in}^\dagger(t')] = \delta(t-t'). \tag{11.1.14}$$

11.1.2 Noise Terms

The terms depending on $b_{\rm in}(t)$, $b_{\rm in}^\dagger(t)$ are to be taken as noise terms. The definition of $b_{\rm in}(t)$ in terms of the the operators of $b_k(t_0)$ ensures that these operators may be freely specified on the same basis as initial conditions. Similarly, we may freely specify the state of the system to be such that, at $t = t_0$, the system and bath density operators both factorize. It is *not* necessary (nor indeed is it possible) to make any further "independence" assumption. In fact it is not necessary to make any particular assumption about the initial state of the system at all for the Langevin equation to be valid in the form (11.1.13). However, the terms $b_{\rm in}(t)$, $b_{\rm in}^\dagger(t)$ can only be reasonably interpreted as noise when the state of the system is initially factorized and the state of $b_{\rm in}(t)$ is incoherent. Normally this will be a thermal state for which we can write

$$\langle b^\dagger(t) b(t')\rangle = \bar{N}(T,\omega_0)\,\delta(t-t'), \tag{11.1.15}$$

$$\langle b(t) b^\dagger(t')\rangle = \big(\bar{N}(T,\omega_0)+1\big)\,\delta(t-t'). \tag{11.1.16}$$

In the case, that, for example $b_{\rm in}(t)$ is in a coherent state, we would have a classical driving field being applied to the system, and other intermediate situations can easily be envisaged.

11.2 The Quantum Langevin Equation in Terms of Outputs

The development of the quantum Langevin equation in the previous section gives an operator driving field $b_{\rm in}(t)$ defined in terms of the values of the field operators at a time t_0, which is implicitly understood to be in the remote past. We can also formulate the methodology in terms of the values of the field operators at a time t_1, implicitly understood to be in the remote future.

Thus, taking $t_1 > t$, we can write, analogously to (11.1.3),

$$b_k(t) = \mathrm{e}^{-\mathrm{i}\omega_k(t-t_1)}\, b_k(t_1) - \kappa_k^* \int_t^{t_1} \mathrm{e}^{-\mathrm{i}\omega_k(t-t')}\, c(t')\, dt', \tag{11.2.1}$$

and then from this derive the quantum Langevin equation driven by output fields

$$\dot{a} = -\frac{i}{\hbar}[a, H_{sys}] - \int_t^{t_1} dt' \left(\bar{\gamma}^*(t-t')c^\dagger(t')[a,c] - \bar{\gamma}(t-t')[a,c^\dagger]c(t')\right)$$
$$+ \bar{f}^\dagger_{out}(t)[a,c] - [a,c^\dagger]\bar{f}_{out}(t), \tag{11.2.2}$$

$$\bar{f}_{out}(t) \equiv \sum_{|\omega_k-\omega_0|<\vartheta} e^{-i\omega_k t}\kappa(\omega_k)b_k(t_1) \equiv e^{-i\omega_0 t} f_{out}(t). \tag{11.2.3}$$

We can take the equations (11.1.3) and (11.2.1), and eliminate $b_k(t)$, to get the relationship

$$f_{out}(t) = f(t) + \int_{t_0}^{t_1} e^{i\omega_0 t'} c(t')\gamma(t-t'). \tag{11.2.4}$$

In the limit that $\gamma(t-t')$ is sharply peaked, we note that $e^{i\omega_0 t'}c(t')$ is slowly varying in t', and $f_{out}(t) \to \sqrt{\Gamma}\, b_{out}(t)$, where the *output field* is defined as

$$b_{out}(t) = \sum_{|\omega_k-\omega_0|<\vartheta} e^{-i\omega_k t} b_k(t_1). \tag{11.2.5}$$

11.2.1 Relationship between Input and Output

From (11.2.4, 11.2.5) we get the fundamental relationship between *input* and *output*

$$b_{out}(t) - b_{in}(t) = \sqrt{\Gamma}\, c(t). \tag{11.2.6}$$

11.2.2 Time-Reversed Quantum Langevin Equation

We can follow the same procedure as in the previous section to derive a quantum Langevin equation, in which the system is driven by the *output fields*

$$\dot{a} = -\frac{i}{\hbar}[a, H_{sys}] + \frac{i}{\hbar}[a, \delta\omega^+ c^\dagger c - \delta\omega^- c c^\dagger]$$
$$+ [a, c^\dagger]\left(\tfrac{1}{2}\Gamma c - \sqrt{\Gamma}\, b_{out}(t)\right) - \left(\tfrac{1}{2}\Gamma c^\dagger - \sqrt{\Gamma}\, b^\dagger_{out}(t)\right)[a,c]. \tag{11.2.7}$$

The relationship between the forward and time reversed versions is most clearly seen by rewriting (11.2.6) as

$$b_{out}(t) - \tfrac{1}{2}\sqrt{\Gamma}\, c(t) = b_{in}(t) + \tfrac{1}{2}\sqrt{\Gamma}\, c(t). \tag{11.2.8}$$

11.2.3 Inputs and Outputs, and Causality

The quantities $b_{\rm in}(t)$ and $b_{\rm out}(t)$ will be interpreted as inputs and outputs to the system. The condition (11.2.6) can be viewed as a boundary condition, relating input, output, and internal modes. If (11.1.13) is solved to give values of the system operators in terms of their past values and those of $b_{\rm in}(t)$, then it is clear that $a(t)$ is independent of $b_{\rm in}(t')$ for $t' > t$; that is, the system variables do not depend on the values of the input in the future. Hence we deduce

$$\left[a(t), b_{\rm in}(t')\right] = 0, \quad t' > t \tag{11.2.9}$$

and using similar reasoning, we deduce from (11.2.7)

$$\left[a(t), b_{\rm out}(t')\right] = 0, \quad t' < t. \tag{11.2.10}$$

Defining the step function $u(t)$ as

$$u(t) = \begin{cases} 1, & t > 0, \\ \frac{1}{2}, & t = 0, \\ 0, & t < 0, \end{cases} \tag{11.2.11}$$

and using (11.2.6), we obtain the specific results

$$\left[a(t), b_{\rm in}(t')\right] = -u(t-t')\sqrt{\Gamma}\left[a(t), c(t')\right], \tag{11.2.12}$$

$$\left[a(t), b_{\rm out}(t')\right] = u(t'-t)\sqrt{\Gamma}\left[a(t), c(t')\right]. \tag{11.2.13}$$

We now have, in principle, a complete specification of a system with input and output. We *specify* the input $b_{\rm in}(t)$, and solve (11.1.13) for $a(t)$. We then compute the output from the known $a(t)$ and $b_{\rm in}(t)$ by use of the boundary condition, (11.2.6).

The commutators (11.2.12, 11.2.13) are an expression of quantum causality—that only the future motion of the system is affected by the present input, and that only the future value of the output is affected by the present values of the system operators.

11.2.4 Applications of the Output-Driven Quantum Langevin Equation

This kind of quantum Langevin equation has not often been applied. However, examples of its application is given in Sect. 28.3.4, and Sect. 28.3.5. There we consider the optimal mapping of a qubit encoded in an input pulse into a quantum memory, and then its readout from the memory into an output field. The application of both input- and output-driven quantum Langevin equations gives an elegant solution the problem.

11.3 Correlation Functions of the Output

The output correlation functions of most interest are those related to photon counting of the output field, and as we shall show in Chap. 18, these are the normally ordered correlation functions of the form

$$\langle b^\dagger_{\rm out}(t_1) b^\dagger_{\rm out}(t_2) \dots b^\dagger_{\rm out}(t_n) b_{\rm out}(t_{n+1}) \dots b_{\rm out}(t_m)\rangle. \tag{11.3.1}$$

It does not in fact matter what time ordering is chosen in (11.3.1) since all the $b_{\text{out}}(t_r)$ commute with each other. Products which are not normally ordered can be related to normally ordered products by use of the commutation relations for the output field, which we have shown are the same as those for the input field.

11.3.1 Output Correlations Functions and System Correlation Functions

The input and output fields are related by the fundamental input-output relationship (11.2.6), which means that the properties of the output will reflect those of the system and of the input. Let us show what form of relationship between correlation functions this takes

As noted above, we may consider with complete generality that the correlation function is of the form (11.3.1) and that the $b^{\dagger}_{\text{out}}(t_r)$ are time *antiordered* (i.e., $t_1 < t_2 < \dots < t_n$) and the $b_{\text{out}}(t_r)$ are *time ordered* (i.e., $t_{n+1} > t_{n+2} > \dots > t_m$).

We may now substitute for $b_{\text{out}}(t_r)$ by the relation (11.2.6)

$$b_{\text{out}}(t_r) = \sqrt{\Gamma}c(t_r) + b_{\text{in}}(t_r). \tag{11.3.2}$$

We note that from (11.2.12, 11.2.13) $c^{\dagger}(t_r)$ will commute with all $b^{\dagger}_{\text{in}}(t_s)$ which occur to the right of $c^{\dagger}(t_r)$, since we have shown $t_r < t_s$. Similarly, $c(t_{r'})$ will commute with all $b_{\text{in}}(t_{s'})$ which occur to the left of $c(t_{r'})$, since we have chosen $t_{r'} > t_{s'}$. Thus we may write the correlation function in a form in which we make the substitution (11.3.2) in (11.3.1), and reorder the terms so that:

i) The $c^{\dagger}(t_r)$ are time *antiordered*, and are to the left of all $c(t_r)$;

ii) The $c(t_{r'})$ are *time ordered*;

iii) The $b^{\dagger}_{\text{in}}(t_s)$ all stand to the left of all other operators; and

iv) The $b_{\text{in}}(t_{s'})$ all stand to the right of all other operators.

We now consider various cases.

a) Input in the Vacuum State: In this case, $b_{\text{in}}(t)\rho_{\text{in}} = \rho_{\text{in}}b^{\dagger}_{\text{in}}(t) = 0$, and we derive

$$\begin{aligned}\langle b^{\dagger}_{\text{out}}(t_1)\, b^{\dagger}_{\text{out}}(t_2)\dots b^{\dagger}_{\text{out}}(t_n) b_{\text{out}}(t_{n+1})\dots b_{\text{out}}(t_m)\rangle = \\ \Gamma^{(m+n)/2}\langle \tilde{T}\left[c^{\dagger}(t_1)c^{\dagger}(t_2)\dots c^{\dagger}(t_n)\right] T\left[c(t_{n+1})\dots c(t_m)\right]\rangle,\end{aligned} \tag{11.3.3}$$

where $\tilde{T}$ is the time-antiordered product and T the time-ordered product.

b) Input is in a Coherent State: In this case

$$b_{\text{in}}(t)\rho_B = \beta(t)\rho_B, \qquad \rho_B b^{\dagger}_{\text{in}}(t) = \beta^*(t)\rho_B, \tag{11.3.4}$$

where $\beta(t)$ is the coherent field amplitude; a c-number. In this case we may replace $b, b^{\dagger}$ by β, β^*, respectively, and since β and β^* are mere numbers, reorder them so that

$$\begin{aligned}
&\langle b_{\rm out}^\dagger(t_1) b_{\rm out}^\dagger(t_2)\dots b_{\rm out}^\dagger(t_n) b_{\rm out}(t_{n+1})\dots b_{\rm out}(t_m)\rangle \\
&= \Big\langle \tilde{T}\Big\{[\sqrt{\Gamma}c^\dagger(t_1)+\beta^*(t_1)]\dots\Big\} T\Big\{[\sqrt{\Gamma}c(t_{n+1})+\beta(t_{n+1})]\dots[\sqrt{\Gamma}c(t_m)+\beta(t_m)]\Big\}\Big\rangle .
\end{aligned} \tag{11.3.5}$$

c) The Input is Quantum White Noise: By carrying out this procedure we are left with the problem of evaluating terms like $\langle c^\dagger(t_1)c^\dagger(t_2)b_{\rm in}(s_1)b_{\rm in}(s_2)\rangle$ when the density matrix is not coherent or vacuum, but represents a more general state.

The case where the input is quantum white noise, of the form given by (9.6.32), is solved in [11.2] and in *Quantum Noise* with the following result:

$$\langle a(t')b_{\rm in}(t)\rangle = \sqrt{\Gamma}\bar{N}u(t'-t)\langle[a(t'),c(t)]\rangle, \tag{11.3.6}$$

$$\langle b_{\rm in}^\dagger(t)a(t')\rangle = \sqrt{\Gamma}\bar{N}u(t'-t)\langle[c^\dagger(t),a(t')]\rangle, \tag{11.3.7}$$

$$\langle a(t')b_{\rm in}^\dagger(t)\rangle = \sqrt{\Gamma}(\bar{N}+1)u(t'-t)\langle[c^\dagger(t),a(t')]\rangle, \tag{11.3.8}$$

$$\langle b_{\rm in}(t)a(t')\rangle = \sqrt{\Gamma}(\bar{N}+1)u(t'-t)\langle[a(t'),c(t)]\rangle, \tag{11.3.9}$$

and we note that these are consistent with the commutation relations (11.2.12, 11.2.13). We can now use the results of Sect. 11.2.3 to derive

$$\langle b_{\rm out}(t)b_{\rm out}(t')\rangle = \Gamma(\bar{N}+1)\langle T[c(t)c(t')]\rangle - \Gamma\bar{N}\langle\tilde{T}[c(t)c(t')]\rangle, \tag{11.3.10}$$

$$\langle b_{\rm out}^\dagger(t)b_{\rm out}(t')\rangle = \Gamma(\bar{N}+1)\langle c^\dagger(t)c(t')\rangle - \Gamma\bar{N}\langle c(t')c^\dagger(t)\rangle + \bar{N}\delta(t-t'), \tag{11.3.11}$$

$$\langle b_{\rm out}^\dagger(t)b_{\rm out}^\dagger(t')\rangle = \Gamma(\bar{N}+1)\langle\tilde{T}[c^\dagger(t)c^\dagger(t')]\rangle - \Gamma\bar{N}\langle T[c^\dagger(t)c^\dagger(t')]\rangle. \tag{11.3.12}$$

Although this is a very elegant result, its interest is more mathematical than physical, since in practice non-vacuum inputs are noisy, but normally have quite finite correlation times, and are thus non-white. Realistic physical cases are of greater complexity, and need individual treatment.

11.4 Quantum Stochastic Differential Equations

As derived in the previous section, the quantum Langevin equation has not been formulated as a truly stochastic equation. The manipulations based upon the Markovian approximations are purely heuristic, and do not use any of the rather subtle techniques of Sect. 9.3–Sect. 9.6 to obtain a stochastic description using the quantum Ito algebra.

Let us use the quantum stochastic formalism we have already developed for the quantum stochastic Schrödinger equation and the evolution operator $\mathcal{V}_T(t,t_0)$, and transform directly to the Heisenberg picture. Then we can use the quantum stochastic differential equations for the evolution operator to develop the appropriate quantum stochastic differential equations for the Heisenberg picture operators. Note that this procedure is an alternative to that developed in *Quantum Noise* and in [11.2], where the stochastic interpretation was done directly in the Heisenberg picture.

11.4.1 Development of the Quantum Stochastic Differential Equation

We can use the evolution operator to move to the Heisenberg picture by defining ing the Heisenberg operators in the usual way

$$a(t) = \mathcal{V}_T^\dagger(t,t_0)a(t_0)\mathcal{V}_T(t,t_0), \tag{11.4.1}$$

$$\boldsymbol{\rho}_H = \mathcal{V}_T^\dagger(t,t_0)\boldsymbol{\rho}(t)\mathcal{V}_T(t,t_0) \equiv \boldsymbol{\rho}(t_0). \tag{11.4.2}$$

This enables us to derive equations of motion for the Heisenberg operators using the usual second order expansion of differential

$$\begin{aligned} da(t) &= d\mathcal{V}_T^\dagger(t,t_0)\,a(t_0)\mathcal{V}_T(t,t_0) + \mathcal{V}_T^\dagger(t,t_0)a(t_0)\,d\mathcal{V}_T(t,t_0) \\ &\qquad + d\mathcal{V}_T^\dagger(t,t_0)\,a(t_0)\,d\mathcal{V}_T(t,t_0). \end{aligned} \tag{11.4.3}$$

This can be used to generate the appropriate quantum stochastic differential equation by taking into account

i) The expression (9.6.25) for $d\mathcal{V}_T(t,t_0)$, which gives the quantum stochastic differential equation for $\mathcal{V}_T(t,t_0)$.

ii) The fact that the increments $dB(t)$, $dB^\dagger(t)$ commute with the evolution operator $\mathcal{V}_T(t,t_0)$,

iii) The Ito rules (9.6.32).

The equation of motion arising from this is the *Heisenberg operator quantum stochastic differential equation*

$$\begin{aligned} da(t) = {}& \frac{\mathrm{i}}{\hbar}\left(H_{\text{eff}}(t)a(t) - a(t)H_{\text{eff}}^\dagger\right)dt \\ &+\Gamma\left(\bar{N}(T,\omega_0)+1\right)c^\dagger(t)a(t)c(t)\,dt + \Gamma\bar{N}(T,\omega_0)c(t)a(t)c^\dagger(t)\,dt. \\ &+\sqrt{\Gamma}[c^\dagger(t),a(t)]\,dB(t) - \sqrt{\Gamma}[c(t),a(t)]\,dB^\dagger(t). \end{aligned} \tag{11.4.4}$$

11.4.2 Alternative Form

Expanding the definition (9.6.26) of the effective Hamiltonian, the quantum stochastic differential equation can also be written in the form

$$
\begin{aligned}
da(t) = & -\frac{\mathrm{i}}{\hbar}\left[a(t), H_{\mathrm{sys}}(t)\right] dt \\
& -\frac{\mathrm{i}}{\hbar}\left[a(t)\,,\,\delta\omega^{+} c^{\dagger}(t)c(t) - \delta\omega^{-} c(t)c^{\dagger}(t)\right] dt \\
& -\sqrt{\Gamma}\left[a(t)\,, c^{\dagger}(t)\right] dB(t) + \sqrt{\Gamma}\left[a(t)\,, c(t)\right] dB^{\dagger}(t) \\
& +\tfrac{1}{2}\Gamma\left(\bar{N}(T,\omega_0)+1\right)\left(2c^{\dagger}(t)a(t)c(t) - a(t)c^{\dagger}(t)c(t) - c^{\dagger}(t)c(t)a(t)\right) dt \\
& +\tfrac{1}{2}\Gamma\bar{N}(T,\omega_0)\left(2c(t)a(t)c^{\dagger}(t) - a(t)c(t)c^{\dagger}(t) - c(t)c^{\dagger}(t)a(t)\right) dt.
\end{aligned}
\tag{11.4.5}
$$

11.4.3 Correspondence with the Quantum Langevin Equation

Making the substitution $dB(t) \to b_{\mathrm{in}}(t)\,dt$ into this second form of the quantum stochastic differential equation above gives a form in which the first three lines match exactly with the quantum Langevin equation (11.1.13), and in which the last two lines match at $T = 0$. The extra terms which come at finite temperature are thermal noise terms.

a) Stratonovich and Ito Forms: The basic difference between the two forms arises because the quantum stochastic differential equation (11.4.4) is an Ito equation, while the quantum Langevin equation (11.1.13) is a Stratonovich equation. It can easily be checked that if a and b both obey the quantum Langevin equation, then the correct equation of notion for ab comes from ordinary *non-commuting* calculus:

$$d(ab) = a(db) + (da)b. \tag{11.4.6}$$

On the other hand, (11.4.4) is an Ito quantum stochastic differential equation, for which the Ito rules are

$$d(ab) = a(db) + (da)b + (da)(db), \tag{11.4.7}$$

and then one uses the the Ito rules (9.6.32) to evaluate such products explicitly.

b) Formal Relationship between Input Fields and Stochastic Increments: The two mappings between the physical electromagnetic field and the noise operators are given by (11.1.12) for b_{in} and by (9.3.1) for the quantum noise increment $\Delta B(t)$. These mean that we can write either of

$$\Delta B(t) = \int_t^{t+\Delta t} b_{\mathrm{in}}(t')\,dt', \tag{11.4.8}$$

$$dB(t) = b_{\mathrm{in}}(t)\,dt. \tag{11.4.9}$$

Exercise 11.1 Damped Harmonic Oscillator: Set

$$H_{\rm sys} \to \hbar\Omega a^\dagger a, \qquad c \to a, \qquad [a, a^\dagger] = 1. \tag{11.4.10}$$

Show that the quantum Langevin equation becomes

$$\dot a = -\left(\mathrm{i}\Omega + \tfrac{1}{2}\Gamma\right) a - \sqrt{\Gamma}\, b_{\rm in}(t). \tag{11.4.11}$$

Notice that this represents a *damped* motion for $\langle a\rangle$ independent of the statistics of $b_{\rm in}(t)$, because of the linearity of the equation in the variables a and $b_{\rm in}(t)$.

Exercise 11.2 Two-Level System: Set

$$H_{\rm sys} \to \tfrac{1}{2}\hbar\Omega\sigma_z, \qquad c \to \sigma^-. \tag{11.4.12}$$

Show that the quantum Langevin equations become

$$\left.\begin{aligned}
\dot\sigma^- &= -\left(\mathrm{i}\Omega + \tfrac{1}{2}\Gamma\right)\sigma^- + \sqrt{\Gamma}\sigma_z b_{\rm in}(t),\\
\dot\sigma^+ &= \left(\mathrm{i}\Omega - \tfrac{1}{2}\Gamma\right)\sigma^+ + \sqrt{\Gamma}\sigma_z b^\dagger_{\rm in}(t),\\
\dot\sigma_z &= -\Gamma(1+\sigma_z) - 2\sqrt{\Gamma}\left\{\sigma^+ b_{\rm in}(t) + b^\dagger_{\rm in}(t)\sigma^-\right\}.
\end{aligned}\right\} \tag{11.4.13}$$

These equations are non-linear in $\boldsymbol{\sigma}$ and $b_{\rm in}(t)$—hence we cannot express, for example, $\langle\sigma^+ b_{\rm in}(t)\rangle$ as a function of the $\langle\boldsymbol{\sigma}\rangle$. However, by making white noise assumptions and formulating the equivalent quantum stochastic differential equation as in as in (11.4.5), the averages $\langle\sigma_z\rangle$ and $\langle\sigma^\pm\rangle$ can be computed. Show how to do this. Notice that we cannot conclude that the damping is independent of the statistics of the bath, as in the case of the harmonic oscillator.

Exercise 11.3 Input and Output for a Damped Cavity Mode: We take the Hamiltonian and equation of motion for the damped cavity mode—illustrated in Fig. 11.1—from Ex. 11.1. The solution of the Langevin equation (11.4.11) is

$$a(t) = a(0)\mathrm{e}^{-(\mathrm{i}\Omega+\frac{1}{2}\Gamma)t} - \sqrt{\Gamma}\int_0^t ds\, \mathrm{e}^{-(\mathrm{i}\Omega+\frac{1}{2}\Gamma)(t-s)}\, b_{\rm in}(s). \tag{11.4.14}$$

i) Show directly from this solution that

$$\left.\begin{aligned}
\left[a^\dagger(t), b_{\rm in}(t')\right] &= 0, & t < t',\\
\left[a^\dagger(t), b_{\rm in}(t)\right] &= -\tfrac{1}{2}\sqrt{\Gamma}, & \\
\left[a^\dagger(t), b_{\rm in}(t')\right] &= -\sqrt{\Gamma}\mathrm{e}^{-(\mathrm{i}\Omega+\frac{1}{2}\Gamma)(t-t')}, & t > t'.
\end{aligned}\right\} \tag{11.4.15}$$

ii) Using the solution for $a(t)$, show that the commutator is given by

$$\left[a(t), a^\dagger(s)\right] = \mathrm{e}^{-\frac{1}{2}\Gamma|t-s|}. \tag{11.4.16}$$

iii) Using the time-reversed Langevin Equation

$$\dot a(t) = \left(-\mathrm{i}\Omega + \tfrac{1}{2}\Gamma\right)a(t) - \sqrt{\Gamma}\, b_{\rm out}(t) \tag{11.4.17}$$

find the solution in terms of a *final condition* at a future time T

$$a(t) = a(T)\mathrm{e}^{-(\mathrm{i}\Omega-\frac{1}{2}\Gamma)(t-T)} + \sqrt{\Gamma}\int_t^T ds\, \mathrm{e}^{-(\mathrm{i}\Omega-\frac{1}{2}\Gamma)(t-s)}\, b_{\rm out}(s). \tag{11.4.18}$$

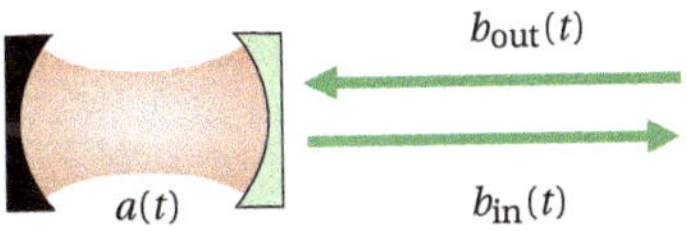

Fig. 11.1. Input and output from a damped single-sided cavity.

iv) We define a Fourier transform of any function by

$$\tilde{f}(\omega) \equiv \frac{1}{2\pi}\int_{-\infty}^{\infty} e^{i\omega t} f(t)\, dt, \tag{11.4.19}$$

and use the Langevin equation to show that

$$\tilde{a}(\omega) = \frac{\sqrt{\Gamma}}{i(\omega-\Omega)-\frac{1}{2}\Gamma}\tilde{b}_{in}(\omega). \tag{11.4.20}$$

v) Relate the "out" and "in" fields by (11.2.6), giving

$$b_{out}(t) = b_{in}(t) + \sqrt{\Gamma}a(t), \tag{11.4.21}$$

and thus in terms of the Fourier transforms

$$\tilde{b}_{out}(\omega) = \tilde{b}_{in}(\omega) + \sqrt{\Gamma}\tilde{a}(\omega), \tag{11.4.22}$$

$$= \tilde{b}_{in}(\omega) + \frac{\Gamma}{i(\omega-\Omega)-\frac{1}{2}\Gamma}\tilde{b}_{in}(\omega) \tag{11.4.23}$$

$$= \frac{i(\omega-\Omega)+\frac{1}{2}\Gamma}{i(\omega-\Omega)-\frac{1}{2}\Gamma}\tilde{b}_{in}(\omega) \equiv e^{i\delta(\omega)}\tilde{b}_{in}(\omega), \tag{11.4.24}$$

with $\delta(\omega)$ a phase shift, thus preserving the commutation relations

$$\left[\tilde{b}_{out}(\omega), \tilde{b}^{\dagger}_{out}(\omega')\right] = \left[\tilde{b}_{in}(\omega), \tilde{b}^{\dagger}_{in}(\omega')\right] = \delta(\omega-\omega'). \tag{11.4.25}$$

11.4.4 Phase Shift on Reflection From a Cavity

The phase shift $\delta(\omega)$ in (11.4.24) has the properties

$$e^{i\delta(\omega)} \longrightarrow \begin{cases} -1, & \omega = \Omega, & \text{on resonance,} \\ 1, & |\omega-\Omega| \gg \Gamma, & \text{far detuned.} \end{cases} \tag{11.4.26}$$

This simple property has quite far-reaching implications. The phase reversal on resonance requires the buildup of a significant intra-cavity optical field, and this corresponds to the solution (11.4.20) for the internal mode operator, which exhibits a resonance denominator.

The absence of a phase reversal off resonance corresponds to simple reflection of the surface of the cavity, and happens whenever the cavity is detuned, and no internal field build up happens. More generally, anything which prevents the build up of an internal field will yield the same result; for example, if the cavity contains an atom which can interact significantly with the field, this will detune the system, and the phase change will not occur.

Thus, the observation of the phase change is confirmation of the absence of such an atom. From this, one can deduce the presence or absence of an atom in the cavity.

11.5 Several Inputs and Outputs

In Sect. 9.4.6 we considered the quantum stochastic differential equation for a system coupled to several independent light fields. It is quite straightforward to generalize the formalism of this chapter in the same way. We choose the Hamiltonians for a system is coupled to several heat baths of the form that

$$H_B = \sum_{l,k} \hbar\omega_{k,l} b^\dagger_{k,l} b_{k,l}, \tag{11.5.1}$$

$$H_{\rm int} = i\hbar \sum_{k,l} \kappa_{k,l}[b^\dagger_{l,k} c_l - c^\dagger_l b_{l,k}], \tag{11.5.2}$$

and in this case the quantum Langevin equation becomes

$$\begin{aligned}\dot a = -\frac{i}{\hbar}[a, H_{\rm sys}] - \sum_l [a, c^\dagger_l]\left\{\tfrac{1}{2}\Gamma_l c_l + \sqrt{\Gamma_l}\, b_{l,\rm in}(t)\right\} \\ + \sum_l \left\{\tfrac{1}{2}\Gamma_l c^\dagger_l + \sqrt{\Gamma_l}\, b^\dagger_{l,\rm in}(t)\right\}[a, c_l]. \end{aligned}\tag{11.5.3}$$

The relations between the inputs and outputs are of course then

$$b_{l,\rm out}(t) = b_{l,\rm in}(t) + \sqrt{\Gamma_l}\, c_l(t). \tag{11.5.4}$$

Exercise 11.4 Damped Cavity Mode with Two Input Ports: Systems with two input ports can be realized in various ways, and two possibilities are illustrated in Fig. 11.2.

For simplicity we introduce the notation $\gamma \equiv \gamma_1 + \gamma_2$.

i) Show that the correct Langevin equations are

$$\begin{aligned}\dot a(t) &= -(i\Omega + \tfrac{1}{2}\gamma)a(t) - \sum_i \sqrt{\gamma_i}\, b_{i,\rm in}(t), &(11.5.5)\\ &= -(i\Omega - \tfrac{1}{2}\gamma)a(t) - \sum_i \sqrt{\gamma_i}\, b_{i,\rm out}(t). &(11.5.6)\end{aligned}$$

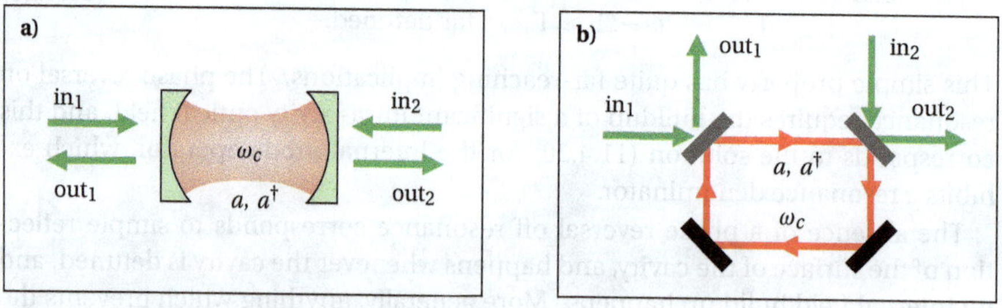

Fig. 11.2. Two implementations of a resonator with inputs in two ports. **a)** A Fabry–Pérot cavity, with partially transmitting mirrors at both ends; **b)** A ring cavity in which a single clockwise circulating mode is coupled via the two partially transmitting mirrors.

ii) Show the Fourier transform solution can be written

$$b_{i,\mathrm{out}}(\omega) = \sum_j (M(\omega))_{ij}\, b_{j,\mathrm{in}}(\omega), \tag{11.5.7}$$

in which the matrix elements are

$$M_{ij}(\omega) = \delta_{ij} + \frac{\sqrt{\gamma_i \gamma_j}}{\mathrm{i}(\omega - \Omega) - \frac{1}{2}(\gamma_1 + \gamma_2)}. \tag{11.5.8}$$

iii) Show that this is a unitary matrix.

iv) Show that the matrix can be written in full as

$$M(\omega) = \frac{\begin{pmatrix} \mathrm{i}(\omega - \Omega) - \frac{1}{2}(\gamma_2 - \gamma_1) & \sqrt{\gamma_1\gamma_2} \\ \sqrt{\gamma_1\gamma_2} & \mathrm{i}(\omega - \Omega) - \frac{1}{2}(\gamma_1 - \gamma_2) \end{pmatrix}}{\mathrm{i}(\omega - \Omega) - \frac{1}{2}(\gamma_1 + \gamma_2)}. \tag{11.5.9}$$

v) The output fields are

$$\tilde{b}_{1,\mathrm{out}}(\omega) = \frac{\left(\frac{1}{2}(\gamma_1 - \gamma_2) + \mathrm{i}(\omega - \Omega)\right)\tilde{b}_{1,\mathrm{in}}(\omega) + \sqrt{\gamma_1\gamma_2}\,\tilde{b}_{2,\mathrm{in}}(\omega)}{\mathrm{i}(\omega - \Omega) - \frac{1}{2}(\gamma_1 + \gamma_2)}, \tag{11.5.10}$$

$$\tilde{b}_{2,\mathrm{out}}(\omega) = \frac{\left(\frac{1}{2}(\gamma_2 - \gamma_1) + \mathrm{i}(\omega - \Omega)\right)\tilde{b}_{2,\mathrm{in}}(\omega) + \sqrt{\gamma_1\gamma_2}\,\tilde{b}_{1,\mathrm{in}}(\omega)}{\mathrm{i}(\omega - \Omega) - \frac{1}{2}(\gamma_1 + \gamma_2)}. \tag{11.5.11}$$

vi) *A Far-Detuned Cavity*: In the case that $|\omega - \Omega| \gg \gamma$, we find

$$\tilde{b}_{1,\mathrm{out}}(\omega) = \tilde{b}_{1,\mathrm{in}}(\omega), \qquad \tilde{b}_{2,\mathrm{out}}(\omega) = \tilde{b}_{2,\mathrm{out}}(\omega). \tag{11.5.12}$$

Both inputs are reflected with no phase change.

vii) *Behaviour on Resonance*: On resonance $\omega = \Omega$, and

$$\tilde{b}_{1,\mathrm{out}}(\omega) = \frac{(\gamma_2 - \gamma_1)\tilde{b}_{1,\mathrm{in}}(\omega) - 2\sqrt{\gamma_1\gamma_2}\,\tilde{b}_{2,\mathrm{in}}(\omega)}{\gamma_1 + \gamma_2}, \tag{11.5.13}$$

$$\tilde{b}_{2,\mathrm{out}}(\omega) = \frac{(\gamma_1 - \gamma_2)\tilde{b}_{2,\mathrm{in}}(\omega) - 2\sqrt{\gamma_1\gamma_2}\,\tilde{b}_{1,\mathrm{in}}(\omega)}{\gamma_1 + \gamma_2}. \tag{11.5.14}$$

There are then two special cases:

a) *Single-Sided Cavity*: We set $\gamma_1 = 0$, then

$$\tilde{b}_{1,\mathrm{out}}(\omega) = \tilde{b}_{1,\mathrm{in}}(\omega), \qquad \tilde{b}_{2,\mathrm{out}}(\omega) = -\tilde{b}_{2,\mathrm{out}}(\omega). \tag{11.5.15}$$

The mode which does not couple to the cavity mode is reflected without any phase change; the mode that does couple to the cavity is reflected and its phase is reversed.

b) *Equal Coupling—Impedance Matching*: We set $\gamma_1 = \gamma_2$; then

$$\tilde{b}_{1,\mathrm{out}}(\omega) = -\tilde{b}_{2,\mathrm{in}}(\omega), \qquad \tilde{b}_{2,\mathrm{out}}(\omega) = -\tilde{b}_{1,\mathrm{in}}(\omega). \tag{11.5.16}$$

The input from each mode is transferred to the output of the other mode, and the phase is reversed in both cases. The equality of the damping constants is equivalent to impedance matching.

viii) Check that because M is unitary, the commutation relations are preserved;

$$\left[\tilde{b}_{i,\mathrm{out}}(\omega), \tilde{b}^{\dagger}_{j,\mathrm{out}}(\omega')\right] = \left[\tilde{b}_{i,\mathrm{in}}(\omega), \tilde{b}^{\dagger}_{j,\mathrm{in}}(\omega')\right] = \delta_{ij}\delta(\omega-\omega'). \tag{11.5.17}$$

ix) *Transformation of the Spectrum Matrix*: We define input and output spectra by

$$\langle \tilde{b}^{\dagger}_{j,\mathrm{in}}(\omega)\tilde{b}_{i,\mathrm{in}}(\omega')\rangle \equiv S_{ij,\mathrm{in}}(\omega)\delta(\omega-\omega'), \tag{11.5.18}$$

$$\langle \tilde{b}^{\dagger}_{j,\mathrm{out}}(\omega)\tilde{b}_{i,\mathrm{out}}(\omega')\rangle \equiv S_{ij,\mathrm{out}}(\omega)\delta(\omega-\omega'). \tag{11.5.19}$$

Using the transfer function as in (11.5.7), the spectrum matrix transforms as

$$S_{\mathrm{out}}(\omega) = M^{\dagger}(\omega)S_{\mathrm{in}}(\omega)M(\omega). \tag{11.5.20}$$

11.6 Fermionic Input-Output Theory

The formulation of input-output theory in this chapter assumes the system is driven by a Bosonic field, and this field is normally assumed to be an optical field. The generalization to other Bose fields is not very difficult, but the generalization to Fermionic fields is non-trivial. The main problem that arises is quite simple—the equations of motion for system operators are different, depending on whether a system operator is viewed as *commuting* or *anticommuting* with the Fermionic heat bath operators.

This technical problem is overcome in [11.4], where it is shown how we may define "restricted" system operators so as to commute with all bath operators, and that these internal system operators all obey a quantum Langevin equation of the same form.

The equations of motion in the original operators are, for the two level atom, linear and thus exactly soluble. Thus, we have a description of a two level system interacting with a Fermi bath which is essentially the same as that of a harmonic oscillator interacting with a Bosonic heat bath. The two level atom behaves like a Fermion coming to equilibrium with all the other Fermions.

It is possible to develop Fermionic quantum stochastic integration, and the corresponding Ito and Stratonovich formulations of quantum stochastic differential equations, and finally, from these, to derive the master equation in the expected form.

12. Cascaded Quantum Systems

The generalization of the input-output formalism of Chap. 11 to the case where the output of one system is fed into the input of another system is highly relevant to the idea of quantum communication. We can envisage a quantum processor which produces an output in the form of a pulse of light in an appropriate quantum state, and another quantum processor which absorbs this pulse and the quantum information encoded in it—indeed this has already been achieved experimentally [12.1, 12.2]. Systems linked in this way are known as *cascaded quantum systems*, a name which emphasizes that the output of the first system is fed into that of the second system, but that the reverse process does not happen.

The requisite quantum stochastic formalism was developed independently by *Kolobov* and *Sokolov* [12.3], *Carmichael* [12.4] and *Gardiner* [12.5], as a way of understanding how non-classical light sources could be used to drive quantum optical systems. The two main issues considered were:

i) How do we write a physically acceptable formalism which allows coupling from system 1 to system 2 *without* allowing coupling in the reverse direction. This is achievable in the laboratory with a unidirectional coupler—indeed the isolation of the laser used to drive an optical system, from reflections off the driven system is a common experimental issue.

ii) How does one specify a light field with arbitrary statistics? In principle this involves the specification of all possible correlation functions, but this is obviously impractical.

    The answer provided by the cascaded quantum systems approach is physically very natural—the best way of specifying the statistics of a light field is to model the system which produces it. Using this specification, the cascaded quantum systems approach shows how to couple the resulting output light into the system under study, and in [12.6] it was demonstrated that the cascaded systems approach provides a very practical way of simulating such atomic systems.

12.1 Coupling Equations

Consider a quantum system which can be decomposed into two subsystems. Using the notation of Chap. 11, a driving field $b_{\text{in}}(1, t)$ drives the first system, and gives rise to an output $b_{out}(1,t)$ which, after a propagation delay τ, becomes the input field $b_{in}(2,t)$ to the second system, as illustrated in Fig. 12.1. Now let a_1, a_2 be operators for the two systems, and let $H_{sys} \equiv H_{sys}(1) + H_{sys}(2)$ be the sum of their otherwise independent system Hamiltonians.

The quantum Langevin equations, as in (11.1.13), in this case take the form

$$\begin{aligned}\dot{a}_1 &= -\frac{i}{\hbar}[a_1, H_{sys}] \\ &\quad -[a_1, c_1^\dagger]\left(\tfrac{1}{2}\Gamma_1 c_1 + \sqrt{\Gamma_1}\, b_{in}(1,t)\right) + \left(\tfrac{1}{2}\Gamma_1 c_1^\dagger + \sqrt{\Gamma_1}\, b_{in}^\dagger(1,t)\right)[a_1, c_1], \qquad (12.1.1)\\ \dot{a}_2 &= -\frac{i}{\hbar}[a_2, H_{sys}] \\ &\quad -[a_2, c_2^\dagger]\left(\tfrac{1}{2}\Gamma_2 c_2 + \sqrt{\Gamma_2}\, b_{in}(2,t)\right) + \left(\tfrac{1}{2}\Gamma_2 c_2^\dagger + \sqrt{\Gamma_2}\, b_{in}^\dagger(2,t)\right)[a_2, c_2]. \qquad (12.1.2)\end{aligned}$$

The output field from the first system, as in (11.2.6) is given by

$$b_{out}(1,t) = b_{in}(1,t) + \sqrt{\Gamma_1}\, c_1(t). \qquad (12.1.3)$$

If it takes a time τ for light to travel from system 1 to system 2, then we can get the effect of feeding the output of system 1 into the input of system 2 by writing

$$b_{in}(2,t) = b_{out}(1,t-\tau) = b_{in}(1,t-\tau) + \sqrt{\Gamma_1}\, c_1(t-\tau). \qquad (12.1.4)$$

If a is an operator from either of the systems, the resulting two quantum Langevin equations can be written as one equation for a (using the abbreviated notation $b_{in}(1,t) \to b_{in}(t)$):

$$\begin{aligned}\dot{a} &= -\frac{i}{\hbar}[a, H_{sys}] \\ &\quad -[a, c_1^\dagger]\left(\tfrac{1}{2}\Gamma_1 c_1 + \sqrt{\Gamma_1}\, b_{in}(t)\right) + \left(\tfrac{1}{2}\Gamma_1 c_1^\dagger + \sqrt{\Gamma_1}\, b_{in}^\dagger(t)\right)[a, c_1] \\ &\quad -[a, c_2^\dagger]\left(\tfrac{1}{2}\Gamma_2 c_2 + \sqrt{\Gamma_1\Gamma_2}\, c_1(t-\tau) + \sqrt{\Gamma_2}\, b_{in}(t-\tau)\right) \\ &\quad +\left(\tfrac{1}{2}\Gamma_2 c_2^\dagger + \sqrt{\Gamma_1\Gamma_2}\, c_1^\dagger(t-\tau) + \sqrt{\Gamma_2}\, b_{in}^\dagger(t-\tau)\right)[a, c_2]. \qquad (12.1.5)\end{aligned}$$

The major technical difficulty is the fact that operators at the two times t and $t-\tau$ both turn up. Because we are considering only the case in which the driving is one way, it is clear that τ must be positive, but otherwise may be chosen arbitrarily—its only effect is to shift the origin of the time axis for the second atom. If we let $\tau \to 0+$, the quantum Langevin equation for an operator a (which may relate to either system or to both of them) takes the form

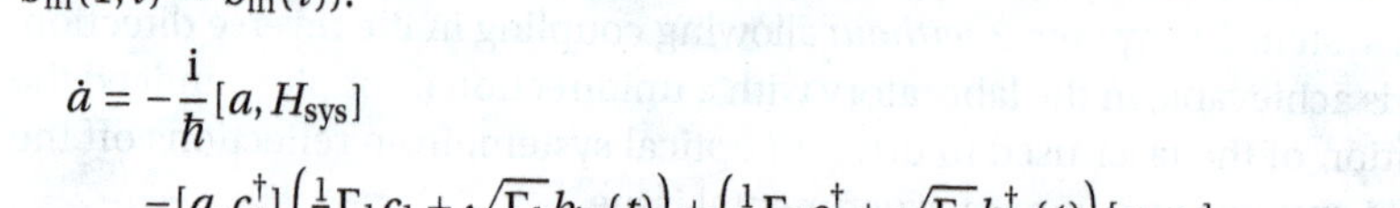

Fig. 12.1. Schematic diagram of system 2 being driven by the output from system 1.

$$
\begin{aligned}
\dot{a} = &-\frac{\mathrm{i}}{\hbar}[a, H_{\mathrm{sys}}] \\
&-[a, c_1^\dagger]\left(\tfrac{1}{2}\Gamma_1 c_1 + \sqrt{\Gamma_1}\, b_{\mathrm{in}}(t)\right) + \left(\tfrac{1}{2}\Gamma_1 c_1^\dagger + \sqrt{\Gamma_1}\, b_{\mathrm{in}}^\dagger(t)\right)[a, c_1] \\
&-[a, c_2^\dagger]\left(\tfrac{1}{2}\Gamma_2 c_2 + \sqrt{\Gamma_2}\, b_{\mathrm{in}}(t)\right) + \left(\tfrac{1}{2}\Gamma_2 c_2^\dagger + \sqrt{\Gamma_2}\, b_{\mathrm{in}}^\dagger(t)\right)[a, c_2] \\
&-[a, c_2^\dagger]\sqrt{\Gamma_1\Gamma_2}\, c_1 + \sqrt{\Gamma_1\Gamma_2}\, c_1^\dagger [a, c_2].
\end{aligned}
\tag{12.1.6}
$$

In *Quantum Noise* we show, by defining time-advanced operators for the second system by $\hat{a}_2(t) \equiv a_2(t+\tau)$, that this equation is valid for any τ, with the interpretation that all operators in the second system are time-advanced operators.

12.1.1 Interpretation of the Quantum Langevin Equation

The quantum Langevin equation (12.1.6) for the combined cascaded system is of a very special form, which we shall now examine.

a) Inputs and Outputs: The output of the combined system is found by using (12.1.3) in the limit $\tau \to 0+$,

$$
\begin{aligned}
b_{\mathrm{out}}(t) &= b_{\mathrm{out}}(2, t) \\
&= b_{\mathrm{in}}(2, t) + \sqrt{\Gamma_2}\, c_2(t) \\
&= b_{\mathrm{in}}(t) + \sqrt{\Gamma_1}\, c_1(t) + \sqrt{\Gamma_2}\, c_2(t).
\end{aligned}
\tag{12.1.7}
$$

This form implies that the effective coupling of the combined cascaded system to the input field is by the operator $\sqrt{\Gamma_1} c_1 + \sqrt{\Gamma_2} c_2$.

b) Standard Form of the Cascaded System Quantum Langevin Equation: We can rewrite the quantum Langevin equation (12.1.6) as a quantum Langevin equation of the standard form (11.1.13)

$$
\begin{aligned}
\dot{a} = &-\frac{\mathrm{i}}{\hbar}\left[a, H_{\mathrm{sys}} + \tfrac{1}{2}\mathrm{i}\hbar\sqrt{\Gamma_1\Gamma_2}\left(c_1^\dagger c_2 - c_2^\dagger c_1\right)\right] \\
&-\left[a, \sqrt{\Gamma_1}\, c_1^\dagger + \sqrt{\Gamma_2}\, c_2^\dagger\right]\left(\tfrac{1}{2}\left(\sqrt{\Gamma_1}\, c_1 + \sqrt{\Gamma_2}\, c_2\right) + b_{\mathrm{in}}(t)\right) \\
&+\left(\tfrac{1}{2}\left(\sqrt{\Gamma_1}\, c_1^\dagger + \sqrt{\Gamma_2}\, c_2^\dagger\right) + b_{\mathrm{in}}^\dagger(t)\right)\left[a, \sqrt{\Gamma_1}\, c_1 + \sqrt{\Gamma_2}\, c_2\right].
\end{aligned}
\tag{12.1.8}
$$

By rewriting the equation in this form, it can be seen that the coupling can be viewed as arising from:

i) A *coherent* radiation term from the combined operator $\sqrt{\Gamma_1} c_1 + \sqrt{\Gamma_2} c_2$, given by the second two lines.

ii) An additional Hamiltonian term $\frac{1}{2}\mathrm{i}\hbar\sqrt{\Gamma_1\Gamma_2}\left(c_1^\dagger c_2 - c_2^\dagger c_1\right)$ which exchanges the excitation between the two atoms.

Under the exchange $1 \leftrightarrow 2$ the radiation term is symmetric, while the Hamiltonian term is antisymmetric—the asymmetry of the process arises from the interference between these.

c) Interpretation as a Single System: The combined cascaded system therefore behaves like a single system, with a single coupling to the radiation field given by $\sqrt{\Gamma_1}c_1 + \sqrt{\Gamma_2}c_2$, and this is the way in which all of the quantum stochastic formalism must be implemented.

12.2 Quantum Stochastic Formalism

Although the quantum Langevin equation gives an elegant description of the physics involved, it is necessary to convert this to the appropriate quantum stochastic equations to use it in practice. For simplicity we will consider here only the case in which the first input field is the vacuum, which will be sufficiently general for the purposes of this book, since coherent driving fields can always be absorbed into the definition of the Hamiltonian H_{sys}, as shown in Sect. 9.5. The equations for the case of a finite temperature quantum white noise input are treated in *Quantum Noise.*

Thus, we will write

$$b_{\mathrm{in}}(t)\,dt = dB(t), \tag{12.2.1}$$

with

$$dB(t)dB^\dagger(t) = dt, \qquad dB(t)^2 = dB^\dagger(t)^2 = dB(t)^\dagger dB(t) = 0. \tag{12.2.2}$$

12.2.1 Quantum Stochastic Differential Equation

Using the methods we used in Sect. 11.4 to derive (11.4.4), the appropriate form of the quantum stochastic differential equation follows directly from the form (12.1.8) of the quantum Langevin equation. To make the structure more easily visible, let us define the operator

$$C \equiv \sqrt{\Gamma_1}c_1 + \sqrt{\Gamma_2}c_2. \tag{12.2.3}$$

The quantum stochastic differential equation is

$$\begin{aligned} da = &-\frac{\mathrm{i}}{\hbar}\left[a,\, H_{\mathrm{sys}} + \tfrac{1}{2}\mathrm{i}\hbar\sqrt{\Gamma_1\Gamma_2}\left(c_1^\dagger c_2 - c_2^\dagger c_1\right)\right] dt \\ &+\tfrac{1}{2}\left(2C^\dagger aC - C^\dagger Ca - aC^\dagger C\right) dt \\ &-\left[a, C^\dagger\right] dB(t) + \left[a, C\right] dB^\dagger(t). \end{aligned} \tag{12.2.4}$$

12.2.2 Master Equation

The corresponding master equation is best written in terms of c_1 and c_1, and takes the form

$$\frac{d\rho}{dt} = \frac{\mathrm{i}}{\hbar}[\rho, H_{\rm sys}] + \tfrac{1}{2}\Gamma_1\left(2c_1\rho c_1^\dagger - \rho c_1^\dagger c_1 - c_1^\dagger c_1\rho\right) + \tfrac{1}{2}\Gamma_2\left(2c_2\rho c_2^\dagger - \rho c_2^\dagger c_2 - c_2^\dagger c_2\rho\right)$$
$$-\sqrt{\Gamma_1\Gamma_2}\left([c_2^\dagger, c_1\rho] + [\rho c_1^\dagger, c_2]\right). \tag{12.2.5}$$

12.2.3 Quantum Stochastic Schrödinger Equation

To write the quantum stochastic Schrödinger equation for cascaded systems we use the compact notation introduced in Sect. 9.4.2. In the case of the quantum stochastic Schrödinger equation (12.2.3, 12.2.4), the effective Hamiltonian, analogous to that defined in (9.4.6), can be written in either of the forms

$$H_{\rm eff} \equiv H_{\rm sys} + \tfrac{1}{2}\mathrm{i}\hbar\sqrt{\Gamma_1\Gamma_2}(c_1^\dagger c_2 - c_2^\dagger c_1) - \tfrac{1}{2}\mathrm{i}\hbar C^\dagger C, \tag{12.2.6}$$
$$= H_{\rm sys} + \tfrac{1}{2}\mathrm{i}\hbar\left(\Gamma_1 c_1^\dagger c_1 + \Gamma_2 c_2^\dagger c_2 + 2\sqrt{\Gamma_1\Gamma_2}c_2^\dagger c_1\right). \tag{12.2.7}$$

We then define the operators K and J by

$$\mathsf{K} \equiv -\frac{\mathrm{i}}{\hbar}H_{\rm eff}, \tag{12.2.8}$$
$$\mathsf{J} \equiv C^\dagger = \sqrt{\Gamma_1}c_1^\dagger + \sqrt{\Gamma_2}c_2^\dagger. \tag{12.2.9}$$

Corresponding to (12.2.4), the quantum stochastic differential equation is then written as

$$d|\Psi, t\rangle = \left(\mathsf{K}\,dt + \mathsf{J}\,dB^\dagger(t)\right)|\Psi, t\rangle. \tag{12.2.10}$$

12.3 Applications

12.3.1 Driving a Quantum System with Light of Arbitrary Statistics

Gardiner and *Parkins* [12.6] (see also *Quantum Noise* Chap. 11 and [12.7] Chap. 19) applied the cascaded systems approach to model the driving of several kinds of quantum systems by thermal light, squeezed light, coherent light and antibunched light. It is important to note that all of these generate *non-Markovian* equations of motion for the driven atom, but that the equations of motion for the full cascaded system are Markovian.

Thus the cascaded systems approach provides a complete method for handling non-Markovian inputs, provided one has an appropriate theoretical model for the production of the non-Markovian input. In practice, this is almost always the case. It is important to note that, while it is not difficult to conceive of non-Markovian inputs, the specification of any particular non-Markovian input does present a major problem. Only in the case of a quantum Gaussian input, is it sufficient to specify a small number of correlation functions. In the case of a non-Gaussian driving field, for example antibunched light, there is no similar simple specification.

The cascaded systems approach shows that the best way to model such an input is *by modelling the system which produces the input*. By modelling this system, we can produce the best possible specification of the light being produced as a driving field. In other word, the parameters needed to describe the non-Markovian driving field are those of the system producing light. They are definitely *not* provided by any collection of moments and correlation functions. Even in the case of Gaussian light, [12.6] shows that the the cascaded systems approach provides the most practical method of modelling the driven system.

12.3.2 Cascaded Quantum Networks

There is no limit to the number of systems which can be linked by one-way cascaded coupling, and recent work on such *cascaded networks* has provided some very interesting results. *Stannigel, Rabl, Ramos, Pichler, Daley* and *Zoller* [12.8–12.10] have investigated the dissipative dynamics and the formation of entangled states in driven cascaded quantum networks, where multiple systems are coupled to a common unidirectional bath. Under certain conditions emission and coherent reabsorption of radiation drives the whole network into a pure stationary state with non-trivial quantum correlations between the individual nodes.

For cascaded two-level systems an explicit preparation scheme allows one to tune the whole network through "bright" and "dark" states associated with different multipartite entanglement patterns. In a complementary setting consisting of cascaded nonlinear cavities, two cavity modes can be driven into a non-Gaussian entangled dark state.

They also show how to make an atomic implementation of a system composed of the coupled spins and the coupling field with a two-species mixture of cold quantum gases. The coupling field is represented by a spin-orbit coupled one-dimensional quasicondensate of atoms in a magnetized phase, while the spins are identified with motional states of a separate species of atoms in an optical lattice. Under certain conditions a pure steady state arises, in which pairs of neighboring spins form dimers that decouple from the remainder of the chain.

13. Dissipative Dynamics of Driven Atoms

This chapter presents the effects of dissipation, such as spontaneous emission, on the equations governing the elementary techniques for coherent manipulation of two-level and three-level systems presented in Chap. 5 and Chap. 6. The aim here is to present only the basic formalism and results—applications occur extensively throughout the remainder of the book. The results on the two-level system complement those presented in *Bk. I: Ch.13* and *Ch.14*.

13.1 Formulation of the Optical Bloch Equations for a Two-Level System

The optical Bloch equations are the equations of motion for matrix elements of the reduced density operator of an atom, driven by a classical field, and interacting with the vacuum modes of the electromagnetic field.

13.1.1 System Hamiltonian

The system of interest is the driven two-level atom system itself, whose Hilbert space consists of the vectors spanned by $\{|g\rangle, |e\rangle\}$. We will use the two equivalent notations for the complete set of operators in this Hilbert space

$$\sigma^+ = |e\rangle\langle g|, \tag{13.1.1}$$

$$\sigma^- = |g\rangle\langle e|, \tag{13.1.2}$$

$$\sigma_z = |e\rangle\langle e| - |g\rangle\langle g|, \tag{13.1.3}$$

$$1 = |e\rangle\langle e| + |g\rangle\langle g|. \tag{13.1.4}$$

The system Hamiltonian can be written

$$H_{\text{sys}} = \hbar\omega_{eg}|e\rangle\langle e| - \sigma^+ \boldsymbol{d}_{eg}\cdot\boldsymbol{E}_{\text{cl}}^{(+)}(0,t) - \sigma^- \boldsymbol{d}_{eg}^*\cdot\boldsymbol{E}_{\text{cl}}^{(-)}(0,t), \tag{13.1.5}$$

where the classical driving field is provided by a laser with frequency ω:

$$\boldsymbol{E}_{\text{cl}}^{(+)}(\boldsymbol{x}=0,t) = \boldsymbol{\epsilon}_{\text{cl}}\mathcal{E}(t)\mathrm{e}^{-\mathrm{i}\omega t}. \tag{13.1.6}$$

Here $\boldsymbol{\epsilon}_{\text{cl}}$ is the polarization vector of the driving field, which can in practice be any of plane, circularly, or elliptically polarized.

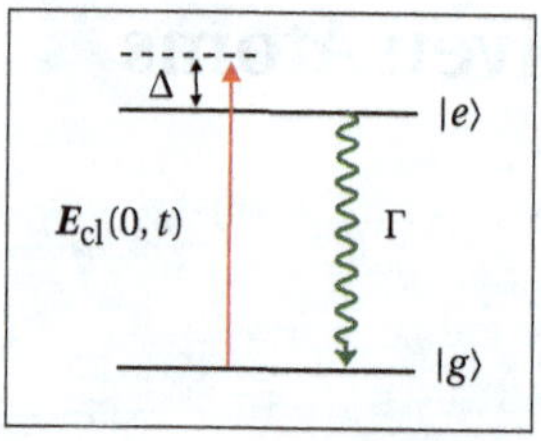

Fig. 13.1. The level structure, driving field and decay of the two-level atom system Hamiltonian (13.1.5).

13.1.2 The Master Equation

The master equation itself has the very familiar form which we have already found in *Bk. I: Sect.14.2*—the level structure is illustrated in Fig. 13.1. In terms of our formulation in Chap. 10, it is obtained from (10.1.7) or (10.1.8) by setting

$$\omega_0 \to \omega_{eg}, \qquad c \to \sigma^-, \qquad c^\dagger \to \sigma^+, \qquad \bar{N}(\omega_0, T) \to 0. \tag{13.1.7}$$

The master equation can then be conveniently written in two ways:

i) *System Hamiltonian and Lindblad Damping*

$$\frac{d}{dt}\rho_A(t) = -\frac{i}{\hbar}\left[H_{\text{sys}} + \hbar\delta\omega\,|e\rangle\langle e|,\, \rho_A(t)\right] + \tfrac{1}{2}\Gamma\left(2\sigma^-\rho_A(t)\sigma^+ - \sigma^+\sigma^-\rho_A(t) - \rho_A(t)\sigma^+\sigma^-\right), \tag{13.1.8}$$

$$H_{\text{sys}} = \hbar\omega_{eg}|e\rangle\langle e| - \boldsymbol{d}_{eg}\cdot\boldsymbol{E}_{\text{cl}}^{(+)}(0,t)^*\sigma^- - \boldsymbol{d}_{eg}\cdot\boldsymbol{E}_{\text{cl}}^{(+)}(0,t)\sigma^+, \tag{13.1.9}$$

$$\boldsymbol{E}_{\text{cl}}^{(+)}(0,t) = \boldsymbol{\epsilon}_{\text{cl}}\mathcal{E}(t). \tag{13.1.10}$$

Here we have used the same notation for the polarization and amplitude of the classical driving field as was introduced in (4.1.1).

ii) *Effective Hamiltonian plus Recycling Term*

$$\frac{d}{dt}\rho_A(t) = -\frac{i}{\hbar}\left(H_{\text{eff}}\rho_A(t) - \rho_A(t)H_{\text{eff}}^\dagger\right) + \mathcal{J}\rho_A(t), \tag{13.1.11}$$

Effective Hamiltonian: $$H_{\text{eff}} \equiv H_{\text{sys}} + \hbar\left(\delta\omega - \tfrac{1}{2}i\Gamma|e\rangle\langle e|\right), \tag{13.1.12}$$

Recycling Term: $$\mathcal{J}\rho_A(t) \equiv \Gamma\sigma^-\rho_A(t)\sigma^+ \equiv \Gamma\rho_{ee}|g\rangle\langle g|. \tag{13.1.13}$$

The recycling term is a projection operator onto the ground state, multiplied by the spontaneous emission rate.

This first form (13.1.8) displays the optical Bloch equations as a master equation of the Lindblad form, such as is described in *Bk. I: Sect.13.1.3*. As noted there, the Lindblad form is the most general equation for a reduced density operator of a system which is local in time, and which is compatible with the properties of a density operator

a) Transforming away the Optical Frequencies: We can make a transformation to a frame rotating at the *laser frequency* ω by defining a density operator $\bar{\rho}_A$ in this frame through

$$\bar{\rho}_A = \mathrm{e}^{\mathrm{i}\omega|e\rangle\langle e|t}\rho_A \mathrm{e}^{-\mathrm{i}\omega|e\rangle\langle e|t}, \tag{13.1.14}$$

which is equivalent to

$$\left.\begin{aligned} \rho_{A,ee} &= \bar{\rho}_{A,ee}, & \rho_{A,ge} &= \bar{\rho}_{A,ge}\mathrm{e}^{\mathrm{i}\omega t}, \\ \rho_{A,gg} &= \bar{\rho}_{A,gg}, & \rho_{A,eg} &= \bar{\rho}_{A,eg}\mathrm{e}^{-\mathrm{i}\omega t}. \end{aligned}\right\} \tag{13.1.15}$$

This transformation to the frame rotating at the laser frequency has two effects:

i) It eliminates the time dependence at frequency ω from the laser field, leaving only the relatively slow modulation, $\mathcal{E}(t)$, as the field is turned on and off.

ii) The frequencies ω and ω_{eg} normally differ only slightly, and the transformation then replaces terms involving ω_{eg} with terms involving $-\Delta$, where $\Delta \equiv \omega - \omega_{eg} - \delta\omega$, a relatively slow frequency.

b) The Optical Bloch Equations: The resulting master equation, whose individual components are normally known as *the optical Bloch equations,* is a master equation of the same form as in (13.1.8–13.1.13), but with no frequencies in the optical range:

$$\frac{d\bar{\rho}_A}{dt} = -\frac{\mathrm{i}}{\hbar}\bar{H}_{\text{eff}}\bar{\rho}_A + \frac{\mathrm{i}}{\hbar}\bar{\rho}_A\bar{H}^{\dagger}_{\text{eff}} + \mathcal{J}\bar{\rho}_A \equiv L_{\text{Bloch}}\,\bar{\rho}_A, \tag{13.1.16}$$

with the effective Hamiltonian

$$H_{\text{eff}} \to \bar{H}_{\text{eff}} = \hbar\left(-\Delta - \tfrac{1}{2}\mathrm{i}\Gamma\right)|e\rangle\langle e| - \tfrac{1}{2}\hbar\Omega_R(t)^*\sigma^- - \tfrac{1}{2}\hbar\Omega_R(t)\sigma^+, \tag{13.1.17}$$

where

$$\Delta \equiv \omega - \omega_{eg} - \delta\omega\,, \tag{13.1.18}$$

$$\Omega_R(t) \equiv \frac{2\boldsymbol{d}_{eg}\cdot\boldsymbol{\epsilon}_{\text{cl}}\,\mathcal{E}(t)}{\hbar}\,. \tag{13.1.19}$$

13.2 Explicit Form and Solutions of the Optical Bloch Equations

It is often convenient to express the optical Bloch equations as equations for the density matrix elements, rather the in terms of the spin averages, as was done in *Bk. I: Sect.14.2.* These then take a form which is more easily generalizable to atoms with more than two levels:

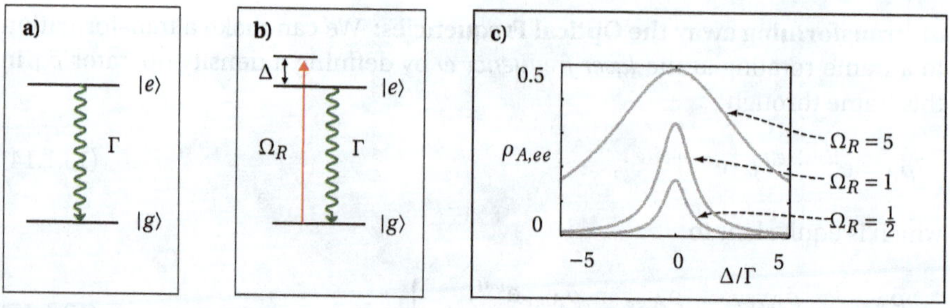

Fig. 13.2. **a)** Atom decaying to the ground state with decay rate Γ and without a driving field; **b)** Atom interacting with an electromagnetic field with Rabi frequency Ω_R, detuned by Δ; **c)** Excited state population for various Rabi frequencies, showing power broadening for large values of Ω_R.

$$\frac{d}{dt}\begin{pmatrix}\bar\rho_{A,eg}\\ \bar\rho_{A,ge}\\ \bar\rho_{A,ee}\\ \bar\rho_{A,gg}\end{pmatrix} = \begin{pmatrix} i\Delta-\frac{1}{2}\Gamma & 0 & -\frac{1}{2}i\Omega_R & \frac{1}{2}i\Omega_R \\ 0 & -i\Delta-\frac{1}{2}\Gamma & \frac{1}{2}i\Omega_R^* & -\frac{1}{2}i\Omega_R^* \\ -\frac{1}{2}i\Omega_R^* & \frac{1}{2}i\Omega_R & -\Gamma & 0 \\ \frac{1}{2}i\Omega_R^* & -\frac{1}{2}i\Omega_R & \Gamma & 0 \end{pmatrix}\begin{pmatrix}\bar\rho_{A,eg}\\ \bar\rho_{A,ge}\\ \bar\rho_{A,ee}\\ \bar\rho_{A,gg}\end{pmatrix}. \tag{13.2.1}$$

These equations can be solved, and there are a number of aspects of these solutions which are useful in applications.

13.2.1 The Zero Driving Field Case

The populations decay with the spontaneous emission rate

$$\rho_{A,ee}(t) = \rho_{A,ee}(0)e^{-\Gamma t}, \tag{13.2.2}$$

$$\rho_{A,gg}(t) = 1-\rho_{A,ee}(t) = 1-\rho_{A,ee}(0)e^{-\Gamma t}, \tag{13.2.3}$$

and coherences damp out

$$\rho_{A,eg}(t) = \rho_{A,eg}(0)e^{-(i\omega_{eg}-\frac{1}{2}\Gamma)t}, \tag{13.2.4}$$

$$\rho^*_{A,eg}(t) = \rho_{A,eg}(t). \tag{13.2.5}$$

13.2.2 Stationary Solution in the Presence of a Coherent Driving Field

When the Rabi frequency Ω_R is time-independent, the solution of the optical Bloch equations given by (13.2.1) for $\bar\rho_A$ approaches a stationary state corresponding to a dynamical equilibrium between spontaneous emission and driving.

a) Populations: The diagonal matrix elements are the populations of the two states, and are given by

$$\rho_{A,ee} = \frac{\frac{1}{4}|\Omega_R|^2}{\Delta^2 + \frac{1}{4}\Gamma^2 + \frac{1}{2}|\Omega_R|^2} = 1 - \rho_{A,gg}. \tag{13.2.6}$$

This shows:

i) *Saturation*: The excited population $\rho_{A,ee} \to \frac{1}{2}$ as the Rabi frequency Ω_R becomes very large.

ii) *Resonance Behaviour and Power Broadening*: Expressed as a function of the detuning Δ, $\rho_{A,ee}$ has a resonance form, with a peak at $\Delta = 0$. The resonance width is $\sqrt{\frac{1}{4}\Gamma^2 + \frac{1}{2}|\Omega_R|^2}$, which increases with Ω_R^2, which is proportional to the power of the driving field. This is known as *power broadening* of the resonance.

b) Coherences: The off-diagonal matrix elements are known as the *coherences*, and are given by

$$\rho_{A,eg}(t) = \bar{\rho}_{A,eg}\,\mathrm{e}^{-\mathrm{i}\omega t}, \tag{13.2.7}$$

$$\bar{\rho}_{A,eg} = -\tfrac{1}{2}\Omega_R \frac{\Delta + \frac{1}{2}\mathrm{i}\Gamma}{\Delta^2 + \frac{1}{4}\Gamma^2 + \frac{1}{2}|\Omega_R|^2}. \tag{13.2.8}$$

This shows the saturation behaviour corresponding to that for the populations. The real part of $\bar{\rho}_{A,eg}$ is known as the *dispersive part*, and the imaginary part as the *absorptive part.*

13.2.3 Electric Polarization and Susceptibility

As formulated in Sect. 4.1.1, the coherences are directly related to the mean of the electric dipole moment by

$$\langle \boldsymbol{d}(t)\rangle = \mathrm{Tr}_A\{\boldsymbol{d}\rho_A(t)\} = \boldsymbol{d}_{eg}\rho_{eg}(t) + \boldsymbol{d}_{ge}\rho_{ge}(t). \tag{13.2.9}$$

The *electric polarization* of a uniform gas of atoms with a density of n_A atoms per unit volume is given by the mean density of dipole moments, namely

$$\boldsymbol{P} \equiv n_A\langle \boldsymbol{d}(t)\rangle. \tag{13.2.10}$$

13.2.4 Time-Dependent Solutions

The time-dependent solutions of the optical Bloch equations are best treated using the method described in *Bk. I: Sect.14.2.* This is formulated in terms of the spin averages $\boldsymbol{S}(t) \equiv \mathrm{Tr}_A\{\boldsymbol{\sigma}\rho_A(t)\}$, and their stationary value

$$\bar{\boldsymbol{S}} = -\left(\Delta^2 + \tfrac{1}{4}\Gamma^2 + \tfrac{1}{2}\Omega_R^2\right)^{-1}\begin{pmatrix}\Delta\Omega_R \\ -\frac{1}{2}\Gamma\Omega_R \\ \frac{1}{4}\Gamma^2 + \Delta^2\end{pmatrix}. \tag{13.2.11}$$

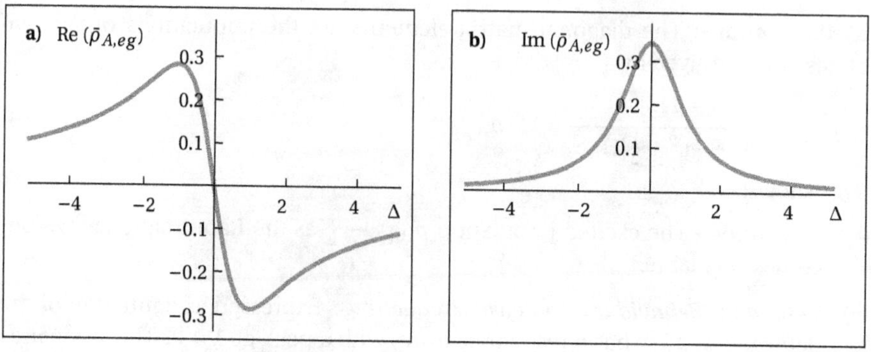

Fig. 13.3. Dispersive **a)** and absorptive **b)** parts of the coherence $\bar{\rho}_{A,eg}$ as a function of detuning.

In the situation under consideration here, the equations of motion (13.2.1) can be written symbolically as

$$\dot{\boldsymbol{S}}(t) = A(\boldsymbol{S}(t) - \bar{\boldsymbol{S}}), \tag{13.2.12}$$

with

$$A = \begin{pmatrix} -\frac{1}{2}\Gamma & \Delta & 0 \\ -\Delta & -\frac{1}{2}\Gamma & \Omega_R \\ 0 & -\Omega_R & -\Gamma \end{pmatrix}, \tag{13.2.13}$$

and solutions given by

$$\boldsymbol{S}(t) = \mathrm{e}^{At}\left(\boldsymbol{S}(0) - \bar{\boldsymbol{S}}\right) + \bar{\boldsymbol{S}}. \tag{13.2.14}$$

Apart from minor notational changes, the only difference from the treatment in *Book I* is the presence of the detuning Δ, which, however, prevents any simple analytic solutions.

To construct the solution for the density matrix we first calculate $\boldsymbol{S}(0)$ from the initial value $\rho_A(0)$, solve the equation of motion (13.2.13), and then use the formula $\rho_A(t) = \frac{1}{2}(1 + \boldsymbol{S}(t)\cdot\boldsymbol{\sigma})$.

13.3 Dissipative Dynamics of the Λ-System

It is important to show that the concept of a dark state is quite general and robust. Our treatment in Chap. 6 did not include the influence of dissipation—that is, in this case, spontaneous emission. Our first task is, therefore, to show that this can be accounted for correctly, and that the dark state is still a valid concept, and that the idea—as introduced in Sect. 6.3.1—of STIRAP [13.1–13.3] is valid and useful when spontaneous emission is included.

This section will, as an additional task, establish the notation and terminology we will need for the description of the dark-state quantum memory.

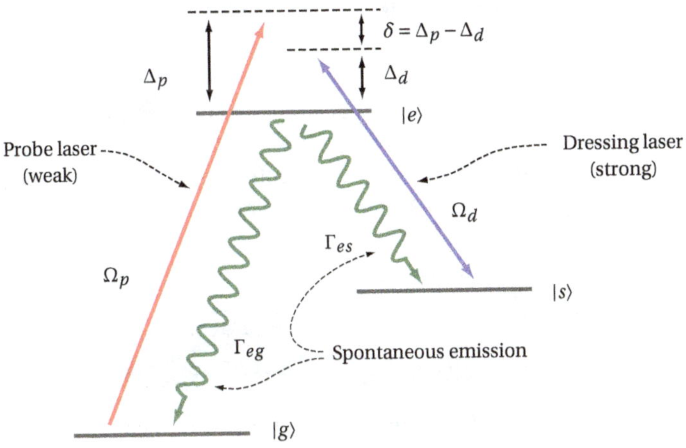

Fig. 13.4. The Λ-configuration being considered, and notation for the states, transitions, driving fields and detunings.

13.3.1 Parameters for the Three-Level System

We will therefore generalize the ideas of Chap. 6 to include the correct master equation treatment of the interaction between transitions $e \longleftrightarrow g$ and $e \longleftrightarrow s$. The energy level structure, transitions, frequencies and detunings are as illustrated in Fig. 13.4. The notation we shall use for this chapter is:

i) There are two stable ground states $|g\rangle$ and $|s\rangle$.

ii) There is one excited state $|e\rangle$, which decays to the ground states by spontaneous emission with decay constants Γ_{eg} and Γ_{es}.

iii) We use the notation

$$\sigma_{ab} \equiv |a\rangle\langle b|, \tag{13.3.1}$$

in defining the Hamiltonians relevant to this system. This notation is analogous to that used in (13.1.1–13.1.4)—however, in the case of three or more levels, there is no clear choice for an analogue of σ_z, so instead we use σ_{ee}, σ_{gg} and σ_{ss}, which of course are related to each other by $\sigma_{ee} + \sigma_{gg} + \sigma_{ss} = 1$.

iv) The ground state $|g\rangle$ is coupled to the excited state $|e\rangle$ by a *weak* probe laser with frequency ω_p, and with Rabi frequency Ω_p.

v) The ground state $|s\rangle$ is coupled to the excited state $|e\rangle$ by a *strong* dressing laser with frequency ω_d, and with Rabi frequency Ω_d.

vi) There are two detunings

$$\Delta_p \equiv \omega_p - \omega_{eg}, \tag{13.3.2}$$

$$\Delta_d \equiv \omega_d - \omega_{es}, \tag{13.3.3}$$

and the Raman detuning is

$$\delta \equiv \Delta_p - \Delta_d = \omega_p - \omega_d - \omega_{sg}. \tag{13.3.4}$$

vii) We can add a dissipative dephasing term connecting the two ground states, with decay constant Γ_{gs} to account collisional dephasing. However, such dephasing will tend to destroy the electromagnetically induced transparency, and will in practice be minimized in experiments.

13.3.2 System Hamiltonian

The system Hamiltonian is essentially the same as that given in (6.1.1)

$$H = \sum_{i=g,s,e} \hbar\omega_i\sigma_{ii} + \tfrac{1}{2}\hbar\left(\Omega_p\sigma_{eg}e^{-i\omega_p t} + \Omega_d\sigma_{es}e^{-i\omega_d t} + \text{h.c.}\right), \tag{13.3.5}$$

where we write $\sigma_{ij} = |i\rangle\langle j|$, and the Rabi frequencies have been chosen to be real. When written in a rotating frame corresponding the choice of coefficients (6.1.7), it takes the form

$$H = -\hbar\Delta_p\sigma_{ee} + \hbar\left(\Delta_p - \Delta_d\right)\sigma_{ss} + \tfrac{1}{2}\hbar\left(\Omega_p\sigma_{eg} + \Omega_d\sigma_{es} + \text{h.c.}\right). \tag{13.3.6}$$

We have studied coherent dynamics of this a Λ-System in Chap. 6, and in particular, in Sect. 6.2.1, we considered the dark states, which occur when the Raman detuning $\delta = 0$.

a) Hamiltonian Matrix and Rabi Frequency: In a matrix form, in which the rows and columns are labelled in the sequence (g, e, s), the Hamiltonian (13.3.6) can be written

$$H = \hbar\begin{pmatrix} 0 & \frac{1}{2}\Omega_p & 0 \\ \frac{1}{2}\Omega_p^* & -\Delta_p & \frac{1}{2}\Omega_d \\ 0 & \frac{1}{2}\Omega_d^* & -\delta \end{pmatrix}. \tag{13.3.7}$$

This corresponds to the equations of motion (6.1.8), with the Rabi frequencies $\Omega_a \to \Omega_p$ and $\Omega_b \to \Omega_d$. The relationship between the electric fields and the Rabi frequencies follows from (5.1.6, 5.1.9), and takes the form

$$\Omega_p = -\frac{2\boldsymbol{d}_{eg}\cdot\boldsymbol{\epsilon}_p\mathcal{E}_p}{\hbar}, \tag{13.3.8}$$

and similarly for the dressing field.

b) Eigenstates: For exact Raman resonance, that is $\Delta_p = \Delta_d \equiv \Delta$, and hence $\delta = 0$, we have shown in Sect. 6.2.1 that the eigenvalues of this Hamiltonian are

$$E_0 = 0, \tag{13.3.9}$$

$$E_\pm = \tfrac{1}{2}\hbar\left(-\Delta \pm \sqrt{\Delta^2 + |\Omega_p|^2 + |\Omega_d^2|}\right). \tag{13.3.10}$$

The corresponding eigenstates can be conveniently written using angles θ and ϕ defined by

$$\tan\theta = \frac{|\Omega_p|}{|\Omega_d|}, \qquad \tan 2\phi = \frac{\sqrt{|\Omega_p|^2 + |\Omega_d|^2}}{\Delta}. \tag{13.3.11}$$

Let us also define the phases of the Rabi frequencies by

$$\Omega_p = |\Omega_p| e^{i\varphi_p}, \qquad \Omega_d = |\Omega_d| e^{i\varphi_d}. \tag{13.3.12}$$

Using this parametrization, the results of (6.2.11) are equivalent to

$$|E_+\rangle = \sin\phi\left(\sin\theta e^{i\varphi_p}|g\rangle + \cos\theta e^{-i\varphi_d}|s\rangle\right) + \cos\phi|e\rangle, \tag{13.3.13}$$

$$|E_-\rangle = \cos\phi\left(\sin\theta e^{i\varphi_p}|g\rangle + \cos\theta e^{-i\varphi_d}|s\rangle\right) - \sin\phi|e\rangle, \tag{13.3.14}$$

$$|E_0\rangle = \cos\theta e^{-i\varphi_d}|g\rangle - \sin\theta e^{i\varphi_p}|s\rangle \equiv \frac{\Omega_d^*}{\sqrt{|\Omega_p|^2 + |\Omega_d|^2}}|g\rangle - \frac{\Omega_p}{\sqrt{|\Omega_p|^2 + |\Omega_d|^2}}|s\rangle. \tag{13.3.15}$$

c) The Dark State: The state $|E_0\rangle$ does not evolve with time, and has no admixture of the excited state, and thus cannot interact with the laser field; it is the *dark state*, and we normally write

$$|D\rangle \equiv |E_0\rangle. \tag{13.3.16}$$

d) The Bright State: This is a convenient terminology for linear combination of $|s\rangle$ and $|g\rangle$ orthogonal to the dark state, namely, the bright state is defined as

$$|B\rangle = \frac{\Omega_p}{\sqrt{|\Omega_p|^2 + |\Omega_d|^2}}|g\rangle + \frac{\Omega_d^*}{\sqrt{|\Omega_p|^2 + |\Omega_d|^2}}|s\rangle. \tag{13.3.17}$$

13.3.3 Master Equation

We will write the master equation in the form involving the non-Hermitian effective Hamiltonian, and in an appropriate rotating frame, as

$$\begin{aligned} H_{\text{eff}} = &-\hbar\left(\Delta_p + \tfrac{1}{2}i\Gamma_e\right)\sigma_{ee} + \hbar\left(-\Delta_p + \Delta_d\right)\sigma_{ss} \\ &+ \tfrac{1}{2}\hbar\left(\Omega_p\sigma_{eg} + \Omega_p^*\sigma_{ge}\right) + \tfrac{1}{2}\hbar\left(\Omega_d\sigma_{es} + \Omega_d^*\sigma_{se}\right). \end{aligned} \tag{13.3.18}$$

Here $\Gamma_e = \Gamma_{eg} + \Gamma_{es}$ is the total decay rate from the upper state, and δ is the Raman detuning, as defined in (13.3.4). The master equation then becomes

$$\dot{\rho} = -\frac{i}{\hbar}\left(H_{\text{eff}}\rho - \rho H_{\text{eff}}^\dagger\right) + \Gamma_{eg}\sigma_{ge}\rho\sigma_{eg} + \Gamma_{es}\sigma_{se}\rho\sigma_{es}. \tag{13.3.19}$$

a) The Dark State as the Stationary Solution of the Master Equation: Because the dark state has no component of the excited state $|e\rangle$, it has the properties

$$\sigma_{ge}|E_0\rangle = \sigma_{se}|E_0\rangle = \sigma_{ee}|E_0\rangle = 0. \tag{13.3.20}$$

From these properties, it follows that the pure state $\rho_{ss} \equiv |E_0\rangle\langle E_0|$ is the stationary state of the master equation, since all terms involving dissipation vanish when acting on the dark state. This is quite natural, since the only dissipation corresponds to radiative decay from the excited state $|e\rangle$, itself absent from the dark state.

b) Interfering Pathways: The dark state happens when the transition *rates* $s \leftrightarrow e$ and $g \leftrightarrow e$ are equal, but the transition *amplitudes* have opposite phase. Thus, the total transition rate is zero, as a result of the cancellation of the amplitudes for the two pathways.

c) Optical Pumping: Thus, any initial density matrix will be irreversibly approach a pure state, namely the dark state

$$\rho(t) \xrightarrow{t\to\infty} \rho_{ss} \equiv |E_0\rangle\langle E_0|. \tag{13.3.21}$$

Although mathematically this is a damping process, physically it is best seen as a process of optical pumping of the atom by the laser fields into this dark state.

d) Adiabatic Population Transfer along the Dark State—STIRAP: In Sect. 6.3.1 we showed how, by adiabatically varying the relative sizes of the two Rabi frequencies, it is possible change the relative contributions of the states $|g\rangle$ and $|s\rangle$ to the dark state without ever involving the excited state $|e\rangle$. The advantage achieved is that no spontaneous emission will be involved, since this can only proceed from the state $|e\rangle$. The arguments given in that section did not include any dissipation, and therefore did not properly include the effects of spontaneous emission.

However, now that we have included the dissipative processes, we see that the same dark state $|E_0\rangle$ occurs, and the adiabatic transfer is possible in exactly the same way; that is STIRAP works even in the presence of spontaneous emission, giving the transfer

$$|g\rangle \xrightarrow{\Omega_p \ll \Omega_d} |E_0\rangle \xrightarrow{\Omega_p \gg \Omega_d} -|s\rangle. \tag{13.3.22}$$

e) Comparison between Optical Pumping and STIRAP: Let us consider atoms initially prepared in the ground state $|g\rangle$, with the dressing laser on and the probe laser off. We note that initially $|g\rangle \equiv |E_0\rangle$.

i) Suppose we turn on the probe laser *instantaneously*—this populates the excited state $|e\rangle$, and leads to an optical pumping into the dark state $|E_0\rangle$ at a rate involving the *dissipative* time scale Γ_e.

ii) Alternatively, let us turn on the probe laser *adiabatically* on a time scale $1/\Omega_d$. This pulls the state $|g\rangle$ adiabatically into the dark state $|g\rangle \to |E_0\rangle$. This can be significantly faster than the time scale of optical pumping, since $1/\Omega_d \ll 1/\Gamma_e$ is easily achievable in practice.

14. Multi-Atom Optical Bloch Equations

When an ensemble of atoms interacts with a light field, each atom will produce radiation which then interacts with the other atoms, a problem that was first treated by *Lehmberg* [14.1, 14.2] in 1970. The quantum stochastic formalism we developed in Chap. 9 provides a very logical and explicit framework for the description of this situation, with appropriate modifications, and this chapter will be devoted to development of the appropriately modified formalism.

To achieve this, we will build directly on the foundations we set up in Chap. 9. Thus, we will consider an ensemble of N two-level atoms coupled to the quantized radiation field and driven by laser light, as illustrated in Fig. 14.1. Using the quantum stochastic formalism, we will derive the appropriate quantum Langevin and optical Bloch equations. In particular we will discuss the following effects:

i) The interaction between atoms arising from induced dipole moments, known as the dipole-dipole interaction.

ii) Spontaneous emission, and collective effects such as superradiance and subradiance.

The distances $r_{i,j}$ between the atoms vary, but have a typical value which we shall call $\bar{r}$, and the behaviour depends on the size of the parameter $\zeta \equiv k_{eg}\bar{r} \equiv 2\pi\bar{r}/\lambda$. If $\zeta \gg 1$, the atoms behave independently, and the formalism of Chap. 9 gives a satisfactory picture. When $\zeta \approx 1$ or less the atoms affect each other, in the way which we will now explain.

14.1 Coupling to Several Non-Independent Light Fields

When we have several identical atoms at positions $\boldsymbol{x}_i$, each atom interacts with the electromagnetic field at its position $\boldsymbol{x}_i$. For sufficiently large separations $|\boldsymbol{x}_i - \boldsymbol{x}_j|$,

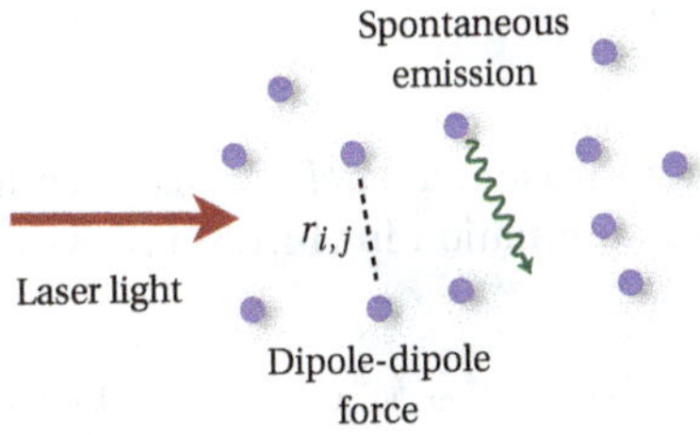

Fig. 14.1. An ensemble of atoms irradiated with laser light experience spontaneous emission, and a force arising from induced dipole moments.

the field operators $\boldsymbol{E}(\boldsymbol{x}_i, t)$ and $\boldsymbol{E}(\boldsymbol{x}_j, t)$ would be expected to act independently, but at shorter distances this would become less valid. The procedure we formulated in Sect. 9.2.3 is quite easily generalized to deal with this situation as follows.

14.1.1 Hamiltonian and Noise Operators

The electromagnetic field is described for all space by the operators b_k, and the system will be composed of several atoms, each with a transition operator σ_i^- and the *same* transition frequency ω_{eg}. The interaction Hamiltonian will take a form analogous to that of (9.1.4, 9.2.8);

$$H^{(\vartheta)}_{\text{int},I}(t) = \sum_{\substack{i,k \\ |\omega_k-\omega_{eg}|<\vartheta}} \mathrm{i}\hbar\left(\kappa_k^* b_k^\dagger e^{\mathrm{i}(\omega_k-\omega_{eg})t-\mathrm{i}\boldsymbol{k}\cdot\boldsymbol{x}_i}\sigma_i^- - \kappa_k b_k e^{-\mathrm{i}(\omega_k-\omega_{eg})t+\mathrm{i}\boldsymbol{k}\cdot\boldsymbol{x}_i}\sigma_i^+\right), \tag{14.1.1}$$

$$\equiv \mathrm{i}\hbar\sum_i\left(f_i^\dagger(t)\sigma_i^- - \sigma_i^+ f_i(t)\right), \tag{14.1.2}$$

in which the noise operators are defined analogously to (9.2.9)

$$f_i(t) = \sum_{\substack{i,k \\ |\omega_k-\omega_{eg}|<\vartheta}} \kappa_k b_k e^{-\mathrm{i}(\omega_k-\omega_{eg})t+\mathrm{i}\boldsymbol{k}\cdot\boldsymbol{x}_i}. \tag{14.1.3}$$

These operators have the commutation relation

$$\left[f_i(t), f_j^\dagger(t')\right] = \gamma_{i,j}(t-t'), \tag{14.1.4}$$

with

$$\gamma_{i,j}(\tau) = \sum_{\substack{i,k \\ |\omega_k-\omega_{eg}|<\vartheta}} |\kappa_k|^2 e^{-\mathrm{i}(\omega_k-\omega_{eg})\tau+\mathrm{i}\boldsymbol{k}\cdot(\boldsymbol{x}_i-\boldsymbol{x}_j)}. \tag{14.1.5}$$

14.1.2 Decay Constants and Energy Shifts

The reasoning of Sect. 9.2.3 is easy to generalize, and leads to the evaluation of the quantities

$$\int_0^{\Delta t} dt_1 \int_0^{t_1} dt_2\, \gamma_{i,j}(t_1-t_2) \equiv \left(\tfrac{1}{2}\Gamma_{i,j} + \mathrm{i}\delta\omega_{i,j}\right)\Delta t. \tag{14.1.6}$$

We perform the detailed evaluation of $\Gamma_{i,j}$ and $\delta\omega_{i,j}$ below.

14.1.3 Evaluation of Damping Constants

The evaluation proceeds in several steps.

i) We first take the form (14.1.5), and evaluate κ_k using the definitions given in *Bk. I: (11.1.43)* and *Bk. I: (12.1.28)*. Thus the summation in (14.1.5) involves a sum over polarizations, so that

$$\sum_\lambda |\kappa_k|^2 = \frac{\omega_{\boldsymbol{k}}}{2\hbar\epsilon_0 V}\sum_\lambda |\boldsymbol{d}_{eg}.\hat{\boldsymbol{e}}^\lambda|^2 = \frac{\omega_{\boldsymbol{k}}}{2\hbar\epsilon_0 V}\left(|\boldsymbol{d}_{eg}|^2 - |\boldsymbol{d}_{eg}.\hat{\boldsymbol{k}}|^2\right), \tag{14.1.7}$$

where $\hat{\boldsymbol{k}}$ is a unit vector in the direction of $\boldsymbol{k}$.

ii) We now make the conversion from a sum to an integral by writing

$$\frac{1}{V}\sum_{\boldsymbol{k}} \longrightarrow \frac{1}{(2\pi)^3}\int d^3\boldsymbol{k} \longrightarrow \frac{1}{(2\pi c)^3}\int \omega^2\, d\omega \int d\Omega_{\boldsymbol{k}}, \tag{14.1.8}$$

where $\omega = \omega_{\boldsymbol{k}} = c|\boldsymbol{k}|$.

iii) We now need the angular integral

$$\int d\Omega_{\boldsymbol{k}} \left(|\boldsymbol{d}_{eg}|^2 - |\boldsymbol{d}_{eg}\cdot\hat{\boldsymbol{k}}|^2\right) e^{iz\hat{\boldsymbol{k}}\cdot\hat{\boldsymbol{r}}_{ij}}, \tag{14.1.9}$$

in which we have used the compact notation

$$\boldsymbol{r}_{ij} \equiv \boldsymbol{x}_i - \boldsymbol{x}_j, \tag{14.1.10}$$

$$z \equiv |\boldsymbol{k}|\,|\boldsymbol{r}_{ij}| = \omega|\boldsymbol{r}_{ij}|/c, \tag{14.1.11}$$

and $\hat{\boldsymbol{r}}_{ij}$ is a unit vector in the direction of $\boldsymbol{r}_{ij}$.

iv) To evaluate the angular integral, we introduce $\hat{\boldsymbol{d}}_{eg}$ as a unit vector in the direction of $\boldsymbol{d}_{eg}$, and note that we can rewrite the expressions in (14.1.9) using the Legendre polynomials $P_l(x)$. Thus we write for the first term

$$|\boldsymbol{d}_{eg}|^2 - |\boldsymbol{d}_{eg}\cdot\hat{\boldsymbol{k}}|^2 = \tfrac{2}{3}|\boldsymbol{d}_{eg}|^2\left(P_0\big(\hat{\boldsymbol{k}}\cdot\hat{\boldsymbol{d}}_{eg}\big) - P_2\big(\hat{\boldsymbol{k}}\cdot\hat{\boldsymbol{d}}_{eg}\big)\right). \tag{14.1.12}$$

For the exponential term we use the Rayleigh expansion of a plane wave

$$e^{iz\hat{\boldsymbol{k}}\cdot\hat{\boldsymbol{r}}} = \sum_{l=0}^{\infty}(2l+1)i^l j_l(z)P_l\big(\hat{\boldsymbol{k}}\cdot\hat{\boldsymbol{r}}\big), \tag{14.1.13}$$

After these substitutions, we can evaluate the integral using the spherical harmonic addition theorem, which we write in the form

$$\int d\Omega_{\boldsymbol{k}}\, P_l\big(\hat{\boldsymbol{k}}\cdot\hat{\boldsymbol{r}}\big)P_{l'}\big(\hat{\boldsymbol{k}}\cdot\hat{\boldsymbol{d}}_{eg}\big) = \frac{4\pi}{2l+1}\delta_{l,l'}P_l\big(\hat{\boldsymbol{r}}\cdot\hat{\boldsymbol{d}}_{eg}\big). \tag{14.1.14}$$

v) Using all of these formulae, we can now say

$$\gamma_{i,j}(\tau) = \frac{|\boldsymbol{d}_{eg}|^2}{6\pi^2 c^3}\int_{\omega_{eg}-\vartheta}^{\omega_{eg}+\vartheta}\frac{\omega^3\, d\omega}{\hbar\epsilon_0}\Big(j_0(z) + P_2\big(\hat{\boldsymbol{r}}\cdot\hat{\boldsymbol{d}}_{eg}\big)j_2(z)\Big)e^{-i(\omega-\omega_{eg})\tau}. \tag{14.1.15}$$

vi) We now need to evaluate the double time integral in (14.1.6), and to do this we use the results of (9.2.22, 9.2.23), which amount to the substitution

$$\int_0^{\Delta t} dt_1 \int_0^{t_1} dt_2\, e^{-i(\omega-\omega_{eg})(t_1-t_2)} \longrightarrow \left(\pi\delta(\omega-\omega_{eg}) - \frac{iP}{\omega-\omega_{eg}}\right)\Delta t, \quad \text{for } \Delta t \gg 1/\vartheta. \tag{14.1.16}$$

vii) The delta function part of this equation gives the decay constants, and we get straightforwardly

$$\Gamma_{i,j} = \frac{|\boldsymbol{d}_{eg}|^2 k_{eg}^3}{3\hbar\pi\epsilon_0}\Big(j_0\big(k_{eg}r\big) + P_2\big(\hat{\boldsymbol{r}}\cdot\hat{\boldsymbol{d}}_{eg}\big)\, j_2\big(k_{eg}r\big)\Big), \tag{14.1.17}$$

$$\text{where}\quad k_{eg} \equiv \frac{2\pi}{\lambda_{eg}} \equiv \frac{\omega_{eg}}{c}. \tag{14.1.18}$$

Setting $r = 0$ gives the decay constant of a single atom

$$\Gamma_{i,i} = \Gamma \equiv \frac{|\boldsymbol{d}_{eg}|^2 k_{eg}^3}{3\hbar\pi\epsilon_0}. \tag{14.1.19}$$

14.1.4 Evaluation of Lineshifts

To evaluate the second, imaginary, part of (14.1.15) requires more care. The range of ω is very small compared to the value of ω_{eg}, and we can make the approximation $\omega \to \omega_{eg}$ everywhere except in the exponentials. The argument of the spherical Bessel functions is $\omega r/c$, which is the ratio of the distance between atoms to the optical wavelength. The change of this value over the range $\omega_{eg} \pm \vartheta$ needs to be assessed.

i) We therefore make the replacements, using the spherical Bessel function properties in Appendix 14.A:

$$j_0(\omega r/c) \longrightarrow \frac{\sin(\omega r/c)}{\omega_{eg} r/c}, \tag{14.1.20}$$

$$j_2(\omega r/c) \longrightarrow \frac{3\sin(\omega r/c)}{(\omega_{eg} r/c)^3} - \frac{\sin(\omega r/c)}{\omega_{eg} r/c} - \frac{3\cos(\omega r/c)}{(\omega_{eg} r/c)^2}. \tag{14.1.21}$$

The integrals that occur are then

$$\mathrm{P}\int_{\omega_{eg}-\vartheta}^{\omega_{eg}+\vartheta} \frac{d\omega\,\cos(\omega r/c)}{\omega-\omega_{eg}} = -2\mathrm{Si}(\vartheta r/c)\sin(k_{eg} r) \approx -\pi\sin(k_{eg} r), \tag{14.1.22}$$

$$\mathrm{P}\int_{\omega_{eg}-\vartheta}^{\omega_{eg}+\vartheta} \frac{d\omega\,\sin(\omega r/c)}{\omega-\omega_{eg}} = 2\mathrm{Si}(\vartheta r/c)\cos(k_{eg} r) \approx \pi\cos(k_{eg} r). \tag{14.1.23}$$

Here we have used the function, defined in [14.3],

$$\mathrm{Si}(x) = \int_0^x \frac{\sin z\,dz}{z} \approx \tfrac{1}{2}\pi \text{ when } x > 2. \tag{14.1.24}$$

Putting these results together, the effect of the formulae (14.1.22, 14.1.23) is to substitute spherical Neumann functions for spherical Bessel functions in the result for the damping constants; precisely, after taking account of the factor of π and the signs, we get

$$\delta\omega_{i,j} = \tfrac{1}{2}\Gamma\Big(n_0\left(k_{eg}r\right) + P_2\big(\hat{\boldsymbol{r}}\cdot\hat{\boldsymbol{d}}_{eg}\big)n_2\left(k_{eg}r\right)\Big), \qquad i \neq j. \tag{14.1.25}$$

Here we use the usual physicist's notation $n_l(z)$ for the spherical Neumann function as opposed to the standard mathematical notation $y_l(z)$ as used, for example, by [14.3].

ii) The approximation (14.1.24) means that we can use these results provided $\vartheta r/c > 2$. This can be interpreted to mean that there is a characteristic frequency $f_{\text{transit}} \equiv r/c$ associated with the transit of light from one atom to another, and that the bandwidth ϑ of the frequencies included in our theoretical description, must be sufficiently large to encompass all frequencies up to $2f_{\text{transit}}$. On the other hand, the value of ϑ must be small compared to ω_{eg} for the Markov approximation and the rotating-wave approximation to be valid, so this restriction will mean that in practice that the interatomic distances must satisfy $|\boldsymbol{x}_i - \boldsymbol{x}_j| \gg c/\vartheta$.

This is actually a purely technical issue. The dipole force between atoms too close to each other for the above calculation to be valid has to be calculated by a different means, and will effectively be included in the elimination of other levels which is necessary to produce the form of Hamiltonian and interaction we set up in Sect. 9.1.

iii) The case when $i = j$ is the self reaction of an atom on itself, and since $r = 0$, the result (14.1.25) above is not valid. The energy shift $\omega_{i,i}$ can be calculated by setting $r = 0$ in (14.1.15), and proceeding through the same calculation as with $r \neq 0$. It is finite, but depends strongly on ϑ, and in fact it is given by

$$\delta\omega_{i,i} = -\tfrac{1}{2}\Gamma\left(\frac{\vartheta}{\omega_{eg}} + \frac{4\vartheta^3}{\omega_{eg}^3}\right). \tag{14.1.26}$$

This is a contribution to the Lamb shift, and it is very much smaller than the typical values of $\delta\omega_{i,j}$. It can be absorbed in a redefinition of the transition frequency—that is, the value of ω_{eg} used in the model should be corrected in advance to yield the correct observed frequency.

Exercise 14.1 Lineshift Calculations: Check the validity of the approximations used in the equations (14.1.20–14.1.24).

Exercise 14.2 The Lamb Shift: Derive the energy shift formula (14.1.26).

14.2 Formulation of the Quantum Stochastic Equations

Having derived the decay constants and energy shifts, we can now carry out the same procedure as in Chap. 9 to get the appropriate form of quantum stochastic algebra, which the leads to the quantum stochastic Schrödinger equation, and then to the master equation.

14.2.1 Correlated Quantum Ito Increments

We now define the quantum Ito increments; they are most conveniently defined in the form

$$\Delta F_i(t) = \int_t^{t+\Delta t} f_i(t')\,dt', \tag{14.2.1}$$

and these obey an Ito commutator algebra—analogous to that given by (9.3.8)—of the form

$$\left[\Delta F_i(t), \Delta F_j^\dagger(t)\right] = \tfrac{1}{2}(\Gamma_{i,j}+\Gamma_{j,i})\,\Delta t. \tag{14.2.2}$$

Exercise 14.3 Proof of Ito Algebra: Using the same methods as in Ex. 9.1, and noting that $\gamma_{i,j}(t-t') = \gamma_{j,i}(t'-t)^*$, prove the Ito equation (14.2.2).

14.2.2 Quantum Stochastic Differential Equation

The procedure to derive the quantum stochastic Schrödinger equation now follows that used in Sect. 9.2.3 to derive the single atom quantum stochastic Schrödinger equation. The result, in the system picture, takes the form analogous to (9.4.5, 9.4.6)

$$d|\Psi_{\text{sys}}, t\rangle = \left(-\frac{\mathrm{i}}{\hbar}H_{\text{eff}}\,dt + \sum_i \sigma_i^-\,dF_i^\dagger(t)\right)|\Psi_{\text{sys}}, t\rangle. \tag{14.2.3}$$

Here H_{eff}, the *effective Hamiltonian*, which is non-Hermitian, is defined in this case as

$$H_{\text{eff}} \equiv H_{\text{sys}} + \hbar\sum_{i,j}\left(\delta\omega_{i,j} - \tfrac{1}{2}\mathrm{i}\Gamma_{i,j}\right)\sigma_i^+\sigma_j^-. \tag{14.2.4}$$

a) Inclusion of a Driving Field: A classical driving field $\boldsymbol{E}_{\rm cl}(\boldsymbol{x},t)$ is included in the same way as for a single atom, and this amounts to the replacement

$$H_{\rm sys} \to H_{\rm sys} - \sum_i \left(\boldsymbol{d}_{eg} \cdot \boldsymbol{E}_{\rm cl}^{(+)}(\boldsymbol{x}_i,t)\sigma_i^+ + \boldsymbol{d}_{eg} \cdot \boldsymbol{E}_{\rm cl}^{(-)}(\boldsymbol{x}_i,t)\sigma_i^- \right). \tag{14.2.5}$$

14.3 The Many Atom Master Equation

The reduced atomic density matrix defined, as in (10.1.1), by

$$\rho_N(t) \equiv \mathrm{Tr}_{\rm EM}\left\{ |\Psi_{\rm sys}(t)\rangle\langle\Psi_{\rm sys}(t)| \right\}, \tag{14.3.1}$$

is now in a 2^N dimensional space, and the master equation becomes

$$\frac{d}{dt}\rho_N(t) = -\frac{\mathrm{i}}{\hbar}\Big[H_{\rm sys} + \sum_{\substack{i,j\\ i\neq j}} \hbar\delta\omega_{i,j}\sigma_i^+\sigma_j^- \,,\, \rho_N(t) \Big] + \tfrac{1}{2}\sum_{i,j}\Gamma_{i,j}\left(2\sigma_i^-\rho_N(t)\sigma_j^+ - \sigma_j^+\sigma_i^-\rho_N(t) - \rho_N(t)\sigma_j^+\sigma_i^- \right), \tag{14.3.2}$$

$$H_{\rm sys} = \sum_i \left(\hbar\omega_{eg}\sigma_i^+\sigma_i^- - \boldsymbol{d}_{eg}\cdot\boldsymbol{E}_{\rm cl}^{(+)}(\boldsymbol{x}_i,t)\sigma_i^+ - \boldsymbol{d}_{eg}\cdot\boldsymbol{E}_{\rm cl}^{(-)}(\boldsymbol{x}_i,t)\sigma_i^- \right). \tag{14.3.3}$$

14.3.1 Spontaneous Emission Rates and Dipole-Dipole Forces

The master equation contains terms in which the spontaneous emission from an atom is modified by the presence of other atoms. The similar modification to the line-shifts leads to a force between atoms, as a result of the dipole moments induced by the optical field, known as the *dipole-dipole force*.

a) Spontaneous Emission Rates: These are given by (14.1.17). Explicitly writing out the spherical Bessel functions in terms of sines and cosines, the spontaneous emission rates can be written in terms of Γ, the single atom decay constant given by (14.1.19), in the form

$$\Gamma_{ij} = \Gamma F_{ij}(\xi)\big|_{\xi=k_{eg}r_{ij}}, \tag{14.3.4}$$

in which we define

$$F_{ij}(\xi) = \frac{3}{2}\left\{ \left[1 - \left(\hat{\boldsymbol{d}}_{eg}\cdot\hat{\boldsymbol{r}}_{ij}\right)^2 \right]\frac{\sin\xi}{\xi} + \left[1 - 3\left(\hat{\boldsymbol{d}}_{eg}\cdot\hat{\boldsymbol{r}}_{ij}\right)^2 \right]\left(\frac{\cos\xi}{\xi^2} - \frac{\sin\xi}{\xi^3} \right) \right\}. \tag{14.3.5}$$

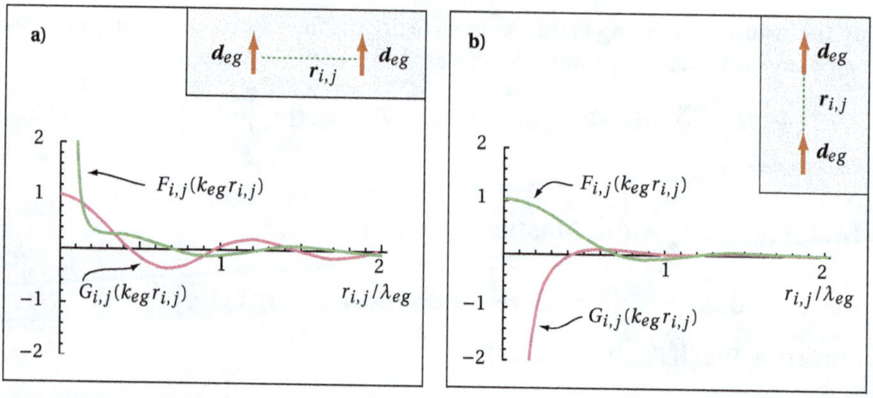

Fig. 14.2. The functions determining the spontaneous emission rates $\Gamma_{i,j}$ (brown), and the dipole-dipole shifts $\delta\omega_{i,j}$ (green); **a)** Parallel orientation; **b)** Perpendicular orientation.

b) Dipole-Dipole Shifts: These are given by (14.1.25), and can similarly be rewritten in the form

$$\delta\omega_{ij} = -\frac{\Gamma}{(k_{eg}r_{ij})^3}\mathcal{P}\int_{-\infty}^{+\infty}\frac{d\xi}{2\pi}\frac{\xi^3 F_{ij}(\xi)}{\xi - k_{eg}r_{ij}} = \Gamma G_{ij}(\xi)\big|_{\xi=k_{eg}r_{ij}}, \tag{14.3.6}$$

$$G_{ij}(\xi) = \tfrac{3}{4}\left\{-\left[1-\left(\hat{\boldsymbol{d}}_{eg}\cdot\hat{\boldsymbol{r}}_{ij}\right)^2\right]\frac{\cos\xi}{\xi} + \left[1-3\left(\hat{\boldsymbol{d}}_{eg}\cdot\hat{\boldsymbol{r}}_{ij}\right)^2\right]\left[\frac{\sin\xi}{\xi^2}+\frac{\cos\xi}{\xi^3}\right]\right\}. \tag{14.3.7}$$

As noted in Sect. 14.1.4, the terms with $i = j$ are given by (14.1.26), and are omitted in the expressions for dipole-dipole shifts, since they have already been absorbed in a redefinition of the value of ω_{eg}.

c) Dipole-Dipole Force: The expressions (14.3.6, 14.3.7) show that the dipole-dipole shift depends on the distances between the atoms, and therefore there is a corresponding force between the atoms, known as the *dipole-dipole force.*

14.3.2 Special Cases

In the limiting cases of distances between the atoms much smaller and much larger than the wavelength, $r_{ij} \ll \lambda$ and $r_{ij} \gg \lambda$, the expressions above are easily evaluated and interpreted.

a) Atoms Far Apart from Each Other: One expects that if the atoms are sufficiently separated, they behave independently, and this is indeed what happens. For $k_{eg}r_{ij} \equiv 2\pi r_{ij}/\lambda_{eg} \gg 1$ we find the decay constants become

$$\Gamma_{ij} = \Gamma F_{ij}(k_{eg}r_{ij}) \xrightarrow{k_{eg}r_{ij}\to\infty} \Gamma\delta_{ij}, \tag{14.3.8}$$

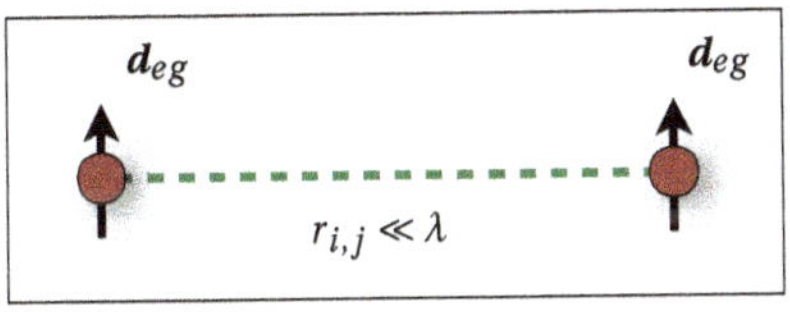

Fig. 14.3. The static dipole-dipole interaction

and the level shifts become

$$\delta\omega_{ij} = \Gamma G_{ij}(k_{eg} r_{ij}) \xrightarrow{k_{eg} r_{ij} \to \infty} 0. \tag{14.3.9}$$

The master equation then describes non-interacting atoms coupled to independent reservoirs, and interacting with the classical driving field at their individual locations; namely

$$\dot{\rho}(t) = -\frac{i}{\hbar}\sum_{i=1}^{N}\left[h_i, \rho(t)\right] + \tfrac{1}{2}\sum_i \Gamma\left\{2\sigma_i^- \rho(t)\sigma_i^+ - \sigma_i^+\sigma_i^-\rho(t) - \rho(t)\sigma_i^+\sigma_i^-\right\}, \tag{14.3.10}$$

$$h_i \equiv \hbar\omega_{eg}\sigma_i^+\sigma_i^- - \boldsymbol{d}_{eg}\cdot \boldsymbol{E}_{\rm cl}^{(+)}(\boldsymbol{x}_i, t)\sigma_i^+ - \boldsymbol{d}_{eg}\cdot \boldsymbol{E}_{\rm cl}^{(-)}(\boldsymbol{x}_i, t)\sigma_i^- . \tag{14.3.11}$$

b) Atoms Very Close to Each Other—Superradiance: When all of the atoms are much closer than a wavelength of light, they act together coherently. The precise condition is

$$k_{eg} r_{ij} \equiv 2\pi r_{ij}/\lambda_{eg} \ll 1, \tag{14.3.12}$$

and then we find:

i) *Emission rates*:

$$\Gamma_{ij} \xrightarrow{k_{eg} r_{ij} \to 0} \Gamma, \tag{14.3.13}$$

that is, all of the emission rates approach the single atom emission rate.

ii) *Energy shifts*: At the same time, the energy shifts approach the static dipole-dipole interaction, that is

$$\hbar\delta\omega_{ij} \xrightarrow{k_{eg} r_{ij} \to 0} \tfrac{3}{4}\hbar\Gamma \frac{\hat{\boldsymbol{d}}_{eg}\cdot\hat{\boldsymbol{d}}_{eg} - 3(\hat{\boldsymbol{d}}_{eg}\cdot\hat{\boldsymbol{r}}_{ij})(\hat{\boldsymbol{r}}_{ij}\cdot\hat{\boldsymbol{d}}_{eg})}{(k_{eg} r_{ij})^3}, \tag{14.3.14}$$

$$= \frac{1}{4\pi\epsilon_0}\frac{\boldsymbol{d}_{eg}\cdot\boldsymbol{d}_{eg} - 3(\boldsymbol{d}_{eg}\cdot\hat{\boldsymbol{r}}_{ij})(\hat{\boldsymbol{r}}_{ij}\cdot\boldsymbol{d}_{eg})}{r_{ij}^3}. \tag{14.3.15}$$

This is proportional to $1/r_{ij}^3$ as $r_{ij} \to 0$— however, these results all require the atoms not to overlap, that is, the results will only be valid as long as $r_{ij} \gg a_B$.

iii) *Electric fields*: The closeness condition (14.3.12) means that we can set the driving fields at the various positions to be equal to that evaluated at $\bar{\boldsymbol{x}}$, the mean position of the atoms; thus we will write

$$\boldsymbol{E}^{\pm}(\boldsymbol{x}_i, t) \longrightarrow \boldsymbol{E}^{\pm}(\bar{\boldsymbol{x}}, t). \tag{14.3.16}$$

iv) *The Master Equation*: The master equation (14.3.2, 14.3.3) then becomes

$$\frac{d}{dt}\rho_N(t) = -\frac{i}{\hbar}\Big[H_{\text{sys}} + \sum_{\substack{i,j \\ i\neq j}} \hbar\delta\omega_{i,j}\sigma_i^+\sigma_j^- \,,\, \rho_N(t)\Big]$$
$$+2\Gamma\left(2S^-\rho_N(t)S^+ - S^+S^-\rho_N(t) - \rho_N(t)S^+S^-\right), \tag{14.3.17}$$

$$H_{\text{sys}} = \tfrac{1}{2}\hbar\omega_{eg}(2S_z - N) - 2\boldsymbol{d}_{eg}\cdot\boldsymbol{E}_{\text{cl}}^{(+)}(\bar{\boldsymbol{x}},t)S^+ - 2\boldsymbol{d}_{eg}\cdot\boldsymbol{E}_{\text{cl}}^{(-)}(\bar{\boldsymbol{x}},t)S^-. \tag{14.3.18}$$

This describes a *collective decay* involving the collective spin operators

$$\boldsymbol{S} = \tfrac{1}{2}\sum_i \boldsymbol{\sigma}_i. \tag{14.3.19}$$

14.3.3 Superradiance

The fact that the decay of N atoms could happen collectively was first pointed out by *Dicke* in 1954 [14.4], who coined the term *superradiance* to describe the phenomenon. The elementary phenomenon occurs when the atoms are close, as specified above in (14.3.12), but nevertheless sufficiently far apart for the dipole-dipole term—that term containing $\delta\omega_{ij}$ in the master equation—to be negligible. However, it will be easier to investigate the solutions of the master equation in which this term is neglected first, and then consider what the effect of the dipole-dipole term would have on these solutions.

Here we will give only the most elementary sketch of the phenomenon of superradiance, for the case in which all atoms are within a wavelength of each other. The behaviour is also very interesting for larger systems, that is when the sample has many atoms within a wavelength of each other, but also extends over a larger volume. A very thorough presentation of all aspects of superradiance has been given by *Gross* and *Haroche* [14.5].

a) Solutions of the Superradiance Master Equation: The collective spin operators obey the same commutation relations as the angular momentum operators $J_\pm$, J_0 defined in Appendix 8.A, so we can classify the states of the atoms in the same way as angular momentum states, that is, by the eigenvalue $S(S+1)$ of $\boldsymbol{S}^2$ and the eigenvalue M of S_0, so that we write the states as $|S,M\rangle$.

The simplest case happens when we set $\boldsymbol{E}_{\text{cl}}(\bar{\boldsymbol{x}},t) = 0$, and consider the time evolution of the occupation probabilities

$$P_N(S,M) \equiv \langle SM|\rho_N|S,M\rangle. \tag{14.3.20}$$

From the master equation (14.3.17), the equation of motion for $P_N(S,M,t)$ is

$$\dot{P}_N(S,M) = 2\Gamma\Big(T(S,M+1)P_N(S,M+1,t) - T(S,M)P_N(S,M,t)\Big), \tag{14.3.21}$$

$$T(S,M) \equiv (S-M+1)(S+M), \tag{14.3.22}$$

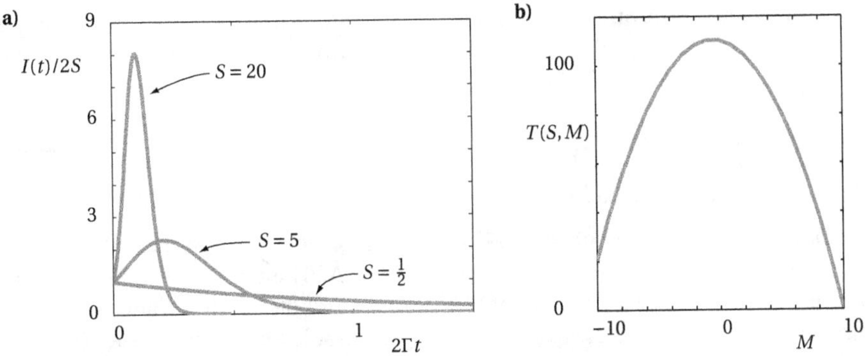

Fig. 14.4. Superradiant decay. **a)** The normalized radiated intensity for various values of S, with the initial state corresponding to $M = S$, as in (14.3.24). **b)** The transition probability $T(S, M)$ for $S = 20$ and as a function of M; this function peaks at $M = \frac{1}{2}$, leading to the peak in the radiated intensity as the system passes through states of reducing M.

where, as previously noted, we have neglected the dipole-dipole forces.

There is no simple analytic solution to this equation, but numerical solutions are easy. The most illuminating thing to present is the *light intensity*, that is mean rate at which photons are being emitted, and this is given by the mean rate of transitions $M \to M-1$, which is

$$I(t) \equiv 2\Gamma \sum_{M=-S}^{S} T(S,M)\, P_N(S,M,t). \tag{14.3.23}$$

The solutions for $I(t)/2S$ are given in Fig. 14.4, in the case that the initial condition is

$$P_N(S,M,0) = \delta_{S,M}. \tag{14.3.24}$$

If $S = \frac{1}{2}N$, this corresponds to all atoms being in the excited state, for $S < \frac{1}{2}N$ the number of excited atoms is $n = S + \frac{1}{2}N$.

Notice the following points:

i) The value of N does not enter into the equation (14.3.20), except to the extent that we know that $S \leqslant \frac{1}{2}N$.

ii) As seen in Fig. 14.4 a, the decay is characterized by a pulse of light which rises to a peak, and then decays.

iii) This kind of decay is a signature of the co-operative nature of the decay. It arises directly from the way the transition probability $(S - M + 1)(S + M)$ rises as M decreases, as shown in Fig. 14.4 b.

iv) The decay is exponential only for the case of $S = \frac{1}{2}$, corresponding to a single atom.

b) Time Scale of the Superradiant Pulse: The numerical solutions show that the pulse occurs more rapidly for larger S, and one can empirically deduce from the numerical results the typical time scale is

$$T_{\text{superradiant}} \approx \frac{1}{2\Gamma(2S+1)}. \tag{14.3.25}$$

This can also be deduced from a semiclassical approximation, see Ex. 14.4.

c) Estimate of the Effect of Dipole-Dipole Terms: Let us suppose that the spread in values of $\delta\omega_{i,j}$ has a value which we shall call $\bar{\Delta}$. Then any dephasing will take place on a time scale of the order of $1/\bar{\Delta}$. Since the whole superradiance process is finished within the time $T_{\text{superradiant}}$, the dipole-dipole terms will have a negligible effect when

$$T_{\text{superradiant}}\bar{\Delta} = \frac{\bar{\Delta}}{2\Gamma(2S+1)} \ll 1. \tag{14.3.26}$$

An estimate of the spread can (in principle) be made from the formula (14.3.6), for the case that all dipoles are oriented parallel to each other (to give the correct initial state for superradiance) and the atoms are distributed randomly within a known volume.

Exercise 14.4 Semiclassical Approximation: Let us define the average value of M as $\bar{M}(t) \equiv \sum M P_N(S,M,t)$. For a sufficiently large number of atoms we can expect the probability distribution to retain its sharply peaked character during the decay, and therefore we can make a factorization approximation to the averages, i.e. we can write $\langle M^2\rangle \to \bar{M}^2$.

Using (14.3.21, 14.3.22), show this leads to the equation of motion

$$\dot{\bar{M}}(t) = -2\Gamma\big(S-\bar{M}(t)+1\big)\big(S+\bar{M}(t)\big). \tag{14.3.27}$$

Show that the solution to this equation with the initial condition $\bar{M}(t=0)=S$ is

$$\bar{M}(t) = S+1-\frac{1+2S}{1+2S\exp\big(-2(2S+1)\Gamma t\big)}, \tag{14.3.28}$$

which verifies the estimate of $T_{\text{superradiant}}$ in (14.3.25).

How accurate is this semiclassical solution when compared to the results of Fig. 14.4?

Exercise 14.5 Example Spontaneous Emission from a Pair of Atoms: Write out and solve the master equation for two atoms. Use the quantities $Q_{ij}(t) \equiv \operatorname{Tr}\left\{\sigma_i^+\sigma_j^-\rho(t)\right\}$, and show that we get the equations of motion

$$\dot{Q}_{11} = -\Gamma Q_{11} - \left(\tfrac{1}{2}\Gamma_{12} + i\delta\omega_{12}\right)Q_{12} - \left(\tfrac{1}{2}\Gamma_{12} - i\delta\omega_{12}\right)Q_{21}, \tag{14.3.29}$$

$$\dot{Q}_{12} = -\Gamma Q_{12} - \left(\tfrac{1}{2}\Gamma_{12} + i\delta\omega_{12}\right)Q_{11} - \left(\tfrac{1}{2}\Gamma_{12} - i\delta\omega_{12}\right)Q_{22} + 2\Gamma\langle\sigma_1^+\sigma_1^-\sigma_2^+\sigma_2^-\rangle, \tag{14.3.30}$$

$$\frac{d}{dt}\langle\sigma_1^+\sigma_1^-\sigma_2^+\sigma_2^-\rangle = -2\Gamma\langle\sigma_1^+\sigma_1^-\sigma_2^+\sigma_2^-\rangle, \qquad \text{etc.} \tag{14.3.31}$$

Exercise 14.6 Example Spontaneous Emission from a Pair of Atoms—Use of Coupled States: It is more convenient, and provides more insight to introduce a new basis in terms of the *triplet* states

$$|S=1, M=+1\rangle = |ee\rangle, \tag{14.3.32}$$

$$|S=1, M=0\rangle = \frac{1}{\sqrt{2}}\big(|eg\rangle + |ge\rangle\big), \tag{14.3.33}$$

$$|S=1, M=-1\rangle = |gg\rangle, \tag{14.3.34}$$

and the *singlet states*

$$|S=0, M=0\rangle = \frac{1}{\sqrt{2}}\big(|eg\rangle - |ge\rangle\big). \tag{14.3.35}$$

Show that the master equation becomes

$$\dot{\rho}_{11,11} = -2\Gamma\rho_{11,11}, \tag{14.3.36}$$

$$\dot{\rho}_{10,10} = \big(\Gamma+\Gamma_{12}\big)\rho_{11,11} - \big(\Gamma+\Gamma_{12}\big)\rho_{10,10}, \tag{14.3.37}$$

$$\dot{\rho}_{00,00} = \big(\Gamma-\Gamma_{12}\big)\rho_{11,11} - \big(\Gamma-\Gamma_{12}\big)\rho_{00,00}, \tag{14.3.38}$$

$$\dot{\rho}_{10,00} = = -2\big(i\delta\omega_{12}+\Gamma\big)\rho_{10,00}. \tag{14.3.39}$$

14.A Appendix: Spherical Bessel Functions

The spherical Bessel functions are fully described in [14.3], and we illustrate their behaviour in Fig. 14.5. The properties used in this chapter are:

i) $j_l(z)$ is *regular* at the origin, and in fact for small l we can give explicit formulae

$$j_0(z) = \frac{\sin z}{z}, \tag{14.A.1}$$

$$j_1(z) = \frac{\sin z}{z^2} - \frac{\cos z}{z}, \tag{14.A.2}$$

$$j_2(z) = \left(\frac{3}{z^3} - \frac{1}{z}\right)\sin z - \frac{3}{z^2}\cos z. \tag{14.A.3}$$

ii) $n_l(z)$ is *singular* at the origin

$$n_0(z) = -\frac{\cos z}{z}, \tag{14.A.4}$$

$$n_1(z) = -\frac{\cos z}{z^2} - \frac{\sin z}{z}, \tag{14.A.5}$$

$$n_2(z) = -\left(\frac{3}{z^3} - \frac{1}{z}\right)\cos z - \frac{3}{z^2}\sin z. \tag{14.A.6}$$

iii) Note that the spherical Neumann functions are obtained from the spherical Bessel functions by the substitutions

$$\sin z \longrightarrow -\cos z, \tag{14.A.7}$$

$$\cos z \longrightarrow \sin z. \tag{14.A.8}$$

iv) For small z

$$j_l(z) \approx \frac{z^l}{1\cdot 3\cdot 5\ldots(2l+1)}, \qquad n_l(z) \approx -\frac{1\cdot 3\cdot 5\ldots(2l-1)}{z^{l+1}}. \tag{14.A.9}$$

v) For large z

$$j_l(z) \approx \frac{\sin(z - l\pi/2)}{z}, \qquad n_l(z) \approx -\frac{\cos(z - l\pi/2)}{z}. \tag{14.A.10}$$

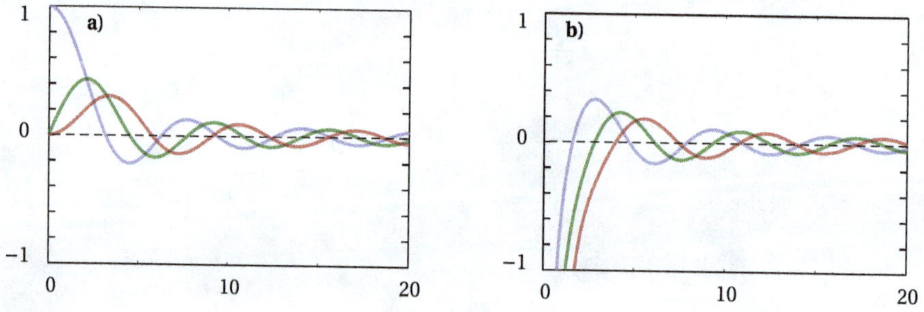

Fig. 14.5. Plots of **a)** The spherical Bessel functions $j_l(z)$ for $l = 0, 1, 2$, and **b)** The spherical Neumann functions $n_l(z)$ for $l = 0, 1, 2$.

IV LASER COOLING

PART IV: LASER COOLING

Laser cooling is the central technology that underpins all of the physics of trapping and manipulating ultra-cold atoms and ions. The process of laser cooling involves the application of carefully designed laser fields to a cloud of atoms, with the directions and frequencies of the laser fields being chosen so as to force the atoms to transfer their kinetic energy into atomic excitation energy, which is then radiated into the electromagnetic field.

This part of the book covers three aspects of the theory of laser cooling:

i) In Chap. 15, the quantum stochastic framework is modified to a form appropriate to laser cooling. As formulated in Part III, the quantum stochastic framework implicitly assumed that atoms are held in a fixed position. Here we show how to include the recoil effects when light is emitted or absorbed. This necessarily means that the atomic Hamiltonian has to include the kinetic term, as was done in Chap. 7. It is also necessary to separate the quantum noise increments into components corresponding to a definite direction of propagation.

   The final result is the *laser cooling master equation* of Sect. 15.4, which governs the motion of a two-level atom moving under the influence of a classical optical field, as would be provided by a laser.

ii) Chap. 16 is devoted to the development of a systematic approximation procedure for solving the laser cooling master equation for the case of an untrapped vapour. The procedure is a perturbation theory in the parameter $\bar{\eta} \equiv \hbar k_L / \Delta p$, where $\hbar k_L$ is the momentum of a single laser photon, and Δp is a measure of the width of the momentum distribution of the atoms.

   The main result of the procedure is a validation of the results of the simple stochastic model we considered in *Bk. I: Sect.14.4.4*, and in particular the result that the minimum temperature achievable by this process is given in terms of the decay constant Γ of the atomic transition by the *Doppler limit* temperature $T_{\text{Cool}} \approx \hbar\Gamma / k_B$.

iii) In Chap. 17 we develop the theory for the case of a trapped ion, the case of most importance for the construction of quantum devices based on ion traps, which will be considered in more detail in Part VII. Here the trapped motion of the ion is fully quantized, and a similar perturbation procedure yields a master equation for transitions between these energy levels.

   However, in the case of an ion trap the laser frequency can be tuned to yield the process of *sideband cooling*, by means of which the ion can be cooled to a very much lower temperature than the Doppler limit—in fact it is possible to achieve *ground-state cooling*, in which the temperature is so low that the ion is in the ground-state with 100% probability.

15. Quantum Stochastic Equations for Laser Cooling of Atoms

In Part III of this book we have given a rather formal description of the theory of quantum Markov processes, in which the electromagnetic field is formalized as quantum white noise, or possibly a set of quantum noises corresponding to different modes or different transition frequencies. This chapter is devoted to exploring the application of the methodology to the case of the two-level atom driven by a coherent light field. What we present here is a reformulation in the context of the quantum stochastic Schrödinger equation of a methodology introduced in [15.1], and summarized in *Quantum Noise Chap. 11.*

The intention of this chapter is to show how the formalism of Part III is used in a more realistic system, and in this case, there will be the following additional features which we wish account for:

i) The three-dimensional nature of the radiated light field has not so far been explicitly accounted for. This will lead us to express the quantum noise increment $dB(t)$ as a sum of different increments $dB_{\boldsymbol{n},\lambda}(t)$ corresponding to the light field being radiated in a definite direction specified by the unit vector $\boldsymbol{n}$, and with a particular polarization state λ.

ii) Furthermore, by formulating the procedure in terms of noise operators, each corresponding to light propagating in a definite direction $\boldsymbol{n}$, it becomes possible to include the recoil of the atom induced by the momentum transfer in the direction $\boldsymbol{n}$ to the radiation field. In this chapter we will include this recoil effect in the quantum stochastic equations, and the remainder of this part of the book will explore the applications of the resulting formalism to laser cooling.

iii) The strength of the particular transition is also specified by the the matrix element of the dipole moment, and we will also want to make sure that this is directly related to the quantum stochastic differential equation formalism.

The generalization of the formalism does not represent a major change, but is rather detailed. The main task is to formulate the quantization of the radiation field in terms of photons travelling in the direction $\boldsymbol{n}$ and with a particular energy, rather than the more conventional treatment in terms of the momentum $\hbar\boldsymbol{k}$ that we gave in *Bk. I: Sect.11.1.3.* After that, the appropriate generalization of the quantum stochastic differential equation proceeds analogously to the procedure of Chap. 9.

15.1 Notation and Hamiltonians

In this case it is best to start with values of $\boldsymbol{k}$ with a continuous range, and not normalized to a box, so that the creation and destruction operator commutation relations are

$$\left[b_{\boldsymbol{k},\lambda}, b^{\dagger}_{\boldsymbol{k}',\lambda'}\right] = \delta_{\lambda,\lambda'}\delta(\boldsymbol{k}-\boldsymbol{k}'). \tag{15.1.1}$$

We want alternative operators which would satisfy the commutation relations

$$\left[\mathcal{B}_{k,\boldsymbol{n},\lambda}, \mathcal{B}^{\dagger}_{k',\boldsymbol{n}',\lambda'}\right] = \delta_{\lambda,\lambda'}\delta(k-k')\delta(\boldsymbol{n},\boldsymbol{n}'), \tag{15.1.2}$$

where the solid angle delta function $\delta(\boldsymbol{n},\boldsymbol{n}')$ satisfies

$$\int d\Omega_{\boldsymbol{n}}\, F(\boldsymbol{n}')\delta(\boldsymbol{n},\boldsymbol{n}') = F(\boldsymbol{n}), \qquad \text{for any function } F(\boldsymbol{n}). \tag{15.1.3}$$

Setting $\boldsymbol{k} = k\boldsymbol{n}$, we can write

$$d^3\boldsymbol{k} = k^2\, dk\, d\Omega_{\boldsymbol{n}}, \tag{15.1.4}$$

so that

$$\delta(\boldsymbol{k}-\boldsymbol{k}') = \frac{1}{k^2}\delta(k-k')\delta(\boldsymbol{n},\boldsymbol{n}'). \tag{15.1.5}$$

This means that the relationship between our new notation and the conventional notation will be

$$\mathcal{B}_{k,\boldsymbol{n},\lambda} = \frac{1}{k} b_{\boldsymbol{k},\lambda}, \qquad \text{with} \quad \boldsymbol{k} = k\boldsymbol{n}. \tag{15.1.6}$$

15.1.1 Electromagnetic Field and Hamiltonian

Correspondingly, the Hamiltonian for the radiation field can be written

$$H_{\text{EM}} = \sum_{\lambda}\int dk\, d\Omega_{\boldsymbol{n}}\, \hbar\omega_k \mathcal{B}^{\dagger}_{k,\boldsymbol{n},\lambda}\mathcal{B}_{k,\boldsymbol{n},\lambda}. \tag{15.1.7}$$

Corresponding to the form given in *Bk. I: (11.1.48)*. The Schrödinger picture electric field operator can now be written

$$\boldsymbol{E}(\boldsymbol{x}) = \mathrm{i}\sum_{\lambda}\int dk\, d\Omega_{\boldsymbol{n}}\sqrt{\frac{\hbar c k^3}{2\epsilon_0}}\left(\mathcal{B}_{k,\boldsymbol{n},\lambda}U_{k,\boldsymbol{n},\lambda}(\boldsymbol{x}) - \mathcal{B}^{\dagger}_{k,\boldsymbol{n},\lambda}U^{*}_{k,\boldsymbol{n},\lambda}(\boldsymbol{x})\right), \tag{15.1.8}$$

$$U_{k,\boldsymbol{n},\lambda}(\boldsymbol{x}) = \frac{1}{\sqrt{(2\pi)^3}}\epsilon_{\boldsymbol{n},\lambda}\mathrm{e}^{ik\boldsymbol{n}\cdot\boldsymbol{x}}. \tag{15.1.9}$$

15.1.2 System Hamiltonian

The system of interest is the driven two-level atom system itself, as in Fig. 15.1, whose Hilbert space consists of the vectors spanned by $\{|g\rangle, |e\rangle\}$. We will use the

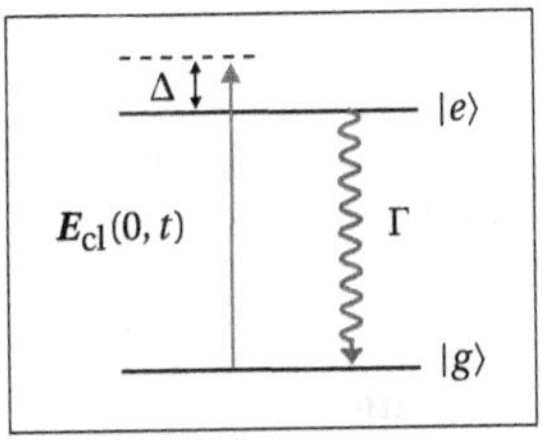

Fig. 15.1. The level structure, driving field and decay of the two-level atom system Hamiltonian (15.1.14).

two equivalent notations for the complete set of operators in this Hilbert space

$$\sigma^+ = |e\rangle\langle g|, \tag{15.1.10}$$

$$\sigma^- = |g\rangle\langle e|, \tag{15.1.11}$$

$$\sigma_z = |e\rangle\langle e| - |g\rangle\langle g|, \tag{15.1.12}$$

$$1 = |e\rangle\langle e| + |g\rangle\langle g|. \tag{15.1.13}$$

The system Hamiltonian can be written

$$H_{\text{sys}} = \hbar\omega_{eg}|e\rangle\langle e| - \sigma^+ \boldsymbol{d}_{eg}\cdot\boldsymbol{E}^{(+)}_{\text{cl}}(0,t) - \sigma^- \boldsymbol{d}^*_{eg}\cdot\boldsymbol{E}^{(-)}_{\text{cl}}(0,t), \tag{15.1.14}$$

where the classical driving field is provided by a laser with frequency ω:

$$\boldsymbol{E}^{(+)}_{\text{cl}}(\boldsymbol{x}=0,t) = \boldsymbol{\epsilon}_{\text{cl}}\mathcal{E}(t)\mathrm{e}^{-\mathrm{i}\omega t}. \tag{15.1.15}$$

Here $\boldsymbol{\epsilon}$ is the polarization vector of the driving field, which can in practice be any of plane, circularly, or elliptically polarized.

15.1.3 Interaction Hamiltonian

The interaction Hamiltonian between the radiation field and the atom (in the rotating-wave approximation and the electric dipole approximation) then becomes

$$H_{\text{Int}} = -\sigma^+ \boldsymbol{d}_{eg}\cdot\boldsymbol{E}^{(+)}(0) - \sigma^- \boldsymbol{d}^*_{eg}\cdot\boldsymbol{E}^{(-)}(0), \tag{15.1.16}$$

$$= -\mathrm{i}\hbar\sum_{\lambda}\int dk\,d\Omega\Big(\kappa(k,\boldsymbol{n},\lambda)\sigma^+\mathcal{B}_{k,\boldsymbol{n},\lambda} - \kappa^*(k,\boldsymbol{n},\lambda)\sigma^-\mathcal{B}^\dagger_{k,\boldsymbol{n},\lambda}\Big), \tag{15.1.17}$$

in which

$$\kappa(k,\boldsymbol{n},\lambda) = \sqrt{\frac{ck^3}{2(2\pi)^3\hbar\epsilon_0}}\,\boldsymbol{d}_{eg}\cdot\boldsymbol{\epsilon}_{\boldsymbol{n},\lambda}. \tag{15.1.18}$$

15.2 Quantum Stochastic Differential Equation Formalism

In this section we will generalize the quantum stochastic differential equation formalism developed in Chap. 9 to the situation here, in which the direction of the emitted photon is explicitly tracked.

15.2.1 Noise Operators

Corresponding to the definition (9.2.9) we introduce the noise operators, defined by integrating over the k-region limited to the neighbourhood of ω_{eg}/c

$$\mathcal{D}(\omega_{eg},\vartheta) \equiv |k/c - \omega_{eg}| < \vartheta, \tag{15.2.1}$$

as introduced in Sect. 9.1. The relevant noise operators are defined by

$$f_{\boldsymbol{n},\lambda}(t) \equiv \int_{\mathcal{D}(\omega_{eg},\vartheta)} dk\, \mathcal{B}_{k,\boldsymbol{n},\lambda}\, \mathrm{e}^{-\mathrm{i}(\omega_k-\omega_{eg})t}, \tag{15.2.2}$$

which have the commutation relations

$$\left[f_{\boldsymbol{n},\lambda}(t), f^{\dagger}_{\boldsymbol{n}',\lambda'}(t')\right] = \delta(\boldsymbol{n},\boldsymbol{n}')\delta_{\lambda,\lambda'}\gamma_{\boldsymbol{n},\lambda}(t-t'), \tag{15.2.3}$$

in which

$$\gamma_{\boldsymbol{n},\lambda}(\tau) \equiv \int_{\mathcal{D}(\omega_{eg},\vartheta)} dk\, |\kappa(k,\boldsymbol{n},\lambda)|^2\, \mathrm{e}^{-\mathrm{i}(\omega_k-\omega_{eg})\tau}. \tag{15.2.4}$$

15.2.2 Total Noise Operator

The total noise operator plays the same rôle as $f(t)$ in Sect. 9.2.1, so we use the same notation, but define $f(t)$ here by

$$f(t) = \sum_{\lambda}\int d\Omega_{\boldsymbol{n}}\, f_{\boldsymbol{n},\lambda}(t), \tag{15.2.5}$$

and this has commutation relations of exactly the same form as (9.2.11)

$$[f(t), f^{\dagger}(t')] = \gamma(t-t'), \tag{15.2.6}$$

in which

$$\gamma(\tau) = \sum_{\lambda}\int_{\mathcal{D}(\omega_{eg},\vartheta)} d\Omega_{\boldsymbol{n}}\, \gamma_{\lambda,\boldsymbol{n}}(\tau). \tag{15.2.7}$$

15.2.3 Density of States

In Sect. 9.2.1 we rewrote the expression for $\gamma(\tau)$ in terms of a frequency integral, a density of states and an average value of $|\kappa(\omega)|^2$. Here we can simply write $\omega = kc$ in (15.2.4) and similar equations, and no explicit density of states appears; rather, we find that $|\kappa(k,\boldsymbol{n},\lambda)|^2$, as defined by (15.1.18), has an explicit factor ω^3, so that it makes some sense to isolate this factor explicitly by introducing a different notation. Therefore let us write

$$|\kappa(k,\boldsymbol{n},\lambda)|^2 dk = \omega^3 |K(\omega,\boldsymbol{n},\lambda)|^2 d\omega, \tag{15.2.8}$$

so that

$$|K(\omega,\boldsymbol{n},\lambda)|^2 \equiv \frac{\left|\boldsymbol{d}_{eg}\cdot\boldsymbol{\epsilon}_{\boldsymbol{n},\lambda}\right|^2}{2(2\pi)^3\epsilon_0\hbar c^3}, \tag{15.2.9}$$

and we can then rewrite (15.2.4) as

$$\gamma_{n,\lambda}(\tau) = \int_{\mathcal{D}(\omega_{eg},\vartheta)} d\omega\,\omega^3 |K(\omega,\boldsymbol{n},\lambda)|^2 e^{-i(\omega_k-\omega_{eg})\tau}. \tag{15.2.10}$$

Since the integral is only over the narrow band $\mathcal{D}(\omega_{eg},\vartheta)$, the function $\gamma_{n,\lambda}(\tau)$ is approximately proportional to ω_{eg}^3.

15.2.4 Quantum Noise Increments

The procedure to follow now is essentially the same as that we used in Sect. 9.2.3–Sect. 9.3.4. The Markov approximation of Sect. 9.2.3d takes the form

$$\int_0^{\Delta t} dt_1 \int_0^{t_1} dt_2 \gamma_{n,\lambda}(t_1 - t_2)$$

$$\longrightarrow \left(\pi\omega_{eg}^3 |K(\omega_{eg},\boldsymbol{n},\lambda)|^2 - iP\int_{\mathcal{D}(\omega_{eg},\vartheta)} \frac{\omega^3\, d\omega}{\omega-\omega_{eg}} |K(\omega,\boldsymbol{n},\lambda)|^2 \right) \Delta t, \qquad \text{for } \Delta t \gg 1/\vartheta, \tag{15.2.11}$$

$$\equiv \left(\tfrac{1}{2}\Gamma_{n,\lambda} + i\delta\omega_{n,\lambda}\right). \tag{15.2.12}$$

The linewidth $\Gamma_{n,\lambda}$ and lineshift $\delta\omega_{n,\lambda}$ are thus separated into individual components corresponding to the different directions and polarizations of the radiated photons.

a) Zero Temperature Ito Algebra: The quantum Ito increments are defined analogously to (9.3.1)

$$\Delta B_{n,\lambda}(t) \equiv \frac{1}{\sqrt{\Gamma_{n,\lambda}}} \int_t^{t+\Delta t} f_{n,\lambda}(t')\, dt', \tag{15.2.13}$$

$$\longrightarrow dB_{n,\lambda}(t) \qquad \text{in the limit of } \Delta t \text{ sufficiently small.} \tag{15.2.14}$$

The resulting Ito algebra is essentially that of (9.4.31), with the only non-zero product being

$$dB_{n,\lambda}(t) dB^\dagger_{n',\lambda'}(t) = \delta(\boldsymbol{n},\boldsymbol{n}')\delta_{\lambda,\lambda'}\, dt. \tag{15.2.15}$$

b) Interpretation: The radiation field operator in the the solid angle $d\Omega_n$ with polarization λ is given by the product $dB_{n,\lambda}(t)\, d\Omega_n$. This means that we can introduce a photon counting formalism in which photons of a definite polarization and radiated in a definite direction can be individually counted.

15.2.5 Quantum Stochastic Schrödinger Equation

The procedure of Sect. 9.4.2 leads to a quantum stochastic Schrödinger equation in the form

$$d|\Psi_{\mathrm{sys}},t\rangle = \left(-\frac{\mathrm{i}}{\hbar}H_{\mathrm{eff}}\,dt + \sigma^- \sum_\lambda \int d\Omega_n \sqrt{\Gamma_{n,\lambda}}\, dB^\dagger_{n,\lambda}(t)\right)|\Psi_{\mathrm{sys}},t\rangle, \qquad (15.2.16)$$

$$H_{\mathrm{eff}} = H_{\mathrm{sys}} + \sum_\lambda \int d\Omega_n \left(\hbar\,\delta\omega_{n,\lambda} - \tfrac{1}{2}\mathrm{i}\hbar\Gamma_{n,\lambda}\right)\sigma^+\sigma^-. \qquad (15.2.17)$$

15.2.6 Equivalence to a Single Set of Noise Increments

If we write

$$\Gamma = \sum_\lambda \int d\Omega_n\, \Gamma_{n,\lambda}, \qquad (15.2.18)$$

$$\delta\omega = \sum_\lambda \int d\Omega_n\, \delta\omega_{n,\lambda}, \qquad (15.2.19)$$

$$dB(t) = \frac{1}{\sqrt{\Gamma}}\sum_\lambda \int d\Omega_n\, dB_{n,\lambda}(t), \qquad (15.2.20)$$

then:

i) The quantity $dB(t)$ defined this way generates the single noise quantum Ito algebra (9.3.21).

ii) The quantum stochastic differential equation (15.2.16) is identical with the single noise version (9.3.19).

iii) However, the physics is not identical, in that this formalism gives us access to the full angular distribution of photon counting, as well as the dependence on polarization.

15.3 Recoil Effects

The model of a driven two-level system so far formulated does not include the translational degrees of freedom of the atom. In this section we will extend the Hamiltonian of Chap. 7 to include a radiatively damped two-level atom driven by a classical light field, and this time include the motion of the atom and mechanical effects arising from induced and spontaneous radiative processes. We will develop a full set of quantum stochastic equations to describe this system. Later chapters will explore methods of solving these equations, and also generalize to systems with more than two levels. Taken together, these will provide a description of a variety of ways in which laser cooling can be achieved.

15.3.1 The Model

Our goal in this chapter is to derive a master equation for the reduced system density operator

$$\rho_A(t) = \mathrm{Tr}_{\mathrm{EM}}\left\{|\Psi(t)\rangle\langle\Psi(t)|\right\}. \tag{15.3.1}$$

We will closely follow the derivation of the optical Bloch equations given in the previous chapter. Including kinetic and potential energy terms, the Hamiltonian for the system is

$$H = H_{\mathrm{Atom}} + H_{\mathrm{EM}} + H_{\mathrm{int}}, \tag{15.3.2}$$

in which the three terms describe respectively the atom, the electromagnetic field, and the interaction between them, as we shall now describe in detail.

15.3.2 The Atom

a) Atomic Degrees of Freedom: These include:

i) *Motional Degrees of Freedom*: These are associated with the motion of the centre of mass of the atom, represented by the position operator $\boldsymbol{X}$ and the momentum operator $\boldsymbol{X}$.

ii) *Internal Degrees of Freedom*: These are associated with the states of the electrons in the atom, and are often called electronic degrees of freedom. We represent them in general by the quantum states $|i\rangle$. In the case of the two-level atom there are only two of these $|g\rangle$ and $|e\rangle$, the ground and excited states.

b) Atomic Part of the Hamiltonian: This is formed from the system Hamiltonian (13.1.5) by including, in addition to the energy of the atomic energy levels, the kinetic energy of the centre of mass and an optional trapping potential. Furthermore, the interaction energy induced by the classical driving field is now obtained by evaluating the field at the centre of mass operator position X. The atomic Hamiltonian therefore has the form

$$H_{\mathrm{Atom}} = \frac{\boldsymbol{P}^2}{2M} + V(\boldsymbol{X}) + \hbar\omega_{eg}|e\rangle\langle e| - \left(\boldsymbol{d}_{eg}\cdot\boldsymbol{E}^{(+)}_{\mathrm{cl}}(\boldsymbol{X},t)\sigma^+ + \boldsymbol{d}^*_{eg}\cdot\boldsymbol{E}^{(-)}_{\mathrm{cl}}(\boldsymbol{X},t)\sigma^-\right). \tag{15.3.3}$$

i) *The classical electric field* $\boldsymbol{E}_{\mathrm{cl}}$ is in principle arbitrary, but is normally either a travelling wave $\boldsymbol{E}^{(+)}_{\mathrm{cl}}(\boldsymbol{x},t) = \boldsymbol{\epsilon}_{\mathrm{cl}}\mathcal{E}e^{\mathrm{i}(\boldsymbol{k}_L\cdot\boldsymbol{x}-\omega_L t)}$, or the sum of a number of such travelling waves.

ii) *The trapping potential* is usually a harmonic potential

$$V(\boldsymbol{X}) = \tfrac{1}{2}M\left(\nu_x^2X^2 + \nu_y^2Y^2 + \nu_z^2Z^2\right). \tag{15.3.4}$$

iii) This Hamiltonian acts in the atomic Hilbert space $\mathcal{H}_A = L(R^3)\otimes\left\{|g\rangle,|e\rangle\right\}$, which is the product space of square integrable centre of mass wave packets and internal atomic states.

15.3.3 The Radiation Field

This is the electromagnetic Hamiltonian, written as in (15.1.7)

$$H_{\mathrm{EM}} = \sum_{\lambda}\int dk\, d\Omega_{\boldsymbol{n}}\, \hbar\omega_k B^\dagger_{k,\boldsymbol{n},\lambda} B_{k,\boldsymbol{n},\lambda}. \tag{15.3.5}$$

15.3.4 The Interaction Hamiltonian

The interaction Hamiltonian evaluates the electric field at the centre of mass operator position $\boldsymbol{X}$, and thus is

$$H_{\mathrm{int}} = -\boldsymbol{d}_{eg}\cdot \boldsymbol{E}^{(+)}(\boldsymbol{X},t)\sigma^+ - \sigma^-\boldsymbol{d}^*_{eg}\cdot \boldsymbol{E}^{(-)}(\boldsymbol{X},t). \tag{15.3.6}$$

This is a natural generalization of the form we used in Sect. 15.1.3 of the previous chapter, and can be written, in the notation of that section, as

$$H_{\mathrm{Int}} = -\mathrm{i}\hbar\sum_{\lambda}\int dk\, d\Omega\Big(\kappa(k,\boldsymbol{n},\lambda)\mathrm{e}^{\mathrm{i}k\boldsymbol{n}\cdot\boldsymbol{X}}\sigma^+ B_{k,\boldsymbol{n},\lambda} - \kappa^*(k,\boldsymbol{n},\lambda)\mathrm{e}^{-\mathrm{i}k\boldsymbol{n}\cdot\boldsymbol{X}}\sigma^- B^\dagger_{k,\boldsymbol{n},\lambda}\Big). \tag{15.3.7}$$

The behaviour of this interaction term is very significantly different from (15.1.17), where the motion of the atom is assumed to be negligible. The two main aspects of the difference are:

i) Since $\mathrm{e}^{-\mathrm{i}k\boldsymbol{n}\cdot\boldsymbol{X}}$ is an operator which does not commute with the centre of mass momentum $\boldsymbol{P}$, the interaction with the electromagnetic field changes the momentum and the kinetic energy of the atom.

ii) This interaction with the centre of mass degrees of freedom is different for every direction $\boldsymbol{n}$ of propagation of the photon.

15.3.5 Quantum Stochastic Schrödinger Equation Including Atomic Motion

The formulation of the quantum stochastic differential equation formalism requires care in order to account correctly for the terms $\mathrm{e}^{\pm\mathrm{i}k\boldsymbol{n}\cdot\boldsymbol{X}}$, since in the transition from the concept of a quantum field to that of a quantum noise operator, the integrals of these terms over k must be accounted for.

a) Noise Operators: The relevant noise operator terms corresponding to (15.2.2) are now

$$\mathcal{F}_{\boldsymbol{n},\lambda}(t) \equiv \int_{\mathcal{D}(\omega_{eg},\vartheta)} dk\, B_{k,\boldsymbol{n},\lambda}\mathrm{e}^{\mathrm{i}k\boldsymbol{n}\cdot\boldsymbol{X}}\mathrm{e}^{-\mathrm{i}(\omega_k-\omega_{eg})t}. \tag{15.3.8}$$

The commutation relations are, despite the factor $\mathrm{e}^{\mathrm{i}k\boldsymbol{n}\cdot\boldsymbol{X}}$, the same as those for $f_{n,\lambda}(t)$, and since $\boldsymbol{X}$ commutes with the photon operators, so that

$$\Big[\mathcal{F}_{\boldsymbol{n},\lambda}(t),\mathcal{F}^\dagger_{\boldsymbol{n},\lambda}(t')\Big] = \delta(\boldsymbol{n},\boldsymbol{n}')\delta_{\lambda,\lambda'}\int dk\,|\kappa(k,\boldsymbol{n},\lambda)|^2\,\mathrm{e}^{-\mathrm{i}(\omega_k-\omega_{eg})(t-t')}$$
$$\times\Big[\mathrm{e}^{\mathrm{i}k\boldsymbol{n}\cdot\boldsymbol{X}}B_{k,\boldsymbol{n},\lambda}\,,\, B^\dagger_{k,\boldsymbol{n},\lambda}\mathrm{e}^{-\mathrm{i}k\boldsymbol{n}\cdot\boldsymbol{X}}\Big], \tag{15.3.9}$$

$$= \delta(\boldsymbol{n},\boldsymbol{n}')\delta_{\lambda,\lambda'}\gamma_{\boldsymbol{n},\lambda}(t-t'). \tag{15.3.10}$$

Here $\gamma_{n,\lambda}(\tau)$ is as previously defined in (15.2.4).

b) Momentum Transferred by Noise Operators: Although the commutation relation is not changed by the inclusion of the term $e^{\pm ik\boldsymbol{n}\cdot\boldsymbol{X}}$, the operators $\mathcal{F}_{n,\lambda}(t)$ are not the same as the operators $f_{n,\lambda}(t)$ defined in (15.2.2), which do not have such a factor. It is this factor which generates the exchange of momentum between the electromagnetic field and the atom.

c) Momentum-Jump Operators: In order to get a compact description, we introduce the *momentum jump operators,* which we define as

$$\sigma^{\pm}(\boldsymbol{n}) \equiv e^{\pm ik_{eg}\boldsymbol{n}\cdot\boldsymbol{X}}\sigma^{\pm}, \qquad \text{where } k_{eg} \equiv \omega_{eg}/c. \tag{15.3.11}$$

for which we have the algebraic relationships

$$\sigma^{+}(\boldsymbol{n})\sigma^{-}(\boldsymbol{n}) = \sigma^{+}\sigma^{-} = \tfrac{1}{2}(1+\sigma_z), \tag{15.3.12}$$

$$\sigma^{-}(\boldsymbol{n})\sigma^{+}(\boldsymbol{n}) = \sigma^{-}\sigma^{+} = \tfrac{1}{2}(1-\sigma_z), \tag{15.3.13}$$

$$\left[\sigma^{\pm}(\boldsymbol{n}),\boldsymbol{P}\right] = \pm\hbar k_{eg}\boldsymbol{n}\sigma^{\pm}(\boldsymbol{n}). \tag{15.3.14}$$

The operator $\sigma^{+}(\boldsymbol{n})$ raises the atom from the ground state and increases its momentum by $\hbar k_{eg}\boldsymbol{n}$, the momentum of the photon absorbed, while the operator $\sigma^{+}(\boldsymbol{n})$ lowers the atom to its ground state, reducing its momentum by $\hbar k_{eg}\boldsymbol{n}$, the momentum of the photon emitted.

d) Form of the Quantum Stochastic Schrödinger Equation: This is similar to the form found in Sect. 15.2.5, and uses the same quantum Ito increments as formulated in (15.2.13–15.2.15), but appropriate replacements arising from (15.3.11–15.3.14) must also be introduced. Making these replacements, the quantum stochastic Schrödinger equation takes the form

$$d|\Psi_{\text{sys}},t\rangle = \left(-\frac{i}{\hbar}H_{\text{eff}}\,dt + \sum_{\lambda}\int d\Omega_{\boldsymbol{n}}\sqrt{\Gamma_{\boldsymbol{n},\lambda}}\,\sigma^{-}(\boldsymbol{n})\,dB^{\dagger}_{\boldsymbol{n},\lambda}(t)\right)|\Psi_{\text{sys}},t\rangle, \tag{15.3.15}$$

$$H_{\text{eff}} = H_{\text{Atom}} + \sum_{\lambda}\int d\Omega_{\boldsymbol{n}}\left(\hbar\delta\omega_{\boldsymbol{n},\lambda} - \tfrac{1}{2}i\hbar\Gamma_{\boldsymbol{n},\lambda}\right)\sigma^{+}\sigma^{-}, \tag{15.3.16}$$

$$\equiv H_{\text{Atom}} + \hbar\left(\delta\omega - \tfrac{1}{2}i\Gamma\right)\sigma^{+}\sigma^{-}. \tag{15.3.17}$$

The parameters $\Gamma_{\boldsymbol{n},\lambda}$ and $\delta\omega_{\boldsymbol{n},\lambda}$ are the same as in Chap. 13, and are as defined by the equations (15.2.12, 15.2.8, 15.1.18). In the last equation $\delta\omega$ and Γ are as previously defined in (15.2.18, 15.2.19).

e) The Amount of Momentum Transferred: In the definition (15.3.8) of the noise operators $\mathcal{F}_{n,\lambda}(t)$ the exponential factor depends on k, a variable of integration, but in the definition of the momentum jump operators, and therefore also on the quantum stochastic differential equation (15.3.15) the dependence is on k_{eg}, the value of k appropriate to a photon emitted in the transition $e \longrightarrow g$. Let us explain how this comes about.

The relevant object to consider is the term involving $dB^\dagger_{n,\lambda}(t)$ in the quantum stochastic differential equation in (15.3.15), and what we will want to show is that

$$\int_0^{\Delta t} dt \int_0^\infty dk B^\dagger_{n,\lambda} e^{i(\omega_k-\omega_{eg})t} e^{-ikn\cdot X} \longrightarrow e^{-ik_{eg}n\cdot X} \Delta B^\dagger_{n,\lambda}(t). \tag{15.3.18}$$

The validity of this limit is heuristically obvious, since, because of the time exponential, the k-integral only gives a significant contribution in a neighbourhood centred on the value $\omega_{eg}c$, and of width of order of magnitude $c/\Delta t$; hence we can replace $k \to k_{eg}$ in the space exponential, thus giving the result.

15.4 The Laser Cooling Master Equation

This is evaluated using the same procedure as in Chap. 13, and takes the form

$$\frac{d\rho_A}{dt} = -\frac{i}{\hbar}(H_{\text{eff}}\rho_A - \rho_A H^\dagger_{\text{eff}}) + \Gamma \int d\Omega_n \, \Phi(n)\, \sigma^-(n)\rho_A\sigma^+(n), \tag{15.4.1}$$

$$\Phi(n) \equiv \frac{1}{\Gamma}\sum_\lambda \Gamma_{n,\lambda}\,. \tag{15.4.2}$$

There is also the alternative expression

$$\frac{d\rho_A}{dt} = -\frac{i}{\hbar}\left[H_{\text{Atom}} + \delta\omega\,|e\rangle\langle e|\,,\, \rho_A\right] + \tfrac{1}{2}\Gamma\left(2\int d\Omega_n \Phi(n)\sigma^-(n)\rho_A\sigma^+(n) - \sigma^+\sigma^-\rho_A - \rho_A\sigma^+\sigma^-\right), \tag{15.4.3}$$

Remember that from the definition (15.3.3) the classical driving field has already been included in H_{Atom} thus:

$$H_{\text{Atom}} = \frac{p^2}{2M} + V(X) + \hbar\omega_{eg}|e\rangle\langle e| - \left(d_{eg}\cdot E^{(+)}_{\text{cl}}(X,t)\sigma^+ + d^*_{eg}\cdot E^{(-)}_{\text{cl}}(X,t)\sigma^-\right). \tag{15.4.4}$$

The solutions of this master equation in the context of laser cooling will be the subject matter of the remainder of this part of the book.

a) Transforming away the Optical Frequencies: When the classical driving field has a single optical frequency, that is

$$E^{(+)}_{\text{cl}}(x,t) = \epsilon_{\text{cl}}\mathcal{E}(x)e^{-i\omega t}, \tag{15.4.5}$$

exactly the same procedure as in Sect. 13.1.2 can be used to give a master equation in a frame rotating at the laser frequency, resulting in the form in which the master equation is normally used

$$\frac{d\bar{\rho}_A}{dt} = -\frac{\mathrm{i}}{\hbar}\left[\bar{H}_{\text{Atom}} + \Delta\,|e\rangle\langle e|\,,\,\bar{\rho}_A\right] + \tfrac{1}{2}\Gamma\left(2\int d\Omega_{\boldsymbol{n}}\,\Phi(\boldsymbol{n})\sigma^-(\boldsymbol{n})\bar{\rho}_A\sigma^+(\boldsymbol{n}) - \sigma^+\sigma^-\bar{\rho}_A - \bar{\rho}_A\sigma^+\sigma^-\right), \tag{15.4.6}$$

$$\bar{H}_{\text{Atom}} \equiv \frac{\boldsymbol{p}^2}{2M} + V(\boldsymbol{X}) - \left(\boldsymbol{d}_{eg}\cdot\boldsymbol{\epsilon}_{\text{cl}}\mathcal{E}(\boldsymbol{X})\sigma^+ + \left(\boldsymbol{d}_{eg}\cdot\boldsymbol{\epsilon}_{\text{cl}}\mathcal{E}(\boldsymbol{X})\right)^*\sigma^-\right). \tag{15.4.7}$$

b) Applying the Laser Cooling Master Equation: This equation describes the mechanical effects of the interaction of an atom with the quantized electromagnetic field, but its practical use is normally confined to the case where the spontaneous emission process is very much more rapid than the motion of the atom. The next two chapters show how adiabatic elimination of the internal atomic degrees of freedom is used to effect a further simplification, and thus provide a practical description of the cooling process.

16. Laser Cooling of Untrapped Atoms

The solution of the laser cooling master equation that we derived in the previous chapter can only be achieved by approximation methods, and these are chosen to suit the particular problem under study. However, there is one practical principle that guides the methods chosen, and that is that the cooling will be most effective when the noise terms are small. Thus we will analyse the equations to determine when this happens, and use approximation methods based on an expansion in an appropriate small parameter. The main ideas in this chapter follow those of [16.1–16.4], while the book by *Metcalf* and *van der Straten* [16.5] provides an introduction to and an overview of the practical aspects of laser cooling,

The simplified model of laser cooling that was given in *Bk. I: Sect.14.4*, which we used to get an estimate of the lowest temperature achieved by Doppler cooling, depends on the smallness of the ratio of two momentum scales:

i) *The Momentum of a Single Resonant Photon*: This is $\hbar k_{eg} \approx \hbar k_L = h/\lambda$.

ii) *The Width Δp of the Probability Distribution for the Centre of Mass Momentum of the Atom*: This provides a momentum scale over which this momentum distribution changes significantly.

We will therefore define a dimensionless parameter $\bar{\eta} \equiv \hbar k_{eg}/\Delta p$, and develop a procedure which is valid when $\bar{\eta}$ is small.

In the case of a tightly trapped atom, Δp is inversely proportional to the spatial size of the ground state wavefunction, and $\bar{\eta}$ becomes the Lamb–Dicke parameter η, as will be defined in Sect. 23.1.5. This is in fact a rather special case, in which the state of the atom is determined by only a few trap levels. In the general case of a dilute gas of atoms, the width of the momentum distribution is not directly related the width of the position distribution. For this kind of situation a description in terms of the Wigner distribution function is more relevant.

16.1 Wigner Representation of the Atomic Density Matrix

Let us return to the laser cooling master equation for the two-level system as given in (15.4.6, 15.4.7), and develop methods of solving these equations in the case that there is no trapping potential, or possibly only a very weak and broad trap. The *Wigner distribution function*, which we have introduced in *Bk. I: Sect.16.3*, provides the framework we need to describe the system in terms of momentum and position simultaneously, in a way quite closely related to the classical methods of kinetic theory.

16.1.1 Internal and Centre of Mass Degrees of Freedom

We do need to consider the internal structure of the atom, since cooling is implemented by means of transitions between the internal energy levels. This means that it will be necessary to generalize the definition *Bk. I: (16.3.1)* so as to map the density operator $\rho_A(t)$ to include the internal degrees of freedom; thus we define a *matrix Wigner distribution function* by

$$W(\boldsymbol{x},\boldsymbol{p},t) = \frac{1}{(2\pi\hbar)^3}\int d^3u\,\langle \boldsymbol{x}+\tfrac{1}{2}\boldsymbol{u}|\,\rho_A(t)\,|\boldsymbol{x}-\tfrac{1}{2}\boldsymbol{u}\rangle\, e^{-i\boldsymbol{p}\cdot\boldsymbol{u}/\hbar}. \tag{16.1.1}$$

Note that $W(\boldsymbol{x},\boldsymbol{p},t)$ can be viewed as a *density operator* with respect to the internal atomic degrees of freedom, and a *Wigner distribution function* with respect to the centre of mass degrees of freedom.

a) Distribution for the Centre of Mass Motion: Although the internal degrees of freedom are an essential part of the cooling process, the quantity of interest is the kinetic temperature, and this is determined by the momentum distribution function. Thus we need the trace over the atomic degrees of freedom, g and e

$$N(\boldsymbol{x},\boldsymbol{p},t) \equiv \mathrm{Tr}_{\mathrm{At}}\left\{W(\boldsymbol{x},\boldsymbol{p},t)\right\} = W_{gg}(\boldsymbol{x},\boldsymbol{p},t) + W_{ee}(\boldsymbol{x},\boldsymbol{p},t). \tag{16.1.2}$$

This is the quantity of main interest in laser cooling theory. and our main aim is to get a Fokker–Planck equation in momentum space for $N(\boldsymbol{x},\boldsymbol{p},t)$, from which we can determine the stationary distribution function, and therefore the temperature achievable by the cooling process.

16.1.2 Evolution Equation for the Wigner Distribution Function

The first step is to use the operator mappings in *Bk. I: Sect.16.3.2* to convert the master equation (15.4.6) for $\bar\rho_A(t)$ to an equation for $W(\boldsymbol{x},\boldsymbol{p},t)$. We obtain

$$\begin{aligned}\frac{\partial}{\partial t}W(\boldsymbol{x},\boldsymbol{p},t) &= -\frac{\boldsymbol{p}}{m}\cdot\nabla_{\boldsymbol{x}}W(\boldsymbol{x},\boldsymbol{p},t) + i\Delta\left[|e\rangle\langle e|, W(\boldsymbol{x},\boldsymbol{p},t)\right]\\ &\quad+\frac{i}{\hbar}\int d^3\boldsymbol{k}\, e^{i\boldsymbol{k}\cdot\boldsymbol{x}}\left[W\big(\boldsymbol{x},\boldsymbol{p}+\tfrac{1}{2}\hbar\boldsymbol{k},t\big)\tilde V_{\text{A-L}}(\boldsymbol{k}) - \tilde V_{\text{A-L}}(\boldsymbol{k})W\big(\boldsymbol{x},\boldsymbol{p}-\tfrac{1}{2}\hbar\boldsymbol{k},t\big)\right]\\ &\quad-\tfrac{1}{2}\Gamma\left[\sigma^+\sigma^- W\big(\boldsymbol{x},\boldsymbol{p},t\big) + W\big(\boldsymbol{x},\boldsymbol{p},t\big)\sigma^+\sigma^-\right]\\ &\quad+\Gamma\int d\Omega_{\boldsymbol{n}}\,\Phi(\boldsymbol{n})\,\sigma^- W\big(\boldsymbol{x},\boldsymbol{p}+\hbar k_{eg}\boldsymbol{n},t\big)\sigma^+ .\end{aligned} \tag{16.1.3}$$

We have written this equation in terms of the Fourier transform of the laser-atom interaction

$$V_{\text{A-L}}(\boldsymbol{X}) \equiv -\left(\boldsymbol{d}_{eg}\cdot\boldsymbol{e}_{\text{cl}}\mathcal{E}(\boldsymbol{X})\sigma^+ + \left(\boldsymbol{d}_{eg}\cdot\boldsymbol{e}_{\text{cl}}\mathcal{E}(\boldsymbol{X})\right)^*\sigma^-\right), \tag{16.1.4}$$

$$\equiv \int d^3\boldsymbol{k}\, e^{i\boldsymbol{k}\cdot\boldsymbol{X}}\,\tilde V_{\text{A-L}}(\boldsymbol{k}). \tag{16.1.5}$$

Here of course $\boldsymbol{X}$ is the operator for the position of the centre of mass, so that $e^{i\boldsymbol{k}\cdot\boldsymbol{X}}$ is a *momentum kick operator*, that is, an operator which increases the centre of mass momentum by $\hbar\boldsymbol{k}$.

a) Derivation of the Equation of Motion for the Wigner Function: This proceeds as follows:

i) The first line uses the methods of *Bk. I: Sect.16.3.3*—for simplicity here, we have omitted the trapping potential term.

ii) The second line uses the mappings *Bk. I: (16.3.33)–(16.3.36)*.

iii) In the fourth line we explicitly write out the definition (15.3.11) of $\sigma^{\pm}(\boldsymbol{n})$, and use the mappings *Bk. I: (16.3.33)–(16.3.36)*.

b) Interpretation: We note the following points:

i) The first line contains the kinetic energy and evolution due to the free atomic Hamiltonian.

ii) The second line is the laser-atom interaction, and includes the associated momentum transfer $\hbar\boldsymbol{k}$.

iii) The last two lines are dissipative terms; note that the recycling term is associated with a momentum transfer $\hbar k_{eg}\boldsymbol{n}$ to the centre of mass, in the direction of the unit vector $\boldsymbol{n}$, which is distributed according to the angular distribution $\Phi(\boldsymbol{n})$ of spontaneous emission.

iv) The magnitude of the momentum transfer from the laser-atom interaction as in ii) is governed by the laser frequency, while the momentum transfer from the dissipative terms as in iii) is governed by the atomic transition frequency. This is the fundamental mechanism of energy transfer into and out of the centre of mass motion.

16.2 Doppler Cooling of a Two-Level System Using a Travelling Light Wave

In the general case, laser cooling involves multiple laser beams and atoms in which the energy levels involved have non-zero angular momentum, leading to degenerate excited and ground states. The resultant equations are quite complicated, although the fundamental processes are essentially the same as those of the simplest case.

Therefore we will first formulate laser cooling for the simplest situation, that of a single incident laser beam as illustrated in Fig. 16.1, for which the atom-laser interaction in a rotating frame is

$$V_{\text{A-L}}(z) = -\left(\boldsymbol{d}_{eg}\cdot\boldsymbol{e}_{\text{cl}}\mathcal{E}\,e^{ik_Lz}\sigma^+ + \sigma^-(\boldsymbol{d}_{eg}\cdot\boldsymbol{e}_{\text{cl}}\mathcal{E})^*\,e^{-ik_Lz}\right), \tag{16.2.1}$$

$$\equiv -\tfrac{1}{2}\hbar\Omega_R\left(e^{ik_Lz}\sigma^+ + e^{-ik_Lz}\sigma^-\right). \tag{16.2.2}$$

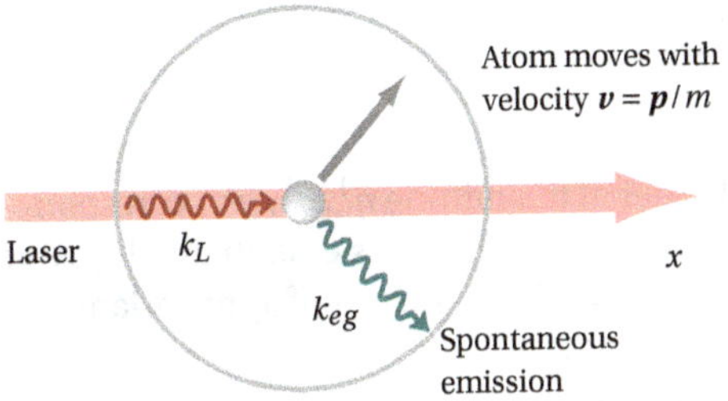

Fig. 16.1. A single laser beam travelling along the x-axis, and with wavenumber k_L is incident on an atom. Spontaneous emission from the excited level occurs in all directions, with wavenumber k_{eg}. For laser cooling, k_L is chosen very slightly less than k_{eg}.

Here, for simplicity, we have chosen the phase of $\mathcal{E}$ so that the Rabi frequency $\Omega_R \equiv 2\boldsymbol{d}_{eg} \cdot \boldsymbol{e}_{\rm cl}\mathcal{E}/\hbar$ is real and positive.

16.2.1 Equation for the Wigner Function

If we insert this particular form of the laser-atom interaction into the general equation (16.1.3) for the matrix Wigner function, we get the basic equation for Doppler cooling of a two-level atom by a single travelling wave:

$$
\begin{aligned}
\frac{\partial}{\partial t} W(\boldsymbol{x},\boldsymbol{p},t) = & -\frac{\boldsymbol{p}}{m}\cdot\nabla_{\boldsymbol{x}} W(\boldsymbol{x},\boldsymbol{p},t) + \mathrm{i}\Delta\big[|e\rangle\langle e|\,,\, W(\boldsymbol{x},\boldsymbol{p},t)\big] \\
& + \tfrac{1}{2}\mathrm{i}\Omega_R\Big(\mathrm{e}^{\mathrm{i}k_L z}\sigma^+ W\big(\boldsymbol{x},\boldsymbol{p}-\tfrac{1}{2}\hbar\boldsymbol{k}_L,t\big) + \sigma^-\mathrm{e}^{-\mathrm{i}k_L z} W\big(\boldsymbol{x},\boldsymbol{p}+\tfrac{1}{2}\hbar\boldsymbol{k}_L,t\big)\Big) \\
& - \tfrac{1}{2}\mathrm{i}\Omega_R\Big(W\big(\boldsymbol{x},\boldsymbol{p}+\tfrac{1}{2}\hbar\boldsymbol{k}_L,t\big)\mathrm{e}^{\mathrm{i}k_L z}\sigma^+ + W\big(\boldsymbol{x},\boldsymbol{p}-\tfrac{1}{2}\hbar\boldsymbol{k}_L,t\big)\sigma^-\mathrm{e}^{-\mathrm{i}k_L x}\Big) \\
& - \tfrac{1}{2}\Gamma\Big(\sigma^+\sigma^- W\big(\boldsymbol{x},\boldsymbol{p},t\big) + W\big(\boldsymbol{x},\boldsymbol{p},t\big)\sigma^+\sigma^-\Big) \\
& + \Gamma\int d\Omega_{\boldsymbol{n}}\,\Phi(\boldsymbol{n})\,\sigma^- W\big(\boldsymbol{x},\boldsymbol{p}+\hbar k_{eg}\boldsymbol{n},t\big)\sigma^+\,,
\end{aligned}
\qquad (16.2.3)
$$

$$
\boldsymbol{k}_L \equiv (0,0,k_L). \qquad (16.2.4)
$$

This equation forms the basis from which we will develop an approximation method, which will ultimately yield the desired Fokker–Planck equation.

16.2.2 Evolution Operators and Stationary Solutions

We will formulate a methodology based on perturbation theory, and executed using projection operator techniques, very like those used in *Bk. I: Sect.13.2.1* in deriving the quantum-optical master equation.

Thus, in the case of laser cooling in a travelling wave incident on a two-level system, the equation of motion (16.2.3) can be written in terms of the sum of five operators as

$$\frac{\partial W}{\partial t} = (\mathcal{L}_{0a} + \mathcal{L}_{0b} + \mathcal{L}_{0c} + \mathcal{L}_1 + \mathcal{L}_2)\,W. \tag{16.2.5}$$

We will show that these operators are of order of magnitude $\bar{\eta}^0, \bar{\eta}^1, \bar{\eta}^2$, corresponding to the value of the subscript. The perturbation theory is essentially the same as that used in adiabatic elimination, and depends on the use of appropriate projectors.

The evolution operators have the explicit forms:

$$\mathcal{L}_{0a} W \equiv -\frac{\boldsymbol{p}\cdot\nabla}{m}, \tag{16.2.6}$$

$$\mathcal{L}_{0b} W \equiv \mathrm{i}\Delta\big[\,|e\rangle\langle e|, W\big] - \tfrac{1}{2}\Gamma\left(\sigma^+\sigma^- W + W\sigma^+\sigma^- - 2\sigma^- W\sigma^+\right), \tag{16.2.7}$$

$$\mathcal{L}_{0c} W \equiv \tfrac{1}{2}\mathrm{i}\Omega_R\left(\mathrm{e}^{ikz}\left[\sigma^+, W\right] + \mathrm{e}^{-ikz}\left[\sigma^-, W\right]\right), \tag{16.2.8}$$

$$\mathcal{L}_1 W \equiv \tfrac{1}{2}\mathrm{i}\Omega_R\left(\mathrm{e}^{ikz}\left(\sigma^+ D^- W - D^+ W\sigma^+\right) + \mathrm{e}^{-ikz}\left(\sigma^- D^+ W - D^- W\sigma^-\right)\right), \tag{16.2.9}$$

$$\mathcal{L}_2 W \equiv \Gamma\int d\Omega_{\boldsymbol{n}}\,\phi(\boldsymbol{n})\,\sigma^-\big(T(\boldsymbol{n})W\big)\sigma^+. \tag{16.2.10}$$

In the definition of $\mathcal{L}_1$ and $\mathcal{L}_2$ we have used operators defined by

$$D^{\pm} W(\boldsymbol{x},\boldsymbol{p}) \equiv W(\boldsymbol{x},\boldsymbol{p} \pm \tfrac{1}{2}\hbar\boldsymbol{k}_L) - W(\boldsymbol{x},\boldsymbol{p}), \tag{16.2.11}$$

$$T(\boldsymbol{n}) W(\boldsymbol{x},\boldsymbol{p}) \equiv W(\boldsymbol{x},\boldsymbol{p} + \hbar k_{eg}\boldsymbol{n}) - W(\boldsymbol{x},\boldsymbol{p}). \tag{16.2.12}$$

a) The Small *k* Approximation: The wavenumbers k_{eg} and k_L are almost equal, and we will want to develop a perturbation method which uses the fact that the momentum kicks of their order of magnitude are very small compared to the typical momentum of the atom. The criterion we need is that such a momentum kick changes the Wigner function by only a small amount. To put this explicitly, we will define the quantity

$$\bar{\eta} \equiv \frac{\hbar k_{eg}}{\Delta p}, \tag{16.2.13}$$

where Δp is the typical momentum scale over which the Wigner function changes. This means that the operators $D^{\pm}$ and $T(\boldsymbol{n})$ are of order of magnitude $\bar{\eta}$, while $\int d\Omega_{\boldsymbol{n}}\,\phi(\boldsymbol{n})\,T(\boldsymbol{n})$ is of order $\bar{\eta}^2$, since $\phi(\boldsymbol{n})$ is a symmetric function of $\boldsymbol{n}$.

b) Kramers–Moyal Expansion: It is most convenient to use a derivative approximation for these operators, which yields an approximation similar to the Kramers–Moyal expansion introduced in *Bk. I: Sect.6.1.1*; thus we can write approximate expressions to the lowest non-vanishing order. For $D^{\pm}$ we get

$$D^{\pm} \approx \pm\tfrac{1}{2}\hbar\boldsymbol{k}_L\cdot\nabla_{\boldsymbol{p}}. \tag{16.2.14}$$

For the other expression we assume that $\phi(\boldsymbol{n})$ has inversion symmetry, so that the first non-vanishing term is of second order, namely

$$\int d\Omega_{\boldsymbol{n}}\,\phi(\boldsymbol{n})T(\boldsymbol{n}) \approx \tfrac{1}{2}(\hbar k_{eg})^2 \sum_{i,j} \alpha_{ij} \frac{\partial^2}{\partial p_i \partial p_j}\,, \tag{16.2.15}$$

$$\text{where} \qquad \alpha_{ij} \equiv \int d\Omega_{\boldsymbol{n}}\,\phi(\boldsymbol{n}) n_i n_j. \tag{16.2.16}$$

In practice only the diagonal elements of α_{ij} are normally non-zero.

The expression of the jump operators in terms of derivatives makes for a considerable simplification of the laser cooling equations. Without it we would be forced to consider momentum jumps of magnitude k_{eg} in all directions in three dimensions, and of magnitude k_L in the positive and negative z-directions. However, this simplification is not *essential* for the approximation procedures we will use; rather it is *convenient and simple.*

The validity of the approximation requires that the momentum distribution of the laser-cooled atom is sufficiently broad, even at the coldest temperature achieved. As noted in *Bk. I: (14.4.29)*, the width of the momentum distribution is then given by

$$\Delta p \to \sqrt{2m\hbar\Gamma}\,, \tag{16.2.17}$$

so a sufficiently narrow linewidth Γ can in principle lead to a situation in which $k_{eg} \sim \Delta p$. But in that case the parameter $\bar{\eta} \sim 1$, and the whole approximation procedure breaks down. Thus, the derivative approximation is expected to be valid whenever the expansion in $\bar{\eta}$ is valid.

c) Orders of Magnitude Operators in Powers of $\bar{\eta}$: The evolution operators in (16.2.5–16.2.10) can be classified as follows:

i) *Terms of Order* $\bar{\eta}^0$: $\mathcal{L}_{0a}$ — Transport,
$\mathcal{L}_{0b}$ — Spontaneous emission,
$\mathcal{L}_{0c}$ — Coherent driving.

ii) *Terms of Order* $\bar{\eta}^1$: $\mathcal{L}_1$ — Momentum transfer and fluctuations arising from driving.

iii) *Terms of Order* $\bar{\eta}^2$: $\mathcal{L}_2$ — Momentum fluctuations arising from spontaneous emission.

d) Stationary Solutions at Order Zero: The terms of order $\bar{\eta}^0$ separate naturally into two parts, corresponding to optical driving, and spatial transport.

i) *Optical Driving*: The equation

$$\frac{\partial \bar{\rho}(\boldsymbol{x},\boldsymbol{p})}{\partial t} = \big(\mathcal{L}_{0b} + \mathcal{L}_{0c}\big)\bar{\rho}(\boldsymbol{x},\boldsymbol{p}) \equiv \mathcal{L}_{\mathrm{OBE}}(\boldsymbol{x})\bar{\rho}(\boldsymbol{x},\boldsymbol{p}), \tag{16.2.18}$$

corresponds to a version of optical Bloch equations in the interaction picture, as formulated in (13.1.16–13.1.19). The driving field, being proportional to $\exp(\mathrm{i}k_L z)$, provides an explicit space dependence, but there is no explicit dependence on momentum. Furthermore, there is no coupling between the equations for different $\boldsymbol{x}$ and $\boldsymbol{p}$.

The stationary solutions of this equation are therefore as given in (13.2.6, 13.2.8) with the substitution $\Omega_R \to \Omega_R \exp(\mathrm{i}kz)$, and thus take the form

$$\bar{\rho}(\boldsymbol{x},\boldsymbol{p})_s \equiv \begin{pmatrix} p_{ee} & p_{eg}\mathrm{e}^{\mathrm{i}k_L z} \\ p_{eg}^*\mathrm{e}^{-\mathrm{i}k_L z} & p_{gg} \end{pmatrix}, \tag{16.2.19}$$

$$= U(k_L z)\begin{pmatrix} p_{ee} & p_{eg} \\ p_{eg}^* & p_{gg} \end{pmatrix} U^\dagger(k_L z), \tag{16.2.20}$$

$$\text{where} \quad U(k_L z) = \exp(\mathrm{i}k_L z\sigma_z). \tag{16.2.21}$$

ii) *Spatial Transport*: The other operator of order $\bar{\eta}^0$ is the transport operator, arising from the kinetic energy

$$\mathcal{L}_{\text{Kin}} \equiv -\boldsymbol{v}\cdot\nabla \equiv -\frac{\boldsymbol{p}\cdot\nabla}{m}. \tag{16.2.22}$$

The corresponding stationary solutions will be discussed in Sect. 16.2.3.

iii) *Elimination of spatial dependence in the case of a plane travelling wave*: The unitary transformation (16.2.21) can be used to eliminate all spatial dependence in the equations of motion (16.2.5–16.2.10) by defining

$$W_L(\boldsymbol{x},\boldsymbol{p},t) \equiv U^\dagger(k_L z)W(\boldsymbol{x},\boldsymbol{p},t)U(k_L z), \tag{16.2.23}$$

for which the equations of motion are obtained from (16.2.5–16.2.10) by the substitutions

$$W \to W_L, \qquad \mathrm{e}^{\pm\mathrm{i}k_L z} \to 1, \qquad \Delta \to \Delta - \boldsymbol{k}_L\cdot\boldsymbol{v}. \tag{16.2.24}$$

iv) *Stationary Solution of the Full Equations in the Case of a Travelling Plane Wave*: This means that the stationary solution $W_s(\boldsymbol{x},\boldsymbol{p})$ of the evolution equation to order $\bar{\eta}^0$, that is

$$\frac{\partial W}{\partial t} = \mathcal{L}_0 W \equiv (\mathcal{L}_{\text{Kin}} + \mathcal{L}_{\text{OBE}})W, \tag{16.2.25}$$

is given by the formulae (13.2.6, 13.2.8), with the substitution $\Delta \to \Delta - \boldsymbol{k}_L\cdot\boldsymbol{v}$ from (16.2.24), hence

$$W_{s,ee} = 1 - W_{s,gg} = \frac{\frac{1}{4}|\Omega_R|^2}{(\Delta - \boldsymbol{k}_L\cdot\boldsymbol{v})^2 + \frac{1}{4}\Gamma^2 + \frac{1}{2}|\Omega_R|^2}, \tag{16.2.26}$$

$$W_{s,eg} = W_{s,ge}^* = -\tfrac{1}{2}\Omega_R\mathrm{e}^{\mathrm{i}k_L z}\frac{(\Delta - \boldsymbol{k}_L\cdot\boldsymbol{v}) + \frac{1}{2}\mathrm{i}\Gamma}{(\Delta - \boldsymbol{k}_L\cdot\boldsymbol{v})^2 + \frac{1}{4}\Gamma^2 + \frac{1}{2}|\Omega_R|^2}. \tag{16.2.27}$$

Exercise 16.1 Derivation of the Doppler Shift: Explicitly prove that the unitary transformation (16.2.23) leads to (16.2.24).

16.2.3 Projectors

In laser cooling the quantity of primary interest is the phase-space distribution function, since this determines the kinetic temperature, and the spatial distribution of the atoms. The occupations of the internal states are not of immediate interest. This means that the quantity of interest will be the phase space distribution function, obtained by tracing over the internal degrees of freedom:

$$N(\boldsymbol{x},\boldsymbol{p},t) \equiv \mathrm{Tr}_{\mathrm{Atom}}\left\{W(\boldsymbol{x},\boldsymbol{p},t)\right\}. \tag{16.2.28}$$

The definitions (16.2.6–16.2.8) show that the traces of the terms involving $\mathcal{L}_{0b}$ and $\mathcal{L}_{0a}$ vanish, so the equation of motion for the phase space distribution function to order $\bar{\eta}^0$ is simply the flow equation

$$\frac{\partial N(\boldsymbol{x},\boldsymbol{p},t)}{\partial t} = \boldsymbol{v}\cdot\nabla N(\boldsymbol{x},\boldsymbol{p},t). \tag{16.2.29}$$

This means that to order $\bar{\eta}^0$ the light field exerts no force on the atom, whose velocity is therefore constant. The stationary solutions of this equation are functions $N_s(\boldsymbol{x},\boldsymbol{p})$ such that

$$N_s(\boldsymbol{x}+\tau\boldsymbol{v},\boldsymbol{p}) = N_s(\boldsymbol{x},\boldsymbol{p}), \quad \text{for all } \tau. \tag{16.2.30}$$

a) Definition of the Projector: The projector we need will project onto the stationary solution of the evolution equation to order $\bar{\eta}^0$, as defined in (16.2.25), and we will write this in the form

$$\mathcal{P}W(\boldsymbol{x},\boldsymbol{p},t) = W_s(\boldsymbol{x},\boldsymbol{p})\,\overline{N(\boldsymbol{x},\boldsymbol{p},t)}. \tag{16.2.31}$$

In this equation the quantities are defined by

i) $W_s(\boldsymbol{x},\boldsymbol{p})$ is a stationary solution of the zero order evolution equation (16.2.25), so that

$$\mathcal{L}_0 W_s(\boldsymbol{x},\boldsymbol{p}) = 0, \tag{16.2.32}$$

$$\mathrm{Tr}_{\mathrm{Atom}}\left\{W_s(\boldsymbol{x},\boldsymbol{p})\right\} = 1. \tag{16.2.33}$$

ii) The quantity $\overline{N(\boldsymbol{x},\boldsymbol{p},t)}$ contains the information we need, and is defined by

$$\overline{N(\boldsymbol{x},\boldsymbol{p},t)} = \lim_{T\to\infty}\left\{\frac{1}{2T}\int_{-T}^{T} ds\,\mathrm{Tr}_{\mathrm{Atom}}\left\{W(\boldsymbol{x}+s\boldsymbol{v},\boldsymbol{p},t)\right\}\right\}. \tag{16.2.34}$$

iii) Defined this way, $\overline{N(\boldsymbol{x}+\tau\boldsymbol{v},\boldsymbol{p},t)} = \overline{N(\boldsymbol{x},\boldsymbol{p},t)}$, so that $\boldsymbol{v}\cdot\nabla\,\overline{N(\boldsymbol{x},\boldsymbol{p},t)} = 0$. As a result, the definition (16.2.31) of the projector allows us to conclude that $\mathcal{L}_0\mathcal{P} = \mathcal{P}\mathcal{L}_0 = 0$, as is required for the method.

Exercise 16.2 Projector Properties: Satisfy yourself that this definition yields these mandatory projector properties.

b) Projector Properties and the Radiation Pressure Force: Here we list a number of technical properties of the projectors, and put the evolution operator in a form to which we can apply the techniques of adiabatic elimination.

i) As we have just noted, $\mathcal{L}_0\mathcal{P} = \mathcal{P}\mathcal{L}_0 = 0$.

ii) *Radiation Pressure Force*: However $\mathcal{P}\mathcal{L}_1\mathcal{P} \equiv \bar{\mathcal{L}}_1 \neq 0$, and we will therefore define

$$\bar{\mathcal{L}}_1 \equiv \mathcal{P}\mathcal{L}_1\mathcal{P}. \tag{16.2.35}$$

We can then use the derivative approximation (16.2.14) to compute that

$$\bar{\mathcal{L}}_1\mathcal{P}W(\boldsymbol{x},\boldsymbol{p}) = W_s(\boldsymbol{x},\boldsymbol{p})\left(\tfrac{1}{2}\hbar\Omega_R k_L\frac{\partial}{\partial p_z}\left\{\langle S_y(\boldsymbol{p})\rangle_s\overline{N(\boldsymbol{x},\boldsymbol{p})}\right\}\right), \tag{16.2.36}$$

where we have defined

$$\langle S_y(\boldsymbol{p})\rangle_s \equiv \mathrm{Tr}_{\mathrm{Atom}}\left\{\left(\mathrm{e}^{\mathrm{i}k_L z}\sigma^+ - \mathrm{e}^{-\mathrm{i}k_L z}\sigma^-\right)W_s(\boldsymbol{x},\boldsymbol{p})\right\}, \tag{16.2.37}$$

and it is straightforward to check that this result is indeed independent of $\boldsymbol{x}$, justifying the notational dependence on $\boldsymbol{p}$ alone.

It is now convenient to define the *radiation pressure force operator* by

$$F(\boldsymbol{x},\boldsymbol{p}) \equiv -\tfrac{1}{2}\mathrm{i}\hbar\Omega_R k_L\left(\mathrm{e}^{\mathrm{i}k_L z}\sigma^+ - \mathrm{e}^{-\mathrm{i}k_L z}\sigma^-\right). \tag{16.2.38}$$

We can then write $\bar{\mathcal{L}}_1$ as

$$\bar{\mathcal{L}}_1 W(\boldsymbol{x},\boldsymbol{p}) = W_s(\boldsymbol{x},\boldsymbol{p})\left(-\frac{\partial}{\partial p_z}\left\{\bar{F}(\boldsymbol{x},\boldsymbol{p})\overline{N(\boldsymbol{x},\boldsymbol{p})}\right\}\right), \tag{16.2.39}$$

where we have defined the *mean radiation pressure force* by

$$\bar{F}(\boldsymbol{x},\boldsymbol{p}) \equiv \mathrm{Tr}_{\mathrm{Atom}}\{F(\boldsymbol{x},\boldsymbol{p})W_s(\boldsymbol{x},\boldsymbol{p})\}, \tag{16.2.40}$$

$$= \hbar k_L\Gamma\frac{\tfrac{1}{4}\Omega_R^2}{(\Delta - \boldsymbol{k}_L\cdot\boldsymbol{v})^2 + \tfrac{1}{4}\Gamma^2 + \tfrac{1}{2}|\Omega_R|^2}, \tag{16.2.41}$$

and we have evaluated the last line using (16.2.27).

iii) Note that:

a) The force operator $F(\boldsymbol{x},\boldsymbol{p})$ is independent of $\boldsymbol{p}$, but does depend on $\boldsymbol{x}$ because of the factors $\exp(\pm i k_L z)$.

b) In contrast, the dependence on z in the force operator $F(\boldsymbol{x},\boldsymbol{p})$ is no longer present in the mean force $\bar{F}(\boldsymbol{x},\boldsymbol{p})$, which is in fact independent of $\boldsymbol{x}$, but now depends on $\boldsymbol{p}$ because of the term $\boldsymbol{k}_L\cdot\boldsymbol{v}$.

c) The mean force is the same as found in the simple model of laser cooling given in *Bk. I: (14.4.13)* of *Bk. I: Sect.14.4.2*, apart from the dependence on the Rabi frequency in the denominator, which was neglected in the simple treatment.

iv) *Radiation Pressure Fluctuations Arising from the Driving Field*: We now separate the operator $\mathcal{L}_1$ into two parts, one part arising from the mean radiation pressure force and the other part arising from the fluctuations about the mean. We first write $\mathcal{L}_1$ in terms of the force operator as

$$\mathcal{L}_1 W = -\tfrac{1}{2}\frac{\partial}{\partial p_z}[F(\boldsymbol{x},\boldsymbol{p}),W]_+ , \tag{16.2.42}$$

and then define

$$\delta\mathcal{L}_1 \equiv \mathcal{L}_1 - \bar{\mathcal{L}}_1, \tag{16.2.43}$$

which represents the fluctuations. Thus we find that

$$\bar{\mathcal{L}}_1\mathcal{P} = \mathcal{P}\bar{\mathcal{L}}_1, \tag{16.2.44}$$

$$\mathcal{P}\delta\mathcal{L}_1\mathcal{P} = 0. \tag{16.2.45}$$

v) *Radiation Pressure Fluctuations Arising from Spontaneous Emission*: Finally we note that we can also separate $\mathcal{L}_2$ in the same way, so that

$$\bar{\mathcal{L}}_2 \equiv \mathcal{P}\mathcal{L}_2\mathcal{P}, \tag{16.2.46}$$

$$\delta\mathcal{L}_2 \equiv \mathcal{L}_2 - \bar{\mathcal{L}}_2. \tag{16.2.47}$$

Since $\mathcal{P}\delta\mathcal{L}_2\mathcal{P} = 0$, the fluctuation operator $\delta\mathcal{L}_2$ amounts to a correction to the $\delta\mathcal{L}_1$ term, but is of order of magnitude $\bar{\eta}^2$, and we will show that this produces no effects to order $\bar{\eta}^2$.

It is relatively straightforward to evaluate $\bar{\mathcal{L}}_2$ using the definition (16.2.10), the derivative approximations (16.2.15, 16.2.16), and assuming α_{ij} is diagonal. The result is

$$\bar{\mathcal{L}}_2 W(\boldsymbol{x},\boldsymbol{p},t) = W_s(\boldsymbol{x},\boldsymbol{p})\left(\tfrac{1}{2}\Gamma(\hbar k_{eg})^2\sum_i \alpha_{ii}\frac{\partial^2}{\partial p_i^2}\left\{W_{s,ee}(\boldsymbol{p})\overline{N(\boldsymbol{x},\boldsymbol{p},t)}\right\}\right). \tag{16.2.48}$$

vi) *Resolution of the Evolution Operator*: The equation of motion is then grouped as

$$\frac{\partial W}{\partial t} = \{\mathcal{L}_{\text{cool}} + \mathcal{L}_{\text{fluct}}\} W, \tag{16.2.49}$$

in which we have defined the cooling and fluctuation parts as

$$\mathcal{L}_{\text{cool}} \equiv \mathcal{L}_0 + \bar{\mathcal{L}}_1 + \bar{\mathcal{L}}_2, \tag{16.2.50}$$

$$\mathcal{L}_{\text{fluct}} \equiv \delta\mathcal{L}_1 + \delta\mathcal{L}_2, \tag{16.2.51}$$

which satisfy

$$\mathcal{P}\mathcal{L}_{\text{cool}} = \mathcal{L}_{\text{cool}}\mathcal{P} = 0, \qquad \mathcal{P}\mathcal{L}_{\text{fluct}}\mathcal{P} = 0. \tag{16.2.52}$$

16.2.4 Adiabatic Elimination Procedure

If we define

$$v(\boldsymbol{x},\boldsymbol{p},t) \equiv \mathcal{P}W(\boldsymbol{x},\boldsymbol{p},t) = W_s(\boldsymbol{x},\boldsymbol{p})\overline{N(\boldsymbol{x},\boldsymbol{p},t)}, \tag{16.2.53}$$

then the elimination procedure required is essentially that used to derive the quantum optical master equation in *Bk. I: Sect.13.2*, and in the limit of small $\bar{\eta}$ yields to order $\bar{\eta}^2$

$$\frac{\partial v(t)}{\partial t} = \left(\bar{\mathcal{L}}_1 + \bar{\mathcal{L}}_2\right) v(t) + \mathcal{P}\mathcal{L}_{\text{fluct}} \int_0^t dt' \exp(\mathcal{L}_{\text{cool}} t')\mathcal{L}_{\text{fluct}}\, v(t-t'), \tag{16.2.54}$$

$$\approx \left(\bar{\mathcal{L}}_1 + \bar{\mathcal{L}}_2\right) v(t) + \mathcal{P}\,\delta\mathcal{L}_1 \int_0^t dt' \exp(\mathcal{L}_0 t')\,\delta\mathcal{L}_1 v(t-t'). \tag{16.2.55}$$

Since the motion of $v(t)$ will be of order $\bar{\eta}^1$, while that of $\mathcal{L}_0$ is of order $\bar{\eta}^0$, we can approximate $v(t-t')$ by $v(t)$, and note that on the time scale of order τ^1, the upper limit of the time integral is essentially infinite. This gives us the final Markovian equation

$$\frac{\partial v(t)}{\partial t} = \left(\bar{\mathcal{L}}_1 + \bar{\mathcal{L}}_2 + \mathcal{P}\,\delta\mathcal{L}_1 \int_0^\infty dt' \exp(\mathcal{L}_0 t')\,\delta\mathcal{L}_1\right) v(t). \tag{16.2.56}$$

a) Evaluation of Terms Involving $\delta\mathcal{L}_1$: The term involving the evolution operator is quite intricate, but fortunately it is of order $\bar{\eta}^2$, which enables us to make a number of simplifying approximations. We can write explicitly

$$\mathcal{L}_1 v(t) = -\tfrac{1}{2}\frac{\partial}{\partial p_z}\left(\left[F(\boldsymbol{x},\boldsymbol{p}), W_s(\boldsymbol{x},\boldsymbol{p})\right]_+ \overline{N(\boldsymbol{x},\boldsymbol{p},t)}\right), \tag{16.2.57}$$

$$\bar{\mathcal{L}}_1 v(t) = W_s(\boldsymbol{x},\boldsymbol{p})\left(-\frac{\partial}{\partial p_z}\left\{\bar{F}(\boldsymbol{x},\boldsymbol{p})\overline{N(\boldsymbol{x},\boldsymbol{p},t)}\right\}\right). \tag{16.2.58}$$

We can simplify the procedure considerably by moving the $\partial/\partial p_z$ operator to the right, so that it acts only on $\bar{N}(\boldsymbol{x},\boldsymbol{p},t)$. The terms arising from derivatives of $W_s(\boldsymbol{x},\boldsymbol{p})$ and $\bar{F}W_s(\boldsymbol{x},\boldsymbol{p})$ can be neglected, and no such term arises from $F(\boldsymbol{x},\boldsymbol{p})$, since it does not in fact depend on $\boldsymbol{p}$. We can neglect these terms since they will only contribute to the drift term of the Fokker–Planck equation in $N(\boldsymbol{x},\boldsymbol{p},t)$ we shall eventually derive, and are of order $\bar{\eta}^2$. They are therefore negligible compared to the main term in the drift, that is, the radiation pressure force, which is of order $\bar{\eta}^1$.

The result is that we can write

$$\delta\mathcal{L}_1 v(t) \;\to\; -\tfrac{1}{2}\left[\delta F(\boldsymbol{x},\boldsymbol{p})\,,\,W_s(\boldsymbol{x},\boldsymbol{p})\right]_+ \frac{\partial \overline{N(\boldsymbol{x},\boldsymbol{p},t)}}{\partial p_z}\,, \tag{16.2.59}$$

in which $\quad \delta F(\boldsymbol{x},\boldsymbol{p}) \equiv F(\boldsymbol{x},\boldsymbol{p}) - \bar{F}(\boldsymbol{x},\boldsymbol{p}).$ (16.2.60)

b) Application of the Evolution Operator: Looking at the term $\exp(\mathcal{L}_0 t')\delta\mathcal{L}_1 v(t)$ in (16.2.56), notice that the projection operator commutes with the evolution operator, and that

$$\mathcal{P}\delta\bar{\mathcal{L}}_1\delta\mathcal{L}_1\, v(t') = \delta\bar{\mathcal{L}}_1\mathcal{P}\delta\mathcal{L}_1\, v(t') = 0. \tag{16.2.61}$$

From this it follows that we can simplify the evolution equation to

$$\frac{\partial v(t)}{\partial t} = \left(\bar{\mathcal{L}}_1 + \bar{\mathcal{L}}_2 + \mathcal{P}\,\mathcal{L}_1 \int_0^\infty dt' \exp(\mathcal{L}_0 t')\,\delta\mathcal{L}_1\right) v(t), \tag{16.2.62}$$

$$= \left(\bar{\mathcal{L}}_1 + \bar{\mathcal{L}}_2\right) v(t) + W_s(\boldsymbol{x},\boldsymbol{p})\,\tfrac{1}{2}\frac{\partial}{\partial p_z}\left(\mathcal{D}_{zz}(\boldsymbol{x},\boldsymbol{p})\frac{\partial \overline{N(\boldsymbol{x},\boldsymbol{p},t)}}{\partial p_z}\right), \tag{16.2.63}$$

where the diffusion coefficient is defined by

$$\mathcal{D}_{zz}(\boldsymbol{x},\boldsymbol{p}) = \tfrac{1}{4}\int_0^\infty dt'\, \overline{\mathrm{Tr}_{\mathrm{Atom}}\left\{\left[F(\boldsymbol{x},\boldsymbol{p})\,,\,\mathrm{e}^{\mathcal{L}_0 t'}\left[\delta F(\boldsymbol{x},\boldsymbol{p})\,,\,W_s(\boldsymbol{x},\boldsymbol{p})\right]_+\right]_+\right\}}, \tag{16.2.64}$$

$$= \int_0^\infty dt'\,\left\{\tfrac{1}{2}\left\langle\left[F(\boldsymbol{x},\boldsymbol{p},t'),F(\boldsymbol{x},\boldsymbol{p},0)\right]_+\right\rangle_s - \left\langle F(\boldsymbol{x},\boldsymbol{p})\right\rangle_s^2\right\}. \tag{16.2.65}$$

The formulation on the last line in terms of the stationary correlation function follows from the standard relationships between multitime correlation functions and the Markovian evolution operator given in Sect. 10.2.3.

16.2.5 The Laser Cooling Fokker–Planck Equation

Combining together (16.2.39, 16.2.48, 16.2.63) gives a resultant Fokker–Planck equation of the form

$$\begin{aligned}\frac{\partial \overline{N(\boldsymbol{x},\boldsymbol{p},t)}}{\partial t} &= -\frac{\partial}{\partial p_z}\left(\bar{F}(\boldsymbol{x},\boldsymbol{p})\,\overline{N(\boldsymbol{x},\boldsymbol{p})}\right)\\ &\quad+\tfrac{1}{2}\Gamma(\hbar k_{eg})^2\sum_i \alpha_{ii}\frac{\partial^2}{\partial p_i^2}\left(W_{s,ee}(\boldsymbol{p})\,\overline{N(\boldsymbol{x},\boldsymbol{p},t)}\right)\\ &\quad+\tfrac{1}{2}\frac{\partial^2}{\partial p_z^2}\left(\mathcal{D}_{zz}(\boldsymbol{x},\boldsymbol{p})\,\overline{N(\boldsymbol{x},\boldsymbol{p},t)}\right).\end{aligned} \tag{16.2.66}$$

We have moved the inner operator ∂_{p_z} to the left in the last term, since this amounts to a change of order $\bar{\eta}^2$ to the radiation pressure force term, which is negligible, since we have only kept terms of order $\bar{\eta}^1$ when calculating the the radiation pressure force.

16.2.6 Evaluation of the Diffusion Coefficient $\mathfrak{D}_{zz}$

To evaluate the diffusion coefficient $\mathfrak{D}_{zz}(\boldsymbol{x},\boldsymbol{p})$ is no trivial task, although for the particular case of a travelling wave it can be done exactly using the quantum stochastic formalism. This can be shown using the unitary transformation $U(k_L z)$ defined in (16.2.21, 16.2.23), which can be used to eliminate the spatial dependence.

As noted in paragraph iii) of Sect. 16.2.2d, this unitary transformation takes the Wigner function to a function $W_L(\boldsymbol{x},\boldsymbol{p},t)$, whose equations of motion are determined by the substitutions (16.2.24), in which the only term involving the position $\boldsymbol{x}$ is the term $\boldsymbol{v}\cdot\nabla$. If the initial condition is spatially uniform, the Wigner function will develop no spatial dependence as it evolves in time, and can be written $W_L(\boldsymbol{p},t)$. Thus the term $\boldsymbol{v}\cdot\nabla$ in the evolution equation can be dropped, and the notation $N(\boldsymbol{x},\boldsymbol{p},t)$ is no longer necessary. The equations we have developed then become equations for

$$N(\boldsymbol{p},t) \equiv \mathrm{Tr}_{\mathrm{Atom}}\left\{W_L(\boldsymbol{p},t)\right\}, \tag{16.2.67}$$

with the necessary substitutions from (16.2.24).

a) Equations of Motion: These use the transformed quantities:

i) *The Doppler-Shifted Detuning*: We define this by

$$\Delta_v \equiv \Delta - \boldsymbol{k}_L\cdot\boldsymbol{v}. \tag{16.2.68}$$

ii) *The Evolution Operator to Order $\bar{\eta}^0$*: Since the term $\boldsymbol{v}\cdot\nabla$ has been dropped, this becomes the evolution operator for an atom driven by a laser field with a Doppler-shifted detuning, which we write as

$$\begin{aligned}\mathcal{L}_0^L W \equiv{}& \mathrm{i}\Big[\Delta_v|e\rangle\langle e| + \Omega_R\sigma_x\,,\,W\Big]\\ &-\tfrac{1}{2}\Gamma\left(\sigma^+\sigma^- W + W\sigma^+\sigma^- - 2\sigma^- W\sigma^+\right).\end{aligned} \tag{16.2.69}$$

iii) *Stationary Solution to Order $\bar{\eta}^0$*: For this evolution operator, the stationary solution is as given in (13.2.6–13.2.8), that is

$$W^L_{s,ee} = 1 - W^L_{s,gg} = \frac{\frac{1}{4}|\Omega_R|^2}{\Delta_v^2 + \frac{1}{4}\Gamma^2 + \frac{1}{2}|\Omega_R|^2}, \tag{16.2.70}$$

$$W^L_{s,eg} = \left(W^L_{s,ge}\right)^* = -\tfrac{1}{2}\Omega_R\frac{\Delta_v + \frac{1}{2}\mathrm{i}\Gamma}{\Delta_v^2 + \frac{1}{4}\Gamma^2 + \frac{1}{2}|\Omega_R|^2}. \tag{16.2.71}$$

iv) *Radiation Pressure Force Operator*: The force operator is

$$F_L(\boldsymbol{x},\boldsymbol{p}) \equiv \tfrac{1}{2}\Omega_R k_L \sigma_y. \qquad (16.2.72)$$

The mean radiation pressure force is of course unchanged by the unitary transformation, and is given by (16.2.41).

b) The Diffusion Coefficient: Using the correlation function formulae (10.2.17–10.2.18), the equation for the diffusion coefficient (16.2.64, 16.2.65) can now be reduced to

$$\mathcal{D}_{zz}(\boldsymbol{p}) = \tfrac{1}{4}\hbar^2\Omega_R^2 k_L^2 \int_0^\infty dt' \left\{ \tfrac{1}{2}\left\langle [\sigma_y(t'),\sigma_y(0)]_+ \right\rangle_s - \langle \sigma_y \rangle_s^2 \right\}. \qquad (16.2.73)$$

The correlation function can be written in terms of the matrix A_v defined by substituting $\Delta \to \Delta_v$ in the evolution operator A for the optical Bloch equations as defined by (13.2.14) as follows.

i) Note that

$$\tfrac{1}{2}\left\langle [\sigma_y(t'),\sigma_y(0)]_+ \right\rangle_s = \mathrm{Tr}_A\left\{ \sigma_y e^{\mathcal{L}_0^L t'} \tfrac{1}{2}\left[\sigma_y, W_s^L\right]_+ \right\}. \qquad (16.2.74)$$

We can write the stationary density matrix in the form

$$W_s^L = \tfrac{1}{2}(1+\bar{\boldsymbol{S}}\cdot\boldsymbol{\sigma}), \qquad (16.2.75)$$

where the stationary averages are

$$\bar{\boldsymbol{S}} = \frac{1}{\Delta_v^2 + \frac{1}{4}\Gamma^2 + \frac{1}{2}\Omega_R^2} \begin{pmatrix} -\Delta_v\Omega_R \\ \frac{1}{2}\Omega_R\Gamma \\ -\frac{1}{4}\Gamma^2 - \Delta_v^2 \end{pmatrix}. \qquad (16.2.76)$$

ii) We can now apply the methodology of Sect. 13.2.4, and in particular the solution of the optical Bloch equations. Using the Pauli algebra, we can write

$$\tfrac{1}{2}\left[\sigma_y, W_s^L\right]_+ = \tfrac{1}{2}\left(1+\sigma_y/\bar{S}_y\right)\bar{S}_y, \qquad (16.2.77)$$

and this is proportional to a density operator $\rho_W(0) = \frac{1}{2}(1+\boldsymbol{S}(0)\cdot\boldsymbol{\sigma})$, in which

$$\boldsymbol{S}(0) = \left(0,\ 1/\bar{S}_y,\ 0\right). \qquad (16.2.78)$$

iii) Hence, following (13.2.14), we can write

$$e^{\mathcal{L}_0^L t'} \tfrac{1}{2}\left[\sigma_y, W_s^L\right]_+ = \tfrac{1}{2}\left((1+\bar{\boldsymbol{S}}\cdot\boldsymbol{\sigma}) + e^{A_v t'}(\boldsymbol{S}(0)-\bar{\boldsymbol{S}})\right)\bar{S}_y, \qquad (16.2.79)$$

and therefore

$$\mathrm{Tr}_A\left\{ \sigma_y e^{L_0^L t'} \tfrac{1}{2}\left[\sigma_y, W_s^L\right]_+ \right\} = \bar{S}_y^2 + \tfrac{1}{2}\mathrm{Tr}_A\left\{ \sigma_y e^{A_v t'}(\boldsymbol{S}_0 - \bar{\boldsymbol{S}})\cdot\boldsymbol{\sigma} \right\}. \qquad (16.2.80)$$

iv) Substituting into (16.2.73), we can integrate the exponential, and get

$$\mathcal{D}_{zz}(\boldsymbol{p}) = \tfrac{1}{4}\hbar^2\Omega_R^2 k_L^2\left(-\tfrac{1}{2}\bar{S}_y \operatorname{Tr}_A\{\sigma_y A_v^{-1}(\boldsymbol{S}_0 - \bar{\boldsymbol{S}})\cdot\boldsymbol{\sigma}\}\right), \tag{16.2.81}$$

$$= \tfrac{1}{2}\Gamma\hbar^2 k_L^2\left(\frac{\tfrac{1}{4}\Omega_R^2}{\Delta_v^2+\tfrac{1}{4}\Gamma^2+\tfrac{1}{2}\Omega_R^2} - \frac{\tfrac{1}{8}\Omega_R^4\left(\tfrac{1}{4}\Gamma^2+\Delta_v^2\right)}{\left(\Delta_v^2+\tfrac{1}{4}\Gamma^2+\tfrac{1}{2}\Omega_R^2\right)^3}\right), \tag{16.2.82}$$

$$= \tfrac{1}{2}\Gamma\hbar^2 k_L^2\Omega_R^2\frac{\left(\tfrac{1}{4}\Gamma^2+\Delta_v^2\right)^2+\tfrac{1}{4}\Omega_R^4+\tfrac{1}{2}\Omega_R^2\left(\tfrac{1}{4}\Gamma^2+\Delta_v^2\right)}{\left(\Delta_v^2+\tfrac{1}{4}\Gamma^2+\tfrac{1}{2}\Omega_R^2\right)^3}. \tag{16.2.83}$$

v) The last line shows that the diffusion coefficient is always positive. The second line can be written in terms of the stationary excited state population $p_e \equiv W^L_{s,ee}$ as

$$\mathcal{D}_{zz}(\boldsymbol{p}) = \tfrac{1}{2}\Gamma\hbar^2 k_L^2\left(p_e - p_e^2 + 2p_e^3\right). \tag{16.2.84}$$

This equation means that we can write

$$\mathcal{D}_{zz}(\boldsymbol{p}) = \tfrac{1}{2}\Gamma\hbar^2 k_L^2 p_e G(p_e), \quad \text{where } 1 \leqslant G(p_e) \leqslant 2. \tag{16.2.85}$$

Thus, the diffusion coefficient is within a factor of two of $\tfrac{1}{2}\Gamma\hbar^2 k_L^2 p_e$ over the full range of driving fields.

16.3 Summary

When we compare the calculation in this chapter with the calculation given in *Bk. I: Sect.14.4*, the most significant observation is the accuracy of the simplified model given there. As noted in Sect. 16.2.3 b the expression for the radiation pressure is essentially identical with that in the simple treatment.

The fluctuation terms are also much the same; the two terms contributing to the diffusion terms in the Fokker–Planck equation (16.2.66) correspond essentially to the interpretation we gave in *Bk. I: Sect.14.4.4*, in particular the simple expression *Bk. I: (14.4.24)*.

i) The terms in the second line of (16.2.66) come from the momentum kicks provided by the spontaneous emission term already present in the last two lines of the laser cooling equation (16.1.3). Unlike the formulation in *Book I*, the expression is for three-dimensional diffusion.

ii) The term $\mathcal{D}_{zz}$ arises from the fluctuations in radiation pressure, as shown by its expression in terms of radiation pressure correlation function in (16.2.65). As noted in v), the complicated expression (16.2.82, 16.2.83) for it only produces at most a factor of two change from the expression $\mathcal{D}_{zz} \approx \tfrac{1}{2}\Gamma\hbar^2 k_L^2 p_e$.

The conclusion is that the limits on laser cooling of two-level atom atoms will be those we have already deduced in *Book I*, that limit of Doppler cooling corresponds to the *Doppler limit* temperature

$$T_{\text{Cool}} \approx \hbar\Gamma / k_{\text{B}}. \tag{16.3.1}$$

In fact this is not the limit of temperatures attainable by laser cooling. The discovery of *polarization gradient cooling* in atoms with more than two levels showed that temperatures a factor of hundreds less than the Doppler limit could be achieved using laser cooling techniques, and polarization gradient cooling now provides the workhorse for practical laser cooling. A very good summary of the situation when it had become clear in 1989 how these low temperatures had been achieved, can be found in [16.6].

However, we will not discuss polarization gradient cooling in this book, since it belongs rather to the subject matter of cold atoms. It can, however, be treated by techniques similar to those used here, and its theory has already been sketched in *Quantum Noise Chap. 11*. In the next chapter we will develop the subject of laser cooling of trapped ions, which is directly relevant to the subject matter of this book.

17. Laser Cooling of a Trapped Ion

In this chapter we will treat a one-dimensional version of the cooling of a two-level system trapped by a tight harmonic trap. In practice this is normally implemented using an ion trap, which will be formulated in detail in Chap. 23. For our treatment here, it is sufficient to consider a two-level system in a harmonic trap, in which the tightness of confinement is the important adjustable feature. In particular the treatment we will present will depend on the Lamb–Dicke regime, as will be described in Sect. 23.1.5. This is the regime in which the ion is confined by the trap to a region a_0 much smaller than the wave length of the light λ. Quantitatively, this means that the *Lamb–Dicke* parameter $\eta \equiv 2\pi a_0/\lambda \ll 1$.

The treatment of laser cooling we give here is an adaptation into the quantum stochastic Schrödinger equation methodology of the work in [17.1–17.3]. It is in fact based on methods very similar to those of the previous chapter, but modified to take account of the discrete energy level structure imposed by the harmonic trap.

17.1 Formulation of a One-Dimensional Model

The one-dimensional treatment corresponds to a two-level ion moving only along the x-axis, but interacting with a three-dimensional electromagnetic field. This is an approximation to a trap which is very narrow in the y and z directions, making excitation of modes in those directions very difficult.

17.1.1 Hamiltonian Terms

The Hamiltonian for the trapped ion is the sum of three terms

$$H = H_{\text{trap}} + H_{\text{int}} + H_{\text{dip}}, \tag{17.1.1}$$

whose forms are:

a) Trap: This is describes the one-dimensional centre of mass motion of the ion, with the Hamiltonian

$$H_{\text{trap}} = \frac{P^2}{2M} + V_{\text{trap}}(X). \tag{17.1.2}$$

In the case of a harmonic oscillator this becomes

$$H_{\text{trap}} = \frac{P^2}{2M} + \tfrac{1}{2}M\nu^2X^2 \equiv \hbar\nu\left(a^\dagger a + \tfrac{1}{2}\right). \tag{17.1.3}$$

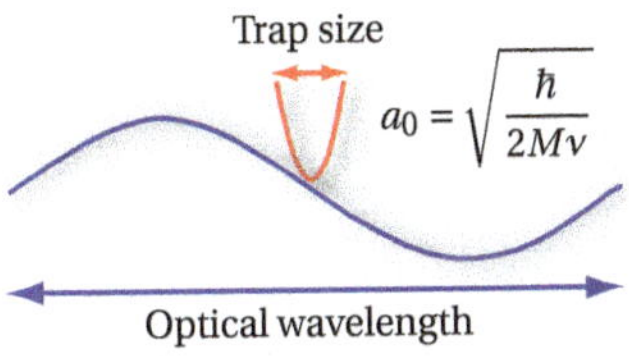

Fig. 17.1. The trap size compared to the optical wavelength in the Lamb–Dicke regime. The illustration is approximately to scale for a typical experimental situation, and corresponds to $\eta \ll 1$.

Here, P and X are the centre of mass momentum and position operators, and a and $a^\dagger$ the corresponding lowering and raising operators.

For a harmonic trap in which only the lowest levels are excited, the typical size of the ion's wavefunction is

$$a_0 = \sqrt{\frac{\hbar}{2M\nu}}\,. \tag{17.1.4}$$

The wavelength to which a_0 is to be compared can be regarded as either that of the laser field or that of the light emitted during spontaneous emission, since these are in practice almost equal. This means that the Lamb–Dicke parameter is defined as

$$\eta = \frac{2\pi a_0}{\lambda_{eg}} = k_{eg} a_0 \approx \frac{2\pi a_0}{\lambda_L} = k_L a_0. \tag{17.1.5}$$

The situation where $\eta \ll 1$ is illustrated in Fig. 17.1.

b) Electronic Degrees of Freedom: For two-level atoms $|g\rangle$, $|e\rangle$ the Hamiltonian $H_{\rm int}$ in a rotating frame is

$$H_{\rm int} = -\tfrac{1}{2}\Delta\sigma_z. \tag{17.1.6}$$

c) Dipole Interaction with the Laser Light: This represents the interaction of the atomic dipole moment with a classical non-operator electric field, evaluated at the operator position X. In our treatment, the Lamb–Dicke condition means that we need only consider the situation in which the atom does not move far from the position $X = 0$. We will use an expansion up to second order in X, and thus write

$$H_{\rm dip}(X) = -\boldsymbol{d}_{eg}\cdot\boldsymbol{\mathcal{E}}(X)\sigma^+ - \boldsymbol{d}_{ge}\cdot\boldsymbol{\mathcal{E}}^*(X)\sigma^-, \tag{17.1.7}$$

$$= H_{\rm dip}(0) - XF - \tfrac{1}{2}X^2F' + \cdots \tag{17.1.8}$$

We will not explicitly specify F and F', other than to say that they arise by expanding the incident classical electromagnetic field $\mathcal{E}(X)$ up to second order in X.

17.1.2 Laser Cooling Terms

The appropriate laser cooling master equation is adapted from the master equations presented in Sect. 15.4 by making the replacements:

$$\boldsymbol{X} \longrightarrow X, \tag{17.1.9}$$

$$e^{-ik_{eg}\boldsymbol{n}\cdot\boldsymbol{X}} \longrightarrow e^{-ik_{eg}Xu}, \quad \text{where } u = \cos\theta. \tag{17.1.10}$$

In particular, we write the recycling term in the form

$$\mathcal{J}\rho_A(t) = \Gamma \int_{-1}^{1} du\, N(u) \left(\sigma^- e^{-ik_{eg}Xu}\right) \rho_A(t) \left(\sigma^+ e^{ik_{eg}Xu}\right). \tag{17.1.11}$$

The quantity $N(u)$ is the angular distribution of the emitted photon, and for a dipolar transition this is given by

$$N(u) = \tfrac{3}{4}\left(1+u^2\right). \tag{17.1.12}$$

This leads to the Liouvillian which describes the process of damping and momentum transfer by spontaneous emission

$$\mathcal{L}^d \rho = \tfrac{1}{2}\Gamma\left(2\sigma^- \tilde{\rho}\sigma^+ - \sigma^+\sigma^-\rho - \rho\sigma^+\sigma^-\right). \tag{17.1.13}$$

Here, for compactness, we have introduced the notation

$$\tilde{\rho} = \int_{-1}^{1} du\, N(u) e^{-ik_{eg}Xu} \rho\, e^{ik_{eg}Xu}. \tag{17.1.14}$$

This term accounts for the momentum transfer by spontaneous emission to the centre of mass motion of the ion when the atomic electron returns from the excited state $|e\rangle$ to the ground state $|g\rangle$.

17.1.3 The Full Master Equation for Cooling of a Trapped Ion

The combination of the Hamiltonian terms and the spontaneous emission Liouvillian is the master equation for the full set of processes:

$$\frac{d\rho}{dt} = -\frac{i}{\hbar}\left[H_{\text{trap}} + H_{\text{int}} + H_{\text{dip}}, \rho\right] + \mathcal{L}^d \rho. \tag{17.1.15}$$

17.2 Perturbation in Terms of the Lamb–Dicke Parameter

In the Lamb–Dicke limit we can make a perturbative expansion of the Liouvillians, and then use the projection operator techniques to eliminate the fast variable, related to the electronic states. The result is an equation of motion for the centre of mass motion of the ion. The technique we shall use is very similar to those introduced in *Bk. I: Sect.13.2.1* in deriving the quantum-optical master equation.

17.2.1 Perturbative Expansion of the Master Equation

a) Lamb–Dicke Expansion of the Dipole Interaction: The expansion (17.1.8) is essentially an expansion in the Lamb–Dicke parameter. This means that $X \sim a_0$, and $\partial_X \mathcal{E} \sim 1/\lambda$, so that

$$H_{\text{dip}}(0) \sim \eta^0, \tag{17.2.1}$$

$$XF \sim \eta^1, \tag{17.2.2}$$

$$\tfrac{1}{2}X^2 F' \sim \eta^2. \tag{17.2.3}$$

Note that $H_{\text{dip}}(0)$, F and F' are operators only referring to the *internal* degrees of freedom of the trapped ion.

b) Lamb–Dicke Expansion of the Dissipative Liouvillian: We expand the exponentials in (17.1.11) in a power series up to second order; thus

$$\tilde{\rho} = \int_{-1}^{1} N(u)\left(1 - \mathrm{i}k_{eg}Xu - \tfrac{1}{2}k_{eg}^2X^2u^2 + \dots\right)\rho\left(1 + \mathrm{i}k_{eg}Xu - \tfrac{1}{2}k_{eg}^2X^2u^2 + \dots\right), \tag{17.2.4}$$

$$\approx \rho + \tfrac{1}{2}k_{eg}^2\left(\int_{-1}^{1} N(u)\,u^2\,du\right)\left(2X\rho X - X^2\rho - \rho X^2\right). \tag{17.2.5}$$

No first order term arises, because $N(u)$ is an even function of u.

Thus, we can write, to second order in η,

$$\mathcal{L}^d = \mathcal{L}_0^d + \eta^2\mathcal{L}_2^d + \dots \tag{17.2.6}$$

in which

$$\mathcal{L}_0^d\rho = \tfrac{1}{2}\Gamma\left(2\sigma^-\rho\sigma_+ - \sigma^+\sigma^-\rho - \rho\sigma^+\sigma^-\right) \quad \sim \eta^0, \tag{17.2.7}$$

$$\eta^2\mathcal{L}_2^d\rho = \tfrac{1}{2}\alpha k_{eg}^2\Gamma\sigma^-\left(2X\rho X - X^2\rho - \rho X^2\right)\sigma^+ \quad \sim \eta^2, \tag{17.2.8}$$

$$\alpha = \tfrac{1}{2}\int_{-1}^{1} u^2 N(u)\,du = \tfrac{2}{5} \quad \text{for the case (17.1.12).} \tag{17.2.9}$$

In these equations

i) $\mathcal{L}_0^d$ is familiar from the optical Bloch equations.

ii) $\mathcal{L}_2^d$ describes the diffusion due to spontaneous emission.

iii) The quadratic dependence on η in the second line arises naturally, since $k_{eg} = 2\pi/\lambda_{eg}$, and $X \sim a_0$.

c) Expansion in Powers of η: We can now write the master equation as an expansion in powers of η up to second order in the following form

$$\frac{d\rho}{dt} = \left(\mathcal{L}_{0E} + \mathcal{L}_{0I} + \eta\mathcal{L}_1 + \eta^2\mathcal{L}_2\right)\rho. \tag{17.2.10}$$

The terms of orders 0, 1 and 2 are:

i) *Terms of order η^0*: At this order the internal dynamics and the external dynamics are quite independent of each other, and there are no mechanical light effects. Thus we can write the evolution as the sum of two terms thus:

$$\mathcal{L}_0\rho \equiv \left(\mathcal{L}_{0E} + \mathcal{L}_{0I}\right)\rho. \tag{17.2.11}$$

a) *The External Degrees of Freedom*, corresponding to undamped motion of the ion in the trap are described by

$$\mathcal{L}_{0E}\rho = -\frac{\mathrm{i}}{\hbar}\left[H_{\text{trap}}, \rho\right]. \tag{17.2.12}$$

b) *The Internal Degrees of Freedom*, corresponding to the electronic levels, are described by

$$\mathcal{L}_{0I}\rho = -\frac{\mathrm{i}}{\hbar}\left[H_{\text{int}} + H_{\text{dip}}(0), \rho\right] + \mathcal{L}_0^d\rho, \tag{17.2.13}$$

and this corresponds to the optical Bloch equations for the dynamics of the two atomic levels of an ion fixed at $X = 0$.

ii) *Terms of order* η^1: These arise from the Hamiltonian term which represents the coupling to the motion of the ion and the laser generated by the dipole force:

$$\eta\mathcal{L}_1\rho = -\frac{i}{\hbar}\left[-FX,\rho\right]. \tag{17.2.14}$$

iii) *Terms of order* η^2: There are two quite different effects at this order; the irreversible diffusion term caused by spontaneous emission, and a Hamiltonian term coming from the derivative of the dipole force:

$$\eta^2\mathcal{L}_2\rho = -\frac{i}{\hbar}\left[\tfrac{1}{2}X^2F',\rho\right] + \eta^2\mathcal{L}_2^d\rho. \tag{17.2.15}$$

17.2.2 Elimination of the Internal Degrees of Freedom

In view of the smallness of η the coupling between the external and internal degrees of freedom (i.e. laser cooling) is slow relative to the internal dynamics. This allows us to eliminate of the internal (atomic) dynamics to obtain an equation of motion for the external (motional) degrees of freedom. This elimination is structurally very similar in nature to that used in *Bk. I: Sect.13.2.1* to derive the quantum master equation. We will adapt this to evaluate this limit, using the correspondences

$$\mathcal{L}_{\text{sys}} \longrightarrow \mathcal{L}_{0E}, \tag{17.2.16}$$

$$\mathcal{L}_{\text{int}} \longrightarrow \eta\mathcal{L}_1, \tag{17.2.17}$$

$$\mathcal{L}_B \longrightarrow \mathcal{L}_{0I}. \tag{17.2.18}$$

There is no term in the master equation derivation corresponding to $\eta^2\mathcal{L}_2$, but we will show that it may easily be handled perturbatively.

a) Projection Operator onto the Null Space of $\mathcal{L}_{0I}$: As in *Bk. I: Sect.13.2.1* we need the projection operator onto null space of $\mathcal{L}_{0I}$, and this can be defined as

$$\mathcal{P}\rho = \rho_{I,s}\otimes\text{Tr}_I\{\rho\} = \lim_{t\to\infty} e^{\mathcal{L}_{0I}t}\rho. \tag{17.2.19}$$

Here $\rho_{I,s}$ is the stationary solution of the optical Bloch equations (17.2.13).

b) Projection Operator Properties: The mandatory projection operator properties *Bk. I: (13.2.10a)–(13.2.10e)* must be satisfied. It is quite obvious that the properties $\mathcal{P}^2 = \mathcal{P}$ and $\mathcal{L}_{0I}\mathcal{P} = \mathcal{P}\mathcal{L}_{0I} = 0$ are satisfied. However, there are two properties which require special attention.

i) For the operator $\mathcal{L}_1$ we find

$$\eta\mathcal{P}\mathcal{L}_1\mathcal{P}\rho = \frac{i}{\hbar}\langle F\rangle_s\left[X,\mathcal{P}\rho\right] \neq 0. \tag{17.2.20}$$

This term represents the mean force exerted on the atom in the stationary state, and it can be absorbed into the harmonic potential, whose minimum is shifted slightly. The result is that we can define

$$\bar{\mathcal{L}}_1 \equiv \mathcal{P}\mathcal{L}_1\mathcal{P} \tag{17.2.21}$$

$$\delta\mathcal{L}_1 = \mathcal{L}_1 - \bar{\mathcal{L}}_1. \tag{17.2.22}$$

Since $\mathcal{P}\bar{\mathcal{L}}_1\mathcal{P} = \bar{\mathcal{L}}_1$, we can treat this term as a correction to $\mathcal{L}_{0,E}$, that is, the mapping (17.2.16) is modified to

$$\mathcal{L}_{\text{sys}} \longrightarrow \mathcal{L}_{0E} + \eta\bar{\mathcal{L}}_1 . \tag{17.2.23}$$

When using these operators, it is normally the projected density operator to which they are applied. For these cases, the definition (17.2.14) of $\mathcal{L}_1$ in terms of the operator F leads to

$$\eta\bar{\mathcal{L}}_1\mathcal{P}\rho = \frac{\mathrm{i}}{\hbar}\langle F\rangle_s\left[X, \mathcal{P}\rho\right], \tag{17.2.24}$$

$$\eta\delta\mathcal{L}_1\rho = \frac{\mathrm{i}}{\hbar}\left[\left(F - \langle F\rangle_s\right)X, \rho\right]. \tag{17.2.25}$$

ii) The additional term $\eta^2\mathcal{L}_2$ does not satisfy the requirement $\mathcal{P}\mathcal{L}_2 = \mathcal{L}_2\mathcal{P}$ since both of the terms in (17.2.15) fail to commute with $\mathcal{P}$.

The solution to this problem is similar; define

$$\bar{\mathcal{L}}_2 \equiv \mathcal{P}\mathcal{L}_2\mathcal{P}, \tag{17.2.26}$$

$$\delta\mathcal{L}_2 \equiv \mathcal{L}_2 - \bar{\mathcal{L}}_2. \tag{17.2.27}$$

Clearly $\mathcal{P}\,\delta\mathcal{L}_2\,\mathcal{P} = \delta\mathcal{L}_2$, so we can absorb $\eta^2\delta\mathcal{L}_2$ into $\eta\mathcal{L}_1$. However, in the limit $\eta \to 0$, this will give a negligible contribution.

Thus, the conclusion is that $\delta\mathcal{L}_2$ can be neglected, and we can therefore make the replacement

$$\mathcal{L}_2 \longrightarrow \bar{\mathcal{L}}_2 \equiv \mathcal{P}\mathcal{L}_2\mathcal{P}. \tag{17.2.28}$$

Since $\mathcal{P}\bar{\mathcal{L}}_2 = \bar{\mathcal{L}}_2\mathcal{P}$, we can treat $\bar{\mathcal{L}}_2$ as a correction to $\mathcal{L}_{0E}$, that is, instead of (17.2.16) or (17.2.23), we now have two corrections, thus

$$\mathcal{L}_{\text{sys}} \longrightarrow \mathcal{L}_{0E} + \eta\bar{\mathcal{L}}_1 + \eta^2\bar{\mathcal{L}}_2 \equiv \mathcal{L}_{\text{trap}}(\eta). \tag{17.2.29}$$

The evolution operator $\mathcal{L}_{\text{trap}}(\eta)$ now includes all of corrections resulting from the presence of the driving field. If the trap is originally harmonic, these corrections simply cause slight changes in position of the minimum and the strength of the potential. Correspondingly, we will define an operator for the corrected potential

$$\mathbf{H}_{\text{trap}}(\eta) \equiv H_{\text{trap}} - \langle F\rangle_s X - \tfrac{1}{2}\langle F'\rangle_s X^2. \tag{17.2.30}$$

c) Master Equation after Elimination of the Internal Degrees of Freedom: As in *Bk. I: Sect. 13.2.1*,the limiting equation of motion is evaluated for the projected quantity, which in this case is

$$\mu(t) \equiv \mathcal{P}\rho(t). \tag{17.2.31}$$

The master equation $\mu(t)$ is therefore of the same form as *Bk. I: (13.2.17)*, and can be written

$$\frac{d\mu(t)}{dt} = \mathcal{L}_{\text{trap}}(\eta)\,\mu(t) - \eta^2 \mathcal{P}\delta\mathcal{L}_1 \int_0^t dt \exp\big(\mathcal{L}_{\text{trap}}(\eta)\tau\big)\delta\mathcal{L}_1\mu(t-\tau). \tag{17.2.32}$$

17.3 Analysis of the Ion Trap Master Equation

The master equation we seek is obtained by inserting the explicit forms of the operators in our result (17.2.32), and by making the approximation analogous to *Bk. I: (13.2.22)* that we made in deriving the quantum optical master equation, namely

$$\mu(t-\tau) \approx \exp\big(-\mathcal{L}_{\text{trap}}(\eta)\tau\big)\mu(t) = e^{i\mathsf{H}_{\text{trap}}(\eta)\tau/\hbar}\mu(t)e^{-i\mathsf{H}_{\text{trap}}(\eta)\tau/\hbar}. \tag{17.3.1}$$

We then follow the procedure we used in *Bk. I: Sect.13.2.1* to derive the quantum-optical master equation, and the result is the sum of four terms:

$$\begin{aligned}\frac{d\mu}{dt} &= -\frac{i}{\hbar}\left[\mathsf{H}_{\text{trap}}(\eta),\mu\right] + \alpha\Gamma k_{eg}^2\langle\sigma_{ee}\rangle_s\left[2X\mu X - X^2\mu - \mu X^2\right] \\ &\quad - \frac{1}{\hbar^2}\int_0^\infty dt\,\langle\delta F(t)\,\delta F(0)\rangle_s\left[X, X(-t)\mu\right] \\ &\quad + \frac{1}{\hbar^2}\int_0^\infty dt\,\langle\delta F(0)\,\delta F(t)\rangle_s\left[X, \mu X(-t)\right].\end{aligned} \tag{17.3.2}$$

Here we have defined:

i) The Heisenberg operators

$$X(-t) = e^{\mathcal{L}_{\text{trap}}(\eta)t}X \equiv e^{-i\mathsf{H}_{\text{trap}}(\eta)t/\hbar}Xe^{i\mathsf{H}_{\text{trap}}(\eta)t/\hbar}. \tag{17.3.3}$$

ii) The stationary *force correlation functions*

$$\langle\delta F(t)\,\delta F(0)\rangle_s = \text{Tr}_I\left\{\delta F e^{\mathcal{L}_{0I}t}\,\delta F\rho_s\right\}, \tag{17.3.4}$$

$$\langle\delta F(0)\,\delta F(t)\rangle_s = \text{Tr}_I\left\{\delta F e^{\mathcal{L}_{0I}t}\rho_s\,\delta F\right\}. \tag{17.3.5}$$

17.3.1 The Harmonic Oscillator Trap

What we have derived so far is independent of the precise nature of the trap. We now explicitly require the trap to be harmonic, which is in practice the most relevant for an ion trap. This means that the trap Hamiltonian becomes of the form

$$\mathsf{H}_{\text{trap}}(\eta) = \hbar\nu\big(a^\dagger a + \tfrac{1}{2}\big), \tag{17.3.6}$$

and the Heisenberg operators (17.3.3) which we shall need are

$$X(-t) = e^{-i\mathbf{H}_{\text{trap}}(\eta)t/\hbar} X e^{i\mathbf{H}_{\text{trap}}(\eta)t/\hbar} = \sqrt{\frac{\hbar}{2M\nu}}\left(a^\dagger e^{-i\nu t} + a\, e^{i\nu t}\right). \tag{17.3.7}$$

a) Spectral Substitution: In order to substitute in the master equation let us first define the *ion spectral function*

$$\mathcal{S}(\omega) \equiv \frac{1}{2M\hbar\nu}\int_0^\infty dt\,\langle \delta F(t)\,\delta F(0)\rangle_s e^{i\omega t}. \tag{17.3.8}$$

We now substitute for X and $X(-t)$ in the master equation (17.3.2), obtaining

$$\begin{aligned}\frac{d\mu}{dt} &= -i\nu\left[a^\dagger a\,,\,\mu\right]\\ &\quad + \alpha\eta^2\Gamma\langle\sigma_{ee}\rangle_s\left\{2(a+a^\dagger)\mu(a+a^\dagger) - (a+a^\dagger)^2\mu - \mu(a+a^\dagger)^2\right\}\\ &\quad - \left[a+a^\dagger\,,\,\left(\mathcal{S}(\nu)a + \mathcal{S}(-\nu)a^\dagger\right)\mu\right]\\ &\quad + \left[a+a^\dagger\,,\,\mu\left(\mathcal{S}^*(-\nu)a + \mathcal{S}^*(\nu)a^\dagger\right)\right].\end{aligned} \tag{17.3.9}$$

Here we have simplified the second line by noting that the Lamb–Dicke parameter can be written $\eta^2 = \hbar k_{eg}^2/2M\nu$.

b) Ion Trap Master Equation in the Rotating Wave Approximation: The terms on lines 2–4 of this equation are all of order of magnitude $\eta^2 \ll 1$ compared to the first line, so that the timescale of the damping will be very many periods of the trap motion. These are the conditions under which the rotating-wave approximation is valid, and thus we may drop all terms involving an unbalanced number of operators a and $a^\dagger$.

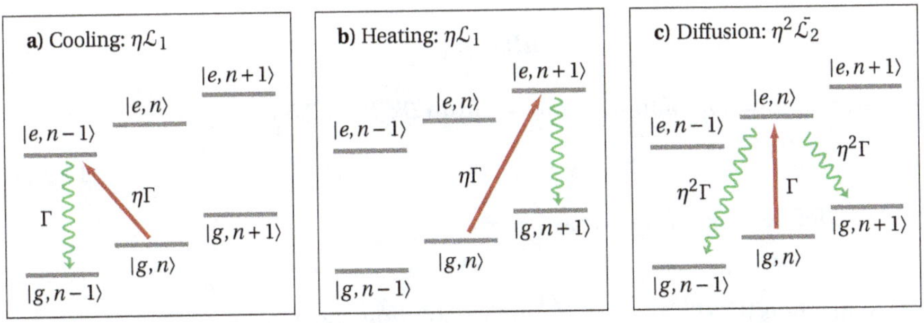

Fig. 17.2. The processes involved in cooling a trapped ion. The cooling process **a)** and the heating process **b)**, are of first order in η, but the resulting cooling takes place as a result of a second order process, so the cooling and heating rates are of order η^2. The diffusion process **c)** is also of order η^2. Sideband cooling arises when the laser can resolve the different transitions, and is tuned to the cooling transition **a)**.

By introducing the notations

$$\mathbf{D} \equiv \alpha\eta^2\Gamma\langle\sigma_{ee}\rangle_s, \tag{17.3.10}$$

$$R^+ \equiv 2\big(\mathrm{Re}\,\mathcal{S}(-\nu)+\mathbf{D}\big), \tag{17.3.11}$$

$$R^- \equiv 2\big(\mathrm{Re}\,\mathcal{S}(\nu)+\mathbf{D}\big), \tag{17.3.12}$$

$$\delta\nu \equiv \mathrm{Im}\,\big(\mathcal{S}(-\nu)+\mathcal{S}(\nu)\big), \tag{17.3.13}$$

the master equation can be written in the simple form

$$\begin{aligned}\frac{d\mu}{dt} &= -\mathrm{i}(\nu+\delta\nu)\left[a^\dagger a,\mu\right] \\ &\quad +\tfrac{1}{2}R^-\left(2a\mu a^\dagger - a^\dagger a\mu - \mu a^\dagger a\right) \\ &\quad +\tfrac{1}{2}R^+\left(2a^\dagger\mu a - aa^\dagger\mu - \mu aa^\dagger\right).\end{aligned} \tag{17.3.14}$$

17.3.2 Solutions of the Ion Trap Master Equation

The master equation (17.3.14) is exactly that of a harmonic oscillator interacting with a heat bath, and we have considered this and its solutions already in *Bk. I: Sect.14.3.1*. We can therefore write down the following solutions.

a) Rate Equation for the Occupation Probabilities: The diagonal matrix elements $P_n \equiv \langle n|\mu|n\rangle$ of the master equation in the number state basis are

$$\frac{dP_n}{dt} = (n+1)R^-P_{n+1} + nR^+P_{n-1} - \big(nR^- + (n+1)R^+\big)P_n\,. \tag{17.3.15}$$

b) Solution of Rate Equation: Here we need only transcribe the results of *Bk. I: Sect.14.3.1* as follows.

i) *Mean Number of Phonons*: This is

$$\bar{n}(t) = \mathrm{Tr}\left\{a^\dagger a\mu(t)\right\} = \sum_{n=0}^{\infty} nP_n(t), \tag{17.3.16}$$

whose equation of motion follows from *Bk. I: (14.3.8)*,

$$\frac{d\bar{n}(t)}{dt} = (R^+ - R^-)\bar{n}(t) + R^+\,. \tag{17.3.17}$$

We define the *cooling rate* as

$$W \equiv R^- - R^+, \tag{17.3.18}$$

and when $W > 0$ there is a stationary solution given by

$$\bar{n}_s = \frac{R^+}{R^- - R^+}\,. \tag{17.3.19}$$

The time dependent solution for $\bar{n}(t)$ is

$$\bar{n}(t) = \bar{n}(0)\mathrm{e}^{-Wt} + \bar{n}_s\left(1-\mathrm{e}^{-2Wt}\right). \tag{17.3.20}$$

ii) *Mean Energy in the Stationary State*: We include the zero point energy of the oscillator, so that this is

$$\bar{E}_s = \hbar\nu\left(\bar{n}_s + \tfrac{1}{2}\right) = \left(\frac{R^+}{R^- - R^+} + \frac{1}{2}\right). \tag{17.3.21}$$

iii) *Stationary Probability Distribution*: This can be found using *Bk. I: (14.3.34), (14.3.36)* and is

$$P_n^s = \left(1 - \frac{R^+}{R^-}\right)\left(\frac{R^+}{R^-}\right)^n. \tag{17.3.22}$$

iv) *Stationary Density Operator*: This is a Bose–Einstein distribution with

$$P_n^s = \frac{1}{Z}\mathrm{e}^{-n\hbar\nu/k_B T_s}, \tag{17.3.23}$$

where the *stationary temperature* T_s is given by

$$\frac{R^+}{R^-} = \frac{\bar{n}_s}{1+\bar{n}_s} = \mathrm{e}^{-\hbar\nu/k_B T_s}. \tag{17.3.24}$$

c) Parameters for a Two-Level System: We will consider only the case of driving by a travelling wave, for which we have the following results.

i) *Internal Hamilton operator*:

$$H_{0I} = H_{\mathrm{int}} + H_{\mathrm{dip}}(0) = -\tfrac{1}{2}\hbar\left(\Delta\sigma_z + \Omega\sigma_x\right). \tag{17.3.25}$$

ii) *Dipole interaction*:

$$H_{\mathrm{dip}}(X) = -\tfrac{1}{2}\hbar\Omega\left(\sigma^+\mathrm{e}^{\mathrm{i}k_L X} + \sigma^-\mathrm{e}^{-\mathrm{i}k_L X}\right). \tag{17.3.26}$$

iii) *Optical Bloch Equations*

$$\frac{d}{dt}\langle\sigma_x\rangle = -\tfrac{1}{2}\Gamma\langle\sigma_x\rangle + \Delta\langle\sigma_y\rangle, \tag{17.3.27}$$

$$\frac{d}{dt}\langle\sigma_y\rangle = -\Delta\langle\sigma_x\rangle - \tfrac{1}{2}\Gamma\langle\sigma_y\rangle + \Omega\langle\sigma_z\rangle, \tag{17.3.28}$$

$$\frac{d}{dt}\langle\sigma_z\rangle = -\Omega\langle\sigma_y\rangle - \Gamma\langle\sigma_z\rangle - \Gamma. \tag{17.3.29}$$

iv) *Steady States*

$$\langle\sigma_x\rangle_s = -\frac{\Omega\Delta}{\Delta^2 + \frac{1}{4}\Gamma^2 + \frac{1}{2}\Omega^2}, \tag{17.3.30}$$

$$\langle\sigma_y\rangle_s = -\frac{\frac{1}{2}\Gamma\Omega}{\Delta^2 + \frac{1}{4}\Gamma^2 + \frac{1}{2}\Omega^2}, \tag{17.3.31}$$

$$\langle\sigma_z\rangle_s = -\frac{\frac{1}{4}\Gamma^2 + \Delta^2}{\Delta^2 + \frac{1}{4}\Gamma^2 + \frac{1}{2}\Omega^2}. \tag{17.3.32}$$

v) *Force Operator*:

$$F = -\tfrac{1}{2}\hbar k_L \Omega \sigma_y, \tag{17.3.33}$$

$$\langle F\rangle_s = \tfrac{1}{2}\hbar k_L \Omega \langle\sigma_y\rangle_s = \frac{\tfrac{1}{4}\hbar k_L \Gamma\Omega^2}{\Delta^2 + \tfrac{1}{4}\Gamma^2 + \tfrac{1}{2}\Omega^2}. \tag{17.3.34}$$

vi) *Diffusion Coefficient*:

$$\mathsf{D} = \tfrac{1}{2}\alpha\eta^2\Gamma\left(1 + \langle\sigma_z\rangle_s\right) = \frac{\tfrac{1}{4}\alpha\eta^2\Gamma\Omega^2}{\Delta^2 + \Gamma^2 + \tfrac{1}{2}\Omega^2}. \tag{17.3.35}$$

vii) *Fluctuation Spectrum*:

$$\mathcal{S}(\omega) = \left(\frac{\eta^2\Omega^2}{4}\right)\int_0^\infty dt\, e^{i\omega t}\left\{\langle\sigma_y(t)\sigma_y(0)\rangle_s - \langle\sigma_y\rangle_s^2\right\}. \tag{17.3.36}$$

Exercise 17.1 Explicit Form of the Spectrum: To calculate the correlation function in (17.3.36) we note that, corresponding to the optical Bloch equations in the form (17.3.27–17.3.29), we can write

$$\mathbf{A} \equiv \begin{pmatrix} -\tfrac{1}{2}\Gamma & \Delta & 0 \\ -\Delta & -\tfrac{1}{2}\Gamma & -\Omega \\ 0 & \Omega & -\Gamma \end{pmatrix}, \qquad \mathbf{c}(t) \equiv \begin{pmatrix} \langle\sigma_x(t)\sigma_y(0)\rangle_s \\ \langle\sigma_y(t)\sigma_y(0)\rangle_s \\ \langle\sigma_z(t)\sigma_y(0)\rangle_s \end{pmatrix}, \qquad \mathbf{s} \equiv \langle\sigma_y\rangle_s \begin{pmatrix} \langle\sigma_x\rangle_s \\ \langle\sigma_y\rangle_s \\ \langle\sigma_z\rangle_s \end{pmatrix}. \tag{17.3.37}$$

The quantum regression theorem, combined with the solution (13.2.12), mean that

$$\mathbf{c}(t) = \exp(\mathbf{A}t)\left(\mathbf{c}(0) - \mathbf{s}\right) + \mathbf{s}. \tag{17.3.38}$$

We can now Fourier transform to get

$$\int_0^\infty dt\, e^{i\nu t}\left(\mathbf{c}(t) - \mathbf{s}\right) = -(\mathbf{A} - i\nu)^{-1}\left(\mathbf{c}(0) - \mathbf{s}\right). \tag{17.3.39}$$

Check that

$$-(\mathbf{A} - i\nu)^{-1} = \frac{\begin{pmatrix} \left(\tfrac{1}{2}\Gamma + i\nu\right)(\Gamma + i\nu) + \Omega^2 & \Delta(\Gamma + i\nu) & -\Delta\Omega \\ -\Delta(\Gamma + i\nu) & \left(\tfrac{1}{2}\Gamma + i\nu\right)(\Gamma + i\nu) & -\left(\tfrac{1}{2}\Gamma + i\nu\right)\Omega \\ -\Delta\Omega & \left(\tfrac{1}{2}\Gamma + i\nu\right)\Omega & \left(\tfrac{1}{2}\Gamma + i\nu\right)^2 + \Omega^2 \end{pmatrix}}{\left(\tfrac{1}{2}\Gamma + i\nu\right)^2(\Gamma + i\nu) + \left(\tfrac{1}{2}\Gamma + i\nu\right)\Omega^2 + \Delta^2(\Gamma + i\nu)}. \tag{17.3.40}$$

Show that

$$\left(\mathbf{c}(0) - \mathbf{s}\right) = \begin{pmatrix} i\langle\sigma_z\rangle_s - \langle\sigma_x\rangle_s\langle\sigma_y\rangle_s \\ 1 - \langle\sigma_y\rangle_s\langle\sigma_y\rangle_s \\ -i\langle\sigma_x\rangle_s - \langle\sigma_z\rangle_s\langle\sigma_y\rangle_s \end{pmatrix} \equiv \begin{pmatrix} c_x \\ c_y \\ c_z \end{pmatrix}. \tag{17.3.41}$$

Hence show that

$$\mathcal{S}(\nu) = \tfrac{1}{4}\eta^2\Omega^2\left(\frac{-\Delta(\Gamma + i\nu)c_x + \left(\tfrac{1}{2}\Gamma + i\nu\right)(\Gamma + i\nu)c_y - \left(\tfrac{1}{2}\Gamma + i\nu\right)\Omega c_z}{\left(\tfrac{1}{2}\Gamma + i\nu\right)^2(\Gamma + i\nu) + \left(\tfrac{1}{2}\Gamma + i\nu\right)\Omega^2 + \Delta^2(\Gamma + i\nu)}\right). \tag{17.3.42}$$

Exercise 17.2 Approximate Spectrum: If $\Omega \ll \Gamma$ and $\Omega \ll \Delta$, show that to lowest order in Ω

$$\big(\mathbf{c}(0)-\mathbf{s}\big) \longrightarrow \begin{pmatrix} -\mathrm{i} \\ 1 \\ 0 \end{pmatrix}, \qquad \mathcal{S}(\nu) \longrightarrow \frac{\frac{1}{4}\eta^2\Omega^2}{\frac{1}{2}\Gamma + \mathrm{i}(\nu - \Delta)}. \tag{17.3.43}$$

This form of the spectrum shows a simple resonance pole at $\nu = \Delta$. The width of the resonance is of course given by the spontaneous emission decay constant $\frac{1}{2}\Delta$. The behaviour for the case when Ω is not negligibly small will be similar, but at sufficiently large Ω the sidebands of resonance fluorescence will become apparent. The behaviour is then best investigated numerically.

17.4 Analysis of the Process of Cooling

For simplicity, we will consider only the case that Ω is sufficiently small, and use the approximate forms derived in Ex. 17.2. In this limit, we can write down the expressions for the cooling coefficients and the diffusion coefficient, based on the general results (17.3.10–17.3.12):

$$\mathsf{D} \longrightarrow \frac{\frac{1}{4}\eta^2\Omega^2\Gamma\alpha}{\Delta^2 + \frac{1}{4}\Gamma^2}, \tag{17.4.1}$$

$$R^{\pm} \longrightarrow \tfrac{1}{2}\eta^2\Omega^2\Gamma\left(\frac{\alpha}{\Delta^2 + \frac{1}{4}\Gamma^2} + \frac{1}{(\Delta \pm \nu)^2 + \frac{1}{4}\Gamma^2}\right). \tag{17.4.2}$$

17.4.1 Doppler Cooling of a Trapped Ion

When the levels are very closely spaced compared to the Γ, the basic behaviour of the results is not very different from that for laser cooling of an untrapped atom. Using the expressions (17.3.19, 17.3.21) for the mean stationary occupation and energy, we can evaluate these quantities in the limit $\nu \ll \Gamma$, where we find

$$\bar{n}_s \approx \tfrac{1}{2}(1+\alpha)\frac{\Gamma}{\nu}\left(\frac{\Gamma}{2\Delta} + \frac{2\Delta}{\Gamma}\right), \tag{17.4.3}$$

which in this limit is a large number.

The stationary temperature reached is found from (17.3.24), in this limit of large $\bar{n}_s$, to be

$$T_{\text{Cool}} \approx \frac{\hbar\Gamma}{2k_B}\left(\frac{\Gamma}{2\Delta} + \frac{2\Delta}{\Gamma}\right) \geqslant \hbar\Gamma/k_B. \tag{17.4.4}$$

This is exactly the same as the *Doppler limit* temperature (16.3.1) for an untrapped atom. This is a reasonable result, since the low ν limit is the limit of a weak trap, and one would expect the result to approach that of the untrapped situation.

17.4.2 Sideband Cooling

When the broadening caused by spontaneous emission is weak—the limit opposite to that of Doppler cooling—the oscillator mode frequency can be resolved, and the cooling process proceeds differently.

We therefore consider the case when the oscillator frequency is sufficiently high that $\nu \gg \Gamma$. The sideband cooling effect arises when we tune so that $\Delta = \nu$, and then $R^- \gg R^+$, because of the resonance effect in the expression for R^-. Using the expressions (17.3.19, 17.3.21) for the mean stationary occupation and energy, we can evaluate these quantities when the laser is tuned to the trap frequency, that is $\Delta = \nu$, and in the limit $\Gamma \ll \nu$, we get

$$\bar{n}_{\rm S} = \left(\frac{\Gamma}{2\nu}\right)^2 \left(\alpha + \tfrac{1}{4}\right) \qquad \ll 1, \tag{17.4.5}$$

and the temperature reached is very low. In fact, we can show that

$$T_{\rm Cool} \approx \frac{\hbar\nu}{k_{\rm B}} \Bigg/ \left|\log\left[\left(\frac{\Gamma}{2\nu}\right)^2 \left(\alpha + \tfrac{1}{4}\right)\right]\right|. \tag{17.4.6}$$

The logarithm in the denominator might be between 1 and 10, and so we find that the cooling temperature is of order of magnitude $\hbar\nu / k_{\rm B}$, very much less than the Doppler limit $\hbar\Gamma / k_{\rm B}$.

In fact, the occupation of all states other than the ground state can be made negligible using sideband cooling—*ground state cooling* is perfectly feasible. This enables the initialization of a trapped ion into the ground state, from which it can be coherently manipulated into any other state—that is, perfect quantum control.

V Continuous Measurement and Quantum Trajectories

Part V: Continuous Measurement and Quantum Trajectories

The focus of this part is the measurement of the output fields produced by a quantum stochastic system. There are two parts to this focus:

i) *Continuous Measurement Theory*: If the light being measured is produced as an output from a system described by the quantum stochastic Schrödinger equation, the measurement of the optical field (i.e. the process of photon counting) induces measurements on the system radiating the optical field. Continuous measurement theory shows how this connection is made, and hence how quantum information in the atom is related to the photon counting process.

   Light is measured by photodetectors, which count photons as they continuously monitor an optical field. Here we show how to relate the quantum properties of the optical field being measured to the correlation functions of the measured photocounts.

ii) *Quantum Trajectories*: The process of photocounting yields a sequence of photocounts at discrete times, and corresponding to each of these counts is a measurement of the system, and the system wavefunction changes instantaneously. This phenomenon is interpreted as a *quantum jump*.

   The reality of quantum jumps provided by this picture is compelling. The full wavefunction, including the electromagnetic field, evolves smoothly, and its role is to provide a way of computing the probabilities of observable events, which are the true reality. The reduced description, where the electromagnetic field is eliminated also involves a smooth evolution, but in this case it is an evolution of the density operator.

   The picture involving quantum jumps gives a smooth evolution of the system wavefunction only *between* the jumps, which occur stochastically following a law deduced from the wavefunction. Thus, a number of sample trajectories can be computed, and averages computed. The jumps correspond to photodetections, mirroring the experimental situation.

18. Continuous Measurement

The concept of a *continuous measurement* arises from the process of counting photons in a light beam, where the photodetector receives light and detects during a time interval of finite length. The process of measurement therefore takes place *continuously*. We have already formulated the basic ideas of continuous measurement in *Bk. I: Ch.19.*

When counting photons all we need to know is the quantum state of the electromagnetic field as it propagates into the detector—the fundamental issue at this level is then how to relate the time-dependent signal generated by the detector to the state of the electromagnetic field. The central counting formula making this connection was formulated in 1958 by *Mandel* [18.1], based on a semiclassical analysis of the process of photon counting, and the full quantum theory of photon counting was derived in 1964 by *Kelley* and *Kleiner* [18.2].

Both *Mandel's formula* and the more general *Kelley–Kleiner formula* give the probability distribution for the number of photons detected in a given time interval T, but do not describe the most general measurement which could be made. More generally, one wishes to know the probability of detecting n photons during infinitesimal time intervals $\Delta t_1, \Delta t_2, \ldots \Delta t_n$, contained within a time interval T. From the knowledge of this probability distribution one can calculate the probability distribution for any conceivable photon counting measurement.

In practice we measure light beams to extract the information they provide about their sources, for example, measuring the light from an ensemble of radiating atoms is used to provide information about the properties of the atoms. More generally, we will want to consider a system which radiates electromagnetic radiation, and an appropriate theoretical description of the radiation process is provided by the quantum stochastic Schrödinger equation.

This chapter will:

i) Give the counting formulae which relate the photodetection process to the properties of the quantum state of the electromagnetic field being measured.

ii) Show how the continuous measurement process on the electromagnetic field is also a measurement process on the system which produces the electromagnetic field, and how such a process is described by the quantum stochastic Schrödinger equation.

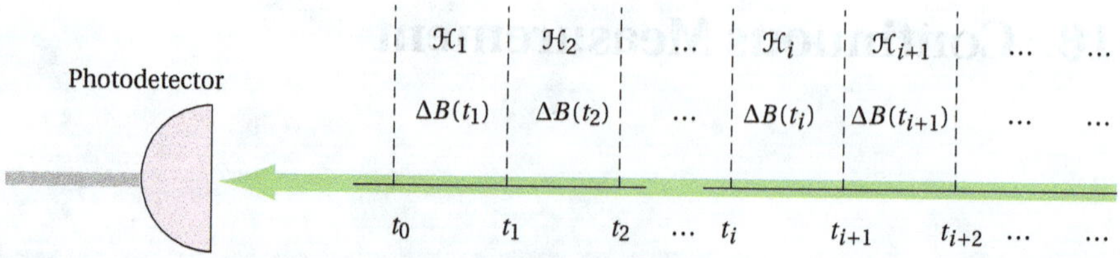

Fig. 18.1. Counting photons: the figure shows the resolution of the Hilbert space into the direct product of Hilbert spaces at each time interval, each with an appropriate set of destruction (and creation) operators. The Hilbert spaces move with the light beam, so that the time t_i corresponds to the time at which photons represented by the particular Hilbert space $\mathcal{H}_i$ are counted by the photodetector.

18.1 Photon Counting

The process of counting photons can be related to our description in Sect. 9.3.2 of a propagating light beam in terms of independent Hilbert spaces at successive time intervals $\Delta t_i \equiv t_{i+1} - t_i$. As illustrated in Fig. 18.1, we can regard the time interval Δt_i as the time at which we measure the number of photons in the Hilbert space $\mathcal{H}_i$.

Unlike the situation in Sect. 9.3.2, which was set up to describe an incoming vacuum field (or that of Sect. 9.6, which describes thermal white noise), the density operator ρ_{in} of the incoming light beam is not necessarily factorizable, thus permitting correlations between photocounts at different times. However, the number operators N_i of photons in the Hilbert spaces do commute with each other, and can be measured independently of each other. This means we can regard the process of photon counting as being that of measuring all of the operators N_i, giving a set of results $n_1, n_2, \dots, n_i, \dots$, whose probability we wish to calculate. The probability of obtaining this set of results can be written in terms of the projectors onto the subspace of $\mathcal{H}_i$ with n_i photons, namely

$$\mathcal{P}_i(n_i) \equiv |n_i, i\rangle\langle n_i, i|, \tag{18.1.1}$$

and this probability is

$$P(n_1, n_2, \dots, n_i, \dots) = \text{Tr}\left\{\rho_{\text{in}}\mathcal{P}_1(n_1)\mathcal{P}_2(n_2)\dots\mathcal{P}_i(n_i)\dots\right\}. \tag{18.1.2}$$

18.1.1 The Normally Ordered Counting Formulae

We start with a preliminary result; the relation between coherent states $|\alpha\rangle$ and number states $|n\rangle$—see *Bk. I: (10.5.2)*—means that

$$\langle\alpha|n\rangle\langle n|\alpha\rangle = \frac{e^{-\alpha^*\alpha}\left(\alpha^*\alpha\right)^n}{n!}. \tag{18.1.3}$$

On the other hand, we have shown in *Bk. I: Sect. 10.5.1 f*, that if $F(a, a^\dagger)$ is a normally ordered function—applying this result, we find $\langle\beta|F(a, a^\dagger)|\alpha\rangle = F(\alpha, \beta^*)$, so that

$$\langle\alpha|:\frac{e^{-a^\dagger a}(a^\dagger a)^n}{n!}:|\alpha\rangle = \frac{e^{-\alpha^*\alpha}(\alpha^*\alpha)^n}{n!}. \tag{18.1.4}$$

Following the reasoning of *Bk. I: Sect.15.1.5*, the diagonal coherent state matrix elements $\langle\alpha|G|\alpha\rangle$ of any operator G determine that operator; in this case, it follows that

$$|n\rangle\langle n| = :\frac{e^{-a^\dagger a}(a^\dagger a)^n}{n!}: \tag{18.1.5}$$

Applying this identity to the counting probability (18.1.2), we can use the creation and destruction operators $A_i^\dagger, A_i$ for the time interval Δt_i, which we introduced in (9.3.12), to write the *normally ordered counting formula*

$$P(n_1, n_2, \dots, n_i, \dots) = \mathrm{Tr}\left\{ :\frac{e^{-\sum_i A_i^\dagger A_i}}{n!}\prod_i\Big(\sum_i A_i^\dagger A_i\Big)^{n_i} : \rho_{\mathrm{in}}\right\}. \tag{18.1.6}$$

Here we have combined the individual normally ordered products into one normally ordered product around all of the operators because the operators in the different Hilbert spaces commute.

18.1.2 Measurement Operators for the Electromagnetic Field

The normally ordered counting formula is very general—it does not involve the infinitesimal nature of the time intervals used constructing the Hilbert spaces $\mathcal{H}_i$. When we take account of the fact that the Δt_i are indeed infinitesimal, we find, as in Sect. 9.3.2, that we need consider only quantum states with either 0 or 1 photon—higher numbers of photons occur only with probabilities proportional to Δt_i^2, and can be neglected.

To make this explicit, we can write

$$A_i \equiv \frac{\Delta B(t_i)}{\sqrt{\Delta t_i}} \equiv b(t_i)\sqrt{\Delta t_i}\,, \tag{18.1.7}$$

where the relationship between the second two is the same as that between the Ito quantum stochastic increment and the input vacuum field as given in (11.4.8, 11.4.9). However, in the situation we are considering here the electromagnetic field does not normally correspond to a vacuum state, and this means that the vacuum Ito rules do not apply. On the other hand, the commutation rule $[b(t), b^\dagger(t')] = \delta(t-t')$ of course does apply.

18.2 Photon Counting Formulae

We can now deduce a number of photon counting formulae. We want to consider the counting of photons in a set of n infinitesimal time intervals $(\tau_1, \tau_1 + d\tau_1]$, $(\tau_2, \tau_2 + d\tau_2], \dots (\tau_n, \tau_n + d\tau_n]$, all contained within the time interval $(0, T]$.

a) The Elementary Counting Probability Density: This represents the *probability density* that one count occurs in each of n time intervals $d\tau_1, d\tau_2, \dots, d\tau_n$, and no counts occur in any other time intervals. This means that in the normally ordered counting formula (18.1.6), n of the n_i are equal to 1, and the others are zero. We then substitute for the A_i, and $A_i^\dagger$, convert sums to integrals, to get the formula for the elementary counting probability density:

$$p_{\mathrm{el}}(\tau_1, \tau_2, \dots, \tau_n) = \frac{1}{n!}\Big\langle : \mathrm{e}^{-\int_0^T I(t)\,dt} I(\tau_1) I(\tau_2) \dots I(\tau_n) : \Big\rangle, \tag{18.2.1}$$

$$\text{where} \quad I(\tau) \equiv b^\dagger(\tau) b(\tau). \tag{18.2.2}$$

The *probability* of the count sequence is of course given by

$$P_{\mathrm{el}}(\tau_1, \tau_2, \dots, \tau_n) = p_{\mathrm{el}}(\tau_1, \tau_2, \dots, \tau_n)\, d\tau_1\, d\tau_2 \dots d\tau_n. \tag{18.2.3}$$

b) Inclusion of Detector Efficiency: The counting formulae we have presented so far represent ideal measurements, in which the photon is detected with 100% efficiency. In practice, detector efficiency is never perfect, and is often very much less than perfect. Hence this is something which must be included in counting formulae.

The simplest way to model an inefficient detector is to use an ideal photodetector, preceded by beam splitter, of intensity transmissivity e, as in Fig. 18.2. The beam splitter mixes the signal $b(t)$ with a vacuum signal $v(t)$, so that the detected beam can be written

$$a(t) = b(t)\sqrt{e} + v(t)\sqrt{1-e}\,. \tag{18.2.4}$$

Since $v(t)$ is in the vacuum state, all of its normally ordered moments are zero, so that the normally ordered moments of the detected intensity $a^\dagger(t)a(t)$ are

$$\langle : \big(a(t)^\dagger a(t)\big)^n : \rangle = e^n \langle : \big(b(t)^\dagger b(t)\big)^n : \rangle. \tag{18.2.5}$$

This gives the counting distributions for an inefficient detector directly in terms of the measured signal $b(t)$. We then get:

Elementary counting probability density for a detector with efficiency e:

$$p_{\mathrm{el}}(\tau_1, \tau_2, \dots, \tau_n) = \frac{e^n}{n!}\Big\langle : \mathrm{e}^{-e\int_0^T I(t)\,dt} I(\tau_1) I(\tau_2) \dots I(\tau_n) : \Big\rangle, \tag{18.2.6}$$

$$\text{where} \quad I(\tau) \equiv b^\dagger(\tau) b(\tau). \tag{18.2.7}$$

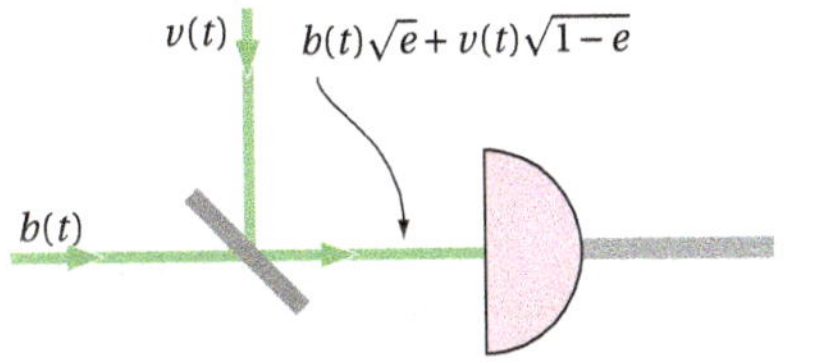

Fig. 18.2. Model of a realistic photodetector as an ideal photodetector preceded by a beam splitter.

c) Coincidence Probability Density: Here we wish to consider the probability density to detect a single count in each one of n time intervals $d\tau_1, d\tau_2, \dots, d\tau_n$, whether or not there are other counts in the other time intervals. This means we sum over all the n_i in the normally ordered counting formula (18.1.6), other than those corresponding to the n time intervals of interest. Since the time intervals are infinitesimal, we keep only terms of first order in the $d\tau_k$, and the result is:

Coincidence counting probability density for a detector with efficiency e:

$$p_{\text{coinc}}(\tau_1, \tau_2, \dots, \tau_n) = \frac{e^n}{n!} \langle : I(\tau_1) I(\tau_2) \dots I(\tau_n) : \rangle. \tag{18.2.8}$$

d) The Kelley–Kleiner Counting Formula: We want to know the probability of counting exactly n photons in the finite time interval $(0, T)$. We get this by summing over all of the ways in which the n photons can be distributed in n time intervals $d\tau_k$, and the result is

Counting probability to detect n photons in the interval $(0, T)$ *for a detector with efficiency e:*

$$P_{\text{K-K}}(n, [0, T]) = \frac{e^n}{n!} \left\langle : \mathrm{e}^{-e \int_0^T I(t)\, dt} \left(\int_0^T I(t)\, dt \right)^n : \right\rangle. \tag{18.2.9}$$

e) Mandel's Counting Formula: Mandel's counting formula arises when the light field can be represented by a P-function, as formulated in *Bk. I: Ch.15*, and *Bk. I: Ch.17.* We must consider a multivariate P-function in this case, using a description like that given in Sect. 9.5. The effect is to make the replacement for the input measured field operator

$$b(t) \longrightarrow \beta(t), \tag{18.2.10}$$

where $\beta(t)$ is a stochastic c-number field, described by a classical probability distribution. The normally ordered products in the Kelley–Kleiner formula map to the corresponding products of the $\beta(t)$ field, and the quantum averages (which require a density operator) become stochastic averages. This gives

Counting probability to detect n photons in the interval $(0, T)$ *for a detector with efficiency* e:

$$P_{\text{Mandel}}(n,[0,T]) = \frac{e^n}{n!}\left\langle \mathrm{e}^{-e\int_0^T |\beta(t)|^2\,dt}\left(\int_0^T |\beta(t)|^2\,dt\right)^n\right\rangle_{\text{Stochastic}}. \tag{18.2.11}$$

18.2.1 Photon Counting Statistics

There are a number of simple cases for which exact formulae are available.

a) Coherent State: If the light is in a coherent state, possibly time-dependent, we get Poissonian statistics

$$P(n,[0,T]) = \frac{\left(e\int_0^T |\beta(t)|^2\,dt\right)^n}{n!}\exp\left(-e\int_0^T |\beta(t)|^2\,dt\right). \tag{18.2.12}$$

This is a Poisson distribution with mean number $\bar{n}(0,T) = e\int_0^T |\beta(t)|^2\,dt$, for any value of detection efficiency e, and for any time dependence of the field $\beta(t)$.

b) Gaussian Light: By Gaussian light, we mean here a light beam with a Gaussian P-function, so that $\beta(t)$ can be regarded as a classical Gaussian field. The general case has not been solved, but if the observation time T is very much smaller than typical correlation time of the field $\beta(t)$, we can treat the field as a time independent Gaussian random variable β, with a known mean and variance; namely

$$\left.\begin{aligned} \langle\beta\rangle_{\text{Stochastic}} &= \bar{B}, & \langle|\beta|^2\rangle_{\text{Stochastic}} &= \bar{B}^2 + \bar{N},\\ \langle\beta^2\rangle_{\text{Stochastic}} &= 0, & \langle(\beta^*)^2\rangle_{\text{Stochastic}} &= 0. \end{aligned}\right\} \tag{18.2.13}$$

The distribution function is

$$P_{\text{Gauss}}(n,[0,T]) = \frac{\mathrm{e}^{-eT|\bar{B}|^2}}{(1+eT\bar{N})}\sum_{m=0}^{n}\frac{(eT|\bar{B}|^2)^m}{m!}\left(\frac{eT\bar{N}}{1+eT\bar{N}}\right)^{n-m}. \tag{18.2.14}$$

In Fig. 18.3 we illustrate the behaviour of this formula for the following situations:

i) *Chaotic Light*: The case when $\bar{B} = 0$, and there is therefore no coherent component, is often called *chaotic light*, and it corresponds to thermalized light. Only the first term in the summation is non-zero, and we get

$$P_{\text{Gauss}}(n,[0,T]) \longrightarrow \frac{(eT\bar{N})^n}{(1+eT\bar{N})^{n+1}}, \tag{18.2.15}$$

a power law distribution in n. This is the kind of result we would expect for light produced by a narrow bandwidth thermal source.

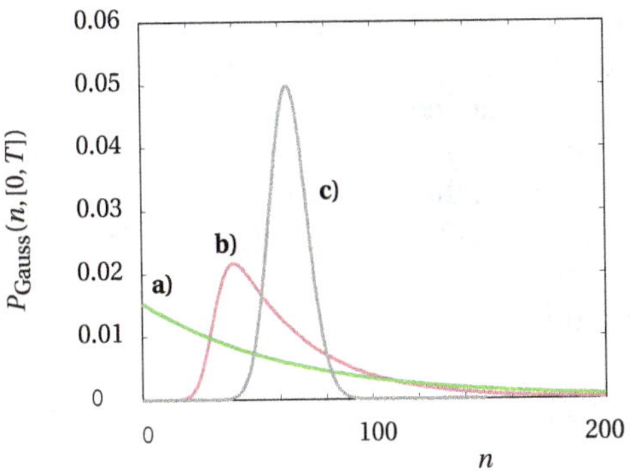

Fig. 18.3. Illustration of the counting probability distributions given by (18.2.14) for the case of: **a)** Chaotic light, with $eT|B|^2 = 0$, $eT\bar{N} = 64$; **b)** Partially coherent light, with $eT|B|^2 = 32$, $eT\bar{N} = 32$; **c)** Coherent light, with $eT|B|^2 = 64$, $eT\bar{N} = 0$.

ii) *Coherent Light*: When $\bar{N} = 0$, the P-function becomes a two-dimensional delta function $\delta^2(\beta - \bar{B}, \beta^* - \bar{B}^*)$, and only the last term in the summation is non-zero. The result is a Poisson distribution

$$P_{\text{Gauss}}(n,[0,T]) \longrightarrow \frac{e^{-eT|\bar{B}|^2}\left(eT|\bar{B}|^2\right)^n}{n!}, \tag{18.2.16}$$

corresponding to the coherent result (18.2.12).

iii) *Partially Coherent Light*: This is the general case, where there is a Gaussian distribution around a definite mean field $\bar{B}$. The full formula provides an interpolation between the coherent and the chaotic cases.

Exercise 18.1 Derivation of the Gaussian Counting Formula: Use the Gaussian P-function

$$P(\beta,\beta^*) = \frac{1}{\pi\bar{N}}\exp\left(\frac{-|\beta-\bar{B}|^2}{\bar{N}}\right), \tag{18.2.17}$$

and the moment generating function

$$G(s) \equiv \sum_{n=0}^{\infty} s^n P_{\text{Mandel}}(n,[0,T]), \tag{18.2.18}$$

to derive the formula (18.2.14).

Exercise 18.2 Mean and Variance of the Photocount: Use the generating function to show that, for the photocount formula (18.2.14), the mean and variance are

$$\langle n(T)\rangle = eT(|B|^2 + \bar{N}), \qquad \text{var}[n(T)] = \langle n(T)\rangle + (eT\bar{N})^2. \tag{18.2.19}$$

The fluctuations are thus always larger than the Poissonian value of $\langle n(T)\rangle$.

c) Number State: Suppose that the light field is in a number state of N photons in a mode with mode function $f_1(t)$ (like, for example, like that for cavity decay in Ex. 9.3, or that for atomic decay in Ex. 9.4). We can choose a complete set of functions $f_i(t)$, which includes this mode function, and write $b(t) = \sum_i f_i(t) A_i$, and the quantum state then consists of N photons in the mode $f_i(t)$, and the vacuum for all the other modes. Let us suppose that T is sufficiently large that the function $f_i(t) = 0$ unless $0 \leqslant t \leqslant T$.

Using this state, the Kelley–Kleiner formula (18.2.9) becomes

$$P_{\text{K-K}}(n,[0,T]) = \frac{e^n}{n!}\left\langle : \mathrm{e}^{-eA_1A_1^\dagger}\left(A_1^\dagger A_1\right)^n : \right\rangle_N, \tag{18.2.20}$$

$$= \frac{e^n}{n!}\left\langle : \mathrm{e}^{-A_1^\dagger A_1}\left(A_1^\dagger A_1\right)^n \mathrm{e}^{(1-e)A_1^\dagger A_1} : \right\rangle_N, \tag{18.2.21}$$

$$= \sum_{p=n}^{\infty}\left\langle : \frac{\mathrm{e}^{-A_1^\dagger A_1}\left(A_1^\dagger A_1\right)^p}{p!} : \right\rangle_N \left(e^n(1-e)^{p-n}\frac{p!}{(p-n)!n!}\right). \tag{18.2.22}$$

We now use the formula (18.1.5) to express this in terms of projectors onto p-photon states, giving

$$P_{\text{K-K}}(n,[0,T]) = \sum_{p=n}^{\infty}\langle |p\rangle\langle p| \rangle_N \left(e^n(1-e)^{p-n}\frac{p!}{(p-n)!n!}\right), \tag{18.2.23}$$

$$= e^n(1-e)^{N-n}\frac{N!}{(N-n)!n!}. \tag{18.2.24}$$

This is a Bernoulli distribution, giving the probability of detecting n photons in an N-photon state.

The mean and variance are given by

$$\langle n(T)\rangle = eN, \qquad \operatorname{var}[N(T)] = Ne(1-e). \tag{18.2.25}$$

These do not depend on T because N, the number of photons in the mode is fixed independently of the mode itself.

18.2.2 Classical and Non-Classical Light

Mandel's counting formula is equivalent to a counting formula for a fluctuating classical optical field $\beta(t)$. Correspondingly, light which can be described by means of such a fluctuating classical optical field is known as *classical light*, and light which cannot be so described is known as *non-classical light*.

Our reasoning above shows that classical light can be described by a P-function which takes on only non-negative values, whereas non-classical light cannot. In the examples of coherent and Gaussian light, the results of Ex. 18.2 show that the fluctuations, as measured by the variance, are greater than the Poissonian value, except for the case of coherent light, which is exactly Poissonian.

In the case of the number state, the fluctuations are always less than Poissonian, but approach a Poissonian value as the detection efficiency e becomes small, and the number N in the state becomes large. The number state is a non-classical state, which cannot be regarded as a fluctuating classical field. Rather, as a state of a fixed number of photons, it directly displays the quantum nature of light.

18.2.3 Photon Counting Correlation Functions

The formula (18.2.8) for the coincidence counting probability density shows that coincidence counting amounts to the measurement of normally ordered correlation functions of the form

$$\mathcal{G}^{(n)}(\tau_1,\tau_2,\dots,\tau_n) \equiv \left\langle b^\dagger(\tau_1)b^\dagger(\tau_2)\dots b^\dagger(\tau_n)b(\tau_n)\dots b(\tau_2)b(\tau_1)\right\rangle. \tag{18.2.26}$$

We can generalize this result, because of the following points:

i) These correlation functions are directly related to the *time-ordered* correlation functions of the system which produces the light, if this is known and if it obeys an appropriate quantum stochastic differential equation, as we have demonstrated in Sect. 11.3.

   In the remainder of this chapter we will show how to relate measurements on the system producing the light directly to the measurements of the optical field, and in Chap. 19 we will show how this can yield a practical numerical simulation method.

ii) The correlation functions of the form (18.2.26) are not the most general kind possible, which would be of the form

$$\mathcal{G}^{(m,n)}(\tau_1,\tau_2,\dots,\tau_m;\tau'_n,\dots,\tau'_2,\tau'_1) \equiv \\ \left\langle b^\dagger(\tau_1)b^\dagger(\tau_2)\dots b^\dagger(\tau_m)b(\tau'_n)\dots b(\tau'_2)b(\tau'_1)\right\rangle. \tag{18.2.27}$$

   We will show in Chap. 21 how it is possible to measure correlation functions other than those of the kind (18.2.26) by using the technique of *homodyne measurement*.

The full description of the state of the electromagnetic field is equivalent to the specification of all such correlation functions, as was shown by *Glauber* [18.3, 18.4], although the measurement of correlation functions of the kind (18.2.27) for values of n and m other than $(1,1)$, $(2,0)$, $(0,2)$ and $(2,2)$ is rare. Nevertheless, the triple intensity correlation function of laser light was measured by *Corti* and *Degiorgio* [18.5] in 1974, corresponding to the $(3,3)$ correlation function.

18.3 Quantum Operations

The solution $|\Psi,t\rangle$ of the quantum stochastic Schrödinger equation describes a state highly entangled between the system and the electromagnetic field. However, the operators which describe the field during any infinitesimal time interval Δt_i are independent of each other, and consequently the measurement processes at each time interval can be formulated independently of each other.

Our aim is to show how the detection of photons—a measurement of the electromagnetic field—induces corresponding measurements to be made on the atom, as a result of the entanglement between the atom and the electromagnetic field induced by the time evolution of the quantum stochastic Schrödinger equation. This is a particular case of the concept of a *quantum operation*, which provides a useful framework for many aspects of measurement theory, and is formulated very clearly in the book of *Nielsen* and *Chuang* [18.6][1].

18.3.1 Definition of a Quantum Operation

Quantum operations are formulated to describe a system-environment model, such as that described by the quantum stochastic Schrödinger equation. There is a system Hilbert space $\mathcal{H}_{\text{sys}}$, and the system is coupled to an environment, with Hilbert space $\mathcal{H}_B$, which is normally infinite dimensional. Together they form a closed quantum system whose time evolution is described by a unitary transformation U in the Hilbert space $\mathcal{H}_{\text{sys}} \otimes \mathcal{H}_B$. Thus, the time evolution of the density operator of the coupled systems can be written

$$\rho_{\text{tot}} \to \rho'_{\text{tot}} = U\rho_{\text{tot}}U^\dagger. \tag{18.3.1}$$

a) The Measurement Model: In our context, we can view the environment as an abstraction of the electromagnetic field, with photons counted at every time interval. The system is described by a reduced density operator, defined by tracing the total density operator ρ_{tot} over the environment

$$\rho_{\text{sys}} = \text{Tr}_B\left\{\rho_{\text{tot}}\right\}. \tag{18.3.2}$$

There is a three-stage process:

i) *Initial Preparation of the System*: By some physical means we prepare the system in a definite state; this initial preparation is equivalent to the preparation of ρ_{tot} in the factorized form

$$\rho_{\text{tot}} = \rho_{\text{sys}} \otimes \rho_B. \tag{18.3.3}$$

ii) *System-Environment Evolution in Time*: The time evolution (18.3.1) then occurs.

iii) *Measurement of the Environment*: A measurement is made on the environment after this time evolution. As a result, the initial system density operator ρ_{sys} is mapped to a new system density operator $\mathcal{E}(\rho_{\text{sys}})$,

$$\rho_{\text{sys}} \longrightarrow \mathcal{E}(\rho_{\text{sys}}) \equiv \text{Tr}_B\left\{U\left(\rho_{\text{sys}} \otimes \rho_B\right)U^\dagger\right\}. \tag{18.3.4}$$

This is the definition of the *quantum operation* $\mathcal{E}(\rho_{\text{sys}})$.

[1] In particular, see Chap. 8.

b) Expression in Terms of the Time Evolution Operator: We will introduce an orthonormal basis set of states $|e_K\rangle$ for the Hilbert space of the bath. For simplicity, we will first consider the case where the environment is in the pure state

$$\rho_B = |e_0\rangle\langle e_0|. \tag{18.3.5}$$

For example, $|e_0\rangle$ could be the vacuum state of the electromagnetic field.

We can now express the evolution of the system density operator using (18.3.4) as

$$\mathcal{E}(\rho_{\text{sys}}) = \sum_K \langle e_K|U\left(\rho_{\text{sys}} \otimes |e_0\rangle\langle e_0|\right)U^\dagger|e_K\rangle, \tag{18.3.6}$$

$$= \sum_K E_K \rho_{\text{sys}} E_K^\dagger, \tag{18.3.7}$$

where the operational elements acting in $\mathcal{H}_{\text{sys}}$ are defined by

$$E_K = \langle e_K|U|e_0\rangle. \tag{18.3.8}$$

From this definition and the fact that $|e_K\rangle$ form a complete orthonormal basis in $\mathcal{H}_B$, it follows that

$$\sum_K E_K^\dagger E_K = \sum_K \langle e_0|U|e_K\rangle\langle e_K|U^\dagger|e_0\rangle = \langle e_0|UU^\dagger|e_0\rangle = 1. \tag{18.3.9}$$

c) Interpretation as Measurement Operators: The set of operators E_K satisfy the properties of measurement operators on the system, such as we have defined in *Bk. I: Sect.18.3.1*. However, the precise nature of what is being measured, and what numerical result is being obtained, can only be specified by using the detailed properties of the evolution operator U.

Imagine that a measurement of the *environment* is performed in the basis $|e_k\rangle$, and the value K is obtained. This can be done using the projection operator $|e_K\rangle\langle e_K|$, but in the case of the electromagnetic field, this would represent a *QND measurement*, that is, a measurement which counts a photon, but does not absorb it. This is not what happens in an absorptive photodetector, but is in principle feasible, and we discuss this further in Sect. 18.3.3.

The measurement operator corresponding to absorptive photon counting is $|e_0\rangle\langle e_K|$, and a more general possibility would be provided by the choice of environment measurement operators

$$F_K(A) \equiv |A(K)\rangle\langle e_K|, \tag{18.3.10}$$

where the states $|A(K)\rangle$ are an arbitrary set of normalized states. With this measurement operator:

i) The state of the *system* after obtaining the result K is

$$\rho_{\text{sys}}(K) \propto \text{Tr}_B\left\{F_K(A)U\left(\rho_{\text{sys}} \otimes |e_0\rangle\langle e_0|\right)U^\dagger F_K^\dagger(A)\right\}, \tag{18.3.11}$$

$$= \langle e_K|U\left(\rho_{\text{sys}} \otimes |e_0\rangle\langle e_0|\right)U^\dagger|e_K\rangle, \tag{18.3.12}$$

$$= E_K \rho_{\text{sys}} E_K^\dagger. \tag{18.3.13}$$

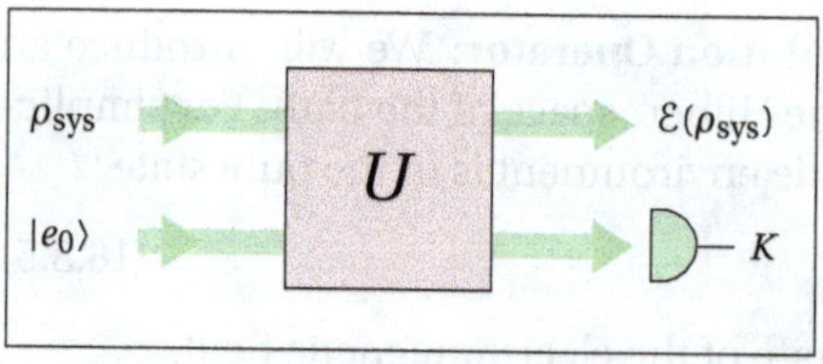

Fig. 18.4. Conceptual illustration of the structure of a quantum operation as given in (18.3.10–18.3.16).

It is important to note that result is actually independent of the choice of state $|A(K)\rangle$. However, the state of the *environment* after the measurement does depend on $|A(K)\rangle$.

ii) The probability for the outcome K is

$$p(K) = \mathrm{Tr}_{sys}\left\{E_K\rho_{sys}E_K^\dagger\right\}. \tag{18.3.14}$$

iii) The system density operator after the measurement K is then obtained by normalizing, so that

$$\rho_{sys}(K) = \frac{E_K\rho_{sys}E_K^\dagger}{p(K)}. \tag{18.3.15}$$

iv) Thus, the quantum operation is

$$\mathcal{E}(\rho_{sys}) = \sum_K p(K)\rho_{sys}(K) = \sum_K E_K\rho_{sys}E_K^\dagger. \tag{18.3.16}$$

In summary, we have constructed a set of measurement operators which act on the *system* density operator, but which measure the value K associated with the *environment.* This is exactly what we aim to do in our theory of photon counting; we wish to detect *photons,* and from these measurements, infer information about the state of the *atom.*

18.3.2 Quantum Operations on the Wavefunction

If the density operator has the form of a pure state, that is $\rho_{sys} = |\varphi\rangle\langle\varphi|$, then the action of the measurement operator on this density operator according to (18.3.15) produces another pure state density operator, that is

$$\rho_{sys}(K) = |\varphi, K\rangle\langle\varphi, K|, \tag{18.3.17}$$

in which the pure state is

$$|\varphi, K\rangle = \frac{E_K|\varphi\rangle}{\sqrt{p(K)}}. \tag{18.3.18}$$

Furthermore, the density operator formula (18.3.16) can be simulated by the following algorithm:

i) By definition, the density operator can always be written as a sum of the pure state density operators corresponding to some quantum states $|\varphi_r\rangle$, thus

$$\rho_{\text{sys}} = \sum_r \mu_r |\varphi_r\rangle\langle\varphi_r|. \tag{18.3.19}$$

ii) Using random numbers, choose a set of N values K_i according to the probability law $p(K)$, as given by (18.3.14).

iii) Construct the sample density operator

$$\bar{\rho}_{\text{sys}}(N) = \sum_r \mu_r \sum_{i=1}^{N} E_{K_i} |\varphi_r\rangle\langle\varphi_r| E^{\dagger}_{K_i}. \tag{18.3.20}$$

iv) The quantum operation is then given by

$$\mathcal{E}(\rho_{\text{sys}}) = \lim_{N\to\infty} \bar{\rho}_{\text{sys}}(N). \tag{18.3.21}$$

This algorithm forms the basis for a very practical method of simulating the time evolution of systems being continuously measured, which we will treat in more detail in Chap. 19, and which has already been introduced in *Bk. I: Ch.19.*

18.3.3 Measurement Operators for the Electromagnetic Field

As we showed in Sect. 9.3.2, the time interval Δt_i is associated with a Hilbert space $\mathcal{H}_i$, in which the quantum state may contain either 0 or 1 photon—higher numbers of photons occur only with probabilities proportional to Δt_i^2, and can be neglected. The Hilbert space is therefore effectively two-dimensional, and can be described by

$$|0,i\rangle = \begin{pmatrix}0\\1\end{pmatrix}, \qquad |1,i\rangle = \begin{pmatrix}1\\0\end{pmatrix}. \tag{18.3.22}$$

The measurement operators can be chosen in many ways, but there are two physically significant choices:

a) Absorption: By far the most important practical choice, here we have the two operators, labelled by $\mathbb{1}$ to indicate detection of one photon, and $\mathbb{0}$ to indicate that no photon is detected:

$$\text{Photon detected and absorbed:} \quad F_i^{\text{abs}}(\mathbb{1}) \equiv \begin{pmatrix}0&0\\1&0\end{pmatrix}, \tag{18.3.23}$$

$$\text{No photon detected:} \quad F_i^{\text{abs}}(\mathbb{0}) \equiv \begin{pmatrix}0&0\\0&1\end{pmatrix}. \tag{18.3.24}$$

b) Quantum Non-Demolition Measurement: This is a measurement which detects a photon, but does not absorb it. This is physically feasible, but does not describe an ordinary photodetector. A very good treatment of the general idea of QND measurements is given in the book by *Walls* and *Milburn* [18.7].

In this case, the appropriate operators are

$$\text{Photon detected and retained:}\quad F_i^{\text{qnd}}(\mathbb{1}) \equiv \begin{pmatrix}1 & 0\\ 0 & 0\end{pmatrix}, \tag{18.3.25}$$

$$\text{No photon detected:}\quad F_i^{\text{qnd}}(\mathbb{O}) \equiv \begin{pmatrix}0 & 0\\ 0 & 1\end{pmatrix}. \tag{18.3.26}$$

Exercise 18.3 Measurement Operators in Terms of Field Operators: Using the Ito algebra, show that we can write the absorption measurement operators as:

$$F^{\text{abs}}(\mathbb{1}) = \frac{\Delta B(t)}{\sqrt{\Delta t}}, \tag{18.3.27}$$

$$F^{\text{abs}}(\mathbb{O}) = 1 - \frac{\Delta^\dagger B(t)\,\Delta B(t)}{\Delta t}. \tag{18.3.28}$$

In the case of a QND measurement, show that

$$F^{\text{qnd}}(\mathbb{1}) = \frac{\Delta B^\dagger(t)\,\Delta B(t)}{\Delta t}, \tag{18.3.29}$$

$$F^{\text{qnd}}(\mathbb{O}) = 1 - \frac{\Delta B^\dagger(t)\,\Delta B(t)}{\Delta t}. \tag{18.3.30}$$

Exercise 18.4 Properties of the Absorption Measurement Operators: The following properties of the operators follow from the definitions (18.3.23, 18.3.24), and are essential for their application:

$$F_i^{\text{abs}}(\mathbb{1})|\text{vac}\rangle = 0, \tag{18.3.31}$$

$$F_i^{\text{abs}}(\mathbb{O})|\text{vac}\rangle = |\text{vac}\rangle, \tag{18.3.32}$$

$$F_i^{\text{abs}}(\mathbb{1})\Delta B^\dagger(t)|\text{vac}\rangle = \sqrt{\Delta t}\,|\text{vac}\rangle. \tag{18.3.33}$$

$$F_i^{\text{abs}}(\mathbb{O})\Delta B^\dagger(t)|\text{vac}\rangle = 0. \tag{18.3.34}$$

Exercise 18.5 Properties of the QND Measurement Operators: Formulate the appropriate properties of the QND operators which correspond to those of the previous exercise.

18.4 Photon Counting Using the Quantum Stochastic Schrödinger Equation

We will now show how the process of continuous measurement of the electromagnetic field, using the measurement operators of Sect. 18.3.3, can induce a measurement process on the system producing the field being measured. The measurement process which results can also be described by the formalism of quantum operations and measurement operators, and provides a physical implementation of the abstract formulation of quantum operations given in Sect. 18.3.1. The measurement operators arising from this procedure are directly related to the interaction Hamiltonian which couples the system to the electromagnetic field.

18.4.1 Compact Notation

Let us introduce a compact notation for the quantum stochastic Schrödinger equation; for the version given in (9.4.5) this is

$$d|\Psi,t\rangle = \mathbf{K}|\Psi,t\rangle\,dt + \mathbf{J}|\Psi,t\rangle\,dB^\dagger(t), \tag{18.4.1}$$

$$\mathbf{K} \equiv -\frac{i}{\hbar}H_{\text{eff}} = -\frac{i}{\hbar}\left(H_{\text{sys}} + \hbar\delta\omega\, c^\dagger c\right) - \tfrac{1}{2}\Gamma\, c^\dagger c, \tag{18.4.2}$$

$$\mathbf{J} \equiv \sqrt{\Gamma}\, c. \tag{18.4.3}$$

The precise form of the operators $\mathbf{K}$ and $\mathbf{J}$ will depend on the details of the system under study, and we will generalize the notation when appropriate for other systems.

The initial condition is chosen as

$$|\Psi,t_0\rangle = |\psi_0\rangle \otimes |\text{vac}\rangle, \tag{18.4.4}$$

and we define the initial system density operator

$$\rho(t_0) \equiv |\psi_0\rangle\langle\psi_0|. \tag{18.4.5}$$

We will use the discretized form of the solution, which follows from the difference equation form (9.3.18) that we used for its derivation. We will use a set of infinitesimally spaced time steps

$$\Delta\tau_i = \tau_{r+1} - \tau_i, \qquad \tau_0 = t_0, \quad t = \tau_n. \tag{18.4.6}$$

Consider now the result of the first time step, which we can write

$$|\Psi,\tau_1\rangle = U_0\,|\Psi,\tau_0\rangle. \tag{18.4.7}$$

The operator U_0 is unitary to order $\Delta\tau_0$, and is defined by

$$U_0 \equiv \left(1 + \mathbf{K}\Delta\tau_0 + \mathbf{J}\Delta B^\dagger(\tau_0)\right). \tag{18.4.8}$$

18.4.2 Formulation of Photon Counting Using Quantum Operations

We can now apply the quantum operation formalism, using the absorptive photon counting operators (18.3.27, 18.3.28) and their properties (18.3.31, 18.3.32). We begin with a consideration of the first time step, and then generalize to many time steps.

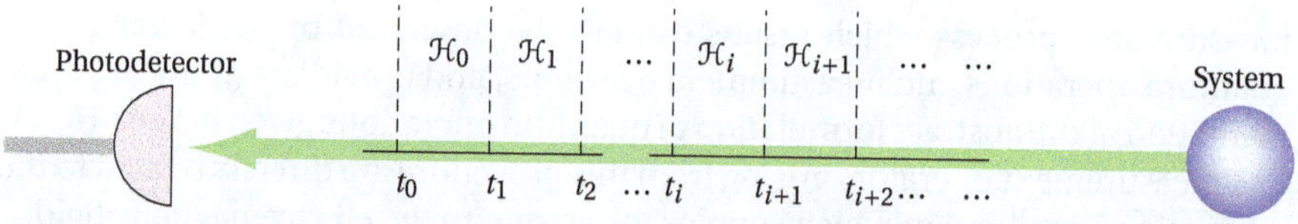

Fig. 18.5. Light emitted from the system and detected by a photodetector. The light produced by the system is described by the same resolution into Hilbert spaces as in Fig. 18.1. The photodetector measures the light produced by the system. Since the optical field is entangled with the state of the system, this induces a measurement on the system, which is described by the corresponding quantum operations (18.4.24, 18.4.25).

a) Measurement Procedure: We will proceed using the following three steps:

i) Evolve the full density operator unitarily for one infinitesimal time step.

ii) Measure the electromagnetic field, using the absorptive measurement operators (18.3.23, 18.3.24) to determine the probability of detecting or not detecting a photon, and to determine the resulting full density operator.

iii) Determine the reduced density operator for the system by tracing over the electromagnetic field. This yields expressions for quantum operations which act only on the system.

b) Unitary Evolution: The unitary evolution of the full density operator for one time step is, keeping only terms up to order $d\tau_0$

$$\rho(\tau_0) \otimes |\mathrm{vac}\rangle\langle\mathrm{vac}| \to U_0 \Big(\rho(\tau_0) \otimes |\mathrm{vac}\rangle\langle\mathrm{vac}|\Big)\, U_0^\dagger, \tag{18.4.9}$$

$$= \Big(\rho(\tau_0) + \big(\mathbf{K}\rho(\tau_0) + \rho(\tau_0)\mathbf{K}^\dagger\big)\, d\tau_0\Big) \otimes |\mathrm{vac}\rangle\langle\mathrm{vac}| + \Big(\mathbf{J}\rho(\tau_0)\mathbf{J}^\dagger\Big) \otimes \Big(dB^\dagger(\tau_0)\,|\mathrm{vac}\rangle\langle\mathrm{vac}|\, dB(\tau_0)\Big). \tag{18.4.10}$$

c) First Time Step: There are two measurement possibilities:

i) *One Photon Counted*: The system-environment density operator is evolved unitarily for one time step. The electromagnetic field is then measured, resulting in a photon count. The system density operator becomes (before normalization)

$$\tilde{\rho}_{\mathrm{sys}}(\mathbb{1}, \tau_1) = \mathrm{Tr}_{\mathrm{EM}}\Big\{F_0^{\mathrm{abs}}(\mathbb{1})\, U_0\Big(\rho(\tau_0) \otimes |\mathrm{vac}\rangle\langle\mathrm{vac}|\Big) U_0^\dagger\, F_0^{\mathrm{abs}}(\mathbb{1})^\dagger\Big\}, \tag{18.4.11}$$

$$= \mathbf{J}\rho(\tau_0)\mathbf{J}^\dagger\, \Delta\tau_0. \tag{18.4.12}$$

The probability of this measurement is

$$\mathrm{Prob}(\mathbb{1}, \tau_1) = \mathrm{Tr}_{\mathrm{sys}}\{\tilde{\rho}_{\mathrm{sys}}(\mathbb{1}, \tau_1)\} = \mathrm{Tr}_{\mathrm{sys}}\Big\{\mathbf{J}\rho(\tau_0)\mathbf{J}^\dagger\Big\}\, \Delta\tau_0. \tag{18.4.13}$$

The normalized system density operator is therefore

$$\rho_{\mathrm{sys}}(\mathbb{1}, \tau_1) = \frac{\tilde{\rho}_{\mathrm{sys}}(\mathbb{1}, \tau_1)}{\mathrm{Prob}(\mathbb{1}, \tau_1)}. \tag{18.4.14}$$

ii) *Zero Photons Counted*: The same time evolution is followed by a measurement of the electromagnetic field resulting in no photon being counted. The system density operator becomes (before normalization)

$$\tilde{\rho}_{\text{sys}}(\mathbb{O},\tau_1) = \text{Tr}_{\text{EM}}\left\{F_0^{\text{abs}}(\mathbb{O})U_0\Big(\rho(\tau_0)\otimes|\text{vac}\rangle\langle\text{vac}|\Big)U_0^\dagger F_0^{\dagger\text{abs}}(\mathbb{O})\right\}, \tag{18.4.15}$$

$$= \rho(\tau_0) + \Big(\mathbf{K}\rho(\tau_0) + \rho(\tau_0)\mathbf{K}^\dagger\Big)\Delta\tau_0, \tag{18.4.16}$$

$$= \mathrm{e}^{\mathbf{K}\Delta\tau_0}\rho(\tau_0)\mathrm{e}^{\mathbf{K}^\dagger\Delta\tau_0}, \quad \text{to order } \Delta\tau_0. \tag{18.4.17}$$

The probability of this measurement is

$$\text{Prob}(\mathbb{O},\tau_1) = \text{Tr}_{\text{sys}}\left\{\mathrm{e}^{\mathbf{K}\Delta\tau_0}\rho(\tau_0)\mathrm{e}^{\mathbf{K}^\dagger\Delta\tau_0}\right\}. \tag{18.4.18}$$

The normalized system density operator is therefore

$$\rho_{\text{sys}}(\mathbb{O},\tau_1) = \frac{\tilde{\rho}_{\text{sys}}(\mathbb{O},\tau_1)}{\text{Prob}(\mathbb{O},\tau_1)}. \tag{18.4.19}$$

d) Photons Detected, but not Counted: In this case we must sum the two previous results, so that density operator becomes

$$\rho_{\text{sys}}(\mathbb{1}+\mathbb{O},\tau_1) = \rho(\tau_0) + \Big(\mathbf{K}\rho(\tau_0) + \rho(\tau_0)\mathbf{K}^\dagger + \mathbf{J}\rho(\tau_0)\mathbf{J}^\dagger\Big)\Delta\tau_0, \tag{18.4.20}$$

$$= \rho(\tau_0) - \frac{\mathrm{i}}{\hbar}\Big[H_{\text{sys}} + \hbar\delta\omega\, c^\dagger c\,,\ \rho(\tau_0)\Big]\Delta\tau_0$$
$$+\tfrac{1}{2}\Gamma\Big(2c\rho(\tau_0)c^\dagger - c^\dagger c\rho(\tau_0) - \rho(\tau_0)c^\dagger c\Big)\Delta\tau_0. \tag{18.4.21}$$

The time evolution corresponds to that given by the master equation.

e) Several Time Steps: Since the Hilbert spaces at each time step are independent of each other, the basic procedure for the next step is the same as for the first time step, the only change being in the initial condition for the density operator at time τ_1, which is given by one of the versions (18.4.14, 18.4.19) of the normalized density operator. It is the same for subsequent steps.

f) Counting for a Finite Time: Using the form (18.4.18), the probability of counting zero photons in the time interval (t_0, t) is

$$\text{Prob}(\mathbb{O},t) = \text{Tr}_{\text{sys}}\left\{\mathrm{e}^{\mathbf{K}(t-t_0)}\rho(t_0)\mathrm{e}^{\mathbf{K}^\dagger(t-t_0)}\right\}, \tag{18.4.22}$$

and the resulting density operator is

$$\rho_{\text{sys}}(\mathbb{O},t) = \frac{\mathrm{e}^{\mathbf{K}(t-t_0)}\rho(t_0)\mathrm{e}^{\mathbf{K}^\dagger(t-t_0)}}{\text{Prob}(\mathbb{O},t)}. \tag{18.4.23}$$

18.4.3 Quantum Operations Induced on the System

We have now shown that the measurement process—counting photons in the electromagnetic Hilbert space—can be expressed entirely within the system Hilbert space. Further, although we have derived operations on the density operator in the form (18.4.12, 18.4.17), these can be expressed as actions on any pure quantum state. To see this, choose the initial density operator to be that of a pure state, for example choose $\rho(\tau_0) = |\varphi\rangle\langle\varphi|$. Both of the operations (18.4.12, 18.4.18) on this state yield a pure state for $\tilde{\rho}_{\rm sys}(\tau_1)$.

Explicitly, we find two measurement operators, corresponding to the mappings (18.4.12, 18.4.17), whose actions on an arbitrary system state $|\varphi\rangle$ are:

i) *Detect No Photon in the Finite Interval* (τ',τ): The measurement operator is $E_{\mathbb{0}}(\tau',\tau)$, defined by

$$E_{\mathbb{0}}(\tau',\tau)|\varphi\rangle = \mathrm{e}^{\mathbf{K}(\tau-\tau')}|\varphi\rangle, \qquad \text{where } \tau > \tau'. \tag{18.4.24}$$

ii) *Detect One Photon in the Infinitesimal Interval* $(\tau,\tau+\Delta\tau)$: The measurement operator is $E_{\mathbb{1}}(\tau)$, defined by

$$E_{\mathbb{1}}(\tau)|\varphi\rangle = \mathbf{J}\sqrt{\Delta\tau}\,|\varphi\rangle. \tag{18.4.25}$$

These operators correspond exactly to those used in *Bk. I: Sect.19.1.1*, when we write them explicitly in terms of the physical operators. In that case we chose (τ',τ) to be the infinitesimal interval $(\tau,\tau+\Delta\tau)$, so that

$$E_{\mathbb{0}}(\tau,\tau+\Delta\tau) = 1 - \frac{\mathrm{i}}{\hbar}\Big(H_{\rm sys} + \hbar\,\delta\omega\, c^\dagger c\Big)\Delta\tau - \tfrac{1}{2}\Gamma c^\dagger c\,\Delta\tau, \tag{18.4.26}$$

$$E_{\mathbb{1}}(\tau) = c\sqrt{\Gamma\,\Delta\tau}\,. \tag{18.4.27}$$

The only difference is the explicit inclusion of the frequency shift generated by the coupling of the system to the electromagnetic field, which is included implicitly in the formulation in *Bk. I: Sect.19.1.1*.

18.4.4 Probability Density of the First Detection Time

a) Conditional Probability: The probability of counting a photon in the interval $(t,t+\Delta t)$, under the condition that no photons were counted in the interval (t_0,t) is given by the formula analogous to (18.4.13),

$$\mathrm{Prob}(\mathbb{1},t) = \mathrm{Tr}_{\rm sys}\Big\{\mathbf{J}\rho_{\rm sys}(\mathbb{0},t)\mathbf{J}^\dagger\Big\}\,\Delta t, \tag{18.4.28}$$

$$= \frac{\mathrm{Tr}_{\rm sys}\Big\{\mathbf{J}\mathrm{e}^{\mathbf{K}(t-t_0)}\rho(t_0)\mathrm{e}^{\mathbf{K}^\dagger(t-t_0)}\mathbf{J}^\dagger\Big\}}{\mathrm{Prob}(\mathbb{0},t)}\,\Delta t. \tag{18.4.29}$$

b) Joint Probability: The joint probability that, after starting counting at time t_0, that no photon is detected in the interval (t_0,t), and that a photon is detected in the interval $(t,t+\Delta t)$ is the product

$$P\big(\mathbb{1},t;\mathbb{0},(t_0,t)\big)\Delta t = \mathrm{Prob}(\mathbb{1},t)\,\mathrm{Prob}(\mathbb{0},t) \tag{18.4.30}$$

$$= \mathrm{Tr}_{\rm sys}\Big\{\mathbf{J}\mathrm{e}^{\mathbf{K}(t-t_0)}\rho(t_0)\mathrm{e}^{\mathbf{K}^\dagger(t-t_0)}\mathbf{J}^\dagger\Big\}\,\Delta t. \tag{18.4.31}$$

This is the probability density for the first count time t, and it is normalized appropriately,

$$\int_{t_0}^{\infty} P\big(\mathbb{1}, t'; \mathbb{0}, (t_0, t')\big)\, dt' = 1. \tag{18.4.32}$$

The proof is straightforward; note that by their definitions (18.4.2, 18.4.3), for any operator μ, the operators $\mathbf{K}$ and $\mathbf{J}$ satisfy

$$\mathrm{Tr}_{\mathrm{sys}}\left\{\mathbf{K}\mu + \mu\mathbf{K}^\dagger + \mathbf{J}\mu\mathbf{J}^\dagger\right\} = 0. \tag{18.4.33}$$

Using this identity, we can eliminate $\mathbf{J}$ to get

$$P\big(\mathbb{1}, t; \mathbb{0}, (t_0, t)\big) = -\frac{d}{dt}\mathrm{Tr}_{\mathrm{sys}}\left\{e^{\mathbf{K}(t-t_0)}\rho(t_0)e^{\mathbf{K}^\dagger(t-t_0)}\right\}, \tag{18.4.34}$$

so that

$$\int_{t_0}^{\infty} P\big(\mathbb{1}, t'; \mathbb{0}, (t_0, t')\big)\, dt' = 1 - \lim_{t\to\infty}\mathrm{Tr}_{\mathrm{sys}}\left\{e^{\mathbf{K}(t-t_0)}\rho(t_0)e^{\mathbf{K}^\dagger(t-t_0)}\right\}. \tag{18.4.35}$$

The condition for the second term to vanish is equivalent to requiring that all possible initial states of the system do decay.

It is conceivable that there are states which do not decay; and if this is physically the case for a particular system, then the theory must be appropriately modified to exclude such states. However, under normal conditions the second term does vanish, and the normalization of the probability density is assured.

c) Superoperator Notation: In order to make compact formulae, we will define the superoperators

$$\mathcal{K}\rho \equiv \mathbf{K}\rho + \rho\mathbf{K}^\dagger, \tag{18.4.36}$$

$$\mathcal{J}\rho \equiv \mathbf{J}^\dagger\rho\mathbf{J}, \tag{18.4.37}$$

$$\mathcal{L} \equiv \mathcal{K} + \mathcal{J}. \tag{18.4.38}$$

The notation is consistent with that introduced in Sect. 10.1.2, and in terms of this notation, (18.4.31) becomes

$$P\big(\mathbb{1}, t; \mathbb{0}, (t_0, t)\big)\Delta t = \mathrm{Tr}_{\mathrm{sys}}\left\{\mathcal{J}e^{\mathcal{K}(t-t_0)}\rho(t_0)\right\}\Delta t. \tag{18.4.39}$$

18.4.5 Resolution in Terms of Photon Counts

When using the photodetector, the statistics of the photocounts will be characterized by an alternation of the two processes described above, that is, instantaneous photocounts alternating with finite periods during which no photon is counted.

18.5 Photon Count Hierarchies

The wavefunction for the system and the electromagnetic field can be resolved into terms corresponding to the creation of definite numbers of photons by writing

$$|\Psi,t\rangle = \sum_{n=0}^{\infty} |\Psi,n,t\rangle, \tag{18.5.1}$$

$$d|\Psi,0,t\rangle = \mathbf{K}|\Psi,0,t\rangle\,dt, \tag{18.5.2}$$

$$d|\Psi,n,t\rangle = \mathbf{K}|\Psi,n,t\rangle\,dt + \mathbf{J}|\Psi,n-1,t\rangle\,dB^\dagger(t), \qquad n = 1,2,3,\ldots \tag{18.5.3}$$

To complete this resolution, we also require:

i) *Initial condition*: At $t = 0$ no photons have been emitted, therefore

$$|\Psi,0,t\rangle = |\psi_0\rangle \otimes |\mathrm{vac}\rangle, \tag{18.5.4}$$

$$|\Psi,n,t\rangle = 0, \qquad n = 1,2,3,\ldots \tag{18.5.5}$$

ii) *Interpretation*: From the quantum stochastic differential equation (18.5.3) it is clear that $|\Psi,n,t\rangle$ has one more photon than $|\Psi,n,t\rangle$ since $dB^\dagger(t)$ adds one more photon. With the initial conditions chosen, it follows that $|\Psi,n,t\rangle$ does correspond to the creation of exactly n photons.

iii) It is obvious that $|\Psi,t\rangle$ as written in (18.5.1) is a solution of the quantum stochastic Schrödinger equation (18.4.1).

18.5.1 Resolution of the System Density Operator

The system density operator $\bar\rho(t)$ is defined, as in Sect. 10.1.1, by the formula $\bar\rho(t) \equiv \mathrm{Tr}_{\mathrm{EM}}\{|\Psi,t\rangle\langle\Psi,t|\}$. If we insert the resolution (18.5.1) in terms of n-photon states, we get

$$\bar\rho(t) = \sum_{n=0}^{\infty} \bar\rho_n(t), \tag{18.5.6}$$

$$\bar\rho_n(t) = \mathrm{Tr}_{\mathrm{EM}}\{|\Psi,n,t\rangle\langle\Psi,n,t|\}. \tag{18.5.7}$$

There are no off-diagonal terms, such as $\mathrm{Tr}_{\mathrm{EM}}\{|\Psi,n,t\rangle\langle\Psi,m,t|\}$, since the many-photon states are orthogonal for $n \neq m$.

Using the quantum Ito algebra in the same way as in Sect. 10.1.1, the hierarchy of wavefunction equations (18.5.3) becomes a hierarchy for the different components of the density operator

$$\bar{\rho}(t) = \sum_{n=0}^{\infty} \bar{\rho}_n(t), \tag{18.5.8}$$

$$d\bar{\rho}_0(t) = \mathcal{K}\bar{\rho}_0(t)\,dt, \tag{18.5.9}$$

$$d\bar{\rho}_n(t) = \mathcal{K}\bar{\rho}_n(t)\,dt + \mathcal{J}\bar{\rho}_{n-1}(t)\,dt, \tag{18.5.10}$$

with the definitions, as in (18.4.36–18.4.38),

$$\mathcal{K}\rho \equiv \mathbf{K}^\dagger\rho + \rho\mathbf{K}, \qquad \mathcal{J}\rho \equiv \mathbf{J}^\dagger\rho\mathbf{J}. \tag{18.5.11}$$

i) *Initial Conditions*: Corresponding to those given in (18.5.4, 18.5.5) for the wavefunction, we require

$$\bar{\rho}_0(0) = \bar{\rho}(0), \tag{18.5.12}$$

$$\bar{\rho}_n(0) = 0, \qquad n = 1,2,3,\ldots \tag{18.5.13}$$

ii) *Comparison of the Hierarchies*: Notice that the two hierarchies (18.5.2, 18.5.3) and (18.5.9, 18.5.10) have an identical structure under the mappings

$$\mathcal{K} \longleftrightarrow \mathbf{K}, \qquad \mathcal{J}\,dt \longleftrightarrow \mathbf{J}\,dB(t). \tag{18.5.14}$$

Their interpretations are, however, subtly different.

iii) *Interpretations*: The wavefunction hierarchy describes components of the wavefunction which are coherent with each other, and the process of time evolution is one of *creating* photons, giving a quantum state with increasing entanglement.

In contrast, the density operator hierarchy describes the process of *detecting* photons, and the different components of the density operator are added incoherently. In essence this is a process of absorption of the photons created by the time evolution, and described by the full wavefunction $|\Psi, t\rangle$.

18.5.2 Comparison with the Classical Poisson Process

The Poisson process provides a classical model of photon counts, whose structure is quite similar to the fully quantum-mechanical formulation, which, however, also takes into account the dynamics of the object emitting the photons.

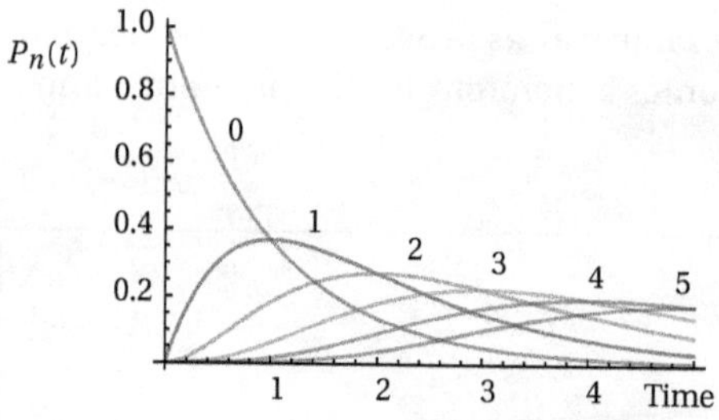

Fig. 18.6. The Poisson process, resolved into the contributions $(\kappa t)^n e^{-\kappa t}/n!$ from the creation of $n = 0, 1, 2, 3, 4, 5$ photons.

a) The Poisson Process: The Poisson process, which we have introduced in *Bk. I: Sect.6.1.2*, has a master equation which takes the very simple form,which we present in the same format as used for density operator in (18.5.8–18.5.10):

$$1 = \sum_{n=0}^{\infty} P_n(t), \tag{18.5.15}$$

$$\dot{P}_0(t) = -\kappa P_0(t), \tag{18.5.16}$$

$$\dot{P}_n(t) = -\kappa P_n(t) + \kappa P_{n-1}(t), \qquad n = 1,2,3,\ldots \tag{18.5.17}$$

It has the solution

$$P_n(t) = \frac{(\kappa t)^n}{n!} e^{-\kappa t}. \tag{18.5.18}$$

This equation hierarchy can be put in an integral form as

$$P_0(t) = e^{-\kappa t}, \tag{18.5.19}$$

$$P_n(t) = \kappa \int_0^t ds\, e^{-\kappa(t-s)} P_{n-1}(s). \tag{18.5.20}$$

The equations we shall derive for the process of continuous measurement have a similar, but more complex structure, as we shall see in Chap. 19.

b) Interpretation of the Density Operator Hierarchy: The equation hierarchy (18.5.9, 18.5.10) can then be seen to have the same kind of structure as the equations (18.5.16, 18.5.17) for the Poisson process, where each term in the hierarchy is fed by the previous term—in fact it is the one-dimensional version of the density operator hierarchy. In the case of the Poisson process the coefficients of the two terms have the same value κ.

In the density operator hierarchy the loss term $\mathcal{K}$ and the recycling term $\mathcal{J}$ are not identical, and the equality of loss and feeding coefficients is replaced by a related result, namely

$$\mathrm{Tr}_{\mathrm{sys}}\{\mathcal{K}\rho\} = -\mathrm{Tr}_{\mathrm{sys}}\{\mathcal{J}\rho\}. \tag{18.5.21}$$

Both this result, and the equality of coefficients in Poisson process, ensure that probability is conserved.

18.6 Unravelling the System Density Operator

The concept of "unravelling" the density operator is defined as a two stage decomposition of the density operator equations:

i) Firstly, identify the terms in which a fixed number of photons n were detected. This we have done, the terms required are $\bar{\rho}_n$, as given in (18.5.7).

ii) For each number n of photons detected, identify the terms which correspond to counting of these photons in the definite time-intervals $(t_1, t_1 + dt_1], (t_2, t_2 + dt_2], \dots (t_n, t_n + dt_n]$. This has not yet been done, and we shall now show how to do it.

The equation hierarchy (18.5.8–18.5.10) with initial conditions (18.5.12, 18.5.13) can be integrated to give

$$\bar{\rho}_0(t) = e^{\mathcal{K}t}\bar{\rho}(0), \tag{18.6.1}$$

$$\bar{\rho}_n(t) = \int_0^t dt'\, e^{\mathcal{K}(t-t')}\mathcal{J}\bar{\rho}_{n-1}(t'), \qquad n = 1,2,3,\dots \tag{18.6.2}$$

From this, an explicit solution for the density operator hierarchy can be written down:

$$\bar{\rho}_n(t) = \sum_{n=1}^{\infty}\int_0^t dt_n \int_0^{t_n} dt_{n-1}\dots\int_0^{t_2} dt_1\, e^{\mathcal{K}(t-t_n)}\mathcal{J}e^{\mathcal{K}(t_n-t_{n-1})}\dots e^{\mathcal{K}(t_2-t_1)}\mathcal{J}e^{\mathcal{K}t_1}\rho(0). \tag{18.6.3}$$

18.6.1 Photon Counting Statistics

We use the notations:

i) The notation p_0^t for probability that there is no count in the time interval $[0, t]$, given that the system was prepared in the state $\rho(0)$ at time 0:

$$p_0^t = \mathrm{Tr}_{\mathrm{sys}}\left\{e^{\mathcal{K}t}\rho(0)\right\}. \tag{18.6.4}$$

ii) The notation $p_0^t(t_1,\dots,t_n)$ is used for the probability density of having a count at each of the times $t_1,\dots,t_n$, and no other counts during the rest of the time interval $(0, t]$:

$$p_0^t(t_1,\dots,t_n) = \mathrm{Tr}_{\mathrm{sys}}\left\{e^{\mathcal{K}(t-t_n)}\mathcal{J}e^{\mathcal{K}(t_n-t_{n-1})}\dots e^{\mathcal{K}(t_2-t_1)}\mathcal{J}e^{\mathcal{K}t_1}\rho(0)\right\}. \tag{18.6.5}$$

iii) The n-photon probabilities are

$$P_0(t) = \mathrm{Tr}_{\mathrm{sys}}\{\bar{\rho}_0(t)\} = p_0^t, \tag{18.6.6}$$

$$P_n(t) = \mathrm{Tr}_{\mathrm{sys}}\{\bar{\rho}_n(t)\}, \tag{18.6.7}$$

$$= \int_0^t dt_n \int_0^{t_n} dt_{n-1}\dots\int_0^{t_2} dt_1\, p_0^t(t_1,\dots,t_n), \qquad n = 1,2,3,\dots. \tag{18.6.8}$$

18.6.2 Density Operator after n Counts

The state of the system for the given count sequence of n counts, one in each of the time intervals $dt_1, dt_2, dt_3, \ldots, dt_n$ is

$$\rho_{\text{sys}}(t|t_1, t_2, \ldots, t_n) = \frac{e^{\mathcal{K}(t-t_n)}\mathcal{J}e^{\mathcal{K}(t_n-t_{n-1})}\ldots\mathcal{J}e^{\mathcal{K}(t_2-t_1)}\mathcal{J}e^{\mathcal{K}(t_1)}\rho(0)}{p_0^t(t_1, \ldots, t_n)}. \tag{18.6.9}$$

Exercise 18.6 The Two-Level System Optical Bloch Equations: For a two-level system we can write $\mathcal{J}\rho(t) = |g\rangle\langle g|\Gamma\rho_{ee}(t)$, and this leads to a hierarchy of optical Bloch equations. Write these out explicitly as a set of equations for the density matrix elements $\rho_n^{ee}, \rho_n^{eg}, \rho_n^{ge}, \rho_n^{gg}$.

19. Quantum Trajectories

The concept of a *quantum trajectory* arises from a consideration of the implications of the Sect. 18.5 and and Sect. 18.6, that the hierarchies of equations for the reduced density operator are equivalent to a stochastic process based on the reduced density operator for the system. The photon counting probabilities and the reduced density operators can be determined from these equations, without there being any need to consider explicitly the full Hilbert space describing the system and the electromagnetic field. Explicitly, the probability densities (18.6.4, 18.6.5) for n-photocounts, and the corresponding density operators (18.6.9) after a sequence of n-photocounts, together provide all that is necessary to describe any counting process based on the original quantum stochastic Schrödinger equation (18.4.1).

In this chapter we will show how to simplify this description, to yield an interpretation which treats the system wavefunction as a smoothly evolving wavefunction, interrupted by *quantum jumps*, which occur randomly according to the photon probability densities (18.6.4). Such a wavefunction has been called a *quantum trajectory*, a terminology introduced by *Carmichael* and *Alsing* [19.1], since the system wavefunction is viewed as a point in the system Hilbert space, whose evolution according to this concept can be regarded as the execution of a trajectory in Hilbert space. Thus, we can simulate a number of wavefunctions $|\psi_a, t\rangle$, corresponding to different realizations (labelled by the index a) of the photocount process. The system density operator can be viewed as the stochastic average of density operators constructed from the individual trajectories, thus if we construct N_{Traj} such wavefunctions, then the system density operator is approximated by

$$\bar{\rho}(t) = \frac{1}{N_{\text{Traj}}}\sum_a |\psi_a, t\rangle\langle\psi_a, t|.$$

The description we arrive at is very much like *Bohr's* original concept of the electron making quantum jumps from one energy level to the other, emitting a photon in the process. Of course Bohr knew nothing about wavefunctions in his original atomic theory, so his jumps were between classical trajectories, whereas the jumps we are considering are between different smooth trajectories in Hilbert space.

19.1 Stochastic Wavefunction Simulations

The density operator hierarchy (18.5.8–18.5.10), and its solutions (18.6.1–18.6.4) have the property that if the initial system density operator represents a pure state $\rho(t_0) = |\psi_0\rangle\langle\psi_0|$, then each of the components $\bar{\rho}_n(t)$ also represents a pure state. Indeed, the density operator hierarchy is reproduced by a hierarchy very like that for the full quantum state, namely

$$\bar{\rho}(t) = \sum_{n=0}^{\infty} |\tilde{\psi}, n, t\rangle\langle\tilde{\psi}, n, t|, \tag{19.1.1}$$

$$\frac{d|\tilde{\psi}, 0, t\rangle}{dt} = \mathbf{K}|\tilde{\psi}, 0, t\rangle, \tag{19.1.2}$$

$$\frac{d|\tilde{\psi}, n, t\rangle}{dt} = \mathbf{K}|\tilde{\psi}, n, t\rangle + \mathbf{J}|\tilde{\psi}, n-1, t\rangle, \qquad n = 1,2,3,\dots \tag{19.1.3}$$

The wavefunctions produced by this procedure are of course not normalized wavefunctions. This is an important feature, since the norm is directly related to a counting probability, and this information is lost if we normalize them. We will codify this feature in two ways:

i) *Terminology—Probability Amplitude*: A wavefunction produced by this procedure will be called a *probability amplitude*. The term wavefunction will be kept for *normalized* wavefunctions.

ii) *Notation*: We shall use the notation $\tilde{\psi}$ to mean that $|\tilde{\psi}\rangle$ is a probability amplitude.

19.1.1 Stochastic Interpretation

Now consider a particular sequence $t_1, t_2, \dots, t_n$ of detections contained *within* the time interval $[t_0, t]$. Associated with this detection sequence there is a *probability amplitude* defined by

$$|\tilde{\psi}(t|t_1, t_2, \dots, t_n)\rangle = \mathrm{e}^{\mathbf{K}(t-t_n)}\mathbf{J}\mathrm{e}^{\mathbf{K}(t_n-t_{n-1})}\dots\mathbf{J}\mathrm{e}^{\mathbf{K}(t_2-t_1)}\mathbf{J}\mathrm{e}^{\mathbf{K}(t_1-t_0)}|\psi_0\rangle, \tag{19.1.4}$$

which occurs with probability density, as defined in (18.6.5), equal to its norm, that is

$$p_{t_0}^t(t_1, t_2, \dots, t_n) = \langle\tilde{\psi}(t|t_1, t_2, \dots, t_n)|\tilde{\psi}(t|t_1, t_2, \dots, t_n)\rangle. \tag{19.1.5}$$

For the *wavefunction* corresponding to the *probability amplitude* $\tilde{\psi}$ we use the symbol ψ—thus

$$|\psi(t|t_1, t_2, \dots, t_n)\rangle \equiv \frac{|\tilde{\psi}(t|t_1, t_2, \dots, t_n)\rangle}{\sqrt{p_{t_0}^t(t_1, t_2, \dots, t_n)}}. \tag{19.1.6}$$

19.1.2 Hierarchy of Equations

Instead of an explicit expression, we can also write the recursive procedure for computing the wavefunction. The formula (19.1.4) results from the following steps:

$$|\tilde{\psi}(t|t_1, t_2, \ldots, t_n)\rangle = \mathrm{e}^{\mathbf{K}(t-t_n)}|\tilde{\psi}_n\rangle, \tag{19.1.7}$$

$$|\tilde{\psi}_n\rangle = \mathbf{J}\mathrm{e}^{\mathbf{K}(t_n-t_{n-1})}|\tilde{\psi}_{n-1}\rangle, \tag{19.1.8}$$

$$\vdots$$

$$|\tilde{\psi}_2\rangle = \mathbf{J}\mathrm{e}^{\mathbf{K}(t_2-t_1)}|\tilde{\psi}_1\rangle, \tag{19.1.9}$$

$$|\tilde{\psi}_1\rangle = \mathbf{J}\mathrm{e}^{\mathbf{K}(t_1-t_0)}|\psi_0\rangle. \tag{19.1.10}$$

This gives all of the wavefunctions for any number r of counts up to n, since

$$|\tilde{\psi}(t|t_1, t_2, \ldots, t_i)\rangle = \mathrm{e}^{\mathbf{K}(t-t_i)}|\tilde{\psi}_i\rangle. \tag{19.1.11}$$

19.1.3 Probabilities

These can be easily computed. The initial wavefunction $|\psi_0\rangle$ is normalized to 1, and therefore:

i) The probability density of the first detection is

$$\mathsf{P}(t_1) = \big\| |\tilde{\psi}_1\rangle \big\|^2. \tag{19.1.12}$$

This is a conditional probability, conditioned on the initial state being $|\psi_0\rangle$.

ii) The probability density of the second detection is also a conditional probability; explicitly, it is the probability of a detection at time t_2 given that a detection occurred at time $t_1 < t_2$. We must therefore first normalize $|\tilde{\psi}_1\rangle$, so that

$$\mathsf{P}(t_2) = \frac{\big\| |\tilde{\psi}_2\rangle \big\|^2}{\big\| |\tilde{\psi}_1\rangle \big\|^2}. \tag{19.1.13}$$

iii) For the r th detection, we get similarly

$$\mathsf{P}(t_i) = \frac{\big\| |\tilde{\psi}_i\rangle \big\|^2}{\big\| |\tilde{\psi}_{r-1}\rangle \big\|^2}. \tag{19.1.14}$$

iv) The joint probability of a detection sequence at the times $t_1, t_2, \ldots t_n$ is given by multiplying all of the conditional probabilities, and is given by the simple formula

$$\mathsf{P}(t_1)\mathsf{P}(t_2)\ldots\mathsf{P}(t_n) = \big\| |\tilde{\psi}_n\rangle \big\|^2. \tag{19.1.15}$$

We see that the probability amplitude $|\tilde{\psi}_n\rangle$, as defined in the hierarchy (19.1.7–19.1.10) contains all of the probabilistic information about the detections at times $t_1, t_2, \ldots t_n$.

19.2 Stochastic Wavefunction Algorithms

We have described a simulation procedure based on these probability densities in *Bk. I: Sect.19.2.2*; however, in practice it is simpler to use a method based on small time intervals. During each time interval there is an infinitesimal probability of a detection, and this is associated with a quantum jump of the system wavefunction, that is $|\tilde{\psi}\rangle \to \mathbf{J}|\tilde{\psi}\rangle$. The explicit procedure is:

i) Choose the initial system wave function $|\psi_0\rangle$, initially normalized to 1.

ii) Set the counter r for the number of quantum jumps equal to zero, $r = 0$.

iii) We choose a time interval Δt which is sufficiently small for the accuracy required.

iv) Compute the probabilities:

$$\text{One photon detected:} \qquad \mathbf{P}_{1}^{(t_0,t_0+\Delta t]} = \big\|\mathbf{J}|\psi_0\rangle\big\|^2 \Delta t \quad \ll 1, \tag{19.2.1}$$

$$\text{Zero photons detected:} \qquad \mathbf{P}_{0}^{(t_0,t_0+\Delta t]} = 1 - \mathbf{P}_{1}^{(t_0,t_0+\Delta t]}. \tag{19.2.2}$$

Using these probabilities, we can make a random choice of either a jump, or no jump.

a) In the first case the system wave function undergoes a quantum jump

$$|\tilde{\psi}, t = t_0 + \Delta t\rangle = \mathbf{J}|\psi_0\rangle, \tag{19.2.3}$$

b) In the second case we propagate the system wave function according to

$$|\tilde{\psi}, t = t_0 + \Delta t\rangle = \mathrm{e}^{\mathbf{K}\Delta t}|\psi_0\rangle. \tag{19.2.4}$$

v) Set

$$|\psi_0\rangle \to \frac{|\tilde{\psi}, t = t_0 + \Delta t\rangle}{\big\||\tilde{\psi}, t = t_0 + \Delta t\rangle\big\|}. \tag{19.2.5}$$

vi) Increase the jump counter $r \to r+1$.

vii) Increase $t_0 \to t_0 + \Delta t$ and we start again.

viii) An approximation for the system density matrix is obtained by repeating these simulations, and taking an average over the different realizations of system wave functions—denoted by $\langle \ldots \rangle_{\mathrm{st}}$—to obtain

$$\bar{\rho}(t) = \Big\langle |\psi, t\rangle\langle\psi, t| \Big\rangle_{\mathrm{st}}. \tag{19.2.6}$$

19.3 Applications

In this section we give a number of the simplest applications of quantum trajectories to atomic and harmonic oscillator systems. The insights provided by a detailed consideration of the consequences of the alternation of quantum jumps and smooth time evolution are quite valuable, and not entirely obvious.

We also refer the reader to Sect. 27.2.4, for a very powerful application of the method in the context of quantum information transmission.

19.3.1 Spontaneous Emission from a Two-Level Atom

We consider a two-level atom described by the ground and excited states $|g\rangle$, $|e\rangle$. The excited state undergoes spontaneous emission, as has already been treated in *Bk. I: Sect. 14.1* using the master equation. Here we apply the photon counting formalism, and observe the system continuously in a counting experiment, in which we subdivide time in steps Δt and check whether a photon was detected or not.

a) Operators: We will work in a rotating frame, writing

$$H_{\text{eff}} = -\tfrac{1}{2}\mathrm{i}\hbar\Gamma\,|e\rangle\langle e|, \qquad c = |e\rangle\langle g|. \tag{19.3.1}$$

Thus

$$\mathbf{K} = -\tfrac{1}{2}\Gamma\,|g\rangle\langle e|, \qquad \mathbf{J} = \sqrt{\Gamma}\,|g\rangle\langle e|. \tag{19.3.2}$$

b) Initial State: We will choose to prepare the atom in the normalized superposition of the ground and excited states:

$$|\psi,0\rangle = c_g|g\rangle + c_e|e\rangle, \qquad |c_e|^2 + |c_g|^2 = 1. \tag{19.3.3}$$

c) Detections:

i) *The Probability of Detecting no Photon During the Time Interval* $(0,t]$: This is

$$\mathbf{P}_{\mathbb{O}}^{(0,t]} = \left\|\mathrm{e}^{\mathbf{K}t/\hbar}|\psi,0\rangle\right\|^2 = |c_g|^2 + |c_e|^2\mathrm{e}^{-\Gamma t}. \tag{19.3.4}$$

The probability of the excited state being populated decreases exponentially during this period, since the photodetector is monitored at each time step, and no photon is ever detected.

a) *The Probability Amplitude at Time t*: This is

$$|\tilde{\psi}(t)\rangle = \mathrm{e}^{\mathbf{K}t}|\psi,0\rangle = c_g|g\rangle + c_e\mathrm{e}^{-\Gamma t/2}|e\rangle. \tag{19.3.5}$$

Thus the amplitude of the ground state component is unchanged, but the amplitude of the excited component decays exponentially.

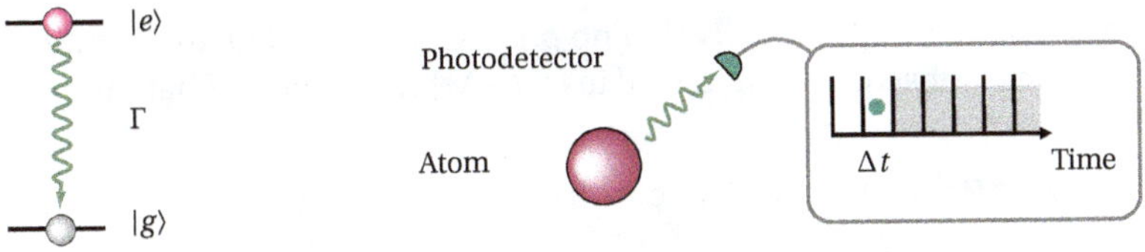

Fig. 19.1. Schematic representation of a photon count (the green spot) in the second time interval, and no photon detected in the previous time interval. The grey area represents the future, in which counting has not yet commenced.

b) If $c_g = 0$, the wavefunction does not change, it is $|e\rangle$ at all times. However, (19.3.4) tells us that this becomes exponentially a less probable result with the progress of time.

c) In the case that $c_g \neq 0$, the wavefunction has the more complex form

$$|\psi(t)\rangle = \frac{e^{\mathbf{K}t}|\psi,0\rangle}{\sqrt{\mathbf{P}_{\mathbb{0}}^{(0,t]}}} = c_g(t)|g\rangle + c_e(t)|e\rangle \xrightarrow{t\to\infty} |g\rangle, \tag{19.3.6}$$

in which

$$c_g(t) \equiv \frac{c_g}{\sqrt{|c_g|^2 + |c_e|^2 e^{-\Gamma t}}}, \qquad c_e(t) = \frac{c_e e^{-\Gamma t/2}}{\sqrt{|c_g|^2 + |c_e|^2 e^{-\Gamma t}}}. \tag{19.3.7}$$

These two coefficients are plotted in Fig. 19.2, which demonstrates how normalizing the state gives a non-exponential behaviour. If there is a very small admixture of the ground state, then it takes some time before the the component of the excited state decays to a comparable size. Thus we see that initially $c_e(t)$ stays close to 1, but this is eventually followed by a behaviour close to an exponential.

ii) *The Probability that a Photon is Detected in the Time Interval* $(t, t+\Delta t]$: This is given in terms of the *wavefunction* at time t. If no photons have been detected up to this time, this is given by (19.3.6), so that

$$\mathbf{P}_{\mathbb{1}}^{(t,t+\Delta t]} = \|\mathbf{J}|\psi,t\rangle\|^2 \,\Delta t, \tag{19.3.8}$$

$$= \Gamma \big\||g\rangle\langle e|\psi,t\rangle\big\|^2 \,\Delta t = \frac{\Gamma|c_e|^2 e^{-\Gamma t}}{|c_g|^2 + |c_e|^2 e^{-\Gamma t}}\,\Delta t. \tag{19.3.9}$$

The detection probability can thus be expressed in terms of physical quantities as

$$\textit{Detection rate} = \textit{Excited state population} \times \textit{Time interval}. \tag{19.3.10}$$

Observation of a photon in $(t, t+\Delta t]$ results in

$$|\psi, t+\Delta t\rangle = \frac{\sqrt{\Gamma}\,|g\rangle\langle e|\psi,t\rangle}{\|\sqrt{\Gamma}\,|g\rangle\langle e|\psi,t\rangle\|} = |g\rangle. \tag{19.3.11}$$

iii) *Joint Probability*: The probability that no photon is detected up to time t and that a photon is subsequently detected in $(t, t+\Delta t]$ is the joint probability, and is given by (18.6.5)

$$p_0^t(t) = \mathbf{P}_{\mathbb{0}}^{(0,t]}\,\mathbf{P}_{\mathbb{1}}^{(t,t+\Delta t]}\,\Delta t = \Gamma|c_e|^2 e^{-\Gamma t}\,\Delta t. \tag{19.3.12}$$

iv) *After Detection of a Photon*: In this case, since $\mathbf{K}|g\rangle = \mathbf{J}|g\rangle = 0$, no further time evolution happens.

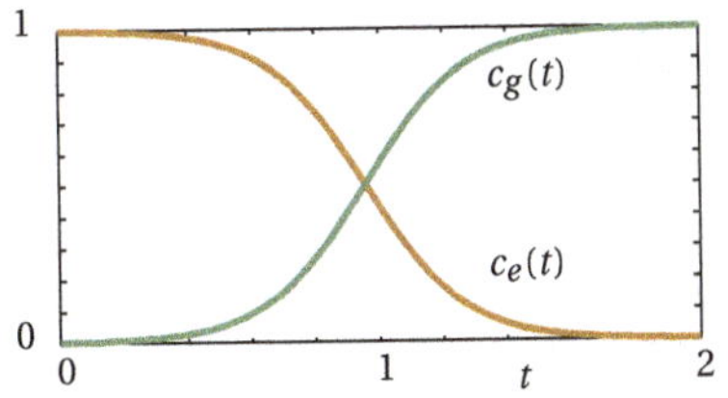

Fig. 19.2. Plot of the amplitudes $c_e(t)$ and $c_g(t)$ as defined in (19.3.7). Here we have chosen $c_e(t=0) = 0.999$, and $\Gamma = 6.5$.

19.3.2 The Driven Two-Level System

We now consider the *driven two level system,* which simultaneously undergoes spontaneous emission. This has been treated using master equation techniques in *Bk. I: Sect.14.2,* and as well in Ex. 9.7. If the laser frequency is ω_L, and the detuning of the laser field form the transition frequency is defined by $\Delta \equiv \omega_L - \omega_{eg}$, then the effective Hamiltonian and the jump operator are (setting Ω_R to be real for convenience)

$$H_{\text{eff}} = \hbar\left(-\Delta - \tfrac{1}{2}\mathrm{i}\Gamma\right)|e\rangle\langle e| - \tfrac{1}{2}\hbar\Omega_R\left(|e\rangle\langle g| + |g\rangle\langle e|\right), \tag{19.3.13}$$

$$c \;= |g\rangle\langle e|. \tag{19.3.14}$$

Thus, the stochastic evolution operators are

$$\mathbf{K} = \left(\mathrm{i}\Delta - \tfrac{1}{2}\Gamma\right)|e\rangle\langle e| + \tfrac{1}{2}\mathrm{i}\Omega_R\left(|e\rangle\langle g| + |g\rangle\langle e|\right), \tag{19.3.15}$$

$$\mathbf{J} = \sqrt{\Gamma}\,|g\rangle\langle e|. \tag{19.3.16}$$

a) Dynamics:

i) For an atom initially prepared the state $|\psi,0\rangle$ the probability of not detecting a photon during $(0,t]$ is

$$\mathbf{P}_{\emptyset}^{(0,t]} = \left\|\mathrm{e}^{\mathbf{K}t}|\psi,0\rangle\right\|^2 = \left\|\mathrm{e}^{-\mathrm{i}H_{\text{eff}}t/\hbar}|\psi,0\rangle\right\|^2. \tag{19.3.17}$$

For the case that $|\psi,0\rangle = |g\rangle$, this probability is plotted in Fig. 19.3 c. The probability is a monotonically decreasing function of time, close to an exponential, but with oscillations related to the detuning.

ii) If no photon is detected the probability amplitude is

$$|\tilde{\psi},t\rangle = \mathrm{e}^{\mathbf{K}t}|\psi,0\rangle = \mathrm{e}^{-\mathrm{i}H_{\text{eff}}t/\hbar}|\psi,0\rangle. \tag{19.3.18}$$

The real part of the excited state amplitude, $\langle e|\tilde{\psi},t\rangle$, is plotted in Fig. 19.3 b, for the case that $|\psi,0\rangle = |g\rangle$. The driving field is seen to repopulate the excited state, which decays to the ground state eventually, because of the damping term. The interpretation is the same as for a decaying atom; if, after a long time, no photon has been detected, the state of the atom is very likely to have no excited component.

iii) The probability that a photon is detected during $(t,t+dt]$ is

$$\mathbf{P}_0^{(t,t+dt]} = \left\|\mathbf{J}|\psi,t\rangle\right\|^2 dt = \Gamma\left\||g\rangle\langle e|\psi,t\rangle\right\|^2 dt. \tag{19.3.19}$$

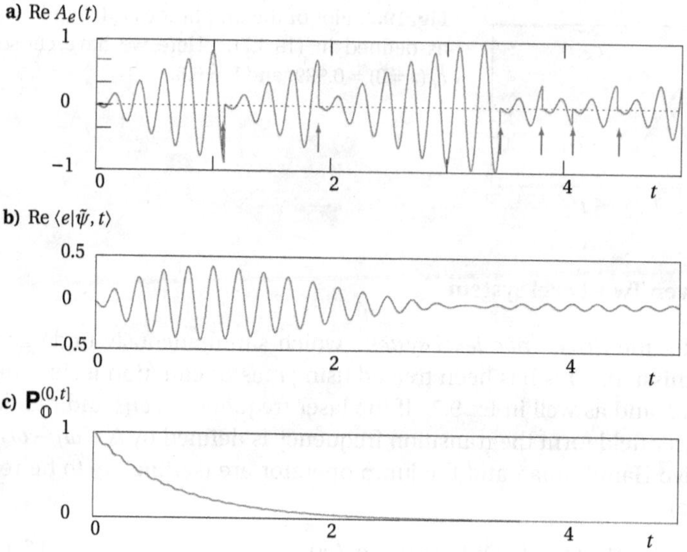

Fig. 19.3. A simulation of the resonance fluorescence of a driven two-level system. **a)** The real part of the excited state amplitude of the *wavefunction* $|\psi, t\rangle$. The wavefunction is given between jumps by normalizing the probability amplitude (19.3.30), and determining the times of the jumps using the probability density (19.3.19) The jumps are indicated by the red arrows. **b)** The real part of excited state amplitude of the probability amplitude as given by (19.3.18). **c)** The probability density (19.3.17) that no photon is detected in $(0, t]$, which is used to determine when a jump will occur.

Detection of a photon is associated with a quantum jump to the state

$$|\psi, t+dt\rangle = \frac{\sqrt{\Gamma}|g\rangle\langle e|\psi, t\rangle}{\left\| \sqrt{\Gamma}|g\rangle\langle e|\psi, t\rangle \right\|} = |g\rangle. \tag{19.3.20}$$

b) The Delay Function: In this system the operator **J** as defined by (19.3.16) always returns the system to the ground state $|g\rangle$, so that after each jump the time evolution of the system is a repetition of that after the previous jump. A classical stochastic process with this property is called a *renewal process* [19.2]

In order to use this fact to our advantage, we define the *delay function* by

$$C(\tau) = \Gamma\langle e|\left(e^{\mathcal{K}\tau}|g\rangle\langle g|\right)|e\rangle, \tag{19.3.21}$$

in terms of which we can write

$$\mathcal{J}e^{\mathcal{K}\tau}|g\rangle\langle g| = \Gamma|g\rangle\langle e|\left(e^{\mathcal{K}\tau}|g\rangle\langle g|\right)|e\rangle\langle g| = C(\tau)|g\rangle\langle g|. \tag{19.3.22}$$

Introducing this into (18.6.5), we see that the probability density for counts at each of the times $t_1, \dots, t_n$ *factorizes*

$$p_0^t(t_1, \dots, t_n) = p_0^{t-t_n}\, C(t_n - t_{n-1}) \dots C(t_1). \tag{19.3.23}$$

This is an expression of the fact that a quantum jump always prepares the atom in the ground state $|g\rangle$, so that there is no dependence on the previous history of the wave function.

This expression assumes the system starts in the ground state; if it starts in some other state, say $|\varphi_0\rangle$, the basic form is exactly the same, but the factor $C(t_1)$ is replaced by

$$C_{\varphi_0}(t_1) \equiv \Gamma\langle e|\left(\mathrm{e}^{\mathcal{K}t_1}|\varphi_0\rangle\langle\varphi_0|\right)|e\rangle. \tag{19.3.24}$$

The delay function is a special case of the formulae (18.4.30, 18.4.30) for the probability that a photon is detected in the time interval $(\tau, \tau + d\tau]$ when the previous photon was detected at time $\tau = 0$. Its normalization follows from the identity

$$\int_0^t C(\tau)d\tau = 1 - p_0^t \longrightarrow 1, \qquad \text{as } t \longrightarrow \infty. \tag{19.3.25}$$

The left-hand side is the probability that *a photon was detected* in $(0, t]$, whereas p_0^t, as defined in (18.6.4), is the probability that *no photon was detected* in the same time interval, clearly their sum must be equal to 1.

The proof is straightforward thus:

$$\frac{d}{dt}p_0^t = \frac{d}{dt}\mathrm{Tr}_{\mathrm{sys}}\left\{\mathrm{e}^{\mathcal{K}t}|g\rangle\langle g|\right\} = \mathrm{Tr}_{\mathrm{sys}}\left\{\mathcal{K}\mathrm{e}^{\mathcal{K}t}|g\rangle\langle g|\right\}, \tag{19.3.26}$$

$$= -\mathrm{Tr}_{\mathrm{sys}}\left\{\mathcal{J}\mathrm{e}^{\mathcal{K}t}|g\rangle\langle g|\right\} = -C(t). \tag{19.3.27}$$

In going from the second line to the third line, we have used

$$\mathrm{Tr}_{\mathrm{sys}}\left\{\mathcal{K}_\rho\right\} = -\mathrm{Tr}_{\mathrm{sys}}\left\{\mathcal{J}\rho\right\}, \tag{19.3.28}$$

valid for any ρ.

The delay function $C(\tau)$ is therefore a normalized probability density satisfying

$$\int_0^\infty C(\tau)d\tau = 1. \tag{19.3.29}$$

The argument here closely parallels that used in Sect. 18.4.4b.

c) Simulating the System: The evolution is described by the probability amplitude

$$|\tilde{\psi}, t\rangle \equiv A_e(t)|e\rangle + A_g(t)|g\rangle, \tag{19.3.30}$$

which consists of a sequence of instantaneous jumps from the ground state $|g\rangle$ to the excited state $|e\rangle$, in between which the system evolves smoothly according to (19.3.18). Since the jumps always return the system to $|g\rangle$, the evolution after a jump is always described by the same function, appropriately displaced in time.

The lengths of times between jumps τ_r have the probability distribution $\mathbf{P}_0^{(0,t]}$, and a sequence of τ_r can be constructed using this probability. The resulting $|\psi, t\rangle$ is then constructed, and is illustrated in Fig. 19.3 a.

19.3.3 The Damped Cavity Mode

The behaviour of a damped cavity mode—that is, a damped harmonic oscillator—is rather different from that of atomic systems. The domination by two-state transitions and their combinations is not present; the primary organizing principle is the infinite tower of uniformly spaced levels, which can be excited coherently. The interest centres mainly on establishing or destroying such coherences. The situation is illustrated in Fig. 19.4.

We work in an interaction picture, in which the stochastic evolution operators become

$$\mathbf{K} \equiv -\tfrac{1}{2}\kappa a^\dagger a = -\mathrm{i}\frac{H_{\mathrm{eff}}}{\hbar}, \tag{19.3.31}$$

$$\mathbf{J} \equiv \sqrt{\kappa}\, a. \tag{19.3.32}$$

a) Time Evolution: The cavity is initially prepared in an arbitrary normalized superposition state

$$|\psi,0\rangle = \sum_{n=0}^{\infty} c_n |n\rangle. \tag{19.3.33}$$

i) The probability that *no* photon from the cavity is detected during $(0,\Delta t]$ is given by

$$\mathbf{P}_{\mathbb{O}}^{(0,t]} = \left\| \mathrm{e}^{\mathbf{K}t}|\psi,0\rangle \right\|^2. \tag{19.3.34}$$

The probability amplitude conditional on having detected no photons is

$$|\tilde{\psi},t\rangle = \mathrm{e}^{\mathbf{K}t}|\psi,0\rangle = \sum_{n=0}^{\infty} c_n \mathrm{e}^{-\kappa n t/2}|n\rangle. \tag{19.3.35}$$

Thus, the decay rate of the n-photon state is proportional to n. As in the case of the two-level system this corresponds to an increase in belief that the actual state is the vacuum state.

ii) The probability that *one* photon from the cavity is detected during $(t, t+\Delta t]$

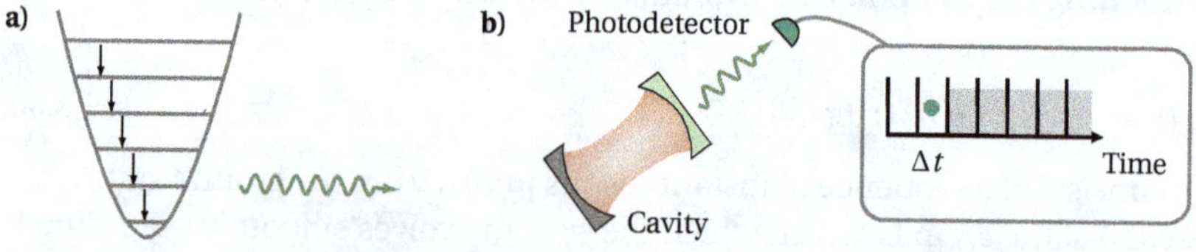

Fig. 19.4. Schematic representation of photon counting from a cavity. **a)** The cavity behaves as a harmonic oscillator, and cascades through the uniformly spaced energy levels, emitting one photon during each jump. **b)** The counting process, represented in the second time interval. The grey area represents the future, in which counting has not yet commenced.

is given by

$$\mathbf{P}_{\mathbb{1}}^{(t,t+\Delta t]} = \kappa \big\| a|\psi,t\rangle \big\|^2 \Delta t = \kappa \left(\sum_{n=0}^{\infty} n|c_n|^2 \mathrm{e}^{-\kappa t} \right) \Delta t, \tag{19.3.36}$$

$$\equiv \kappa \langle n(t) \rangle \, \Delta t. \tag{19.3.37}$$

The probability amplitude after a photon detection is

$$|\tilde{\psi}, t+\Delta t\rangle = \mathbf{J}|\tilde{\psi}, t\rangle = \sqrt{\kappa} a|\tilde{\psi}, t\rangle = \sum_{n=1}^{\infty} c_n \sqrt{\kappa n}\, \mathrm{e}^{-\kappa n t/2} |n-1\rangle, \tag{19.3.38}$$

and the corresponding state is

$$|\psi, t+\Delta t\rangle = \frac{|\tilde{\psi}, t+\Delta t\rangle}{\sqrt{\mathbf{P}_{\mathbb{0}}^{(0,t]} \mathbf{P}_{\mathbb{1}}^{(t,t+\Delta t]} / \Delta t}} . \tag{19.3.39}$$

b) Initial Fock State: In this case the initial state is determined as

$$|\psi, 0\rangle = |n\rangle. \tag{19.3.40}$$

i) The probability amplitude of the cavity state after having not detected any photon in the time interval $(0, t]$ is

$$|\tilde{\psi}, t\rangle = \mathrm{e}^{-\kappa n t/2} |n\rangle, \tag{19.3.41}$$

corresponding to the state

$$|\tilde{\psi}, t\rangle = |n\rangle. \tag{19.3.42}$$

ii) The probability of detection of a photon in $(t, t+\Delta t)$ is given by

$$\mathbf{P}_{\mathbb{1}}^{(t,t+\Delta t]} = \kappa n \Delta t, \tag{19.3.43}$$

and after the detection, the probability amplitude is

$$|\tilde{\psi}(t+\Delta t)\rangle = \sqrt{\kappa n}\, \mathrm{e}^{-\kappa n t/2} |n-1\rangle, \tag{19.3.44}$$

and the corresponding state is

$$|\psi, t+\Delta t\rangle = |n-1\rangle. \tag{19.3.45}$$

The process of detection simply reduces the photon number by one at each detection event.

c) Coherent State: In this case the initial state is

$$|\psi, 0\rangle = |\alpha\rangle = \mathrm{e}^{-|\alpha|^2/2} \sum_{n=0}^{\infty} \frac{\alpha^n}{\sqrt{n!}} |n\rangle. \tag{19.3.46}$$

i) The probability amplitude of the cavity state after having not detected any photon in the time interval $(0, t]$ is

$$|\tilde{\psi}, t\rangle = \mathrm{e}^{\mathbf{K}t} |\alpha\rangle = \exp\left(-\tfrac{1}{2}|\alpha|^2 (1 - \mathrm{e}^{-\kappa t})\right) \big| \mathrm{e}^{-\kappa t} \alpha \big\rangle. \tag{19.3.47}$$

Thus the effect of detecting no photons is to reduce the amplitude of the coherent state, but it remains a coherent state.

The probability of detecting no photons in $(0,t]$ is correspondingly

$$\mathbf{P}_{0}^{(0,t]} = \exp\left(-|\alpha|^2(1-\mathrm{e}^{-\kappa t})\right). \tag{19.3.48}$$

ii) If a photon is detected in the interval $(t, t+\Delta t]$, the probability amplitude is

$$|\tilde{\psi}, t+\Delta t\rangle = \mathbf{J}|\tilde{\psi}, t\rangle = \sqrt{\kappa}\, a|\tilde{\psi}, t\rangle, \tag{19.3.49}$$

$$= \sqrt{\kappa}\mathrm{e}^{-\kappa t}\alpha \exp\left(-\tfrac{1}{2}|\alpha|^2(1-\mathrm{e}^{-\kappa t})\right)\left|\,\mathrm{e}^{-\kappa t}\alpha\right\rangle. \tag{19.3.50}$$

Thus, a quantum jump occurs, and the probability amplitude is changed in both amplitude and phase, but is still proportional to the same normalized wavefunction. Apart from the phase change, the normalized state after the jump is the same coherent state as before the jump.

iii) From the point of view of the normalized wavefunction, this result appears paradoxical—*detecting* a photon does not reduce the coherent state amplitude, while *not detecting* a photon does reduce the coherent state amplitude.

On the other hand, from the point of view of probability amplitudes, which contain probability information that is lost by normalizing, the paradox is not so apparent.

d) Schrödinger Cat State: This kind of state is a superposition of macroscopically distinct quantum states, such as $|\alpha\rangle$ and $|-\alpha\rangle$. The coherent state amplitude α can be as large as one wishes. We therefore take the initial state in the form

$$|\psi, 0\rangle \approx \frac{|\alpha\rangle + |-\alpha\rangle}{\sqrt{2}}. \tag{19.3.51}$$

Here this is written as an approximate equality because we have assumed α is macroscopic, so that we may neglect $\langle\alpha|-\alpha\rangle$ in calculating $\big\||\psi,0\rangle\big\|$—this is not a essential part of the argument.

i) As in the case of the coherent state, not detecting any photon leads to the probability amplitude

$$|\tilde{\psi}, t\rangle = \exp\left(-\tfrac{1}{2}|\alpha|^2(1-\mathrm{e}^{-\kappa t})\right)\frac{|\alpha\mathrm{e}^{-\kappa t/2}\rangle + |-\alpha\mathrm{e}^{-\kappa t/2}\rangle}{\sqrt{2}}, \tag{19.3.52}$$

which thus remains a Schrödinger cat state but with reduced coherent state amplitudes.

ii) On detection, the Schrödinger cat experiences a change in sign of the relative phase between the two components after the quantum jump

$$|\tilde{\psi}, t+\Delta t\rangle = \sqrt{\kappa}\, a \exp\left(-\tfrac{1}{2}|\alpha|^2(1-\mathrm{e}^{-\kappa t})\right)\left(\frac{|\alpha\mathrm{e}^{-\kappa t/2}\rangle + |-\alpha\mathrm{e}^{-\kappa t/2}\rangle}{\sqrt{2}}\right), \tag{19.3.53}$$

$$= \sqrt{\kappa}\alpha\mathrm{e}^{-\kappa t/2} \exp\left(-\tfrac{1}{2}|\alpha|^2(1-\mathrm{e}^{-\kappa t})\right)\left(\frac{|\alpha\mathrm{e}^{-\kappa t/2}\rangle - |-\alpha\mathrm{e}^{-\kappa t/2}\rangle}{\sqrt{2}}\right). \tag{19.3.54}$$

Thus the cat dephases maximally after the time of a *single* photon loss. This is in accordance with the fragility of the interference between the two components of this kind of Schrödinger cat state, which we have shown in *Bk. I: Sect.19.4.2.*

19.4 Quantum Jumps in Three-Level Systems

Experiments with single trapped ions represent an *experimental realization of continuous observation of a single quantum system* in the context of quantum optics. Such a system can be set up using a single trapped ion driven by a coherent laser beam—for example, an ion, such as ^{40}Ca, with two excited states $|e\rangle$ and $|r\rangle$, connected to a common lower level $|g\rangle$ via a *strong* and *weak* transition, as illustrated in Fig. 19.5. Fluorescence photons from the *strong* transition are observed in a photon counting experiment. The observation of quantum jumps in a three-level system is probably one of the best-known examples in quantum optics where quantum jumps are "seen" in experiments—the fluorescence from a single ion can be observed by continuously monitoring the ion with a photodetector.

In this section we will present the theoretical framework that is appropriate for a single quantum system, such as a trapped ion, which is observed continuously by a photodetector.

Excitation of the *weak* transition with a laser will induce a jump to the metastable state $|r\rangle$, where it will remain for a time determined by the the strength of the driving field on that transition and the decay constant Γ_w. This is commonly referred to as "shelving" the atomic electron. While the atom is in the state $|r\rangle$, the fluorescence from the strong transition will be turned off. On returning to the ground state $|g\rangle$, repeated strong transitions to the state $|e\rangle$ will resume, and fluorescence of the corresponding wavelength will be detectable.

Thus, the quantum jumps of the weak transition will correspond to the turning on and off of the strong fluorescence signal.

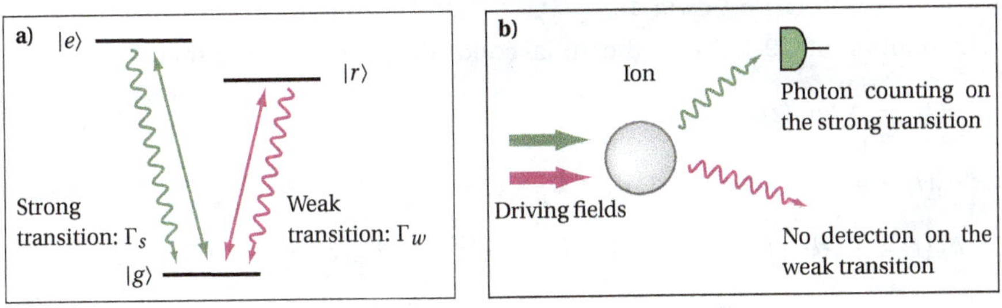

Fig. 19.5. The driven three-level atom. **a)** The two transition processes, one fast (dipole allowed) and one slow (dipole forbidden); **b)** Only the photons from the fast process are monitored.

19.4.1 Theoretical Description

We can treat this theoretically by introducing two sets of quantum noise operators, each coupled to a different transition. A wavefunction simulation can be developed in which both transitions occur, but only the strong transitions are monitored. However, a clearer understanding of the processes involved is obtained by using a density operator formulation, and generalizing the density operator procedures of Sect. 18.5.1–Sect. 18.6.

a) The Master Equation: The three-level system is characterized by:

i) *The Effective Hamiltonian*: This is given by the sum of two terms of the same form, one corresponding to each transition

$$\begin{aligned} H_{\text{eff}} &= -\hbar\left(\Delta_s + \tfrac{1}{2}\mathrm{i}\Gamma_s\right)|e\rangle\langle e| - \tfrac{1}{2}\hbar\Omega_s\left(|g\rangle\langle e| + |e\rangle\langle g|\right) \\ &\quad -\hbar\left(\Delta_w + \tfrac{1}{2}\mathrm{i}\Gamma_w\right)|r\rangle\langle r| - \tfrac{1}{2}\hbar\Omega_w\left(|g\rangle\langle r| + |r\rangle\langle g|\right). \end{aligned} \tag{19.4.1}$$

ii) *Recycling Operators*: There are two recycling operators, which we can write as

$$\mathcal{J}_s\rho = |g\rangle\langle g|\Gamma_s\rho_{ee}\,, \tag{19.4.2}$$

$$\mathcal{J}_w\rho = |g\rangle\langle g|\Gamma_w\rho_{rr}. \tag{19.4.3}$$

iii) For these expressions we use the notations:

a) *Strong Transition*: Decay constant Γ_s, Rabi frequency Ω_s, Detuning Δ_s.

b) *Weak Transition*: Decay constant Γ_w, Rabi frequency Ω_w, Detuning Δ_w.

The master equation is

$$\dot\rho = -\frac{\mathrm{i}}{\hbar}\left(H_{\text{eff}}\rho - \rho H_{\text{eff}}^\dagger\right) + \mathcal{J}_s\rho + \mathcal{J}_w\rho \equiv \mathcal{L}\rho. \tag{19.4.4}$$

b) Counting Photons on Only the Strong Transition: Since we have a photodetector only on the *strong* transition, we can unravel the system density operator in terms of these counts only, in a way similar to that given in (18.6.1–18.6.3).

$$\dot\rho_n = -\frac{\mathrm{i}}{\hbar}(H_{\text{eff}}\rho_n - \rho_n H_{\text{eff}}^\dagger) + \mathcal{J}_s\rho^{(n-1)} + \mathcal{J}_w\rho_n, \tag{19.4.5}$$

$$\equiv (\mathcal{L}_w + \mathcal{K}_s)\rho_n + \mathcal{J}_s\rho_{n-1}. \tag{19.4.6}$$

The solution of (19.4.6) with the initial condition $\rho(0) = |g\rangle\langle g|$ is therefore

$$\rho(t) = \sum_{n=0}^{\infty}\rho_n(t), \tag{19.4.7}$$

$$\rho_0(t) = \mathrm{e}^{(\mathcal{L}_w+\mathcal{K}_s)t}\rho(0), \tag{19.4.8}$$

$$\begin{aligned} \rho_n(t) &= \int_0^t dt_n \int_0^{t_n} dt_{n-1}\ldots\int_0^{t_2} dt_1\,\mathrm{e}^{(\mathcal{L}_w+\mathcal{K}_s)(t-t_n)}\mathcal{J}_s\mathrm{e}^{(\mathcal{L}_w+\mathcal{K}_s)(t_n-t_{n-1})}\ldots \\ &\qquad \times\mathrm{e}^{(\mathcal{L}_w+\mathcal{K}_s)(t_2-t_1)}\mathcal{J}_s\mathrm{e}^{(\mathcal{L}_w+\mathcal{K}_s)t_1}\rho(0). \end{aligned} \tag{19.4.9}$$

This equation is similar to (18.6.1–18.6.3), but with the change $\mathcal{K} \longrightarrow \mathcal{L}_w + \mathcal{K}_s$, which represents the fact that detection happens only on the strong transition.

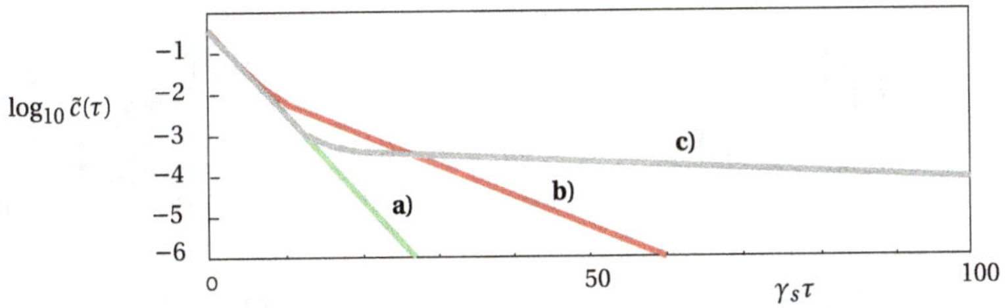

Fig. 19.6. The delay function $C(\tau)$ as a function of $\Gamma_s \tau$. $C(\tau)$ is the conditional probability density that, given a count has occurred on the strong transition $|g\rangle \leftrightarrow |e\rangle$ at time $\tau = 0$, the *next* count on the strong transition will occur at time τ. The three curves show:
a) The metastable state is not excited, $\Omega_w = 0$, and $C(\tau)$ decays essentially exponentially;
b) The lifetime of the metastable state $|r\rangle$ is ten times longer than the lifetime of the rapidly decaying state $|e\rangle$, $W_{gr} = \Gamma_w = 10^{-1}\Gamma_s$.
c) The lifetime of the metastable state $|r\rangle$ is one hundred times longer than the lifetime of the rapidly decaying state $|e\rangle$, $W_{gr} = \Gamma_w = 10^{-2}\Gamma_s$. W_{gr} is an incoherent excitation rate on the weak transition.

c) Counting Probabilities: As in Sect. 18.6, we can interpret the expression for the density operator as a sum of terms corresponding to the detection of exactly n strong transition photons:

i) *The Probability that no Photon is Detected During* $(0, t]$:

$$p_0^t = \mathrm{Tr}_{\mathrm{sys}} \left\{ e^{(\mathcal{L}_w + \mathcal{K}_s)t} |g\rangle\langle g| \right\}. \tag{19.4.10}$$

The initial value of the probability is $p_0^{t=0} = 1$, and because the evolution operator in this expression has no recycling term on the strong transition, the probability decays from to zero, that is $p_0^{t\to\infty} = 0$.

ii) *Probability Density that Exactly n Photons are Detected in the Interval* $(0, t]$: If the detection times for these photons are $t_1, \ldots, t_n$ in the time interval $(0, t]$, this takes the form

$$p_0^t(t_1, \ldots, t_n) = \mathrm{Tr}_{\mathrm{sys}} \Big\{ e^{(\mathcal{L}_w + \mathcal{K}_s)(t - t_n)} \mathcal{J}_s e^{(\mathcal{L}_w + \mathcal{K}_s)(t_n - t_{n-1})} \ldots$$
$$e^{(\mathcal{L}_w + \mathcal{K}_s)(t_2 - t_1)} \mathcal{J}_s e^{(\mathcal{L}_w + \mathcal{K}_s)t_1} |g\rangle\langle g| \Big\}. \tag{19.4.11}$$

19.4.2 Photon Detections in Terms of the Delay Function

As in the case of the two-level systems treated in Sect. 19.3.2, the recycling operators $\mathcal{J}_w$ and $\mathcal{J}_s$ return the system to the ground state. This means that after each photodetection, the process of excitation always starts again from exactly the same state, the process is thus again a renewal process.

The *delay function* in this case is defined slightly differently, because of the existence of two transitions, but it has the same properties as before. Here we define the delay function by

$$C(\tau) = \Gamma_s \langle e| \left(e^{(\mathcal{L}_w + \mathcal{K}_s)\tau} |g\rangle\langle g| \right) |e\rangle, \tag{19.4.12}$$

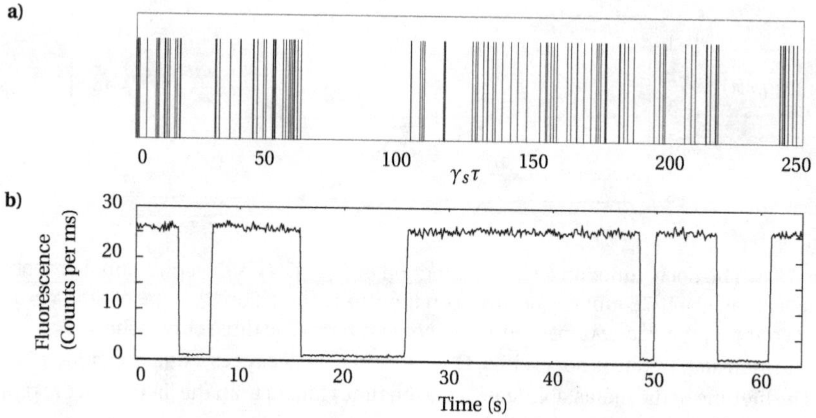

Fig. 19.7. a) Simulation of the photon detection on the strong transition $|g\rangle \leftrightarrow |e\rangle$. A vertical line (of arbitrarily chosen height) corresponds to the detection of one photon. The parameters are $W_{gr} = \Gamma_w = 10^{-1}\Gamma_s$. **b)** Experimental observation of quantum jumps in the intensity of the fluorescence light from a single Ba^+ ion at 493nm. (Data kindly provided by Prof. R Blatt.)

in terms of which we can write

$$\mathcal{J}_s e^{(\mathcal{L}_w+\mathcal{K}_s)\tau}|g\rangle\langle g| = \Gamma_s|g\rangle\langle e|\left(e^{(\mathcal{L}_w+\mathcal{K}_s)\tau}|g\rangle\langle g|\right)|e\rangle\langle g|, \tag{19.4.13}$$

$$= C(\tau)|g\rangle\langle g|. \tag{19.4.14}$$

It is clear that the probability $p_0^t(t_1,\dots,t_n)$ factorizes in exactly the same way as in (19.3.23).

Most importantly, the delay function is again the probability density that, starting from the atom being in the ground state $|g\rangle$, the first photon detected at the time t. Finally, the identities (19.3.25, 19.3.29) continue to be true.

a) Discussion of the Delay Function: In the case of the three-level system under study here, $C(\tau)$ consists of a sum of exponentials with different decay constants, reflecting the different time scales for the excitation and decay on the strong and metastable transition, as shown in Fig. 19.6. The existence of the weak transition leads to a *long time tail* in the delay function.

In a simulation of a photon detection sequence on the strong line this will manifest itself in the *periods of brightness* and *darkness*, as shown in Fig. 19.7.

The atomic density matrix conditional on observing a detection window on the strong line when the *last photon on the strong transition* was detected at time t_r is

$$\rho_c(t) = \frac{e^{(\mathcal{L}-\mathcal{J}_s)(t-t_r)}|g\rangle\langle g|}{\mathrm{Tr_{sys}}\{\dots\}} \longrightarrow |r\rangle\langle r|, \qquad \text{for } \Gamma_s t \gg 1. \tag{19.4.15}$$

That is, *observation of a window* in a single trajectory of counts corresponds to a *preparation of the electron in the metastable state* $|r\rangle$. This state preparation is what is usually referred to as *shelving of the electron.*

VI Phase-Sensitive Quantum Optics

Part VI: Phase-Sensitive Quantum Optics

At optical frequencies it is not possible to follow the evolution of the electric and magnetic field strengths which characterize the electromagnetic field, as it is at radio frequencies. The only practical way to get this kind information is to use the homodyne and heterodyne techniques, in which the field to be measured is mixed with another field, and the difference frequency signal extracted. This then yields information on the *quadrature phases* of the electromagnetic field.

There are many advantages associated with such phase-sensitive measurements. In Chap. 20 we firstly give the theory of homodyne and heterodyne detection, and in a rather extensive appendix, develop the quantum stochastic equations for a system undergoing homodyne measurement. This theory, which is very illuminating, is not often used in practice. However, it is very valuable as an example of the application of a quantized version of the Ito calculus in a form analogous to that used for classical Brownian motion.

In Chap. 21, we give a compact theoretical description of the squeezed light, both of a single mode and of two modes, and of its production and measurement. The theory of *quantum amplifiers* then follows, with special attention to phase-sensitive amplification, which in principle is noise-free when one amplifies one quadrature phase of an optical field. The process of production of squeezed light is in fact equivalent to the amplification of the vacuum fluctuations using an appropriate phase-sensitive amplifier.

20. Homodyne Measurement

Measuring an optical field by direct photodetection reveals only phase-insensitive information, that is, the number of photocounts per second. From a classical point of view, this is information about the intensity of the field, but not its amplitude. To measure amplitude, we need a *phase reference*, and this can be provided by another optical field. In order to provide a stable phase reference, this must be an intense field with minimal fluctuations, as would be provided by a coherent state of high intensity. Under these conditions, we can get very good information about the phase properties of the field to be measured.

The practical detection of light beams using a phase reference is done by mixing the signal to be detected with a strong coherent signal with well-defined frequency, known as a *local oscillator*, before detection. The mixing is done using a beam splitter, and is linear in the fields being mixed, whereas detection is proportional to intensity, so this procedure produces terms which multiply the signal and local oscillator fields. The theory of this kind of detection was first developed by *Yuen* and *Shapiro* [20.1–20.3] who saw that phase-sensitive communication with squeezed light (they used the terminology "two-photon coherent states") presented the possibility of noise reduction in photonic telecommunication systems.

In practice all optical fields have a well-defined frequency, with only a narrow spread around a central frequency. This leads to two kinds of phase-sensitive detection

i) *Homodyne Detection*: Here the local oscillator has the same frequency as that of the detected signal, and the photocount signal which results has a mean frequency of zero, with spread determined by the bandwidth of the signal being measured.

ii) *Heterodyne Detection*: Here the local oscillator has a different frequency from that of the signal, and the photocount signal which results has a frequency equal to the difference of the signal and local oscillator frequencies, with a spread about this determined by the bandwidth of the signal being measured.

The quantities measured as a result of these procedures are known as the *electromagnetic field quadrature phases*. These are essentially the Hermitian and anti-Hermitian parts of the destruction operator for that part of the electromagnetic field which has the frequency determined by the homodyne or heterodyne detection process.

20.1 Procedure for Homodyne and Heterodyne Detection

The basic concept is illustrated in Fig. 20.1. The local oscillator is a stabilized laser field, which is mixed with the signal in a beam splitter, and then detected. The output photocurrent is then multiplied by a $\cos(\omega t)$ function, and the result integrated for a time T, very much larger than ω^{-1}. The photodetector will not normally be 100% efficient, but we will model it as an ideal photodetector preceded by a beam splitter.

20.1.1 General Formulae

The field to be measured is $a(t)$, a quantized optical field. In this section we will not use the interaction picture (or a rotating frame), so that the time dependence of this field is characteristically at an optical frequency. The signal at the photodetector is

$$b(t) = \mathbf{r}\big(\mathcal{F}(t) + v(t)\big) + \mathbf{t}\, a(t). \tag{20.1.1}$$

Here:

i) The beam splitter has transmissivity $\mathbf{t}$ and reflectivity $\mathbf{r}$—these are in principle complex numbers, but for convenience we will often choose them to be *real*.

ii) $\mathcal{F}(t)$ is the *mean value* of the local oscillator field.

iii) $v(t)$ represents the quantized fluctuations about the mean, hence $\langle v(t)\rangle = 0$. If the local oscillator field is in a coherent state, $v(t)$ is identical with vacuum fluctuations.

The quantity being studied is therefore

$$N(\omega, T) = \int_0^T dt\, b^\dagger(t) b(t) \cos(\omega t). \tag{20.1.2}$$

We will consider only the case in which $\cos(\omega t)$ is rather slowly varying, and (20.1.2) is effectively got by the continuous measurement of the photon number.

We now take into account the fact that $\mathbf{r}\mathcal{F}(t)$ is very large, and expanding (20.1.2) to the first two orders only, we find

$$N(\omega, T) = \int_0^T dt \Big\{ |\mathbf{r}|^2 |\mathcal{F}(t)|^2 + \mathbf{r}^* \mathcal{F}^*(t)\big[\mathbf{t}\, a(t) + \mathbf{r} v(t)\big] + \mathbf{r}\mathcal{F}(t)\big[\mathbf{t}^* a^\dagger(t) + \mathbf{r}^* v^\dagger(t)\big]\Big\} \cos(\omega t). \tag{20.1.3}$$

The local oscillator is taken to have a constant amplitude, and a well-defined frequency and phase, so that we can write

$$\mathcal{F}(t) = F e^{-i\omega_L t - i\varphi}, \tag{20.1.4}$$

and let us now choose, for convenience, $\mathbf{r}$ and $\mathbf{t}$ to be *real*. We define a Fourier variable

$$\tilde{a}(\omega) = \frac{1}{T}\int_0^T e^{i\omega t} a(t)\, dt, \tag{20.1.5}$$

with a similar notation for $\tilde{v}(\omega)$. We then find

$$\begin{aligned}N(\omega,T) &= \int_0^T \mathbf{r}^2 F^2 \cos\omega t\,dt \\ &+\frac{\mathbf{r}FT}{2}\mathrm{e}^{\mathrm{i}\varphi}\left\{\mathbf{t}\left(\tilde{a}(\omega_L-\omega)+\tilde{a}(\omega_L+\omega)\right)+\mathbf{r}\left(\tilde{v}(\omega_L-\omega)+\tilde{v}(\omega_L+\omega)\right)\right\} \\ &+\frac{\mathbf{r}FT}{2}\mathrm{e}^{-\mathrm{i}\varphi}\left\{\mathbf{t}\left(\tilde{a}(\omega_L-\omega)^\dagger+\tilde{a}(\omega_L+\omega)^\dagger\right)]+\mathbf{r}\left(\tilde{v}(\omega_L-\omega)^\dagger+\tilde{v}(\omega_L+\omega)^\dagger\right)\right\}.\end{aligned} \tag{20.1.6}$$

a) Mean Detected Signal: We now define the mean detected field by

$$\tilde{\mathcal{E}}(\omega) = \langle\tilde{a}(\omega)\rangle, \tag{20.1.7}$$

so that the mean signal detected is, on the assumption that $T \gg 1/\omega$, and terms of order ω^{-1} are neglected in comparison with terms of order T,

$$\langle N(\omega,T)\rangle = \mathbf{r}^2 F^2 T\delta_{\omega,0} + \mathbf{r}\mathbf{t}FT\,\mathrm{Re}\left\{\mathrm{e}^{\mathrm{i}\varphi}\left[\tilde{\mathcal{E}}(\omega_L-\omega)+\tilde{\mathcal{E}}(\omega_L+\omega)\right]\right\}. \tag{20.1.8}$$

b) Variance of Detected Signal: We are also interested in the variance of $N(\omega,T)$, as a measure of the noise of the detected signal. We have taken $v(t)$ to correspond to the vacuum state, and therefore

$$\left\langle v^\dagger(t)v(t')\right\rangle = 0, \qquad \left\langle v(t)v^\dagger(t')\right\rangle = \delta(t-t'). \tag{20.1.9}$$

The delta function here is, strictly speaking, really a peaked function which can be regarded as a delta function on the time scale of the fluctuations—typically an atomic decay timescale. It is not a delta function on the optical time scale.

The variance of the measured signal can then be computed to be

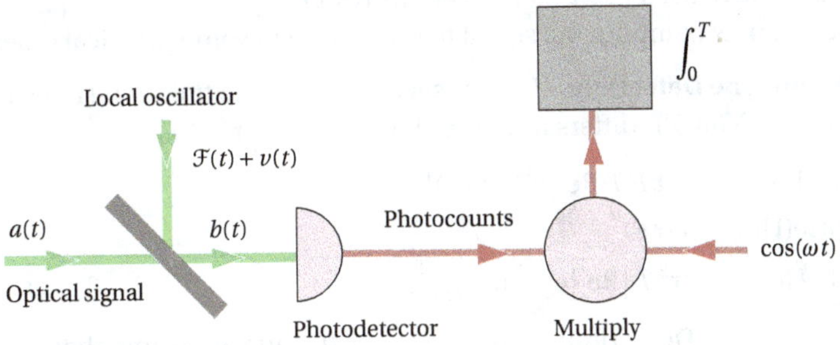

Fig. 20.1. Arrangements for homodyne or heterodyne detection.

$$\mathrm{var}\,[N(\omega,T)] = \left(\frac{\mathbf{r}FT}{2}\right)^2 \left\{\frac{2\mathbf{r}^2}{T}(1+\delta_{\omega,0})\right.$$
$$\left. +\mathbf{t}^2\left\langle\left[e^{i\varphi}\big(\delta\tilde{a}(\omega_L-\omega)+\delta\tilde{a}(\omega_L+\omega)\big)+e^{-i\varphi}\big(\delta\tilde{a}(\omega_L-\omega)^\dagger+\delta\tilde{a}(\omega_L+\omega)^\dagger\big)\right]^2\right\rangle\right\}. \tag{20.1.10}$$

(In this equation, the symbol δ means the difference from the mean value.) This expression for the variance consists of two parts:

i) *Signal Noise*: This is the part arising from the second line, and represents the fluctuations arising from the signal. This is therefore the part of primary interest.

ii) *Local Oscillator Noise*: The first line of the right-hand side is independent of the signal, and it arises from the local oscillator noise $v(t)$. We have taken this to be equivalent to vacuum noise, corresponding to a coherent input, but a more noisy local oscillator is quite possible.

Exercise 20.1 Additional Fluctuations in the Local Oscillator: Explore the consequences of a noise signal with the noise properties

$$\langle v^\dagger(t)v(t')\rangle = \gamma(|t-t'|), \qquad [v(t'),v^\dagger(t)] = \delta(t-t'). \tag{20.1.11}$$

20.1.2 Coherent Signal Detection

If the signal is itself coherent, the signal noise takes on exactly the same form as the vacuum, as in (20.1.9). The result for the variance then becomes

$$\langle[N(\omega,T)-\langle N(\omega,T)\rangle]^2\rangle = \frac{T(\mathbf{r}F)^2}{2}\{1+\delta_{\omega,0}\}, \tag{20.1.12}$$

where we have used $\mathbf{r}^2+\mathbf{t}^2=1$ to get this result.

We can now compute the signal to noise ratio in some practical cases.

a) Homodyne Detection: Here we set $\omega=0$. We take the *signal* to be that part by which the $\langle N(\omega,T)\rangle$ differs from the "background", $\mathbf{r}^2F^2T$.

$$\mathrm{Sig(Ho)} = 2\mathbf{r}\mathbf{t}FT\,\mathrm{Re}\left\{e^{i\varphi}\tilde{\mathcal{E}}(\omega_L)\right\}, \tag{20.1.13}$$
$$\mathrm{Noise(Ho)} = T(\mathbf{r}F)^2, \tag{20.1.14}$$
$$\mathbf{S/N}\mathrm{(Ho)} = 4\mathbf{t}^2T\left[\mathrm{Re}\left\{e^{i\varphi}\tilde{\mathcal{E}}(\omega_L)\right\}\right]^2. \tag{20.1.15}$$

b) Heterodyne Detection: Here we set $\omega\neq 0$, and we assume that $\omega_L-\omega=\omega_{\mathrm{Sig}}$, where ω_{Sig} is the frequency around which the signal has significant power. The background term now vanishes, which is an advantage for this method. It is in

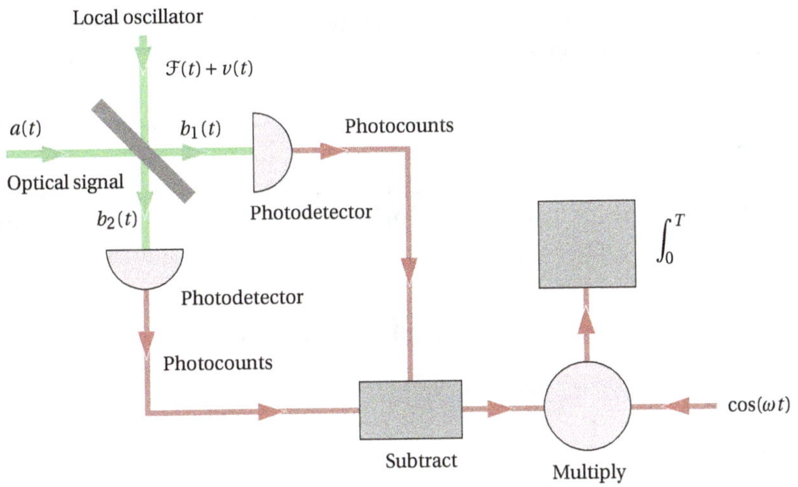

Fig. 20.2. Arrangements for balanced detection.

practice not common to have significant power in both $\omega_L+\omega$ and $\omega_L-\omega$, but it is of course possible. We then find

$$\text{Sig(Het)} = \mathbf{r}\,\mathbf{t}\,F\,T\,\text{Re}\left\{\text{e}^{\text{i}\varphi}\tilde{\mathcal{E}}(\omega_{\text{Sig}})\right\}, \tag{20.1.16}$$

$$\text{Noise(Het)} = \tfrac{1}{2}T(\mathbf{r}F)^2, \tag{20.1.17}$$

$$\mathbf{S/N}\text{(Het)} = 2\mathbf{t}^2 T\left[\text{Re}\left\{\text{e}^{\text{i}\varphi}\tilde{\mathcal{E}}(\omega_{\text{Sig}})\right\}\right]^2. \tag{20.1.18}$$

c) Comment on Signal to Noise Ratios: Note that the signal to noise ratio in heterodyne detection is only half of that for homodyne detection. On the other hand, there is no constant level, as in homodyne detection.

d) Imperfect Photodetectors: If the photodetector has an efficiency of less than 100%, then there is an additional noise source. This can be seen by using the model of the inefficient detector as a perfect detector behind a beam splitter, see Sect. 18.2. Thus the light that reaches the photodetector is

$$c(t) = \sqrt{e}\,b(t)+\sqrt{1-e}\,w(t), \tag{20.1.19}$$

where $w(t)$ is a vacuum white noise. In the limit of $\mathcal{F}(t)$ large, we find that

$$\begin{aligned} N(\omega,T) = \int_0^T dt\,\cos\omega t\Big\{& e\mathbf{r}^2|\mathcal{F}(t)|^2 \\ &+\sqrt{e}\,\mathbf{r}\,\mathcal{F}^*(t)\left[\sqrt{e}\left(\mathbf{t}\,a(t)+\mathbf{r}\,v(t)\right)+\sqrt{1-e}\,w(t)\right] \\ &+\sqrt{e}\mathbf{r}\,\mathcal{F}(t)\left[\sqrt{e}\left(\mathbf{t}\,a(t)+\mathbf{r}\,v(t)\right)+\sqrt{1-e}\,w(t)\right]^\dagger\Big\}. \end{aligned} \tag{20.1.20}$$

By making $\mathcal{F}(t)$ arbitrarily large, and r correspondingly small, the local oscillator noise, which arises from the term with $v(t)$ can be made as small as desired. However the $w(t)$ term, which corresponds to noise directly arising in the inefficient photodetector, can only be eliminated by making the efficiency e as close to 100% as possible.

An arbitrarily large signal to noise ratio is achievable, however, by counting for an arbitrarily long time. But if one is interested in the *noise* itself, which is the case in the study of squeezed light, then great efforts must be made to increase e.

20.1.3 Balanced Homodyne/Heterodyne Detection

In balanced detection, the beam splitter is chosen to have equal reflectivity and transmissivity, i.e. $|\mathbf{r}| = |\mathbf{t}| = 1/\sqrt{2}$, and two detectors are used, as in Fig. 20.2. The photocurrents are then subtracted, and processed as before. In the case that the detectors are 100% efficient, the signal to be measured is

$$N_B(\omega, T) = \int_0^T dt \left\{ b_1^\dagger(t) b_1(t) - b_2^\dagger(t) b_2(t) \right\} \cos\omega t, \tag{20.1.21}$$

with

$$b_1(t) = \frac{\mathcal{F}(t) + v(t) + a(t)}{\sqrt{2}}, \tag{20.1.22}$$

$$b_2(t) = \frac{-\mathcal{F}(t) - v(t) + a(t)}{\sqrt{2}}. \tag{20.1.23}$$

After some manipulation, we find that

$$N_B(\omega, T) = \int_0^T dt \left\{ \mathcal{F}(t)^* a(t) + \mathcal{F}(t) a^\dagger(t) + v^\dagger(t) a(t) + v(t) a^\dagger(t) \right\} \cos\omega t. \tag{20.1.24}$$

Again, one assumes that $\mathcal{F}(t)$ is very large, so that the final two terms can be neglected. However, $\mathcal{F}(t)$ need not be nearly so strong as in unbalanced detection since it does not appear multiplied by $r \ll 1$. Thus balanced detection provides a more satisfactory technique, particularly as the constant term is no longer present. The formulae are essentially the same as those for ordinary detection, with the replacement $\mathbf{r} \to 1, \mathbf{t} \to 1$.

a) Imperfect Detectors: If the photodetectors are not perfect, we have two additional independent noise sources at each detector. Thus, the inputs measured will be

$$c_1(t) = \sqrt{e_1}\, b_1(t) + \sqrt{1 - e_1}\, w_1(t), \tag{20.1.25}$$

$$c_2(t) = \sqrt{e_2}\, b_2(t) + \sqrt{1 - e_2}\, w_2(t). \tag{20.1.26}$$

Here $w_1(t)$ and $w_2(t)$ may be taken as vacuum quantum white noise. In the limit that $\mathcal{F}(t)$ is large (and with $e_1 = e_2 \equiv e$ for simplicity) we find

$$N_B(\omega, T) \to \int_0^T dt \Big\{ e\Big(\mathcal{F}(t)^* a(t) + \mathcal{F}(t) a^\dagger(t)\Big) \\ + \sqrt{\frac{e(1-e)}{2}} \Big(\mathcal{F}(t)^* [w_1(t) + w_2(t)] + \mathcal{F}(t) [w_1(t) + w_2(t)]^\dagger\Big)\Big\} \cos\omega t. \tag{20.1.27}$$

If e is very close to 1, the final terms are negligible, but otherwise they can be significant, as in the case of unbalanced detection.

Exercise 20.2 Effect of an Unbalanced Beam Splitter: To get the full advantage of the technique, one must have a perfectly balanced beam splitter. Analyse the situation in which the beam splitter is unbalanced.

Exercise 20.3 Noise in Balanced Detection: In fact, because there are two *independent* noise sources in balanced detection, the signal to noise ratio is half that of unbalanced detection. Show this.

20.A Appendix: Quantum Stochastic Equations for Homodyne Measurement

In this appendix we will develop the theory of homodyne measurement in a form analogous to that we used for continuous measurement in Chap. 18. In the case of photon counting, the detection of a photon led to definite knowledge of the quantum state of the system, and we were able to develop equations of motion to take account of this. These equations simulated quantum jumps, that is discrete events occurring probabilistically, based on a classical stochastic jump process, in which the system wavefunction changes abruptly.

In the case of homodyne measurement, we will be able to develop a description in which the wavefunction undergoes a continuous stochastic evolution based on the use of the Wiener process. This description gives interesting insights into the nature of homodyne measurement, but it is not of such practical significance as the photon counting methods of Chap. 18.

20.A.1 Ideal Homodyne Measurement

The essence of a homodyne measurement is the mixing of a local oscillator field with the output quantum field from the system of interest, followed by the process of photon counting on the resulting mixed field. The mixing process in its basic form is described by the mixing formula (20.1.1), but we can simplify this by letting $\mathbf{t} \to 1$, $\mathbf{r} \to 0$, and the local oscillator field $\mathcal{F}(t) \to \infty$ in such a way the product

$$\mathbf{r}\mathcal{F}(t) \longrightarrow \bar{\mathcal{F}}(t), \tag{20.A.1}$$

is finite. The quantum fluctuations $v(t)$ become negligible in this limit, and we will call this an *ideal homodyne measurement.*

Thus, in an *ideal homodyne measurement* the field measured is the output field displaced by the c-number field $\bar{\mathcal{F}}(t)$, and can be written as $h(t)$, defined by

$$h(t) = \bar{\mathcal{F}}(t) + b_{\text{out}}(t) = \bar{\mathcal{F}}(t) + b_{\text{in}}(t) + \sqrt{\gamma}c(t), \tag{20.A.2}$$

where $c(t)$ is as usual the operator representing the coupling of the system of interest to the electromagnetic field.

Transfer to a Rotating Frame: For the remainder of this chapter we will move to a frame rotating at the laser frequency, so that we can write

$$\bar{\mathcal{F}}(t) \longrightarrow \bar{F}e^{-i\varphi}, \tag{20.A.3}$$

and make the replacements

$$b_{\text{in}}(t)e^{i\omega_L t} \longrightarrow b_{\text{in}}(t), \tag{20.A.4}$$

$$c(t)e^{i\omega_L t} \longrightarrow b_{\text{out}}(t), \tag{20.A.5}$$

$$h(t)e^{i\omega_L t} \longrightarrow h(t). \tag{20.A.6}$$

Consequently, in the rotating frame, (20.A.2) becomes

$$h(t) = \bar{F}e^{-i\varphi} + b_{\text{out}}(t) = \bar{F}e^{-i\varphi} + b_{\text{in}}(t) + \sqrt{\gamma}c(t). \tag{20.A.7}$$

20.A.2 Quadrature Phases and Homodyne Current

The photon number counted in a time interval dt is

$$\frac{dB^\dagger(t)\,dB(t)}{dt} = h^\dagger(t)h(t)\,dt = \bar{F}^2\,dt + \bar{F}\big(e^{i\varphi}b_{\rm out}(t) + e^{-i\varphi}b^\dagger_{\rm out}(t)\big)\,dt + \dots \tag{20.A.8}$$

Apart from the constant value $|\bar{F}|^2\,dt$, the quantity being continuously measured is proportional to the *electromagnetic field quadrature phase*, which depends on the phase of φ, and which we define by

$$q_\varphi(t) \equiv \frac{e^{-i\varphi}b_{\rm out}(t) + e^{i\varphi}b^\dagger_{\rm out}(t)}{\sqrt{2}}\,. \tag{20.A.9}$$

This leads us to the definition of a quantity called the *homodyne current*,

$$I_h(t) \equiv \sqrt{2}\bar{F}\,q_\varphi(t), \tag{20.A.10}$$

which is directly related to the counts measured—as in (20.A.8)—as a result of the mixing of the signals.

The electromagnetic field quadrature phase operator depends on the phase φ, and homodyne detection is thus a *phase-dependent measurement* of the electromagnetic field. For any given φ, the conjugate quadrature $p_\varphi(t) \equiv q_{\varphi+\pi/2}(t)$, behaves like a momentum canonically conjugate to $q_\varphi(t)$, and measurement of both is required for complete information. Of course this is possible only within the restrictions of the Heisenberg uncertainty principle. In Chap. 21 we will consider the manipulation and measurement of quadrature phase operators in some detail.

20.A.3 Ito Increments for Quadrature Phases

When we considered counting photons in Chap. 18, the detection of each individual photon induced a projection **J** on the quantum state of the system emitting the photons. In homodyne measurement, the quantity measured is the quadrature phase field $q_\varphi(t)$, and the projection required will be of a different kind. This leads us to reformulate the concept of continuous measurement, and introduce a different Ito calculus.

For simplicity let us choose $\varphi = 0$, and define the *quadrature phase Ito increments*

$$dQ(t) \equiv \frac{dB(t) + dB^\dagger(t)}{\sqrt{2}}, \qquad dP(t) \equiv \frac{dB(t) - dB^\dagger(t)}{i\sqrt{2}}, \tag{20.A.11}$$

which have the Ito algebra

$$\left.\begin{aligned} dQ(t)^2 &= dP(t)^2 &&= \tfrac{1}{2}\,dt, \\ dQ(t)\,dP(t) &= -dP(t)\,dQ(t) &&= -\tfrac{1}{2}i\,dt. \end{aligned}\right\} \tag{20.A.12}$$

We will use essentially the same as quadrature phase increments when discussing squeezing in Chap. 21, in particular in Sect. 21.2.2a.

20.A.4 Measurement Operators for Eigenvalues with a Continuous Range

The measurement process will consist of measurements of the increments $\Delta Q(t_i)$ in succeeding time intervals Δt_i. The increments at different time intervals are of course independent of each other. The projection introduced in the case of these variables is, however, quite different from that which is used for photon counting. The eigenvalues $\Delta Q(t)$ are like those of the position operator, and have a continuous range. For such operators the measurements possible consist of determining whether the eigenvalue is within a certain interval.

For such an operator, which we will denote here by X (and which we can think of as being a position operator) with eigenvectors $|x\rangle$, we can introduce a measurement operator defined by

$$E_\epsilon(\bar{x})|x\rangle = \int_{\bar{x}-\epsilon/2}^{\bar{x}+\epsilon/2} |x'\rangle\, dx'. \tag{20.A.13}$$

The resulting state is a superposition of states with position eigenvalues near to the measures value $\bar{x}$, and when ϵ is sufficiently small, this is a kind of measurement.

The probability of obtaining the value $\bar{x}$ is to be interpreted as the probability that the x is within the range $(x-\epsilon/2, x+\epsilon/2)$, that is, if the state of the system is initially $|\psi\rangle$

$$P_\epsilon(\bar{x}) = \int_{\bar{x}-\epsilon/2}^{\bar{x}+\epsilon/2} |\langle x|\psi\rangle|^2\, dx = \langle\psi|E_\epsilon(\bar{x})|\psi\rangle \approx |\langle\bar{x}|\psi\rangle|^2\epsilon. \tag{20.A.14}$$

The final approximate result assumes that $|\langle x|\psi\rangle|$ changes negligibly over the range $(x-\epsilon/2, x+\epsilon/2)$.

The normalized state after measurement is

$$\frac{E_\epsilon(\bar{x})|\psi\rangle}{\sqrt{\langle\psi|E_\epsilon(\bar{x})|\psi\rangle}} \approx \frac{1}{\sqrt{\epsilon}}\int_{\bar{x}-\epsilon/2}^{\bar{x}+\epsilon/2} |x'\rangle\, dx' \equiv \overline{|\bar{x}\rangle}. \tag{20.A.15}$$

The approximation required for this result is that both the phase and the amplitude of $\langle x|\psi\rangle$ change negligibly over the range $(x-\epsilon/2, x+\epsilon/2)$—this is a stronger requirement than that for the previous result.

Exercise 20.4 An Initial State with Large Momentum:

i) Consider the situation where $\langle x|\psi\rangle \propto \mathrm{e}^{\mathrm{i}px}$, and $\epsilon p = n\pi$, where n is integral. What kind of state results from the measurement operator defined by (20.A.13)?

ii) A complete set of measurement operators would be

$$E_\epsilon(\bar{x}, n)|x\rangle \equiv \int_{\bar{x}-\epsilon/2}^{\bar{x}+\epsilon/2} \mathrm{e}^{2\pi\mathrm{i}nx'}|x'\rangle\, dx'. \tag{20.A.16}$$

How should these be interpreted physically?

20.A.5 Formulation of the Quantum Stochastic Schrödinger Equation

We now adapt the arguments of Sect. 18.4 to this situation. The initial condition at $t=0$ is of the form

$$|\Psi,0\rangle = |\psi_C,0\rangle|\mathrm{vac}\rangle \equiv |\psi_C,0\rangle|0\rangle_0|0\rangle_{\Delta t}|0\rangle_{2\Delta t}\cdots \tag{20.A.17}$$

where $|\psi_C,0\rangle$ is a normalized system state vector.

a) Propagation of the State Vector: In the first time step, $[0, \Delta t)$, we can write

$$|\Psi, \Delta t\rangle = U(\Delta t)|\Psi, 0\rangle, \tag{20.A.18}$$

$$= \left(1 - \frac{i}{\hbar} H_{\text{eff}} \Delta t + \sqrt{\gamma}\, c\, \Delta B^\dagger(0)\right)|\Psi, 0\rangle. \tag{20.A.19}$$

Here, as in Chap. 9, we use that fact that the initial quantum state of the electromagnetic field is the vacuum to set equal to zero terms involving $\Delta B(0)$.

b) Introduction of Quadrature Phase Ito Increments: We can again use the fact that terms involving $\Delta B(0)$ can be set equal to zero to write

$$\Delta B^\dagger(0)|\Psi, 0\rangle = \sqrt{2}\, \Delta Q(0)|\Psi, 0\rangle, \tag{20.A.20}$$

where the quadrature phase Ito increment is defined as in (20.A.12). Inserting this into the propagation equation, we get

$$|\Psi, \Delta t\rangle = \left(1 - i H_{\text{eff}} \Delta t + \sqrt{2\gamma}\, c\, \Delta Q(0)\right)|\Psi, 0\rangle. \tag{20.A.21}$$

For the other time steps, $[t, t + \Delta t)$, we get

$$|\Psi, t + \Delta t\rangle = \left(1 - i H_{\text{eff}} \Delta t + \sqrt{2\gamma}\, c\, \Delta Q(t)\right)|\Psi, t\rangle, \tag{20.A.22}$$

where we have used the fact that the $\Delta Q(t)$ commute in different time intervals.

c) Quantum Stochastic Schrödinger Equation: We can summarize the above as the Ito quantum stochastic Schrödinger equation

$$d|\Psi, t\rangle \equiv |\Psi, t + dt\rangle - |\Psi, t\rangle = \left(-i H_{\text{eff}}\, dt + \sqrt{2\gamma}\, c\, dQ(t)\right)|\Psi, t\rangle, \tag{20.A.23}$$

with Ito rule, as in (20.A.12), $dQ(t)^2 = \frac{1}{2} dt$.

20.A.6 Master Equation

We derive a quantum stochastic differential equation for the total density operator $W(t) \equiv |\Psi(t)\rangle\langle\Psi(t)|$ and take the trace $\rho(t) \equiv \text{Tr}_B W(t)$, to get

$$d\rho(t) = -i\left(H_{\text{eff}}\rho(t) - \rho(t) H_{\text{eff}}^\dagger\right) dt + 2\gamma \text{Tr}_B \left\{dQ(t)\, c W(t) c^\dagger\, dQ(t)\right\} \tag{20.A.24}$$

$$+ \sqrt{2\gamma}\, \text{Tr}_B \{dQ(t)\, c W(t)\} + \text{Tr}_B \left\{W(t) \sqrt{2\gamma}\, dX(t) c^\dagger\right\}. \tag{20.A.25}$$

Using

$$\text{Tr}_B \{dQ(t) W(t)\, dQ(t)\} = \text{Tr}_B \left\{dQ(t)^2\, W(t)\right\} = \tfrac{1}{2} dt\, \rho(t) \tag{20.A.26}$$

$$\text{Tr}_B \{dQ(t) W(t)\} = 0, \tag{20.A.27}$$

this reduces to the master equation

$$\dot{\rho}(t) = -i\left(H_{\text{eff}}\rho(t) - \rho(t) H_{\text{eff}}^\dagger\right) + \gamma c \rho(t) c^\dagger. \tag{20.A.28}$$

20.A.7 Continuous Measurement of the Quadrature Phase Components

The process of homodyne measurement is equivalent to making repeated quadrature measurements during read out. We will again look at the first time step $[0, \Delta t)$, and then generalize to the later time steps.

a) State after the Measurement: The initial propagation $|\Psi,0\rangle \to |\Psi,\Delta t\rangle$ is given by (20.A.22). After that time evolution, we measure $\Delta Q(0)$ with a resolution ϵ (which should be small), and obtain the measured value Δq_0 using the measurement operator $E_\epsilon(\Delta q_0)$ as defined as previously in (20.A.13). The resulting probability amplitude is

$$|\Psi,\Delta t\rangle \to E_\epsilon(\Delta q_0)|\Psi,\Delta t\rangle, \tag{20.A.29}$$

$$= \left(1 - \mathrm{i}H_{\text{eff}}\Delta t + \sqrt{2\gamma}c\Delta q_0\right)|\psi,0\rangle \left(E_\epsilon(\Delta q_0)|0\rangle_0\right)|0\rangle_{\Delta t}|0\rangle_{2\Delta t}\ldots, \tag{20.A.30}$$

$$= \sqrt{P_\epsilon(\Delta q_0)}\left(1 - \mathrm{i}H_{\text{eff}}\Delta t + \sqrt{2\gamma}c\Delta q_0\right)|\psi,0\rangle\left(\overline{|\Delta q_0\rangle}\,|0\rangle_{\Delta t}|0\rangle_{2\Delta t}\ldots\right), \tag{20.A.31}$$

In these equations:

i) $P_\epsilon(\Delta q_0)$ and $\overline{|\Delta q_0\rangle}$ are as defined in (20.A.14, 20.A.15).

ii) We use the notation

$$|0\rangle_0, \quad |0\rangle_{\Delta t}, \quad |0\rangle_{2\Delta t}\ldots \tag{20.A.32}$$

to mean the vacuum state of the Hilbert space at the time intervals $0, \Delta t, 2\Delta t\ldots$

iii) Note that the projection has enabled the replacement of the operator $\Delta Q(0)$ with the c-number Δq_0 in the second line.

b) Probability Distribution of Quadrature Phase Measurements:

i) *The Probability Density for Eigenvalue* Δq_0: We now want to evaluate the probability distribution $P_\epsilon(\Delta q_0)$ of the measured values of Δq_0 after the quantum state has been evolved by one time step Δt, and measured. From (20.A.14), the probability that the measured value is in the range $(\Delta q_0 - \epsilon/2, \Delta q_0 + \epsilon/2)$ is given by

$$P_\epsilon(\Delta q_0) = \langle\Psi,\Delta t|E_\epsilon(\Delta q_0)|\Psi,\Delta t\rangle, \tag{20.A.33}$$

$$= \langle\psi_c,0|\langle 0|_0\left(1 + \mathrm{i}H^\dagger_{\text{eff}}\Delta t + \sqrt{2\gamma}c^\dagger\Delta q_0\right)$$

$$\times E_\epsilon(\Delta q_0)|\left(1 - \mathrm{i}H_{\text{eff}}\Delta t + \sqrt{2\gamma}c\Delta q_0\right)|0\rangle_0|\psi_c,0\rangle, \tag{20.A.34}$$

$$= \left(1 + \sqrt{2\gamma}\langle\hat{x}\rangle_c\Delta q_0 + \gamma\langle c^\dagger c\rangle_c\left(2\Delta q_0^2 - \Delta t\right)\right)\langle 0|_0 E_\epsilon(\Delta q_0)|0\rangle_0\,.$$

a) Here we have used the notations:

$$x \equiv c + c^\dagger, \tag{20.A.35}$$

$$\langle x\rangle_c = \langle\psi_c,0|x|\psi_c,0\rangle. \tag{20.A.36}$$

b) To evaluate the term $\langle 0|_0 E_\epsilon(\Delta q_0)|0\rangle_0$, we note that $\langle\Delta q_0|0\rangle_0$ is the wavefunction of the vacuum state in terms of the "position" eigenvalue Δq_0 before the evolution by one time step. It is the ground state wavefunction a harmonic oscillator, which is a Gaussian, and in fact

$$\langle\Delta q_0|0\rangle_0 = \frac{1}{(\pi\Delta t)^{1/4}}\mathrm{e}^{-\frac{\Delta q_0^2}{2\Delta t}}. \tag{20.A.37}$$

Following (20.A.14), we can make the approximation

$$\langle 0|_0 E_\epsilon(\Delta q_0)|0\rangle_0 \approx |\langle\Delta q_0|0\rangle_0|^2\,\epsilon, \tag{20.A.38}$$

provided ϵ is sufficiently small.

We can insert this into (20.A.35), and also note that all terms apart from 1 in the parentheses are infinitesimal, so we can write this term as an exponential and combine these with the Gaussian. Removing the factor ϵ, we get the *probability density*

$$p(\Delta q_0) = \frac{1}{(\pi\Delta t)^{1/2}} \exp\left(-\frac{\left(\Delta q_0 - \sqrt{\gamma/2}\,\langle x\rangle_c\Delta t\right)^2 - \delta}{\Delta t}\right), \tag{20.A.39}$$

where δ is a term which vanishes as $\Delta t \to 0$, and is given by

$$\delta = \tfrac{1}{2}\gamma\langle x\rangle_c^2\Delta t^2 + \left(2\Delta q_0^2 - \Delta t\right)\Delta t. \tag{20.A.40}$$

ii) We can therefore conclude that the probability density of Δq_0 after the first time step is, for sufficiently small Δt,

$$p(\Delta q_0) = \frac{1}{(\pi\Delta t)^{1/2}} \exp\left(-\frac{\left(\Delta q_0 - \sqrt{\gamma/2}\,\langle x\rangle_c\Delta t\right)^2}{\Delta t}\right). \tag{20.A.41}$$

iii) *Simulating a Measurement Using the Wiener Process*: The distribution (20.A.41) says that Δq_0 is a classical random variable which we can write in terms of the Wiener process

$$\Delta q_0 = \frac{\sqrt{\gamma}\langle\hat{x}\rangle_c\Delta t + \Delta W_0}{\sqrt{2}}. \tag{20.A.42}$$

Here ΔW_0 is random number whose probability density process is that of a Wiener increment

$$p(\Delta W_0) = \frac{1}{(2\pi\Delta t)^{1/2}} \exp\left(-\frac{\Delta W_0^2}{2\Delta t}\right). \tag{20.A.43}$$

c) Conditional Evolution of the System Wavefunction: This allows us to write the evolution of a system wave function, conditional on the observation of value Δq_0, in the forms:

i) *Probability Amplitude*:

$$|\tilde{\psi}_c, \Delta t\rangle = \left(1 - \mathrm{i}H_{\text{eff}}\Delta t + \sqrt{2\gamma}\,c\,\Delta q(0)\right)|\psi, 0\rangle. \tag{20.A.44}$$

ii) *Normalized Wavefunction*: The normalized wavefunction is defined as

$$|\psi_c, \Delta t\rangle \equiv \frac{|\tilde{\psi}_c, \Delta t\rangle}{\left\||\tilde{\psi}_c, \Delta t\rangle\right\|} = \frac{\left(1 - \mathrm{i}H_{\text{eff}}\Delta t + \sqrt{2\gamma}\,c\,\Delta q_0\right)|\psi_c, 0\rangle}{\sqrt{1 + \sqrt{2\gamma}\,\langle x\rangle_c\,\Delta q_0}}. \tag{20.A.45}$$

Exercise 20.5 Ito Calculus of Δq_0: Notice that the Ito rules for for the Wiener process allow us to say here that $\Delta W_0^2 = \Delta t$ and $\Delta W_0\,\Delta t = 0$.

Show this means that we can use the Ito rules for Δq_0 in the form $\Delta q_0^2 = \frac{1}{2}\Delta t$ and $\Delta q_0\,\Delta t = 0$.

Exercise 20.6 The Norm of the Probability Amplitude: By using (20.A.42) and (20.A.44), and using the Ito rules for Δq_0, show that

$$\langle\tilde{\psi}_c, \Delta t|\tilde{\psi}_c, \Delta t\rangle = 1 + \sqrt{2\gamma}\langle x\rangle_c\,\Delta q_0. \tag{20.A.46}$$

Exercise 20.7 Behaviour of the Norm: Using the Ito rules for Δq_0, show that

$$1\Big/\sqrt{1+\sqrt{2\gamma}\,\langle x\rangle_c\,\Delta q_0} = 1-\sqrt{\frac{\gamma}{2}}\langle x\rangle_c\,\Delta q_0+\frac{3}{8}\gamma\langle x\rangle_c^2\,\Delta t, \tag{20.A.47}$$

so that

$$1\Big/\big\|\,|\tilde\psi_c,\Delta t\rangle\big\| = 1-\gamma\left(\frac{1}{8}\langle x\rangle_c^2+\tfrac{1}{2}c^\dagger c\right)\Delta t+\sqrt{\frac{\gamma}{2}}\left(c-\tfrac{1}{2}\langle x\rangle_c\right)\Delta W_0. \tag{20.A.48}$$

d) Algorithm for Arbitrary Times: The algorithm for the time step $[t, t+\Delta t)$ has the same forms for the first time step, since the Hilbert spaces at different time intervals are independent. It is as follows:

i) *Propagation of the State Vector*:

$$|\Psi, t+\Delta t\rangle = \left(1-\mathrm{i}H_{\text{eff}}\,\Delta t+\sqrt{2\gamma}\,c\,\Delta Q(t)\right)|\Psi, t\rangle. \tag{20.A.49}$$

ii) *Measurement of $\Delta Q(t)$ at Time $t+\Delta t$*: The probability of obtaining the value Δq_t is

$$P(\Delta q_t) = \frac{1}{(\pi\Delta t)^{1/2}}\exp\left(-\frac{\left(\Delta q_t-\sqrt{\gamma/2}\langle x_t\rangle_c\Delta t\right)^2}{\Delta t}\right), \tag{20.A.50}$$

in which

$$\langle x_t\rangle_c \equiv \langle\psi_c, t|x|\psi_c, t\rangle. \tag{20.A.51}$$

iii) *The Measurement Value*: This can be written as

$$\Delta q_t = \frac{\sqrt{\gamma}\langle x_t\rangle_c\Delta t+\Delta W(t)}{\sqrt{2}}, \tag{20.A.52}$$

where $\Delta W(t)$ is a Wiener increment.

iv) *The Normalized State after the Measurement at Time t*: The measurement trajectory $\Delta q_0, \Delta q_{\Delta t}, \Delta q_{2\Delta t}, \dots \Delta q_t$, occurs with probability

$$P(\Delta q_0)P(\Delta q_{\Delta t})P(\Delta q_{2\Delta t})\cdots P(\Delta q_t). \tag{20.A.53}$$

The normalized state after this trajectory of measurements is

$$|\Psi, t\rangle = |\psi_c, t\rangle\,\overline{|\Delta q_0\rangle}\,\overline{|\Delta q_{\Delta t}\rangle}\,\overline{|\Delta q_{2\Delta t}\rangle}\cdots\overline{|\Delta q_t\rangle}_t\,|0\rangle_{t+\Delta t}\,|0\rangle_{t+2\Delta t}\cdots \tag{20.A.54}$$

v) *The Normalized System Wavefunction Conditional on this Measurement Sequence*: This evolves according to

$$|\psi_c, t+\Delta t\rangle = \frac{|\tilde\psi_c, t+\Delta t\rangle}{\big\|\,|\tilde\psi_c, t+\Delta t\rangle\big\|} = \frac{\left(1-\mathrm{i}H_{\text{eff}}\Delta t+\sqrt{2\gamma}\,c\,\Delta q_t\right)|\psi_c, t\rangle}{\sqrt{1+\sqrt{2\gamma}\,\langle x_t\rangle_c\,\Delta q_t}}. \tag{20.A.55}$$

e) Stochastic Schrödinger Equation for Homodyne Measurement: We can take the limit $\Delta t \to dt$, $\Delta W(t) \to dW(t)$ and write the time evolution of the conditional system wave function as a stochastic Schrödinger equation of the diffusion type. It is convenient to write both an equation for an unnormalized system wave function denoted by $|\tilde\psi_c, t\rangle$, and one for the normalized wavefunction $|\psi_c, t\rangle = |\tilde\psi_c, t\rangle\big/\big\|\,|\tilde\psi_c, t\rangle\big\|$.

i) *Probability Amplitude*: The wavefunction evolution involves two coupled equations, one for the wavefunction, and one for the quadrature increment dq_t, namely

$$d|\tilde{\psi}_c,t\rangle = \left(-\mathrm{i}H_{\mathrm{eff}}\,dt + \sqrt{2\gamma}\,c\,dq_t\right)|\tilde{\psi}_c,t\rangle, \tag{20.A.56}$$

$$dq_t = \frac{\sqrt{\gamma}\langle x_t\rangle_c\,dt + dW(t)}{\sqrt{2}}, \tag{20.A.57}$$

$$\langle x_t\rangle_c = \frac{\langle\tilde{\psi}_c,t|x|\tilde{\psi}_c,t\rangle}{\langle\tilde{\psi}_c,t|\tilde{\psi}_c,t\rangle} = \langle\psi_c,t|x|\psi_c,t\rangle. \tag{20.A.58}$$

ii) *Relationship with the Homodyne Current*: We can write the stochastic Schrödinger equation as

$$d|\psi_c,t\rangle = \left(-\mathrm{i}H_{\mathrm{eff}} + \sqrt{\gamma}\,c\,I_h(t)\right)|\psi_c,t\rangle\,dt. \tag{20.A.59}$$

For a given observed homodyne current

$$I_h(t) = \frac{\sqrt{\gamma}\langle x\rangle_c(t) + \xi(t)}{\sqrt{2}}, \tag{20.A.60}$$

integration of the stochastic Schrödinger equation gives us the corresponding system wave function conditioned on the particular trajectory observed. From the *observed* homodyne current, we can use the wavefunction equation (20.A.59) to construct the wavefunction $|\tilde{\psi}_c,t\rangle$.

iii) *Normalization of the Wavefunction*: The exercises Ex. 20.5–Ex. 20.7 can be adapted to show that

$$\frac{1}{\|\,|\tilde{\psi}_c,t+\Delta t\rangle\,\|} = \left(1-\gamma\left(\tfrac{1}{8}\langle x\rangle_c^2 + \tfrac{1}{2}c^\dagger c\right)\Delta t + \sqrt{\frac{\gamma}{2}}\left(c-\tfrac{1}{2}\langle x\rangle_c\right)\Delta W_0\right)\frac{1}{\|\,|\tilde{\psi}_c,t\rangle\,\|}. \tag{20.A.61}$$

iv) *Normalized Wavefunction*: Using the result (20.A.61), the stochastic Schrödinger equation (20.A.56), for the probability amplitude, and Ito rules, we find the stochastic Schrödinger equation for the normalized wavefunction is

$$d|\psi_c,t\rangle = \left\{\left(-\mathrm{i}H_{\mathrm{eff}} + \tfrac{1}{2}\gamma c\langle x_t\rangle_c - \tfrac{1}{8}\gamma\langle x_t\rangle_c^2\right)dt + \sqrt{\gamma}\left(c-\tfrac{1}{2}\langle x_t\rangle_c\right)dW(t)\right\}|\psi_c,t\rangle. \tag{20.A.62}$$

This is also given in *Quantum Noise Chap. 11*, but with a different derivation.

v) *Alternative Form of the Stochastic Schrödinger Equation*:

$$d|\psi_c,t\rangle = \Big\{-\mathrm{i}H_{\mathrm{eff}}\,dt + \gamma c\langle c_t^\dagger\rangle_c\,dt - \tfrac{1}{2}\gamma|\langle c_t\rangle_c|^2\,dt + \sqrt{\gamma}\left(c-\langle c_t\rangle_c\right)dW(t)\Big\}|\psi_c,t\rangle. \tag{20.A.63}$$

Exercise 20.8 Proof of the Alternative Form of the Stochastic Schrödinger Equation: The second form (20.A.63) is obtained by making a phase transformation of the wave function $\psi_c(t)$. Let $\psi_c(t)$ obey the Ito stochastic differential equation

$$d|\psi_c,t\rangle = \left[A\,dt + B\,dW(t)\right)|\psi_c,t\rangle. \tag{20.A.64}$$

Then the transformation

$$e^{i\phi(t)}|\psi_c,t\rangle \longrightarrow |\psi_c,t\rangle, \tag{20.A.65}$$

where

$$d\phi = a(t)\,dt + b(t)\,dW(t), \tag{20.A.66}$$

with real coefficients $a(t)$ and $b(t)$, gives

$$d|\psi_c,t\rangle = \Big[\big(ia - \tfrac{1}{2}b^2 + A + ibB\big)\,dt + (B+ib)\,dW(t)\Big]\,|\psi_c,t\rangle. \tag{20.A.67}$$

Making the choices

$$a = -\tfrac{1}{4}\gamma\langle y\rangle_c\langle x\rangle_c, \qquad b = -\tfrac{1}{2}\sqrt{\gamma}\langle y\rangle_c \tag{20.A.68}$$

and setting $c \to \frac{1}{2}(x+iy)$ we obtain the alternative form (20.A.63).

Exercise 20.9 Stochastic Density Matrix Equations: Define the stochastic density matrix by

$$\rho_c(t) \equiv |\psi_c,t\rangle\langle\psi_c(t)|. \tag{20.A.69}$$

Use the Ito rules for $dW(t)$ and the alternative form (20.A.63) to show that the stochastic Schrödinger equation is equivalent to the stochastic density matrix equation

$$\begin{aligned} d\rho_c(t) &= -iH_{\text{eff}}\rho_c(t)\,dt + i\rho_c(t)H^\dagger_{\text{eff}}\,dt + \gamma c\rho_c(t)c^\dagger\,dt \\ &\quad + \sqrt{\gamma}\big(c - \langle c\rangle_c\big)\,dW(t)\rho_c(t) + \rho_c(t)\sqrt{\gamma}(c^\dagger - \langle c^\dagger\rangle_c)\,dW(t). \end{aligned} \tag{20.A.70}$$

Note that taking stochastic averages over all measurement trajectories we obtain the master equation $\langle\rho_c,t\rangle = \rho(t)$

$$\begin{aligned} \dot\rho(t) &= -iH_{\text{eff}}\rho(t) + i\rho(t)H^\dagger_{\text{eff}} + \gamma c\rho(t)c^\dagger, \\ &\equiv \mathcal{L}\rho(t). \end{aligned} \tag{20.A.71}$$

21. Squeezing, Quantum Correlations and Quantum Amplifiers

Systems of quantum-correlated Bosons arise physically in a variety of contexts, and the most famous of these is described by the *Bogoliubov Hamiltonian,* which was introduced in *Bk. I: Sect.16.1.3.* There we showed that it could be simplified to the consideration of a Hamiltonian of the form

$$H_{a,b} = \hbar v(a^\dagger a + b^\dagger b) + \hbar u(a^\dagger b^\dagger + ba),$$

where a and b are Boson destruction operators. Because this Hamiltonian is a quadratic form in the operators a, $a^\dagger$, b, $b^\dagger$, the equations of motion are linear in this set of four operators, and can be solved exactly—this makes the model very interesting. Physically, the two parts of the Hamiltonian are very different from each other. The first part is simply the energy associated with the individual Bosons, while the second part represents a process of emission or absorption of pairs of Bosons. This second part always arises as a result of an interaction between three or more modes, some of which have a macroscopic occupation. The amplitudes of these modes are then absorbed into the definitions of the Hamiltonian parameters u and v. We discuss this procedure in more detail for the case of an optical system in Sect. 21.2.1.

There are three main applications of this kind of Hamiltonian:

i) *Squeezing*: If we consider the special case in which the modes a and b are the same mode, we get the *squeezing Hamiltonian,* using which *squeezed states of light* can be described. The idea of a squeezed state, in which the quantum noise in a dynamical variable is reduced, while that in the canonically conjugate variable is increased, in such a way as to not violate Heisenberg's uncertainty principle is really quite old. However, the terminology "squeezed state" was introduced by *Hollenhorst* in 1979 [21.1], and his paper also gives some of the history of the subject.

ii) *Quantum-Correlated States*: When the operators a and b are are distinct from each other, one can produce quantum-correlated pairs of Bosons. In fact these are EPR pairs, as discussed in *Bk. I: Sect.1.2.1,* and play an important role in quantum information.

iii) *Quantum Amplifiers*: The solutions of the equations corresponding to this Hamiltonian always have some amplitudes which grow exponentially—which corresponds to *gain*, and some which decay exponentially, corresponding to *loss*. The system so described is an *amplifier* for those modes with gain. The power which appears in the the modes with gain comes from the macroscopically occupied modes, which are not significantly depleted.

This turns out to be a very good way of providing amplification, since the noise level is very low. When the modes a and b are identical, this is called the *degenerate parametric amplifier*, and no noise is added during the amplification process. However, not all of the information in the initial state is amplified.

Otherwise, this is a two-mode amplifier. However, if initially one of the modes is viewed as being in the vacuum state, and the other mode as carrying a signal, this provides an *ideal quantum amplifier*, as described in *Bk. I: Sect.16.2.1* c. This device amplifies all of the information in the initial state, but adds noise as well.

21.1 Single-Mode Squeezing and Quantum Noise Reduction

Quantum noise cannot be eliminated, but it can be manipulated. The first suggestion that this might be practically applied was made by *Caves* [21.2] in 1981, who suggested using *squeezed states* of light in an interferometer used to measure the effects of gravitational radiation on a suspended weight. In this section we will formulate the concept of *squeezing*, and develop the mathematical framework necessary to understand them.

21.1.1 Heisenberg's Uncertainty Principle

In its most precise form, the Heisenberg uncertainty principle states that:

i) If A and B are Hermitian operators with the commutator $[A, B] = \mathrm{i}C$, and

ii) In the quantum state of system, the operators have means $\langle A\rangle$, $\langle B\rangle$ and $\langle C\rangle$, and

iii) If the uncertainties are defined by
$$\Delta A \equiv \sqrt{\langle A^2\rangle - \langle A\rangle^2} \quad \text{etc.,} \tag{21.1.1}$$

iv) Then
$$\Delta A^2\, \Delta B^2 \geqslant \langle [\Delta A, \Delta B]_+\rangle^2 + \tfrac{1}{4}\langle C\rangle^2, \tag{21.1.2}$$
so that the uncertainty principle takes the form

$$\Delta A\, \Delta B \geqslant \tfrac{1}{2}|\langle [A, B]\rangle|. \tag{21.1.3}$$

This result is derived in *Quantum Noise* and most standard quantum mechanics textbooks.

v) *Minimum Uncertainty State*: This is defined to be a state in which the uncertainty product attains the minimum given by (21.1.2), that is

$$\Delta A\,\Delta B = \tfrac{1}{2}\left|\langle [A,B]\rangle\right|. \tag{21.1.4}$$

21.1.2 Quadrature Phases

If $a, a^\dagger$ are destruction and creation operators for a harmonic oscillator, we define the *quadrature phase operators*, X, Y, by

$$a = \frac{X + \mathrm{i}Y}{\sqrt{2}}, \qquad a^\dagger = \frac{X - \mathrm{i}Y}{\sqrt{2}}. \tag{21.1.5}$$

The quadrature phase operators are dimensionless, and are appropriate for any kind of Bosonic creation and destruction operator. In the case of a mechanical harmonic oscillator with mass m and frequency ω, the physical position and momentum operators are

$$x = \sqrt{\frac{\hbar}{\omega m}}\,X, \qquad p = \sqrt{\hbar\omega m}\,Y, \tag{21.1.6}$$

whose commutation relation is

$$[X, Y] = \mathrm{i}. \tag{21.1.7}$$

The Hamiltonian then takes the form

$$H = \tfrac{1}{2}\hbar\omega\left(X^2 + Y^2\right) = \hbar\omega\left(a^\dagger a + \tfrac{1}{2}\right). \tag{21.1.8}$$

The quadrature phase operators are closely related to the electromagnetic field quadrature phase operators introduced in Sect. 20.A.2, and their measurement in optical systems always involves some kind of homodyne procedure.

21.1.3 Defining Squeezing

Before defining squeezing, we have to have some measure of what we consider not to be squeezed, that is, we want to get some idea of what we consider to be an "ordinary state". For this purpose, we define the *standard quantum limit*, a term introduced to describe the limitations on the precision of specification of two canonically conjugate variables.

a) The Standard Quantum Limit: This is defined in terms of the quadrature phase operators introduced above. Heisenberg's uncertainty principle, applied to the quadrature phase operators, says that

$$\Delta X\,\Delta Y \geqslant \tfrac{1}{2}. \tag{21.1.9}$$

The standard quantum limit requires that in the minimum uncertainty states of interest, $\Delta X = \Delta Y$, and hence $\Delta X = \Delta Y = \frac{1}{\sqrt{2}}$. Such is certainly the case in the lowest energy state of the harmonic oscillator, which corresponds to a zero temperature thermal state. Intuitively one would believe that this uncertainty—the fluctuations at absolute zero—would surely represent the irreducible minimum for any "ordinary state".

b) Squeezed States: However, the uncertainty principle itself does not require this; either ΔX or ΔY can be as small as one likes, as long as the product satisfies the uncertainty principle. This means that we can make ΔX as small as we please, at the cost of making ΔY sufficiently large to satisfy the uncertainty relation. A state in which this occurs is known as a *squeezed state.*

The definition can be made quite general, for any pair of dimensionless operators A and B there is an uncertainty principle of the form (21.1.2), and if the state is a minimum uncertainty state satisfying (21.1.4) and $\Delta A \neq \Delta B$, the state is defined to be squeezed. However, when A and B are not canonically conjugate operators the minimum uncertainty is normally state-dependent, and the concept of squeezing becomes less natural.

21.1.4 Squeezed States of the Harmonic Oscillator

Therefore, although squeezing can be defined for a variety of quantum systems, its natural definition is for the operators of a harmonic oscillator. The *ideal squeezed state* of a harmonic oscillator is taken to be a quantum Gaussian state (see *Bk. I: Sect.10.4.2*) with minimum uncertainty product. This is most straightforwardly defined by means of the linear transformation which will transform a vacuum state into one with prescribed means $\langle X\rangle$, $\langle Y\rangle$ and a prescribed covariance matrix. Because of the Gaussian requirement, this is sufficient to specify the squeezed state.

a) The Squeezing Hamiltonian: The special case of the Bogoliubov Hamiltonian when a and b are the same operator, and when $v = 0$, gives what is known as the *squeezing Hamiltonian,* which is normally written in the form

$$H_{\text{squeezing}} = \tfrac{1}{2}\mathrm{i}\hbar\left(\epsilon a^{\dagger 2} - \epsilon^* a^2\right). \tag{21.1.10}$$

Using this Hamiltonian we can create the desired ideal squeezed states. The corresponding Heisenberg equations are

$$\frac{da}{dt} = \frac{\mathrm{i}}{\hbar}[H_{\text{squeezing}}, a] = \epsilon a^\dagger, \tag{21.1.11}$$

$$\frac{da^\dagger}{dt} = \frac{\mathrm{i}}{\hbar}[H_{\text{squeezing}}, a^\dagger] = \epsilon^* a. \tag{21.1.12}$$

If we write

$$\epsilon \equiv |\epsilon| \mathrm{e}^{\mathrm{i}\theta}, \tag{21.1.13}$$

the solution of these equations is

$$a(t) = a(0)\cosh|\epsilon|t + a^\dagger(0)\mathrm{e}^{\mathrm{i}\theta}\sinh|\epsilon|t. \tag{21.1.14}$$

b) Use of the Quadrature Phase Operators: The appropriate quadrature phase operators take account of the phase θ, and are defined by a modification of (21.1.5), which amounts to changing the phase of the harmonic oscillator operators. Thus we define

$$\frac{\tilde{X}(t) + \mathrm{i}\tilde{Y}(t)}{\sqrt{2}} \equiv \mathrm{e}^{-\mathrm{i}\theta/2} a(t). \tag{21.1.15}$$

The Hamiltonian then takes the form

$$H_{\text{squeezing}} = \hbar|\epsilon|[X, Y]_+ . \tag{21.1.16}$$

The equations of motion for these quadrature phase operators and their solutions are very simple:

$$\frac{d\bar{X}(t)}{dt} = |\epsilon|\bar{X}(t), \qquad \frac{d\bar{Y}(t)}{dt} = -|\epsilon|\bar{Y}(t), \tag{21.1.17}$$

whose solutions are

$$\bar{X}(t) = \bar{X}(0)\mathrm{e}^{|\epsilon|t}, \qquad \bar{Y}(t) = \bar{Y}(0)\mathrm{e}^{-|\epsilon|t}. \tag{21.1.18}$$

c) The Squeezed Vacuum: If the initial state is the vacuum $|0\rangle$, then initially

$$\langle\bar{X}(0)\rangle = \langle\bar{Y}(0)\rangle = \langle[\bar{X}(0), \bar{Y}(0)]_+\rangle = 0, \qquad \Delta X(0) = \Delta Y(0) = \frac{1}{\sqrt{2}}. \tag{21.1.19}$$

As time evolves, the mean values $\langle\bar{X}(0)\rangle$ and $\langle\bar{Y}(0)\rangle$, together with the correlation $\langle[\bar{X}(0), \bar{Y}(0)]_+\rangle$, remain zero, while the uncertainties become

$$\Delta X(t) = \frac{1}{\sqrt{2}}\mathrm{e}^{|\epsilon|t}, \qquad \Delta Y(t) = \frac{1}{\sqrt{2}}\mathrm{e}^{-|\epsilon|t}, \tag{21.1.20}$$

and this is exactly what we mean by squeezing. Since the evolution is linear in the creation and destruction operators, the initially quantum Gaussian vacuum remains quantum Gaussian throughout the time evolution.

d) The Squeezing Operator: The time evolution operator corresponding to these solutions satisfies the equation

$$\mathrm{i}\hbar\frac{d}{dt}U_{\text{squeezing}}(t) = H_{\text{squeezing}}(t)U_{\text{squeezing}}(t), \tag{21.1.21}$$

and is explicitly

$$U_{\text{squeezing}}(t) \equiv \exp\left(\tfrac{1}{2}\left(\epsilon a^{\dagger 2} - \epsilon^* a^2\right)t\right). \tag{21.1.22}$$

In the Schrödinger picture the quantum state becomes what we call a *squeezed vacuum*

$$|\text{vac}_{\text{squeezed}}, t\rangle \equiv U_{\text{squeezing}}(t)|0\rangle. \tag{21.1.23}$$

Exercise 21.1 Expression of the Squeezed Vacuum in Terms of Number States: Because $a|0\rangle = 0$, we can write

$$a(-t)|\text{vac}_{\text{squeezed}}, t\rangle = U_{\text{squeezing}}(t)aU^\dagger_{\text{squeezing}}(t)|\text{vac}_{\text{squeezed}}, t\rangle = 0. \tag{21.1.24}$$

Use the expression (21.1.14) to show that this means that

$$|\text{vac}_{\text{squeezed}}, t\rangle = \sqrt{\operatorname{sech}\xi}\sum_{n=0}^{\infty}\sqrt{\frac{\Gamma(n+1/2)}{n!}}\left(\mathrm{e}^{\mathrm{i}\theta}\tanh\xi\right)^n|2n\rangle, \tag{21.1.25}$$

where $\xi = |\epsilon|t$.

21.1.5 Definition of an Ideal Squeezed State

The most general kind of squeezed state must include the possibility of a non-zero mean field, and this can be generated by using two transformations sequentially, namely:

i) The *squeezing operator* $S(\xi)$ is defined by setting $e^{i\theta}|\epsilon|t \to -\xi$ into (21.1.22), that is

$$S(\xi) \equiv \exp\left(\tfrac{1}{2}(\xi^* a^2 - \xi a^{\dagger 2})\right). \tag{21.1.26}$$

Here ξ is a complex number specifying the amount and kind of squeezing, and is most conveniently discussed using the polar representation

$$\xi = r e^{i\theta}. \tag{21.1.27}$$

The quantity r is known as the *squeezing parameter,* since it determines the amount of squeezing; the phase θ determines the definition of the appropriate quadrature phases, as is shown by (21.1.15).

ii) The *displacement operator* $D(\alpha)$ was used in the definition of coherent states in *Bk. I: (10.5.4),* and is explicitly given by

$$D(\alpha) \equiv \exp(\alpha a^\dagger - \alpha^* a). \tag{21.1.28}$$

An *ideal squeezed state,* $|\alpha,\xi\rangle$ is defined by applying these two transformations sequentially, so that

$$|\alpha,\xi\rangle = D(\alpha)S(\xi)|0\rangle. \tag{21.1.29}$$

In this equation $S(\xi)$ generates squeezing, while $D(\alpha)$ generates a non-zero mean value α of the destruction operator a.

a) Properties of Ideal Squeezed States:

i) Ideal squeezed states are minimum uncertainty states for the variables $\bar{X}$ and $\bar{Y}$, since $[\bar{X},\bar{Y}] = i$ and $\Delta\bar{X}\,\Delta\bar{Y} = \frac{1}{2}$.

ii) The principal axes of the error ellipse are rotated an angle $\theta/2$ with respect to those of X and Y.

iii) For $r = 0$ the state is a coherent state, and the variances of the $\bar{X}$ and $\bar{Y}$ operators correspond to the standard quantum limit.

iv) For positive values of r, the variance of the $\bar{X}$ operator is reduced by a factor $\exp(-2r)$, while variance of the $\bar{Y}$ operator is increased by a factor $\exp(2r)$.

b) Properties of the Squeezing Operator:

i) $S^\dagger(\xi) = S^{-1}(\xi) = S(-\xi)$.

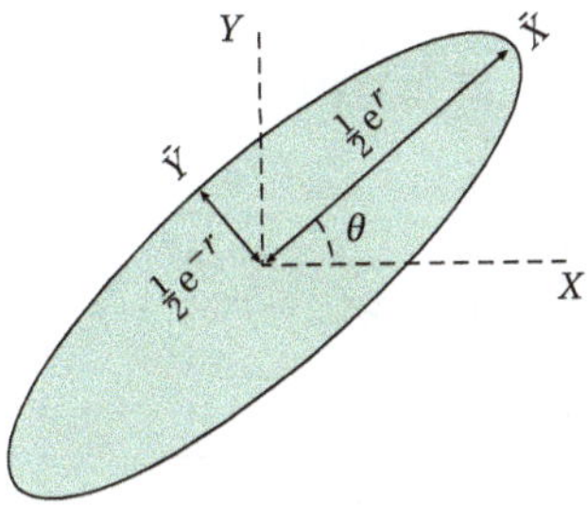

Fig. 21.1. Error ellipse for a squeezed vacuum state.

ii) The transform induced by the squeezing operator on the creation and destruction operators is

$$S^{\dagger}(\xi)aS(\xi) = a\cosh r - a^{\dagger}e^{-i\theta}\sinh r, \tag{21.1.30}$$

$$S^{\dagger}(\xi)a^{\dagger}S(\xi) = a^{\dagger}\cosh r - ae^{-i\theta}\sinh r. \tag{21.1.31}$$

In terms of the rotated quadrature phase components it can be written

$$S^{\dagger}(\xi)\bar{X}S(\xi) = e^{-r}\bar{X}, \tag{21.1.32}$$

$$S^{\dagger}(\xi)\bar{Y}S(\xi) = e^{r}\bar{Y}. \tag{21.1.33}$$

Exercise 21.2 Means and Variances of a Squeezed State: Show that the means for an ideal squeezed state $|\alpha,\xi\rangle$ are given by

$$\langle a\rangle = \alpha, \qquad \langle a^{\dagger}\rangle = \alpha^{*}, \tag{21.1.34}$$

and that the variances in the variables $\bar{X}$ and $\bar{Y}$ are

$$\left.\begin{aligned} \langle\Delta\bar{X}^2\rangle &= \tfrac{1}{2}e^{-2r}, \\ \langle\Delta\bar{Y}^2\rangle &= \tfrac{1}{2}e^{2r}, \\ \langle\Delta\bar{X}\Delta\bar{Y}\rangle &= 0. \end{aligned}\right\} \tag{21.1.35}$$

Graphically, this can be represented as the error ellipse shown in Fig. 21.1.

Exercise 21.3 Electric Field Fluctuations: We can represent a single mode electromagnetic field in a ring cavity in terms of the electric field operator

$$E(x,t) = i\sqrt{\frac{\hbar\omega_c}{2\epsilon_0 LA}}\left(ae^{i(kx-\omega t)} - a^{\dagger}e^{-i(kx-\omega t)}\right). \tag{21.1.36}$$

If this mode is in an ideal squeezed state $|\alpha,\xi\rangle$, show that the electromagnetic field fluctuations are

$$\Delta E^2 = \frac{\hbar\omega_c}{2\epsilon_0 LA}\left\{\Delta\bar{X}^2\sin^2\left(kx-\omega t+\tfrac{1}{2}\theta\right) + \Delta\bar{Y}^2\cos^2\left(kx-\omega t+\tfrac{1}{2}\theta\right)\right\}. \tag{21.1.37}$$

Exercise 21.4 Squeezing in the Field Fluctuations: Fig. 21.2 gives a schematic view of squeezed light in which there is a non-zero mean field, and the fluctuations about the mean field are squeezed. Verify that this is a reasonable representation of the behaviour of the squeezed light.

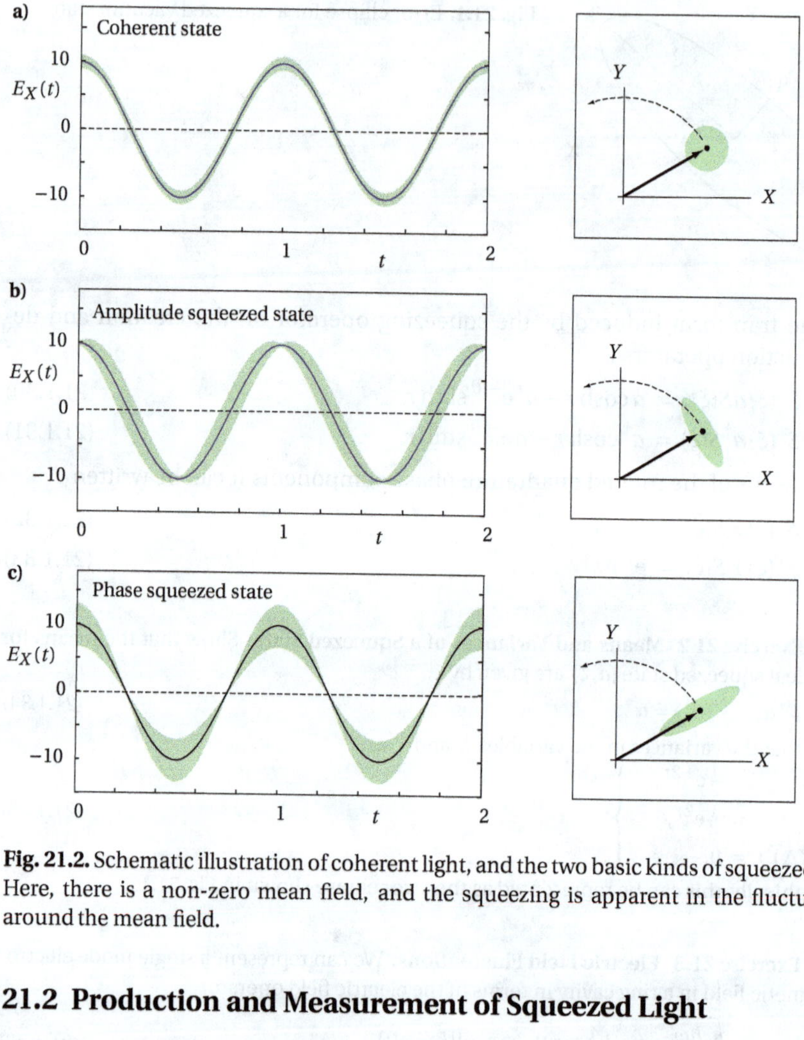

Fig. 21.2. Schematic illustration of coherent light, and the two basic kinds of squeezed light. Here, there is a non-zero mean field, and the squeezing is apparent in the fluctuations around the mean field.

21.2 Production and Measurement of Squeezed Light

The squeezing Hamiltonian, as we have written it in (21.1.10), is an abstraction of a non-linear interaction which may take a range of forms. The first experiment in which squeezed light was made in 1985 [21.3] used sodium vapour as a fourth-order nonlinear medium, in which four-wave mixing was implemented. A more convenient method was implemented [21.4] in 1986, in which the non-linear medium was crystalline MgO : $LiNbO_3$, and a third-order nonlinearity was used. Squeezing was also achieved in a superconductor system in 1990 [21.5], and at that stage it seemed that it was much more difficult to generate squeezing using a Josephson junction than optically. Currently, the situation has changed—superconducting systems, which we consider in Part VIII, may in the future provide a real challenge to optical systems.

21.2.1 The Degenerate Parametric Amplifier

The Hamiltonian for degenerate parametric amplification with a classical pump can be written as

$$H_{\text{sys}} = \hbar\omega_0 a^\dagger a + \tfrac{1}{2}\mathrm{i}\hbar\left[\epsilon \mathrm{e}^{-2\mathrm{i}\omega_0 t}(a^\dagger)^2 - \epsilon^* \mathrm{e}^{2\mathrm{i}\omega_0 t} a^2\right], \tag{21.2.1}$$

where $2\omega_0$ is the frequency of the pump beam and ϵ a measure of the effective pump intensity.

This Hamiltonian is a purely abstract concept, but it can be realized in the laboratory with the use of a strong pump field and an appropriate non-linear crystal.

a) Non-Linear Crystals: The electromagnetic Hamiltonian for the light inside the crystal normally takes a form [21.6] like

$$H_{EM} = \int\left(\frac{\sum_{ij}\varepsilon_{ij}(\boldsymbol{E})E_i E_j}{2} + \frac{\boldsymbol{B}^2}{2\mu_0}\right) d^3\boldsymbol{x}. \tag{21.2.2}$$

The modification to the Hamiltonian in the vacuum—see *Bk. I: (11.1.49)*—does not affect magnetic interaction in such crystals, but the dependence of the dielectric constant $\varepsilon_{ij}(\boldsymbol{E})$ on the electric field gives rise to terms in the Hamiltonian of order 3, 4 or higher in $\boldsymbol{E}$. The tensor form of the dielectric constant is expected in such a crystal, which in practice is normally anisotropic. In particular, this means that $\varepsilon_{ij}(\boldsymbol{E})$ can possess terms of any order in $\boldsymbol{E}$, since there are no considerations of symmetry which would rule this out.

b) Using a Non-Linear Crystal to Make the Squeezing Hamiltonian: It is instructive to look at the experimental setup first used to make squeezed light in this way, and this is illustrated in Fig. 21.3, which has the following features:

i) *Source Laser and Frequency Doubler*: The top left of Fig. 21.3 is an optically pumped ring laser, using the nonlinear Nd:YaG crystal, itself pumped by a more powerful pump laser, which would nowadays be a diode laser, and which is not shown in the diagram. The other crystal, of $Ba_2NaNb_5O_{15}$, is also nonlinear, and acts as a frequency doubler. It converts a small fraction of the laser light, of wavelength 1.06μm (shown as a green line) to a wavelength of 0.53μm (shown as a red dashed line). The ring cavity mirror on the bottom right is essentially transparent at this frequency.

   The output from this device is therefore at frequencies ω_0 and $2\omega_0$, and these two signals are phase coherent with each other.

ii) *Polarizing Beam Splitter*: The main function here is to divert the signal of wavelength 1.06μm for later use in the detection process, leaving only the frequency-doubled signal travel on to the optical parametric oscillator.

iii) *Optical Parametric Oscillator*: This is the central device, consisting of an optical cavity and a non-linear crystal composed of $Mg:LiNbO_3$. This has a third-order nonlinearity. The 0.53μm pump beam enters through the mirror M_1 which is partially transmitting at this wavelength, and is totally reflected by the mirror M_2. This signal is sufficiently strong to be considered classical; on the other hand no signal at $1.06\mu m$ enters, so the field at this wavelength is treated quantum mechanically.

The third order term yields resonant terms which correspond to the Hamiltonian (21.2.1), which correspond to the reversible production of two $1.06\mu m$ photons from on $0.53\mu m$ photon, provided by the classical field. From the point of view of an experimenter, this device is an oscillator, hence the terminology *optical parametric oscillator*. However, it can also be looked upon as an amplifier—a device which amplifies the vacuum fluctuations in one quadrature, and attenuates them in the other quadrature, as we have already seen in Sect. 21.1.4.

iv) *Outputs*: There are now two signals at $1.06\mu m$, the diverted output from the ring laser, and the signal produced by the *parametric down conversion*, and these are phase-coherent. The phase difference θ between the two signals when they meet again can be controlled by adjusting the path length, and this is done by moving the mirror labelled θ over distances of the order of a wavelength.

v) *Balanced Homodyne Detection*: This follows the methods we have described in Sect. 20.1.3.

21.2.2 Quantum Langevin Equation

We can use the formalism of quantum Langevin equations to write an equation of motion for the field mode $a(t)$ inside the cavity in terms of the input $b_{\text{in}}(t)$ entering through the partially reflecting mirror M_1. We work in a rotating frame, writing $\tilde{a}(t) \equiv a(t)e^{-i\omega_0 t}$, and the quantum Langevin equation is

$$\frac{d\tilde{a}(t)}{dt} = \epsilon\tilde{a}(t)^\dagger - \tfrac{1}{2}\gamma\tilde{a}(t) - \sqrt{\gamma}\, b_{\text{in}}(t). \tag{21.2.3}$$

This corresponds exactly to (21.1.12) for the equations of motion arising from the squeezing Hamiltonian, apart from the addition of damping and quantum noise.

To solve this equation we introduce *noise quadrature phase operators* $\bar{q}_{\text{in}}(t)$ and $\bar{p}_{\text{in}}(t)$, defined by

$$\frac{\bar{q}_{\text{in}}(t) + i\bar{p}_{\text{in}}(t)}{\sqrt{2}} \equiv e^{i\theta/2} b_{\text{in}}(t), \tag{21.2.4}$$

in correspondence with the definition of quadrature phase operators in (21.1.15). These satisfy the commutation relation

$$[\bar{q}_{\text{in}}(t), \bar{p}_{\text{in}}(t')] = i\delta(t-t'). \tag{21.2.5}$$

a) Quadrature Phase Ito Calculus: In order to write these as quantum stochastic differential equations, we write

$$\bar{q}_{\text{in}}(t)\,dt = dQ(t), \qquad \bar{p}_{\text{in}}(t)\,dt = dP(t). \tag{21.2.6}$$

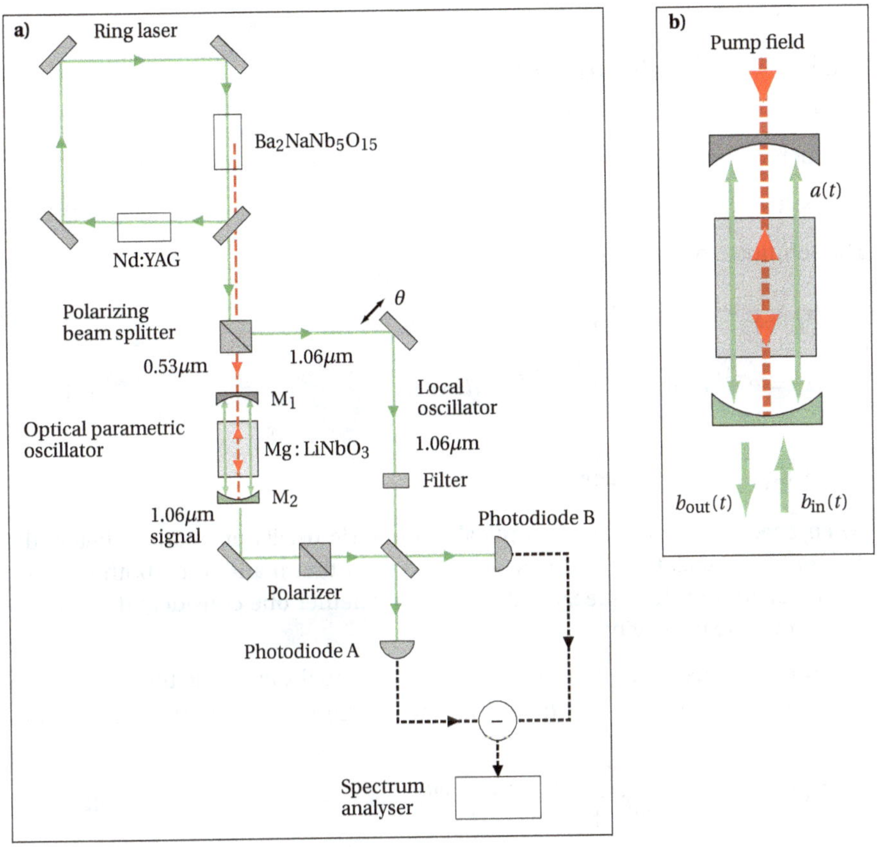

Fig. 21.3. The degenerate parametric amplifier. **a)** The experimental arrangement used by [21.4]. **b)** Enlargement of the essential features, with notation relevant to the description by (21.2.3).

In the case of an input vacuum state, the usual quantum Ito rules of (9.3.21) require that the non-zero products take the form

$$dQ(t)^2 = dP(t)^2 = \tfrac{1}{2}\,dt, \tag{21.2.7}$$

$$dQ(t)\,dP(t) = -dP(t)\,dQ(t) = -\tfrac{1}{2}\mathrm{i}\,dt. \tag{21.2.8}$$

These are identical to the rules given in (20.A.12).

b) Quantum Stochastic Differential Equations and their Solutions: The quantum Langevin equations then are equivalent to the quantum stochastic differential equations for the quadrature phase operators $\bar{X}(t)$ and $\bar{Y}(t)$ we defined in

(21.1.15)

$$d\bar{X}(t) = -\mu\bar{X}(t)\,dt - \sqrt{\gamma}\,dQ(t), \tag{21.2.9}$$

$$d\bar{Y}(t) = -\lambda\bar{Y}(t)\,dt - \sqrt{\gamma}\,dP(t), \tag{21.2.10}$$

in which we have defined

$$\mu = \tfrac{1}{2}\gamma - |\epsilon|, \qquad \lambda = \tfrac{1}{2}\gamma + |\epsilon|. \tag{21.2.11}$$

The solutions to these equations are

$$\bar{X}(t) = e^{-\mu t}\bar{X}(0) - \sqrt{\gamma}\int_0^t e^{-\mu(t-t')}\,dQ(t'), \tag{21.2.12}$$

$$\bar{Y}(t) = e^{-\lambda t}\bar{Y}(0) - \sqrt{\gamma}\int_0^t e^{-\lambda(t-t')}\,dP(t'). \tag{21.2.13}$$

21.2.3 Squeezing Produced

When used as a squeezer, the optical parametric oscillator operates below the threshold of oscillation, that is, with $|\epsilon| < \frac{1}{2}\gamma$, so that μ and λ are both positive. The behaviour of the squeezing depends on whether one considers the field inside or outside the cavity.

a) Squeezing Inside the Cavity: If we take the initial state to be the vacuum, for which $\Delta\bar{X}(0)^2 = \Delta\bar{Y}(0)^2 = \frac{1}{2}$, then from the solutions (21.2.12, 21.2.13), the variances are

$$\Delta\bar{X}(t)^2 = \frac{\gamma}{2(\gamma-2|\epsilon|)} - \frac{|\epsilon|}{\gamma-2|\epsilon|}e^{-(\gamma-2|\epsilon|)t}, \tag{21.2.14}$$

$$\Delta\bar{Y}(t)^2 = \frac{\gamma}{2(\gamma+2|\epsilon|)} + \frac{|\epsilon|}{\gamma+2|\epsilon|}e^{-(\gamma+2|\epsilon|)t}. \tag{21.2.15}$$

Clearly, ΔX increases with time, while ΔY decreases with time, so the qualitative behaviour is that of squeezing. However, the uncertainty product

$$\Delta X\Delta Y = \sqrt{\frac{\gamma^2 - 4\gamma|\epsilon|e^{-\gamma t}\sinh|\epsilon|t - 4|\epsilon|^2 e^{-2\gamma t}}{4(\gamma^2-4|\epsilon|^2)}}, \tag{21.2.16}$$

is an increasing function of time, so that the quality of the squeezing becomes less as time progresses. The stationary values are

$$\Delta X_s = \sqrt{\frac{\gamma}{2(\gamma-2|\epsilon|)}}, \qquad \Delta Y_s = \sqrt{\frac{\gamma}{2(\gamma+2|\epsilon|)}}, \tag{21.2.17}$$

and show that $\Delta Y_s = \frac{1}{2}$ is the minimum possible value of ΔY_s reached, and this is at the threshold of oscillation, $\gamma = 2|\epsilon|$, where $\Delta X_s \to \infty$.

We illustrate the situation in Fig. 21.4; the conclusion is that for any γ, no matter how small, the squeezing achievable in X and Y is only half the value achievable when $\gamma = 0$.

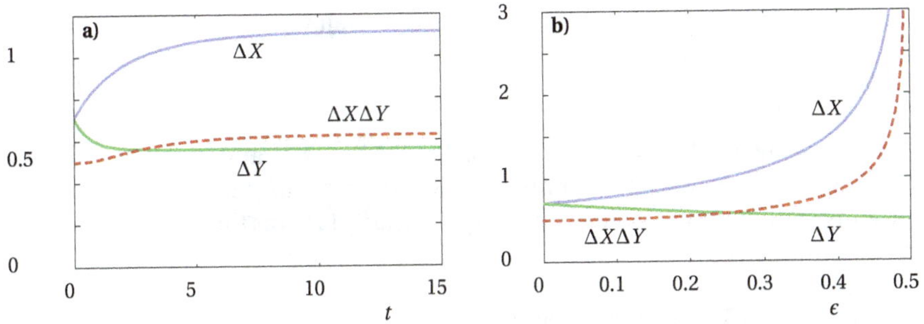

Fig. 21.4. Plots of the uncertainties ΔX and ΔY based on (21.2.14, 21.2.15), in which we set $\gamma = 1$. **a)** The progress of squeezing with time t for $\epsilon = 0.3$. **b)** The stationary values obtainable as a function of the parametric gain ϵ.

b) Squeezing in the Outputs: In terms of the quadrature phase fields, the input-output relation (11.2.6) becomes

$$\bar{q}_{\text{out}}(t) = \bar{q}_{\text{in}}(t) + \sqrt{\gamma}\,\bar{X}(t), \qquad \bar{p}_{\text{out}}(t) = \bar{p}_{\text{in}}(t) + \sqrt{\gamma}\,\bar{Y}(t). \tag{21.2.18}$$

To identify squeezing in the output field, we consider the normalized time integrated output operators defined by

$$Q_{\text{out}} \equiv \frac{1}{\sqrt{T}}\int_0^T q_{\text{out}}(t), \qquad P_{\text{out}} \equiv \frac{1}{\sqrt{T}}\int_0^T p_{\text{out}}(t), \tag{21.2.19}$$

which have canonical commutation relations, thus

$$\left[Q_{\text{out}}, P_{\text{out}}\right] = \mathrm{i}. \tag{21.2.20}$$

We now use the input-output relations (21.2.18) and the solutions (21.2.12) and (21.2.13). After interchanging the orders of integration we find

$$Q_{\text{out}} = \frac{1}{\sqrt{T}}\frac{\sqrt{\gamma}}{\mu}\left(1-\mathrm{e}^{-\mu T}\right)\bar{X}(0) + \frac{1}{\sqrt{T}}\int_0^T \left\{\left(1-\frac{\gamma}{\mu}\right) - \frac{\gamma}{\mu}\mathrm{e}^{-\mu(T-t)}\right\} dQ(t), \tag{21.2.21}$$

$$P_{\text{out}} = \frac{1}{\sqrt{T}}\frac{\sqrt{\gamma}}{\lambda}\left(1-\mathrm{e}^{-\lambda T}\right)\bar{X}(0) + \frac{1}{\sqrt{T}}\int_0^T \left\{\left(1-\frac{\gamma}{\lambda}\right) - \frac{\gamma}{\lambda}\mathrm{e}^{-\lambda(T-t)}\right\} dP(t). \tag{21.2.22}$$

The uncertainties in these can easily be worked out using the Ito rules (21.2.7, 21.2.8), but we can get an understanding of the situation more simply by considering the limit of very large T. In this limit the transient behaviour, characterized by the time scales $1/\mu$ and $1/\lambda$, becomes unimportant. The only terms which do not become negligible are the first two terms in each of the integrals, so that we can say

$$Q_{\text{out}} \longrightarrow \frac{1}{\sqrt{T}}\left(1-\frac{\gamma}{\mu}\right)\int_0^T dQ(t), \qquad P_{\text{out}} \longrightarrow \frac{1}{\sqrt{T}}\left(1-\frac{\gamma}{\lambda}\right)\int_0^T dP(t). \tag{21.2.23}$$

Substituting for μ and λ using (21.2.11), the uncertainties are then given by

$$\Delta Q_{\text{out}} = \frac{1}{\sqrt{2}}\left(\frac{\gamma+2|\epsilon|}{\gamma-2|\epsilon|}\right), \qquad \Delta P_{\text{out}} = \frac{1}{\sqrt{2}}\left(\frac{\gamma-2|\epsilon|}{\gamma+2|\epsilon|}\right). \tag{21.2.24}$$

This represents perfect squeezing, since $\Delta Q_{\text{out}}\Delta P_{\text{out}} = \frac{1}{2}$, and as $\gamma \to 2|\epsilon|$ we find $\Delta P_{\text{out}} \to 0$. Thus, in the large T limit, the output quantum state is a minimum uncertainty state, with squeezing becoming arbitrarily large as the oscillation threshold is approached.

Exercise 21.5 Exact Calculations of Squeezing: If we do not take the limit of large T, transient effects are not negligible.

i) Derive the formulae (21.2.21) and (21.2.22).

ii) Show that the exact results for squeezing are

$$\Delta Q_{\text{out}}^2 = \left(1-\frac{\gamma}{\mu}\right)^2 + \frac{\gamma}{\mu^2 T}\left(1-e^{-\mu T}\right)^2 \langle X(0)^2\rangle$$
$$-\frac{2\gamma}{\mu^2 T}\left(1-\frac{\gamma}{\mu}\right)\left(1-e^{-\mu T}\right) + \frac{\gamma^2}{2\mu^3 T}\left(1-e^{-2\mu T}\right), \tag{21.2.25}$$

with a similar result for ΔP_{out}.

21.2.4 Input-Output View of Squeezing and the Ideal Phase-Sensitive Amplifier

The formulae (21.2.23) are most naturally viewed as a description of an ideal phase-sensitive amplifier; we can define input operators in a way similar to (21.2.19)

$$Q_{\text{in}} \equiv \frac{1}{\sqrt{T}}\int_0^T q_{\text{in}}(t), \qquad P_{\text{in}} \equiv \frac{1}{\sqrt{T}}\int_0^T p_{\text{in}}(t), \tag{21.2.26}$$

and the relationships (21.2.23) can be written

$$Q_{\text{out}} = G Q_{\text{in}}, \qquad P_{\text{out}} = G^{-1} P_{\text{in}}, \tag{21.2.27}$$

where G is defined by

$$G \equiv \frac{\gamma+2|\epsilon|}{\gamma-2|\epsilon|}. \tag{21.2.28}$$

In this form the existence of perfect squeezing in the output fields is seen to be an inevitable consequence of the linearity of the transformation between input and output, and the preservation of commutation relations. The equations of motion, when expressed in terms of quadrature phases as in (21.2.9, 21.2.10), necessarily require the asymptotic output operators as defined in (21.2.23) to be proportional to their input counterparts, and the commutation relations can be preserved only if this relationship takes the form (21.2.27).

a) Perfect Phase-Sensitive Amplification: This form also makes it clear that Q_{out} is an amplified version of Q_{in}, while P_{out} is an attenuated version of P_{in}, and the amplification is perfect—there is no noise impressed on the amplified signal. This is in contrast to the kinds of amplifier we considered in *Bk. I: Sect.16.2*, which were not phase sensitive, and where we found that amplification always adds noise. We will show the connection between these two kinds of amplification in Sect. 21.5.

b) Formulation in Terms of Creation and Destruction Operators: Writing

$$A_{in} \equiv \frac{Q_{in} + iP_{in}}{\sqrt{2}}, \qquad A_{out} \equiv \frac{Q_{out} + iP_{oot}}{\sqrt{2}}, \tag{21.2.29}$$

the amplifier formula (21.2.27) can alternatively be written as

$$A_{out} = \mathcal{M}A_{in} + \mathcal{L}A_{in}^{\dagger}, \qquad \mathcal{M} \equiv \tfrac{1}{2}\left(G + G^{-1}\right), \quad \mathcal{L} \equiv \tfrac{1}{2}\left(G - G^{-1}\right). \tag{21.2.30}$$

21.3 Two-Mode Squeezing and Correlated Quanta

We will now consider the case of a non-degenerate parametric amplifier, where two distinguishable modes of the optical field are involved. We suppose that two systems A and B are localized modes of the electromagnetic field, with frequencies $\omega_{A,B}$ and boson operators a and b respectively. If we use a pump laser at frequency $\omega_A + \omega_B$, and a nonlinear optical crystal which is phase-matched at these wavelengths, the appropriate squeezing Hamiltonian can be written, in the interaction picture, as

$$H_I = i\hbar|\epsilon|\left(a^{\dagger}b^{\dagger} - ab\right). \tag{21.3.1}$$

The Heisenberg equations of motion for this Hamiltonian are

$$\dot{a} = |\epsilon|b^{\dagger}, \qquad \dot{b} = |\epsilon|a^{\dagger}, \tag{21.3.2}$$

with solutions

$$\left.\begin{aligned} a(t) &= \cosh(|\epsilon|t)a(0) + \sinh(|\epsilon|t)b^{\dagger}(0), \\ b(t) &= \cosh(|\epsilon|t)b(0) + \sinh(|\epsilon|t)a^{\dagger}(0). \end{aligned}\right\} \tag{21.3.3}$$

If the initial state is the vacuum of both modes $|0,0\rangle$ this interaction generates two-mode squeezed light, which corresponds to a quantum state in the Schrödinger picture of

$$|\psi\rangle = \sum_{n=0}^{\infty} c_n |n\rangle_A |n\rangle_B, \tag{21.3.4}$$

$$c_n \equiv \tanh^n r \,\mathrm{sech}\, r, \qquad r \equiv |\epsilon|t. \tag{21.3.5}$$

i) *The Squeezing Parameter*: Using a similar convention as for single-mode squeezing, the parameter $r \equiv |\epsilon|t$ is called the squeezing parameter.

ii) The expansion in terms of number states is an example of a *Schmidt decomposition* as described in Sect. 3.1.4. There are an infinite number of Schmidt coefficients, so this state is entangled.

Exercise 21.6 Two-Mode Squeezed State in Terms of Number States: Use a similar method to that used in Ex. 21.1 to prove the formula (21.3.5).

21.3.1 Quadrature Phases

In the case of two-mode squeezing we define quadrature phase variables by

$$a = \frac{X^A + \mathrm{i}Y^A}{\sqrt{2}}, \qquad b = \frac{X^B + \mathrm{i}Y^B}{\sqrt{2}}, \tag{21.3.6}$$

with the commutation relations and Heisenberg uncertainty relations given by

$$[X^A, Y^A] = [X^B, Y^B] = \mathrm{i}, \qquad \Delta X^A \Delta Y^A \geqslant \tfrac{1}{2}, \qquad \Delta X^B \Delta Y^B \geqslant \tfrac{1}{2}. \tag{21.3.7}$$

The time-dependent solutions for these quadrature phases follow from the solutions (21.3.3), namely

$$\left.\begin{aligned} X^A(t) &= \tfrac{1}{2}\big(X^A(0)+X^B(0)\big)\mathrm{e}^r + \tfrac{1}{2}\big(X^A(0)-X^B(0)\big)\mathrm{e}^{-r},\\ X^B(t) &= \tfrac{1}{2}\big(X^A(0)+X^B(0)\big)\mathrm{e}^r - \tfrac{1}{2}\big(X^A(0)-X^B(0)\big)\mathrm{e}^{-r},\\ Y^A(t) &= \tfrac{1}{2}\big(Y^A(0)-Y^B(0)\big)\mathrm{e}^r + \tfrac{1}{2}\big(Y^A(0)+Y^B(0)\big)\mathrm{e}^{-r},\\ Y^B(t) &= -\tfrac{1}{2}\big(Y^A(0)-Y^B(0)\big)\mathrm{e}^r + \tfrac{1}{2}\big(Y^A(0)+Y^B(0)\big)\mathrm{e}^{-r}. \end{aligned}\right\} \tag{21.3.8}$$

a) Sum and Difference Modes: If we define

$$X^{\pm} \equiv \frac{X^A \pm X^B}{\sqrt{2}}, \qquad Y^{\pm} \equiv \frac{Y^A \pm Y^B}{\sqrt{2}}. \tag{21.3.9}$$

The solutions (21.3.8) become like single-mode squeezing

$$X^+(t) = X^+(0)\mathrm{e}^r, \qquad Y^+(t) = Y^+(0)\mathrm{e}^{-r}, \tag{21.3.10}$$

$$X^-(t) = X^-(0)\mathrm{e}^{-r}, \qquad Y^-(t) = Y^-(0)\mathrm{e}^{r}. \tag{21.3.11}$$

The squeezing in the two-mode system is equivalent to single-mode squeezing happening in the opposite quadratures of the sum and difference modes.

b) Quadrature Phase Squeezing: As $r \to \infty$, we can see from (21.3.8) that

$$X^A \longrightarrow X^B, \qquad Y^A \longrightarrow -Y^B. \tag{21.3.12}$$

Thus, the quadrature phases X^A and X^B become perfectly correlated, and the quadrature phases Y^A and Y^B become perfectly anticorrelated.

21.4 Quantum Limited Phase-Insensitive Amplifiers

We want now to consider the effect of coupling to the external electromagnetic field, which need not be the same for the two modes. This provides two models of *phase-insensitive* amplifiers, which are in fact of practical uses.

When we couple to inputs and outputs in the two modes the simple equations of motion (21.3.2) become the quantum Langevin equations

$$\frac{da(t)}{dt} = |\epsilon| b(t)^\dagger - \tfrac{1}{2}\gamma_a a(t) - \sqrt{\gamma_a}\, s_{\text{in}}(t), \tag{21.4.1}$$

$$\frac{db(t)}{dt} = |\epsilon| a(t)^\dagger - \tfrac{1}{2}\gamma_b b(t) - \sqrt{\gamma_b}\, i_{\text{in}}(t), \tag{21.4.2}$$

where $s_{\text{in}}(t)$ and $i_{\text{in}}(t)$ are two independent inputs, with the usual commutation relations $[s_{\text{in}}(t), s_{\text{in}}^\dagger(t')] = [i_{\text{in}}(t), i_{\text{in}}^\dagger(t')] = \delta(t-t')$. The symbols for the two input fields are chosen mnemonically; the *signal* field $s_{\text{in}}(t)$ is expected to carry the input signal, which we may consider amplifying, while the *idler* field $i_{\text{in}}(t)$ will normally be taken to correspond to the vacuum.

Rather than consider the most general case, we will consider the two extreme situations:

i) *Equal Coupling*: The internal modes a and b couple identically to their respective inputs and outputs.

ii) *Only One Mode Coupled*: Only the mode $a(t)$ couples to inputs and outputs.

21.4.1 Identical Couplings to the Input and Output

Here we will set $\gamma_a = \gamma_b \equiv \gamma$, and we will find that the system is equivalent to two sets of degenerate parametric amplifiers.

a) Quantum Langevin Equations in Terms of Sum and Difference Quadrature Phases: In terms of $X^\pm$, $Y^\pm$, these equations become

$$\left.\begin{aligned} \frac{dX^+(t)}{dt} &= -\mu X^+(t) - \sqrt{\gamma}\, q_{\text{in}}^+(t), \\ \frac{dY^+(t)}{dt} &= -\lambda Y^+(t) - \sqrt{\gamma}\, p_{\text{in}}^+(t), \end{aligned}\right\} \tag{21.4.3}$$

$$\left.\begin{aligned} \frac{dX^-(t)}{dt} &= -\lambda X^+(t) - \sqrt{\gamma}\, q_{\text{in}}^-(t), \\ \frac{dY^-(t)}{dt} &= -\mu Y^-(t) - \sqrt{\gamma}\, p_{\text{in}}^-(t), \end{aligned}\right\} \tag{21.4.4}$$

in which μ and λ are defined in the same way as for the degenerate parametric amplifier, namely by (21.2.11).

The inputs here are defined in terms of the signal and idler fields as

$$q_{\text{in}}^\pm(t) \equiv \frac{s_{\text{in}}(t) + s_{\text{in}}^\dagger(t)}{2} \pm \frac{i_{\text{in}}(t) + i_{\text{in}}^\dagger(t)}{2}, \tag{21.4.5}$$

$$p_{\text{in}}^\pm(t) \equiv \frac{s_{\text{in}}(t) - s_{\text{in}}^\dagger(t)}{2\text{i}} \pm \frac{i_{\text{in}}(t) - i_{\text{in}}^\dagger(t)}{2\text{i}}. \tag{21.4.6}$$

These form two sets of independent inputs; that is

$$\left[q_{\text{in}}^+(t)\,,\, p_{\text{in}}^+(t)\right] = \left[q_{\text{in}}^-(t)\,,\, p_{\text{in}}^-(t)\right] = \text{i}\delta(t-t'), \tag{21.4.7}$$

and all other commutators are zero.

b) Interpretation of the Quantum Langevin Equations: The two sets of equations (21.4.3, 21.4.4) are completely independent of each other, and both are identical in form to the quadrature phase quantum stochastic differential equations (21.2.9, 21.2.10) for the degenerate parametric amplifier—but the second pair has exchanged λ and μ. This system is equivalent *mathematically* to two independent degenerate parametric amplifiers, with the gain in the X^+ and the Y^- quadratures.

c) A Quantum-Limited Amplifier: We can take over the kind of treatment we used in Sect. 21.2.4 and apply it separately to the two independent amplifier systems, composed of the symmetric components X^+, Y^+, and the antisymmetric components X^-, Y^-.

We will introduce therefore the input and output operators in the same way as in (21.2.19) and (21.2.26), and will find the input-output relation

$$Q^+_{\text{out}} = GQ^+_{\text{in}}, \qquad P^+_{\text{out}} = G^{-1}P^+_{\text{in}}, \tag{21.4.8}$$

$$Q^-_{\text{out}} = G^{-1}Q^-_{\text{in}}, \qquad P^-_{\text{out}} = GP^-_{\text{in}}, \tag{21.4.9}$$

where the gain is again given by the formula

$$G \equiv \frac{\gamma + 2|\epsilon|}{\gamma - 2|\epsilon|}. \tag{21.4.10}$$

The exchange of $\lambda \leftrightarrow \mu$ in the equations (21.4.3, 21.4.4) leads to the exchange $G \leftrightarrow G^{-1}$ when going from the top two equations to the bottom two equations.

The physical signal and idler input-output quadrature phase operators are to be defined in terms of $s_{\text{in}}(t)$ and $i_{\text{in}}(t)$, and we will define

$$S_{\text{in}} \equiv \frac{1}{\sqrt{T}}\int_0^T s_{\text{in}}(t)\,dt = \frac{S^Q_{\text{in}} + \mathrm{i}S^P_{\text{in}}}{\sqrt{2}}, \tag{21.4.11}$$

$$I_{\text{in}} \equiv \frac{1}{\sqrt{T}}\int_0^T i_{\text{in}}(t)\,dt = \frac{I^Q_{\text{in}} + \mathrm{i}I^P_{\text{in}}}{\sqrt{2}}, \tag{21.4.12}$$

with corresponding definitions for the output operators.

In terms of the signal and idler quadrature phases, the amplifier input-output relationships for the signal become

$$S^Q_{\text{out}} = \mathcal{G}\,S^Q_{\text{in}} + \mathcal{F}\,I^Q_{\text{in}}, \qquad S^P_{\text{out}} = \mathcal{G}\,S^P_{\text{in}} - \mathcal{F}\,I^P_{\text{in}}, \tag{21.4.13}$$

in which

$$\mathcal{G} \equiv \tfrac{1}{2}\left(G + G^{-1}\right), \qquad \mathcal{F} \equiv \tfrac{1}{2}\left(G - G^{-1}\right), \tag{21.4.14}$$

and

$$\mathcal{G} = \frac{\gamma^2 + 4|\epsilon|^2}{\gamma^2 - 4|\epsilon|^2}, \qquad \mathcal{F} = \frac{4\gamma|\epsilon|}{\gamma^2 - 4|\epsilon|^2}, \qquad \mathcal{G}^2 - \mathcal{F}^2 = 1. \tag{21.4.15}$$

d) Signal, Gain and Noise: Clearly $\mathcal{G}$ is a *gain coefficient* and $\mathcal{F}$ is a *noise coefficient*. We can also combine the quadrature phases to write the input-output relation in terms of the creation and destruction operators as

$$S_{\text{out}} = \mathcal{G}\, S_{\text{in}} + \mathcal{F}\, I_{\text{in}}^{\dagger}. \tag{21.4.16}$$

In the model of amplification presented here, the signal input passes into the amplifying medium, is amplified, and then the amplified signal passes into the output. Provided the frequencies involved are sufficiently low this can be a model of a continuously amplified signal.

The component of the idler input is inevitable, and this degrades the fidelity of the amplified signal. In fact we would expect from the results of *Bk. I: Sect.19.4* (even though the system studied is not the same) that this added noise would destroy any kind of coherence associated with superposition states of the input. The amount of noise added in this model is in fact minimal—we shall show in Sect. 21.5 that the relationship between input, amplified output and noise given in (21.4.16) $\mathcal{G}^2 - \mathcal{F}^2 \geqslant 1$, so that this represents the absolute minimum value of added noise.

21.4.2 Input and Output Coupled to Only One Mode

We now consider the situation in which $\gamma_a = 0$ and $\gamma_b = \gamma$. The signal input field $s_{\text{in}}(t)$ no longer plays any part in the equations of motion (21.4.1, 21.4.2), which can be reduced to

$$\frac{d^2 a(t)}{dt^2} = |\epsilon|^2 a(t) - \tfrac{1}{2}\gamma \frac{da(t)}{dt} - \sqrt{\gamma}|\epsilon|\, i_{\text{in}}^{\dagger}(t), \tag{21.4.17}$$

$$b(t) = \frac{1}{|\epsilon|}\frac{da^{\dagger}(t)}{dt}. \tag{21.4.18}$$

a) Approximate Quantum Stochastic Differential Equation: These equations can easily be solved exactly in the frequency domain, but an analysis of behaviour in the low frequency domain is very informative, and can be done by noting that for frequencies $\omega \ll \gamma$ the second order derivative term can be neglected in comparison to the first order derivative, and (21.4.18) can be written in the form of a quantum stochastic differential equation

$$da(t) = \tfrac{1}{2}\Gamma a(t)\, dt - \sqrt{\Gamma}\, dB^{\dagger}(t), \qquad \Gamma \equiv \frac{4|\epsilon|^2}{\gamma}. \tag{21.4.19}$$

Here we have made the mapping to the conventional notation

$$i_{\text{in}}^{\dagger}(t)\, dt \longrightarrow dB^{\dagger}(t). \tag{21.4.20}$$

The idler field $i_{\rm in}(t)$ is taken to be in the vacuum state, so $dB(t)$ and $dB^\dagger(t)$ will obey the usual quantum Ito rules (9.3.21) for a vacuum input.

b) The Ideal Amplifier Quantum Stochastic Differential Equation: This kind of quantum stochastic differential equation, in which the noise input is the field creation operator, does not occur naturally as a result of the procedures we developed in Chap. 9 and Chap. 11. In fact the essential difference is best seen in the Hermitian conjugate equation, in which the *creation* operator is seen to couple to the increment $dB(t)$. This means that the excitation in the cavity mode described by $a(t)$ *increases* when a photon is emitted—a process which reflects the existence of the pump field necessary to produce a Hamiltonian of the form (21.3.1). This kind of equation also arises when considering Raman processes, as we show in Sect. 28.3.5.

c) Master Equation: There will be a corresponding master equation

$$\dot\rho = \tfrac{1}{2}\Gamma\left(2a^\dagger\rho a - \rho a a^\dagger - a a^\dagger\rho\right). \tag{21.4.21}$$

This is a master equation for a system with gain, but no loss—we formulated this in *Bk. I: Sect.14.3.1*, and termed the system an *ideal amplifier* in *Bk. I: Sect.16.2*, where a treatment of its solutions was given using Wigner function formalism. We have also studied it in the context of measurement, gain and decoherence in *Bk. I: Sect.19.4*.

d) Solutions of the Quantum Stochastic Differential Equations: The solution of the quantum stochastic differential equation (21.4.17) can be written

$$a(t) = \mathcal{G}(t)a(0) + \mathcal{F}(t)I^\dagger(t), \tag{21.4.22}$$

$$\mathcal{G}(t) = e^{\Gamma t/2}, \qquad \mathcal{F}(t) = \sqrt{e^{\Gamma t}-1}\,, \qquad \mathcal{G}(t)^2 - \mathcal{F}(t)^2 = 1. \tag{21.4.23}$$

The operator $I^\dagger(t)$ is a noise term

$$I^\dagger(t) \equiv \frac{1}{\sqrt{e^{\Gamma t}-1}}\int_0^t e^{\Gamma(t-t')/2}\,dB^\dagger(t'), \tag{21.4.24}$$

and the Ito rules and independence of $a(0)$ and the increment $dB^\dagger(t)$ for $t>0$ mean that the noise operators satisfy the commutation relation $\left[I(t), I^\dagger(t)\right] = 1$, and that they commute with $a(0)$ and $a^\dagger(0)$.

e) Interpretation: In this model of an amplifier, the initial value $a(0)$ of the mode inside the amplifying medium is amplified for a finite time to produce the value after a time t. The object of interest is the cavity mode $a(t)$. This contrasts with the situation in Sect. 21.4.1, in which the amplification was formulated in terms of the input and output fields to the cavity, and the dynamics inside the cavity was *overdamped*, that is $\gamma > 2|\epsilon|$.

Nevertheless, the equations (21.4.16) and (21.4.23) which relate the amplification process are of exactly the same form.

f) Application to the Atomic Ensemble Quantum Memory: This kind of system arises naturally in the context of atomic ensembles, which we treat in Sect. 28.3.5.

21.5 Quantum Limits on Noise in Linear Amplifiers

In 1982 *Caves* [21.7] introduced a definitive description of the relationship between amplifier noise and the fundamental requirements of quantum mechanics. The starting point is to consider a single-mode system in which the input operators are mapped linearly to the output operators, both of which have Bosonic commutation relations. Thus, the amplifier relates output to input through the formula

$$a_{\text{out}} = \mathcal{M}a_{\text{in}} + \mathcal{L}a_{\text{in}}^\dagger + \mathcal{K}. \tag{21.5.1}$$

Her $\mathcal{L}$ and $\mathcal{M}$ are numerical coefficients, while $\mathcal{K}$ is an operator which does not depend on a_{in} or $a_{\text{in}}^\dagger$. The formula is therefore linear in the creation and destruction operators, or equivalently, the corresponding quadrature phase operators.

The operator $\mathcal{K}$ represents added noise, which is required to be independent of and commute with the signal operators. From this independence, it follows that $\mathcal{K}$ satisfies the commutation relation

$$[\mathcal{K}, \mathcal{K}^\dagger] = 1 - |\mathcal{M}|^2 + |\mathcal{L}|^2. \tag{21.5.2}$$

If this commutator is positive, $\mathcal{K}$ is proportional to a Bosonic destruction operator, if it is negative $\mathcal{K}$ is proportional to a Bosonic creation operator.

This model is consistent with all of the amplifiers we have considered in this chapter, namely the ideal phase-sensitive amplifier of Sect. 21.2.4, as written in the form (21.2.29, 21.2.30), and the two models of an ideal amplifier, namely that described by (21.4.13–21.4.15), and that described by (21.4.22, 21.4.23).

21.5.1 Fundamental Limits on Added Quantum Noise—the Caves Amplifier Noise Bound

If we write $\mathcal{K}$ in terms of its Hermitian quadrature phase components, namely $\mathcal{K} = K_q + \mathrm{i}K_p$, then the uncertainty in $\mathcal{K}$ is represented by

$$\Delta\mathcal{K}^2 \equiv \Delta K_q^2 + \Delta K_p^2 \geqslant 2\Delta K_q \Delta K_p, \tag{21.5.3}$$

$$\geqslant |\langle [K_q, K_p] \rangle| = \tfrac{1}{2}|\langle [\mathcal{K}, \mathcal{K}^\dagger] \rangle|. \tag{21.5.4}$$

In going from the first to the second line we have used the Heisenberg uncertainty principle in the form (21.1.3) for the operators K_q and K_p.

From the commutator (21.5.2) it can be seen that the noise fluctuations satisfy the *Caves amplifier noise bound*:

$$\Delta\mathcal{K}^2 \geqslant \tfrac{1}{2}\left|1 - |\mathcal{M}|^2 + |\mathcal{L}|^2\right|. \tag{21.5.5}$$

From this formula one can deduce a number of consequences:

i) *Condition for a Noise-Free Amplifier*: The noise can only be zero if $|\mathcal{M}|^2 - |\mathcal{L}|^2 = 1$. We can take $\mathcal{M}$ and $\mathcal{L}$ to be real, since any phases can be absorbed into the definitions of the Boson operators—doing so, this yields the ideal phase-sensitive amplifier as in Sect. 21.2.4. Thus, the ideal phase-sensitive amplifier is the unique quantum linear amplifier which does not add noise.

   The disadvantage is that only one quadrature phase is amplified. If one wishes to amplify a quantum signal to a level at which it can be macroscopically measured, this method will only give half of the information. Nevertheless, if we are only interested in measuring one quadrature phase, this can be an enormous advantage.

ii) *The Ideal Phase Insensitive Amplifier*: If $\mathcal{L} = 0$, we have a phase insensitive amplifier if $|\mathcal{M}| > 1$ (otherwise this is an attenuator). The noise then satisfies

$$\frac{\Delta\mathcal{K}^2}{|\mathcal{M}|^2} \geqslant \tfrac{1}{2}\left(1 - \frac{1}{|\mathcal{M}|^2}\right) \longrightarrow \tfrac{1}{2} \quad \text{for sufficiently large } |\mathcal{M}|. \tag{21.5.6}$$

   In this equation $|\mathcal{M}|^2$ represents the *power gain* of the amplifier, while $\Delta\mathcal{K}^2$ represents the noise which is added to the *output*. The left-hand side of this equation therefore represents the size of the noise added to the *input* which would, when amplified, give the same value. This approaches the value 1/2 for any gain sufficiently large to be useful. In practice, this means that a *phase insensitive amplifier adds noise to the input of a size equivalent to at least one half of a quantum.*

iii) *Phase-Conjugating Amplifier*: When $\mathcal{M} = 0$, the amplifier is phase insensitive, in the sense that the magnitude of the gain is the same for both quadrature phases. In this case we find

$$\frac{\Delta\mathcal{K}^2}{|\mathcal{L}|^2} \geqslant \tfrac{1}{2}\left(1 + \frac{1}{|\mathcal{L}|^2}\right) \longrightarrow \tfrac{1}{2} \quad \text{for sufficiently large } |\mathcal{L}|. \tag{21.5.7}$$

   The phase-conjugate amplifier is always noisier than the phase preserving amplifier, but has essentially the same noise performance for sufficiently large gain.

21.5.2 Attenuators and Beam Splitters

There is no restriction that a device described by the linear amplifier formula (21.5.1) should in fact amplify. If the coefficients are small this describes an attenuator, and the same noise bound (21.5.5) applies. The most important case is the *phase-insensitive attenuator*, in which $\mathcal{L} = 0$ and $|\mathcal{M}| < 1$.

This can also be regarded as a beam splitter, or mixer. The noise can be regarded as an input, and represented in the form $\mathcal{K} = \mathcal{N} b_{\text{in}}$, where $\mathcal{N}$ is a coefficient, and

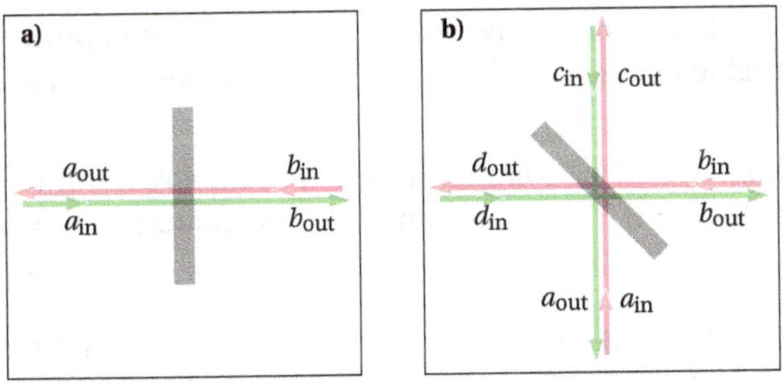

Fig. 21.5. Beam splitter configurations. **a)** The simple two-port configuration, for which the transfer matrix $\mathcal{M}$ is given by (21.5.10). **b)** The more common four-port configuration, in which the transfer matrix is given by (21.5.14). This splits into two sets of input-output relations, and if these are constructed to have the same transfer matrix, this transfer matrix has the same properties as that for the two-port configuration.

the two modes a_{in} and b_{in} are mixed thus:

$$\begin{pmatrix} a_{out} \\ b_{out} \end{pmatrix} = \begin{pmatrix} \mathcal{M} & \mathcal{N} \\ \mathcal{N}' & \mathcal{M}' \end{pmatrix} \equiv \mathcal{M} \begin{pmatrix} a_{in} \\ b_{in} \end{pmatrix}. \tag{21.5.8}$$

The matrix $\mathcal{M}$ must be unitary, and the coefficients $\mathcal{M}'$, $\mathcal{N}'$, which describe the output noise operator, must be determined so as to satisfy any physical constraints which may apply.

a) The Two-Port Beam Splitter: This is illustrated in Fig. 21.5 a. The form of the matrix $\mathcal{M}$ can be fixed as follows:

i) *Unitarity*: The matrix must be unitary, $\mathcal{M}\mathcal{M}^\dagger = 1$.

ii) *Time Reversal*: Electrodynamics satisfies time reversal invariance, and this will also apply to the beam splitter. Time reversal exchanges inputs and outputs, and complex conjugates the coefficients. This means

$$\mathcal{M}^* = \mathcal{M}^\dagger, \quad \text{and hence } \mathcal{M} \text{ is a symmetric matrix.} \tag{21.5.9}$$

iii) Taken together these mean that $\mathcal{M}$ can be written in terms of three coefficients $\mathbf{t}_a$, $\mathbf{t}_b$ and $\mathbf{r}$, in the form

$$\mathcal{M} = \begin{pmatrix} \mathbf{t}_a & \mathbf{r} \\ \mathbf{r} & \mathbf{t}_b \end{pmatrix}, \tag{21.5.10}$$

where the coefficients satisfy the conditions

$$\mathbf{r}^*\mathbf{t}_a + \mathbf{r}\,\mathbf{t}_b^* = 0, \tag{21.5.11}$$

$$|\mathbf{t}_a| = |\mathbf{t}_b| \equiv \mathbf{t}, \tag{21.5.12}$$

$$|\mathbf{r}|^2 + \mathbf{t}^2 = 1. \tag{21.5.13}$$

iv) Because $\mathbf{r}$, $\mathbf{t}_a$ and $\mathbf{t}_b$ are not necessarily real, there are phase changes on reflection and transmission. These cannot be absorbed into the definitions of the operators, since the relative phase between output and input is in principle a physically measurable quantity, and depends on the nature of the partially reflecting mirror being used.

b) Four-Port Configuration: It is more common to use a beam splitter in the four-port configuration of Fig. 21.5 b. The inputs and outputs separate into two disjoint sets, which have input-output relations

$$\begin{pmatrix} c_{\text{out}} \\ d_{\text{out}} \end{pmatrix} = \mathcal{M}_A \begin{pmatrix} a_{\text{in}} \\ b_{\text{in}} \end{pmatrix}, \qquad \begin{pmatrix} a_{\text{out}} \\ b_{\text{out}} \end{pmatrix} = \mathcal{M}_B \begin{pmatrix} c_{\text{in}} \\ d_{\text{in}} \end{pmatrix}. \tag{21.5.14}$$

Both $\mathcal{M}_B$ and $\mathcal{M}_B$ must be unitary, but time reversal gives the relationship

$$\mathcal{M}_B = \mathcal{M}_A^T \equiv \mathcal{M}. \tag{21.5.15}$$

This means that in general the beam splitter is characterized by the matrix $\mathcal{M}$, and the only restriction is that it be unitary.

However, in practice the beam splitter is constructed symmetrically, so that $\mathcal{M}_B = \mathcal{M}_A^T$ and then $\mathcal{M}$ must be symmetric as well as unitary, and this means it has the form given by (21.5.10).

VII Quantum Processing with Atoms, Photons and Phonons

Part VII: Quantum Processing with Atoms, Photons and Phonons

There are two fundamental elementary quantum systems, the harmonic oscillator, and the two-level atom. In this chapter we will describe the behaviour which arises when we couple these two elementary systems together, behaviour which has been fundamental in creating quantum devices. The result of this coupling is described by the *Jaynes–Cummings Model*, whose basic behaviour has already been described in *Bk. I: Sect.12.4*, and here we will develop the description in a form suitable for quantum devices.

The model can be realized physically in a number of ways, and in this part of the book we cover two of the most important physical implementations:

i) *Cavity Quantum Electrodynamics*: In Chap. 22 we cover the optical implementation, in which the two-level system comprises two levels of an atom (which is usually optically trapped), and the harmonic oscillator is provided by an optical resonator, usually in the form of an optical cavity. The interaction between the cavity mode and the two-level atom comes from the dipole interaction with the electromagnetic field.

ii) *Ion Trapping*: Here the two-level system is an ion, is trapped by an ion trap, and the harmonic oscillator is provided by the spatial motion of the ion in the trap. The interaction between the electronic levels of the ion and its centre of mass motion arises when the ion is illuminated with a laser, and comes from the recoil when a photon is emitted or absorbed, in much the same way as in laser cooling. Chap. 23 covers the theory of trapped ions, with a more detailed discussion of the quantum-mechanical details in Appendix 23.A.

Trapped ions have provided the best implementation of a quantum computer, and Chap. 24 is devoted to a presentation of the theory of their operation. Most of the chapter is devoted to the explanations of the operation of appropriate quantum gates. We give a detailed description of both the original Cirac–Zoller gate, and of the Mølmer–Sørensen gate. We also introduce the concept of a geometric phase gate, which is potentially fast and reliable, but as yet experimentally untested.

The three essential features of quantum control stated in Sect. 1.4, *initialization, quantum processing* and *measurement* are all implemented in the ion trap quantum computer. The initialization can be provided by *sideband cooling*, the processing by the *quantum gates*, and the measurement by using the very efficient detection procedure provided by the three-level procedure of Sect. 19.4.

22. Cavity Quantum Electrodynamics

While the idea of an atom interacting with the electromagnetic field is conceptually straightforward, simply aiming a focused laser beam at an atom gives only a comparatively weak coupling between the laser mode and the atom. To achieve the strongest coupling would require a beam matched to the electromagnetic field mode to which the relevant transition couples, and this is normally an electric dipole field. Hence the best coupling would be obtained by creating an *incoming* electric dipole field to impinge on the atom. So far, this has proved to be impractical.

Instead, the interesting strong coupling regime is achieved by improving the matching between the incoming light field and the atom by means of an optical resonator. The incoming light can be made to couple very well to an internal mode of the resonator, whose shape can be chosen to produce very strong coupling to the atom, which is located inside the resonator. This resonator is typically viewed as being some implementation of a Fabry–Pérot cavity, composed of two concave mirrors, as in Fig. 22.1. To be useful the mirrors must have extraordinarily high reflectivity, so that the field which builds up inside the cavity becomes very intense, and the mode must have a very narrow waist where the atom is located. The very high reflectivity mirrors mean that the coupling of the cavity mode to the external modes of the electromagnetic field can be relatively weak. The atom itself still couples significantly to modes other than the cavity mode, but the strength of the coupling to the cavity mode is enormously enhanced compared to the situation of light impinging on an atom in free space.

Because the system exhibits behaviour arising from the quantized nature of the electromagnetic field, the term *cavity quantum electrodynamics*, abbreviated as *CQED* [22.1] has been coined to describe the field. The related field of *circuit quantum electrodynamics* (which we cover in Part VIII), involves superconducting electric circuits and Josephson junctions, and is also often abbreviated as CQED. In this book we will understand the concept of "cavity quantum electrodynamics" to cover both fields, since, from a theoretical point of view, they are essentially identical.

22.1 The Jaynes–Cummings Model

The *Jaynes–Cummings* model [22.2], is normally conceived as a two-level system coupled to a quantized harmonic oscillator mode in the rotating-wave approximation, experimentally realized as in Fig. 22.1 by a single two-level atom interacting with a single mode of a high-Q optical cavity, which provides the harmonic oscillator. The first experimental observation of the effects predicted by the Jaynes–Cummings model were measured by *Rempe, Walther* and *Klein* [22.3] in 1987, using a microwave transition in rubidium, and a superconducting microwave cavity.

When the coupling of both the cavity mode and the atom to other modes is neglected, the dynamics of this system is described by an idealized Hamiltonian

$$H_{JC} \equiv \tfrac{1}{2}\hbar\omega_{eg}\sigma_z + \hbar\omega a^\dagger a + \hbar g(\boldsymbol{x})\left(a^\dagger\sigma^- + a\sigma^+\right). \tag{22.1.1}$$

Here the harmonic oscillator operators $a^\dagger$, a describe the quantized cavity mode of the electromagnetic field, and the Pauli operators $\sigma^\pm$ and σ_z describe the two-level atom. The Hamiltonian H_{JC} is the sum of three terms corresponding to:

i) The atom, simplified to a two-level system;

ii) The electromagnetic field, simplified to a single mode;

iii) The interaction term, which describes atom-field dipole interaction in terms of the coupling parameter $g(\boldsymbol{x})$, whose dependence on $\boldsymbol{x}$ is determined by the shape of the cavity mode.

22.1.1 Implementations of Cavity Quantum Electrodynamics

The rather generic nature of the Hamiltonian H_{JC} has led to its implementation in a wide variety of situations:

i) *Optical Regime Using Free Atoms*: Very small cavities with very high Q can be made using dielectric multilayers, and the atoms, after being laser cooled, can be simply dropped through the cavity, or can be trapped using an optical trap [22.1].

ii) *Rydberg Atoms and Microwaves*: In the microwave regime [22.4] transitions between very highly excited atomic states (known as Rydberg states) are coupled to a microwave field, confined by a cavity constructed from superconducting mirrors, whose losses can be made extraordinarily low.

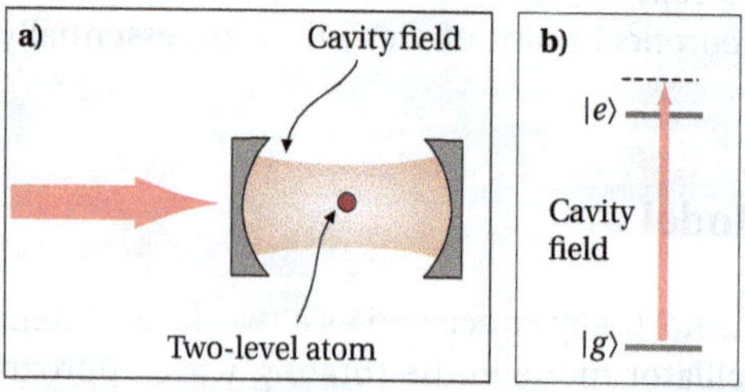

Fig. 22.1. Schematic of cavity QED. **a)** A two-level atom is placed in the centre of a high-Q optical cavity, which confines the cavity field, approximated by a single mode; **b)** The cavity field may be slightly detuned from the transition between the two atomic levels.

iii) *Josephson Junctions and Microwaves*: Here the two-level system is implemented by the quantized energy levels in a Josephson junction system, and a resonator for the microwave field can be made from a superconducting cavity—this is described more fully in Chap. 26.

iv) *Solid-State Systems*: The two-level system can be implemented as a object fixed inside a solid-state substrate. For example, this may be a quantum dot, that is, a very small potential well inside a semiconductor device, coupling to an electromagnetic field (typically in the near infrared regime) inside a nanocavity fabricated as part of the semiconductor device.

   Alternatively, the two-level system may be provided by a rare-earth atom doped into an appropriate substrate, such as YAG, or by a vacancy centre created by implanting nitrogen atoms into diamond.

v) *Ion Traps*: A very useful implementation of the Jaynes–Cummings Hamiltonian can be made using *phonons*, not photons. The two-level system is an *ion*, not a neutral atoms, and the ion is confined by an electromagnetic trap. The operators $a^\dagger$, a are then the phonon creation and destruction operators of the vibrational modes of the trapped ion, as will be shown in Chap. 23.

   The trapped ion implementation of the Jaynes–Cummings Hamiltonian provided the first realistic scenario for a quantum computer, and the only one in which all necessary features of a quantum computer have been experimentally implemented. We will treat this in Chap. 24.

22.1.2 Details of the Jaynes–Cummings Hamiltonian

There are three parts to the Hamiltonian, represented by:

i) *The Two-Level Atom*: This is described by the ground state $|g\rangle$ and the excited state $|e\rangle$, and the Hamiltonian is then written as

$$H_{0A} = \hbar\omega_{eg}|e\rangle\langle e|. \tag{22.1.2}$$

ii) *The Electromagnetic Field*: We describe a single mode radiation field by its electric field

$$\boldsymbol{E}(\boldsymbol{x}) = \mathrm{i}\left(\mathfrak{F}(\boldsymbol{x})a - \mathfrak{F}^*(\boldsymbol{x})a^\dagger\right), \tag{22.1.3}$$

   where $\mathfrak{F}(\boldsymbol{x})$ is the mode function appropriate to the cavity. In the case of an optical beam of area $\mathcal{A}$ inside a Fabry–Pérot cavity using plane mirrors a distance l apart, this would take the form

$$\mathfrak{F}(\boldsymbol{x}) \longrightarrow \sqrt{\frac{\hbar\omega}{\mathcal{A}l\varepsilon_0}}\,\boldsymbol{\epsilon}\sin(\omega x/c)\,. \tag{22.1.4}$$

   However, practical cavities seek to maximize the field at the location of the atom, and have a shape more like that indicated in Fig. 22.1. The Hamiltonian then omits all but this mode, and takes the form

$$H_{0F} = \hbar\omega a^\dagger a, \tag{22.1.5}$$

with the usual harmonic oscillator commutation relation $[a, a^\dagger] = 1$.

iii) *The Interaction between the Field and the Atom*: This is given by the electric dipole moment $\boldsymbol{d}_{eg}$ and the field at the position of the atom, leading to the total Hamiltonian

$$H_{\text{CQED}} = H_0 + H_1, \tag{22.1.6}$$

$$H_0 = \hbar\omega_{eg}|e\rangle\langle e| + \hbar\omega a^\dagger a, \tag{22.1.7}$$

$$H_1 = -\mathrm{i}\left(\boldsymbol{d}_{eg}\cdot\mathfrak{F}(\boldsymbol{x})\, a|e\rangle\langle g| - \boldsymbol{d}_{eg}^*\cdot\mathfrak{F}^*(\boldsymbol{x})\, a^\dagger|g\rangle\langle e|\right). \tag{22.1.8}$$

iv) The form (22.1.6–22.1.8) above matches that of (22.1.1) by noting that:

a) We have chosen the zero of the atom energy at the ground state energy, whereas in (22.1.1) the zero is at the midpoint of the two levels.

b) The real coupling $g(\boldsymbol{x})$ can be obtained by absorbing the phase of $-\mathrm{i}\boldsymbol{d}_{eg}\cdot\mathfrak{F}(\boldsymbol{x})$ into the definition of the creation and destruction operators a, $a^\dagger$. In that case we can say that

$$\hbar g(\boldsymbol{x}) = \left|\boldsymbol{d}_{eg}\cdot\mathfrak{F}(\boldsymbol{x})\right|. \tag{22.1.9}$$

However, for later convenience, we will yet not eliminate the phase in our theoretical development.

c) In principle the phase of the mode function can be spatially dependent, in which case this simplification is not appropriate.

22.1.3 Energy Levels and Eigenstates

The Hilbert space has the basis states

$$|g\rangle\otimes|n\rangle \equiv |g,n\rangle, \quad |e\rangle\otimes|n\rangle \equiv |e,n\rangle, \qquad n = 0,1,2,\ldots \tag{22.1.10}$$

We will use both notations, depending on the emphasis we wish to make in any particular situation.

a) Eigenstates of the Bare Hamiltonian $H_0 = H_{0A} + H_{0F}$: These are written in terms of the field and atom Hamiltonians as follows.

i) The ground state of H_0 is non-degenerate

$$H_0|g,0\rangle = 0. \tag{22.1.11}$$

ii) The atomic transition frequency ω_{eg} and the harmonic oscillator frequency ω are not necessarily equal, and we therefore introduce the *detuning*, defined by

$$\Delta \equiv \omega - \omega_{eg}. \tag{22.1.12}$$

When $\omega \approx \omega_{eg}$, that is, $|\Delta| \ll \omega$, the excited states form pairs $\{|g,n\rangle, |e,n-1\rangle\}$ of nearly degenerate states

$$H_0|g,n\rangle = n\hbar\omega|g,n\rangle, \tag{22.1.13}$$
$$H_0|e,n-1\rangle = \big((n-1)\hbar\omega + \hbar\omega_{eg}\big)|e,n-1\rangle, \tag{22.1.14}$$
$$= (n\hbar\omega - \hbar\Delta)|e,n-1\rangle. \tag{22.1.15}$$

This level structure is illustrated in Fig. 22.2 a.

b) Matrix Form of the Jaynes–Cummings Hamiltonian: The interaction Hamiltonian, as given in (22.1.8), uses the rotating-wave approximation and thus couples only between states in these almost degenerate subspaces. This means that we can diagonalize the Hamiltonian in 2 × 2 blocks which act on the two dimensional subspaces of bare eigenstates $\{|e,n-1\rangle, |g,n\rangle\}$.

These describe all but the ground state $|g,0\rangle$, which is uncoupled. Hence we can write the Hamiltonian in a block diagonal form

$$H_{\text{CQED}} = \begin{pmatrix} 0 & & & & & \\ & H_1 & & & & \\ & & H_2 & & & \\ & & & \ddots & & \\ & & & & H_n & \\ & & & & & \ddots \end{pmatrix}, \tag{22.1.16}$$

where the leading 0 is a null 1 × 1 matrix, and the 2 × 2 submatrices H_n are

$$H_n = n\hbar\omega + \hbar\begin{pmatrix} -\Delta & \frac{1}{2}\Omega_n e^{-i\varphi} \\ \frac{1}{2}\Omega_n e^{i\varphi} & 0 \end{pmatrix}, \tag{22.1.17}$$

in which

$$\tfrac{1}{2}\hbar\Omega_n e^{i\varphi} = -i\boldsymbol{d}_{eg}\cdot\mathcal{F}(\boldsymbol{x})\sqrt{n}, \tag{22.1.18}$$

so that we can write, using the definition (22.1.9) of $g(\boldsymbol{x})$,

$$\Omega_n = \Omega_1\sqrt{n}, \qquad \Omega_1 = 2g(\boldsymbol{x}). \tag{22.1.19}$$

22.1.4 Eigenvalues and Eigenvectors—the Dressed States

Diagonalizing the individual 2 × 2 matrices H_n gives the *dressed states*—the conventional terminology for the eigenstates, as follows. These dressed states are very similar to those described for the driven two-level atom in Sect. 5.3, with the substitution $\Omega_R \longrightarrow \Omega_n$.

a) The Dressed Energy Levels: These are the the energy eigenvalues, given by

$$E_{g,0} = 0, \tag{22.1.20}$$
$$E_{+,n} = n\hbar\omega - \tfrac{1}{2}\hbar\Delta + \tfrac{1}{2}\hbar\sqrt{\Delta^2+\Omega_n^2}, \tag{22.1.21}$$
$$E_{-,n} = n\hbar\omega - \tfrac{1}{2}\hbar\Delta - \tfrac{1}{2}\hbar\sqrt{\Delta^2+\Omega_n^2}. \tag{22.1.22}$$

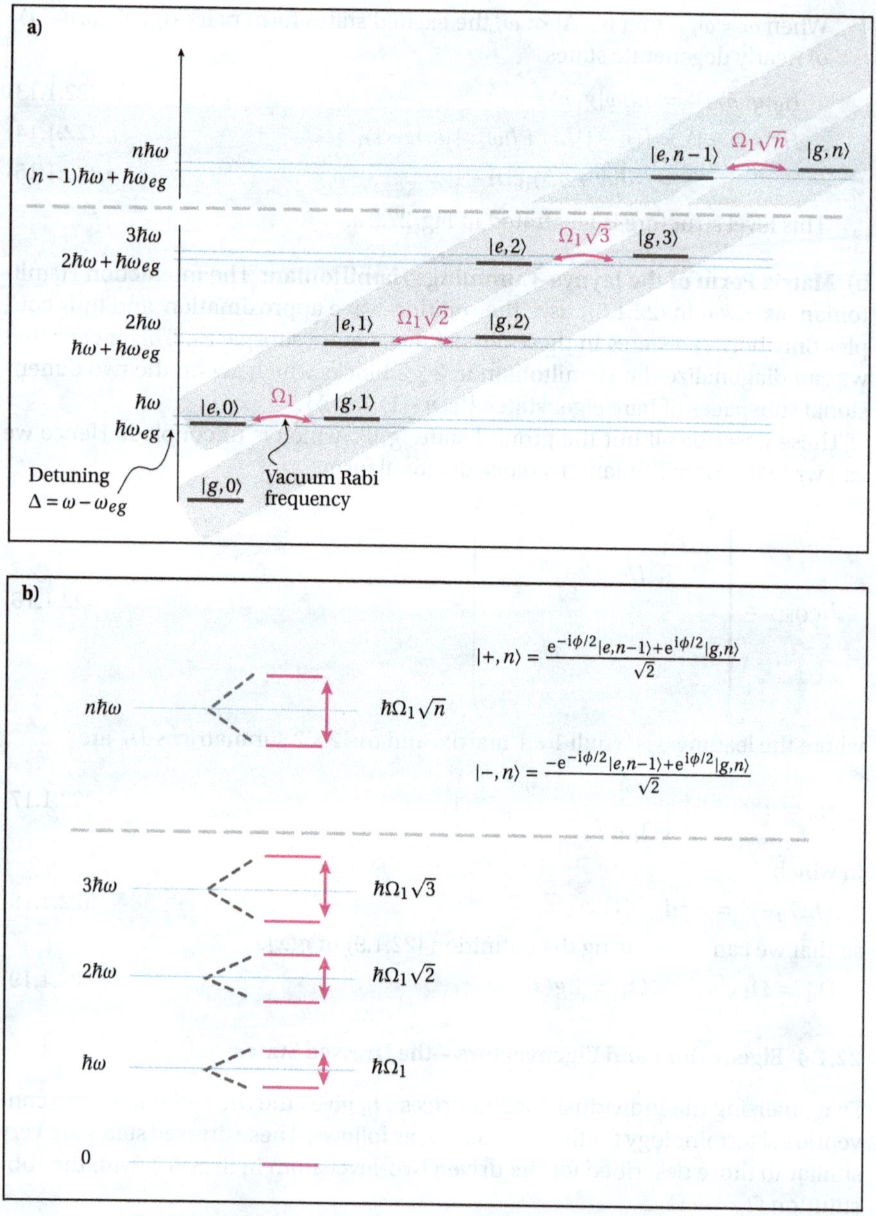

Fig. 22.2. Energy levels of the Jaynes–Cummings Hamiltonian (22.1.6): **a)** The bare energy levels consisting of two towers corresponding to the states $|g\rangle$ and $|e\rangle$ of the atom. The levels are almost degenerate in pairs, which can be strongly coupled by the electromagnetic field; **b)** The dressed energy levels in the case that the detuning $\Delta = 0$. These are even and odd superpositions of the bare energy levels, and these pairs of dressed levels are split by an amount $\hbar\Omega_1\sqrt{n}$.

These are illustrated in Fig. 22.2 b for the case of $\Delta = 0$.

b) The Dressed Eigenstates: These are the eigenvectors

$$|g, 0\rangle, \tag{22.1.23}$$

$$\begin{pmatrix} |+, n\rangle \\ |-, n\rangle \end{pmatrix} = \begin{pmatrix} \cos\frac{1}{2}\theta_n & \sin\frac{1}{2}\theta_n \\ -\sin\frac{1}{2}\theta_n & \cos\frac{1}{2}\theta_n \end{pmatrix} \begin{pmatrix} e^{-i\varphi/2}|e, n-1\rangle \\ e^{i\varphi/2}|g, n\rangle \end{pmatrix}. \tag{22.1.24}$$

In these expressions, the angle θ_n is given by

$$\tan\theta_n = -\frac{\Omega_n}{\Delta}, \quad \sin\theta_n = \frac{\Omega_n}{\sqrt{\Omega_n^2 + \Delta^2}}, \quad \cos\theta_n = -\frac{\Delta}{\sqrt{\Omega_n^2 + \Delta^2}}, \tag{22.1.25}$$

and because $\Omega_n > 0$, these formulae imply that $0 \leqslant \theta_n \leqslant \pi$.

c) Limiting Cases:

i) *Off-Resonance Case*: Consequently, the limiting form of the dressed states as $\Omega_n \to 0$ depends on the sign of Δ, namely

$$\left.\begin{aligned} \Delta > 0: \quad & |+, n\rangle \to e^{i\varphi/2}|g, n\rangle, \quad & |-, n\rangle \to e^{-i\varphi/2}|e, n-1\rangle, \\ \Delta < 0: \quad & |+, n\rangle \to e^{-i\varphi/2}|e, n-1\rangle, \quad & |-, n\rangle \to e^{i\varphi/2}|g, n\rangle. \end{aligned}\right\} \tag{22.1.26}$$

ii) *The On-Resonance Case*: When the atomic transition frequency and the harmonic oscillator frequency coincide, then $\Delta = 0$, and because $\Omega_n > 0$, it follows that $\sin\theta_n = 1$, and therefore

$$\left.\begin{aligned} \theta_n &= \tfrac{1}{2}\pi, \\ E_{\pm,n} &= n\hbar\omega \pm \tfrac{1}{2}\hbar\Omega_n. \end{aligned}\right\} \tag{22.1.27}$$

The dressed states become

$$|\pm, n\rangle = \frac{1}{\sqrt{2}}\left(\pm e^{-i\varphi/2}|e, n-1\rangle + e^{i\varphi/2}|g, n\rangle\right), \qquad n = 1, 2, \ldots. \tag{22.1.28}$$

22.1.5 The Dressed Hamiltonian in the Far-Detuned Limit

The expression (22.1.24) gives the unitary transformation which diagonalizes the Hamilton, and when $|\Delta| \gg \Omega_n$, this can be written in a rather simple form. The number n is the eigenvalue of the operator

$$\mathbf{N} \equiv a^\dagger a + \tfrac{1}{2}(1 + \sigma_z). \tag{22.1.29}$$

Using the dressed energy levels (22.1.21, 22.1.22), it follows that the diagonalized Hamiltonian can be written

$$H_{\text{Dressed}} = \mathbf{N}\hbar\omega - \tfrac{1}{2}\hbar\Delta - \text{sign}(\Delta)\tfrac{1}{2}\hbar\sigma_z\sqrt{\Delta^2 + 4g^2\mathbf{N}}. \tag{22.1.30}$$

In the limit $|\Delta| \gg g$, and provided n is not too large, we can approximate the square root to get (for either positive or negative Δ)

$$H_{\text{Dressed}} \approx \mathbf{N}\hbar\left(\omega - \frac{g^2}{\Delta}\sigma_z\right) - \tfrac{1}{2}\hbar\Delta(1 + \sigma_z). \tag{22.1.31}$$

Now substitute for **N** using (22.1.29), and set $\Delta = \omega - \omega_{eg}$, to get

$$H_{\text{Dressed}} \approx \hbar\omega a^\dagger a + \hbar\left(\omega_{eg} - \frac{g^2}{\Delta}\right)\left(\frac{1+\sigma_z}{2}\right) - \frac{\hbar g^2}{\Delta} a^\dagger a \sigma_z. \tag{22.1.32}$$

a) Interpretation as a QND Hamiltonian: In this form the Hamiltonian is a function of the two commuting operators $a^\dagger a$ and σ_z, each of which can therefore be measured without disturbing the other. The interaction term, the last term in the expression, can be interpreted as:

i) The atomic transition shifted by the combined Lamb and Stark shift of magnitude $(g^2/\Delta)(a^\dagger a + \frac{1}{2})$.

ii) Alternatively we can interpret the Stark shift as a dispersive shift of the cavity transition frequency by $\sigma_z g^2/\Delta$.

iii) This interpretation also appears in [22.5] and [22.6]—note that their definitions of Δ are the negative of ours.

b) Unitary Transformation: We can write $\theta_n \approx 2g\sqrt{n}$, and then the unitary transformation in (22.1.24) can be written as

$$\begin{pmatrix} \cos\frac{1}{2}\theta_n & \sin\frac{1}{2}\theta_n \\ -\sin\frac{1}{2}\theta_n & \cos\frac{1}{2}\theta_n \end{pmatrix} \approx \exp\left(ig\sqrt{n}\sigma_y\right) = \exp\left(\frac{g}{\Delta}\left(a\sigma^+ - a^\dagger\sigma^-\right)\right). \tag{22.1.33}$$

It is also possible to derive the form (22.1.32) by directly implementing the unitary transformation on the Hamiltonian (22.1.17).

c) Relevance to Quantum Computers: This form of the Hamiltonian is of relevance to quantum computing, and in particular to the use of a Josephson junction qubit, as discussed in Sect. 26.3.1.

22.1.6 Rabi Oscillations in the Jaynes–Cummings Model

The dynamics described by each of the 2×2 matrix Hamiltonians H_n is the same as that described in (5.2.9) by a two-level system driven by a coherent field with Rabi frequency $\Omega_R \rightarrow = \Omega_1\sqrt{n}$ and angle $\theta \rightarrow \theta_n$. The physics is rather different, since in this case, no coherent driving field is needed, and the Rabi oscillations describe the oscillation of a single quantum between the excitation in the atom and the cavity field. The absence of any driving field has given rise to the terminology *vacuum Rabi oscillations*, even though there is in fact a non-vanishing optical field. Even in the case of $n = 1$, when the oscillations take place between $|g,1\rangle$ and $|e,0\rangle$, although the electromagnetic vacuum state is involved, there is a component of the $n = 1$ state in both dressed states, giving rise to a non-vanishing electromagnetic field.

Nevertheless, it is conventional to call the quantity Ω_1 the *vacuum Rabi frequency*, and to speak of the resultant oscillations as *vacuum Rabi oscillations*.

a) Nature of the Vacuum Rabi Oscillations: Starting from the atom in the ground state and the field in a Fock state $|g,n\rangle$ the probability that the atom-cavity system is in the state $|e,n-1\rangle$ at time t is

$$\mathcal{P}_{e,n-1\leftarrow g,n}(t) = \left|\langle e,n-1|e^{-iH_{\text{CQED}}t/\hbar}|g,n\rangle\right|^2, \tag{22.1.34}$$

$$= \left|\left\langle e,n-1\right|\left(e^{-iE_{+,n}t/\hbar}|+,n\rangle\langle+,n| + e^{-iE_{-,n}t/\hbar}|-,n\rangle\langle-,n|\right)\left|g,n\right\rangle\right|^2, \tag{22.1.35}$$

$$= \frac{1}{2}\frac{\Omega^2}{\Delta^2+\Omega^2}\left(1-\cos\sqrt{\Delta^2+\Omega_n^2}\,t\right). \tag{22.1.36}$$

Here the detuning $\Delta = \omega - \omega_{eg}$ and Rabi frequency $\Omega_n = 2|g|\sqrt{n+1}/\hbar$ are as defined in (22.1.12, 22.1.19). The behaviour we see is that of the Rabi oscillations that we have already met in *Bk. I: Sect.12.3*, and in Chap. 5.

In particular, an atom prepared in $|e,0\rangle$, the excited atomic state and with photons present, undergoes *vacuum Rabi oscillations* between $|e,0\rangle$ and $|g,1\rangle$ at the *vacuum Rabi frequency* $\Omega_0 = 2|g|/\hbar$.

b) General Initial Field State: If the initial state of the electromagnetic field is not the vacuum, we can describe this situation by a density operator

$$\rho_F = \sum_{n,m=0}^{\infty} |n\rangle\rho_{nm}\langle m|. \tag{22.1.37}$$

The probability of finding an atom in the excited state after it was prepared in its ground state at $t=0$

$$\mathcal{P}_{e\leftarrow g}(t) = \text{Tr}_{A+F}\left\{|e\rangle\langle e|e^{-iH_{\text{CQED}}t/\hbar}\left(|g\rangle\langle g|\otimes\rho_F\right)e^{iH_{\text{CQED}}t/\hbar}\right\}. \tag{22.1.38}$$

Evaluating this expression yields the result

$$\mathcal{P}_{e\leftarrow g}(t) = \sum_{n=0}^{\infty}\mathcal{P}_{e,n-1\leftarrow g,n}(t)\,p_n, \tag{22.1.39}$$

$$\text{where}\quad p_n = \rho_{n,n}. \tag{22.1.40}$$

This result is valid for any initial field density operator, and, because it does not depend on the off-diagonal density matrix elements, is independent of any coherences in the initial field. Therefore the result is true for:

i) A thermal state, for which the density operator is given by the Bose–Einstein distribution

$$\rho_{F,\text{thermal}} = \frac{1}{Z}e^{-H_{0F}/k_B T}, \tag{22.1.41}$$

$$p_{n,\text{thermal}} = \frac{\bar{n}^n}{(1+\bar{n})^{n+1}}. \tag{22.1.42}$$

ii) A coherent state, for which

$$\rho_{F,\text{coherent}} = |\alpha\rangle\langle\alpha|, \tag{22.1.43}$$

$$p_{n,\text{coherent}} = \frac{|\alpha|^2}{n!}e^{-|\alpha|^2}. \tag{22.1.44}$$

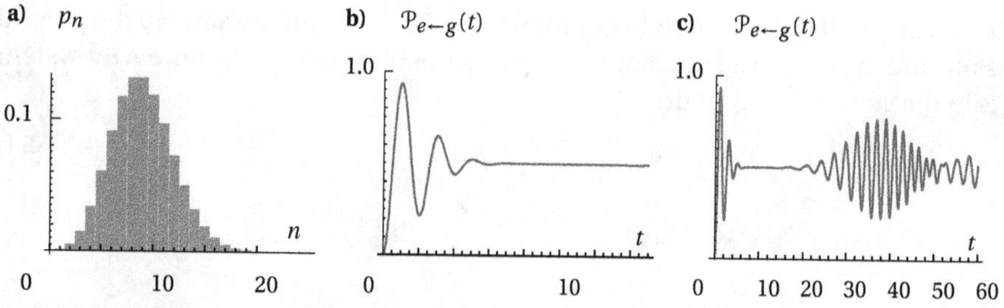

Fig. 22.3. Quantum revivals: **a)** An initial Poisson distribution of photons arising from a coherent state $|\alpha\rangle$ with $\alpha = 3$; **b)** The time dependence of the probability of being in the excited state after a time t for a relatively short time t exhibits a rapid collapse to a steady value; **c)** However, after a sufficiently long time a revival occurs.

c) Quantum Collapses and Revivals: The solutions here have already arisen in *Bk. I: Sect.12.4.1*, and as seen in Fig. 22.3, the coherent state solution exhibits "quantum revivals", which arise because the different frequencies are incommensurate with each other, and soon get out of phase. However, since there are only a finite number which are effectively occupied, eventually the important ones can get in phase again, and the probability has a quantum revival. Here these revivals are seen here in the quantity $\mathcal{P}_{e\leftarrow g}(t)$ defined in (22.1.39). These revivals were first observed by *Rempe, Walther* and *Klein* [22.3] in 1987.

22.2 Interaction with the Environment

The CQED system can interact with the environmental electromagnetic field in two ways:

i) The internal electromagnetic field mode couples to the input and output fields through a partially transparent cavity end mirror.

   There may also be a coherent component to the input field.

ii) The atom can decay by emitting into field modes not confined by the cavity.

22.2.1 Master Equation

These are straightforwardly introduced by writing the master equation

$$\begin{aligned}\frac{d\rho}{dt} &= -\frac{i}{\hbar}[H_{\text{CQED}},\rho] - \sqrt{\kappa}[\mathcal{F}^*(t)a - \mathcal{F}(t)a^\dagger,\rho] \\ &\quad + \tfrac{1}{2}\kappa(2a\rho a^\dagger - \rho a^\dagger a - a^\dagger a\rho) + \tfrac{1}{2}\gamma(2\sigma^-\rho\sigma^+ - \rho\sigma^+\sigma^- - \sigma^+\sigma^-\rho). \qquad (22.2.1)\end{aligned}$$

In these equations the coherent input field $\mathcal{F}(t)$ has been introduced using the method described in Sect. 9.5, and the damping, and the atomic decay constant γ

and cavity decay constant κ are introduced to enable the two loss processes above to a zero temperature environment, as in Chap. 10.

This master equation represents the most elementary form of CQED, in which there is only one atom within the cavity. Generalization to many identical atoms is achieved simply by adding extra atomic terms of the same kind, and possibly including the fact that they may interact with the cavity mode with differing strengths, corresponding to their being located at positions with differing values of the field mode amplitude.

Here we will be mainly concerned with the case where there is only one atom, and where the loss constants κ and γ are small, as is appropriate for a quantum device. This is the region of strong coupling, where as noted by *Kimble* [22.7], most current experimental research is concentrated. More extensive treatments, including the cases of many atoms and weak coupling can be found in [22.8] and [22.9].

22.2.2 The Strong Coupling Condition

Cavity QED can be parametrized by the rates $g(0)$, κ and γ. In order to emphasize the coherent evolution of the system, it is useful to require that the coupling coefficient dominate dissipation:

$$\frac{g(0)}{\gamma} \gg 1, \qquad \frac{g(0)}{\kappa} \gg 1. \tag{22.2.2}$$

This condition is known as the *strong coupling* condition.

Kimble and his coworkers [22.1] have pointed out that:

> It is instructive to explore strong coupling in terms of two dimensionless parameters known as the critical photon and atom numbers. The critical (or saturation) photon number describes the number of photons such that for a cavity of a given geometry the intracavity optical intensity is sufficient to saturate the atomic response
>
> $$n_0 = \frac{\gamma^2}{2g(0)^2}\,. \tag{22.2.3}$$
>
> Similarly, the critical atom number describes the number of strongly coupled atoms necessary to affect appreciably the intracavity field
>
> $$N_0 = \frac{2\kappa\gamma}{g(0)^2}\,. \tag{22.2.4}$$
>
> Many quantum optical systems—lasers for instance, with $\sqrt{n_0} \approx 10^3$–10^4—have large critical parameters and therefore adding or removing one photon or atom does not significantly alter the dynamics as a whole. In these systems, the coherent coupling parameter $g(0)$ is scaled away as processes approach the semi-classical regime. By contrast, the necessary (though not sufficient) criteria for strong coupling are that $n_0 \ll 1$ and $N_0 \ll 1$. This means that in the regime of strong coupling, single quanta

dominate the dynamics of the system such that the interaction between atom and photon can be manifestly nonclassical and nonlinear for single atoms and photons. Strong coupling thereby provides a powerful tool for the study of quantum optics as well as the interaction of the quantized electromagnetic field with matter.

23. Optical Manipulation of Trapped Ions

To study individual atomic systems, it is obviously advantageous to be able to pin the atom at a precise location, and at as low an energy as possible. A neutral atom is not easy to trap, unlike an ion, whose electric charge enables one to exert relatively strong electrical forces, and trap the ion at a well defined location. However, this is not entirely straightforward. One would like a three-dimensional attractive harmonic potential, preferably electrostatic, but Earnshaw's theorem of electrostatics forbids this, so more elaborate stratagems need to be devised.

There are two main kinds of ion trap, both of which are based on an electrostatic quadrupole potential of the kind

$$V(\boldsymbol{x}) = V_0\left(-x^2 - y^2 + 2z^2\right).$$

This potential attracts in the x- and y-directions, but repels in the z-direction, requiring an additional mechanism to provide confinement in the z-direction. The two main solutions to this problem are achieved as follows:

i) *The Paul trap*: Here the static potential is augmented by a radio-frequency quadrupole potential. For slow velocities of the ion, this can be equivalent to a net attractive harmonic potential in all directions. In addition, the ion executes a *micromotion* of small amplitude at the quadrupole radio frequency.

ii) *The Penning trap*: Here the confinement is provided by a magnetic field in the z-direction, and the net effect cannot be described simply by a potential. We will not consider the Penning trap in this book.

This chapter will concentrate on what can be achieved by applying laser light to a two-level ion trapped in the effective harmonic potential provided by the Paul trap. This is a configuration which has proved to be very well adapted to the implementation of quantum computers and quantum processors.

The detailed dynamics of an ion in a Paul trap are covered in Appendix 23.A, where behaviour beyond the approximation by a harmonic potential is covered.

23.1 The Trapped Ion Hamiltonian

The Hamiltonian appropriate for an ion trapped by a harmonic potential, and interacting with an incident laser field, is based on that given in Chap. 7 and Chap. 17, and can be written in the form

$$H = H_T + H_A - \boldsymbol{d}\cdot\boldsymbol{E}_{\rm cl}(\boldsymbol{X},t) - \boldsymbol{d}\cdot\boldsymbol{E}(\boldsymbol{X}), \tag{23.1.1}$$

in which the various terms are:

i) H_T describing the motion of the center of mass of the ion $\boldsymbol{X}$ moving in a harmonic trapping potential.

ii) H_A is the Hamiltonian describing the electronic excitations.

iii) The term $-\boldsymbol{d}\cdot\boldsymbol{E}_{\rm cl}(\boldsymbol{X},t)$ represents the dipole interaction of the atomic electron with the laser field at the center of mass position of the ion, thus coupling the electronic degrees with those of the centre of mass.

iv) The term $-\boldsymbol{d}\cdot\boldsymbol{E}(\boldsymbol{X},t)$ represents the same kind of interaction with the quantized radiation field.

23.1.1 Components of the Trapped Ion Hamiltonian

We will put down the explicit forms of the various components of the Hamiltonian, which are essentially the same as those we used to treat laser cooling of a trapped ion in Chap. 17.

a) Centre of Mass Motion Hamiltonian H_T: We will initially formulate this Hamiltonian in one dimension, and consider the simple case of a purely harmonic trap of angular frequency ν. This has the form

$$H_T = \frac{P^2}{2M} + \tfrac{1}{2}M\nu^2X^2 = \hbar\nu\left(a^\dagger a + \tfrac{1}{2}\right), \tag{23.1.2}$$

with the usual relationship between the operators

$$X = \sqrt{\frac{\hbar}{2M\nu}}\left(a + a^\dagger\right), \qquad P = \mathrm{i}\sqrt{\frac{\hbar M\nu}{2}}\left(a^\dagger - a\right). \tag{23.1.3}$$

b) Quantum States: The eigenstates are the usual number states

$$|n\rangle \equiv \frac{1}{\sqrt{n!}}\left(a^\dagger\right)^n|0\rangle, \tag{23.1.4}$$

$$H_T|n\rangle = \hbar\nu\left(n+\tfrac{1}{2}\right)|n\rangle, \quad n = 0,1,2,\ldots \tag{23.1.5}$$

These quanta, corresponding to vibrational motion, will be called *phonons.*

c) Size of the Ground State: This is determined by the typical distance scale, given in this case by the quantity

$$a_0 \equiv \sqrt{\frac{\hbar}{2M\nu}}. \tag{23.1.6}$$

d) The Experimental Situation:

i) In current experiments, typical trapping frequencies $\nu/2\pi$ are between 1 MHz and 10 MHz.

ii) Using sideband cooling, as discussed in Sect. 17.4.2, it is possible to cool to the vibrational ground state $|0\rangle$, and thus prepare a quantum mechanical pure state of the quantized motion of the ion.

23.1.2 Electronic excitation

We assume that the internal structure of the ion has the form of a two-level system with ground state $|g\rangle$ and excited state $|e\rangle$. The atomic Hamiltonian can then be written as

$$H_A = \hbar\omega_{eg}|e\rangle\langle e|, \qquad \text{where } \omega_{eg} \equiv \omega_e - \omega_g. \tag{23.1.7}$$

a) Two-Level System Description: As usual we define

$$\sigma_z = |e\rangle\langle e| - |g\rangle\langle g|, \tag{23.1.8}$$

$$\sigma_- = |g\rangle\langle e|, \tag{23.1.9}$$

$$\sigma_+ = |e\rangle\langle g|. \tag{23.1.10}$$

The atomic Hamiltonian can be written as

$$H_A = \tfrac{1}{2}\hbar\omega_{eg}(1+\sigma_z) \to \tfrac{1}{2}\hbar\omega_{eg}\sigma_z. \tag{23.1.11}$$

23.1.3 Interaction with Laser Light

The laser field is represented by a classical electric field $\boldsymbol{E}_{\text{cl}}(\boldsymbol{x},t)$, and the Hamiltonian representing the interaction of the laser light with the atomic electron at position $\boldsymbol{X}$ is

$$H_1 = -\boldsymbol{d}\cdot\boldsymbol{E}_{\text{cl}}(\boldsymbol{X},t) \equiv -e\boldsymbol{x}\cdot\boldsymbol{E}_{\text{cl}}(\boldsymbol{X},t). \tag{23.1.12}$$

Note that:

i) We consider the electric field for a *traveling wave* along the x-direction with laser frequency ω and wave vector $k = 2\pi/\lambda = \omega/c$.

$$\boldsymbol{E}(X,t) = \mathcal{E}\,\boldsymbol{e}\,\mathrm{e}^{\mathrm{i}kX-\mathrm{i}\omega t} + \mathcal{E}^*\boldsymbol{e}^*\,\mathrm{e}^{-\mathrm{i}kX+\mathrm{i}\omega t}. \tag{23.1.13}$$

ii) We will consider initially a two-level atom, for which

$$\boldsymbol{d} = \boldsymbol{d}_{eg}|e\rangle\langle g| + \boldsymbol{d}_{ge}|g\rangle\langle e|. \tag{23.1.14}$$

In the rotating-wave approximation this interaction Hamiltonian is

$$H_1 = -\tfrac{1}{2}\hbar\Omega\left(\mathrm{e}^{\mathrm{i}kX-\mathrm{i}\omega t}|e\rangle\langle g| + \mathrm{e}^{-\mathrm{i}kX+\mathrm{i}\omega t}|g\rangle\langle e|\right), \tag{23.1.15}$$

with Rabi frequency $\Omega = 2\mathcal{E}\,\boldsymbol{d}_{eg}\cdot\boldsymbol{e}/\hbar$, which we have taken to be real.

23.1.4 Manipulation of the Quantum State of the Centre of Mass

Excitation by the laser transfers the electron to the excited state. This transition is associated with a momentum transfer $\hbar\boldsymbol{k}$ along the laser propagation direction to the center of mass degrees of freedom. For a momentum eigenstate $|p\rangle$ the action of the interaction operator is

$$|g\rangle\otimes|p\rangle \xrightarrow{H_1} |e\rangle\otimes\mathrm{e}^{\mathrm{i}kX}|p\rangle \equiv |e\rangle\otimes|p+\hbar k\rangle. \tag{23.1.16}$$

Thus, using the laser, we can manipulate the centre of mass motion of the ion.

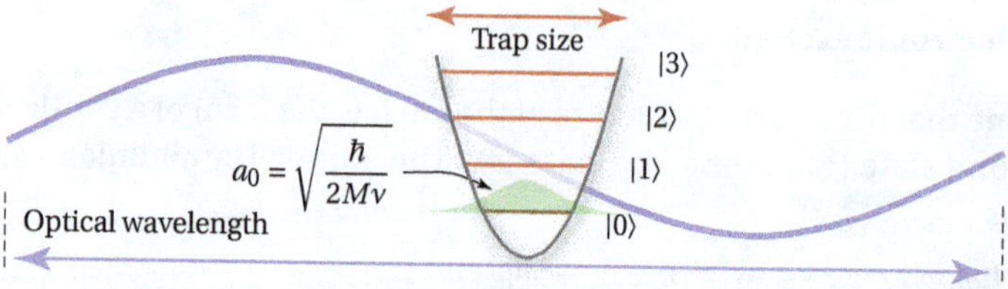

Fig. 23.1. The motional harmonic oscillator levels of an ion trap. The trap size is compared to the optical wavelength in the Lamb–Dicke regime. The illustration is approximately to scale for a typical experimental situation.

23.1.5 The Lamb–Dicke Regime

We define the dimensionless *Lamb–Dicke parameter* as a quantity proportional to the ratio of the spatial dimension of the ground state wavefunction to the wavelength of the laser; precisely, the Lamb–Dicke parameter is defined by

$$\eta = \frac{2\pi a_0}{\lambda}. \tag{23.1.17}$$

The case $\eta \ll 1$ corresponds to a "tight trap", and we call this the Lamb–Dicke regime. For ion traps typically $\eta \approx 0.1$, as illustrated in Fig. 23.1.

i) For ions in an excited trap state n the size of the n-th vibrational wave function will be roughly $\sqrt{n}a_0$.

ii) An alternative interpretation of the Lamb–Dicke parameter can be made in terms of the recoil energy

$$\epsilon_R = \frac{\hbar^2 k^2}{2M}, \tag{23.1.18}$$

which is typically a few kHz in an ion trap. We can then write the Lamb–Dicke parameter as

$$\eta = \sqrt{\frac{\epsilon_R}{\hbar\nu}} \quad \Longrightarrow \quad \eta^2 = \frac{\hbar k^2}{2M\nu}. \tag{23.1.19}$$

that is, the square root of the ratio of the recoil energy to that of a trap quantum.

iii) The interaction Hamiltonian can be rewritten in terms of η as

$$H_1 = -\tfrac{1}{2}\hbar\Omega e^{i\eta(a+a^\dagger)-i\omega t}|e\rangle\langle g| + \text{h.c.} \tag{23.1.20}$$

23.1.6 Total Hamiltonian in the Rotating Frame

The total Hamiltonian can now be written

$$H = H_T + H_A + H_1, \tag{23.1.21}$$

$$\equiv \hbar\nu a^\dagger a + \hbar\omega_{eg}|e\rangle\langle e| - \left(\tfrac{1}{2}\hbar\Omega\, e^{i\eta(a+a^\dagger)-i\omega t}|e\rangle\langle g| + \text{h.c.}\right), \tag{23.1.22}$$

$$\equiv H_0 + H_1. \tag{23.1.23}$$

i) The Hamiltonian has the form $H = H_0 + H_1$ of the sum of a bare and interaction Hamiltonian.

ii) The Hamiltonian H is explicitly time-dependent. However, this time dependence can be transformed away by going to a rotating frame using the unitary transformation

$$|\Psi, t\rangle = e^{i\omega|e\rangle\langle e| t}|\tilde{\Psi}, t\rangle, \tag{23.1.24}$$

so that the Schrödinger equation becomes

$$i\hbar\frac{\partial|\tilde{\Psi}, t\rangle}{\partial t} = \tilde{H}|\tilde{\Psi}, t\rangle, \tag{23.1.25}$$

with the time-independent Hamiltonian

$$\tilde{H} = \hbar\nu a^\dagger a - \hbar\Delta|e\rangle\langle e| - \left(\tfrac{1}{2}\hbar\Omega e^{i\eta(a+a^\dagger)}|e\rangle\langle g| + \text{h.c.}\right), \tag{23.1.26}$$

where $\Delta = \omega - \omega_{eg}$ is the detuning between the laser field and the transition frequency.

23.1.7 Energy Spectrum and Eigenstates of the Bare Hamiltonian

The bare Hamiltonian $H_0 = H_T + H_A$ provides the eigenstates which we will manipulate using the laser field. The eigenstates comprises two systems, corresponding to:

i) A set of states in which the *electronic* state is the ground state, and the motional state an n-quantum state

$$|g\rangle \otimes |n\rangle \equiv |g, n\rangle, \tag{23.1.27}$$

$$H_0|g, n\rangle = \hbar n\nu|g, n\rangle. \tag{23.1.28}$$

ii) A corresponding set in which the *electronic* state is the excited state, and the motional state an n-quantum state excited state

$$|e\rangle \otimes |n\rangle \equiv |e, n\rangle, \tag{23.1.29}$$

$$H_0|e, n\rangle = \hbar(\omega_{eg} + n\nu)|e, n\rangle. \tag{23.1.30}$$

iii) It was pointed out by *Walls, Blockley* and *Risken* [23.1, 23.2] that the structure of the energy level spectrum is identical with that of the Jaynes–Cummings model, as given in Sect. 22.1.3. Here, the vibrational mode of the ion takes on the same role as the optical cavity mode. However, the frequencies involved in trapped ions are very much lower, being typically in the MHz range, much lower than the typical optical frequencies of $\sim 10^{15}$Hz. Thus, in the cavity QED model, the frequency of the harmonic oscillator is in the optical regime, and is almost degenerate with the transition frequency ω_{eg}, leading to the approximate degeneracy of the states $|e, n\rangle$ and $|g, n+1\rangle$.

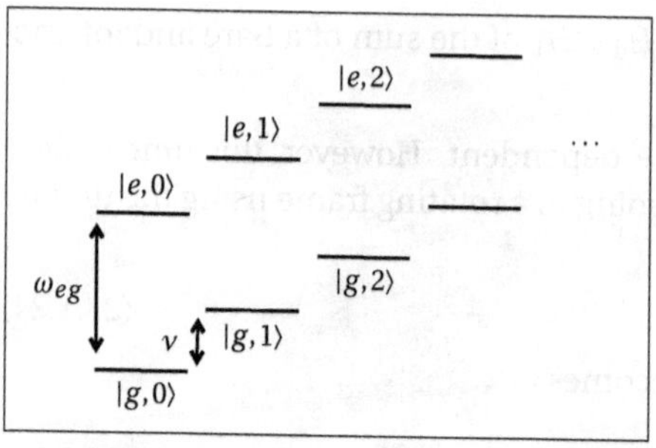

Fig. 23.2. The spectrum of energy levels of a trapped ion, consisting of two branches corresponding to the two electronic levels, each with a full set of associated harmonic oscillator levels.

In the ion trap case the harmonic oscillator frequency ν is very different from the transition frequency ω_{eg}, and the corresponding states are far from degenerate, as illustrated in Fig. 23.2. This then requires the use of laser induced couplings for applications.

23.1.8 Interaction with the Quantized Electromagnetic Field

The most significant additional feature which arises when the quantized electromagnetic field is included is *spontaneous emission*. This is a feature which one wishes to minimize, since all of the applications in quantum processing and quantum computation require the maintenance of coherence for as long as possible.

For a dipole allowed optical transition the spontaneous emission rate is $\Gamma \sim 10$ MHz. In this case spontaneous emission cannot in practice be neglected, since for interaction times T of interest we would find that $\Gamma T \gg 1$. However, the present model can be realized with metastable electronic states, of which the most important example is $^{40}\mathrm{Ca}^{+}$, which we have described in Sect. 19.4, and which is also discussed in Chap. 24. In this case, the spontaneous emission is a very weak effect. Another possibility which can be used to minimize decoherence arising from spontaneous emission is a Raman-coupled system with two long-lived ground states.

23.2 Laser-Induced Couplings in the Lamb–Dicke Regime

In the Lamb–Dicke regime we know that the Lamb–Dicke parameter $\eta \ll 1$, so we can make approximations based on its smallness. Thus we can make the expansion

$$\mathrm{e}^{\mathrm{i}kX} \equiv \mathrm{e}^{\mathrm{i}\eta(a+a^{\dagger})} \approx 1+\mathrm{i}\eta\left(a+a^{\dagger}\right)+\ldots \tag{23.2.1}$$

In the Lamb–Dicke regime, the Hamiltonian can consequently be approximated as

$$\begin{aligned} H &= \hbar\nu a^{\dagger}a+\hbar\omega_{eg}|e\rangle\langle e| \\ &\quad -\tfrac{1}{2}\hbar\left\{\Omega\mathrm{e}^{-\mathrm{i}\omega t}\left(1+\mathrm{i}\eta\left(a+a^{\dagger}\right)+\ldots\right)|e\rangle\langle g|+\mathrm{h.c.}\right\}. \end{aligned} \tag{23.2.2}$$

23.2.1 Approximate Forms in the Lamb–Dicke Regime

Inserting the approximation (23.2.2), we find the following results.

a) Matrix Elements: The transition matrix elements of $\langle e, m|H_1|g, n\rangle$, in leading order in η, are

$$\langle n|e^{ikX}|n\rangle = 1 + O(\eta^2), \tag{23.2.3}$$
$$\langle n+1|e^{ikX}|n\rangle = i\eta\sqrt{n+1} + O(\eta^3), \tag{23.2.4}$$
$$\langle n-1|e^{ikX}|n\rangle = i\eta\sqrt{n} + O(\eta^3). \tag{23.2.5}$$

Notice that the actual expansion parameter is not η itself, but rather it is effectively $\eta\sqrt{n}$, because of the $\sqrt{n}$ which always arises in the off-diagonal matrix elements.

b) Interaction Picture: If we transform to an interaction picture with respect to the trap Hamiltonian $H_T = \hbar\nu\left(a^\dagger a + \frac{1}{2}\right)$, that the wavefunction we use becomes

$$|\Psi_I, t\rangle = e^{iH_T t/\hbar}|\tilde{\Psi}, t\rangle. \tag{23.2.6}$$

This has the interaction picture Hamiltonian

$$H_I = \hbar\omega_{eg}|e\rangle\langle e| - \tfrac{1}{2}\hbar\left\{\left(\Omega e^{-i\omega t} + i\eta\Omega\left(a e^{-i(\omega-\nu)t} + a^\dagger e^{-i(\omega+\nu)t}\right) + \ldots\right)|e\rangle\langle g| + \text{h.c.}\right\}. \tag{23.2.7}$$

c) Excitation Spectrum: When we tune the laser across the atomic resonance we expect:

i) A dominant excitation of the bare transition $|g, n\rangle \to |e, n\rangle$ for $\Delta = \omega - \omega_{eg} \approx 0$ with Rabi frequency Ω.

ii) A red motional sideband $|g, n\rangle \to |e, n-1\rangle$ for $\Delta \approx \omega - \omega_{eg} \approx -\nu$ with Rabi frequency $\eta\Omega$. Note that there is no excitation on the red sideband for the ground state $|g\rangle|0\rangle$, since there is no lower state.

iii) A blue motional sideband $|g, n\rangle \to |e, n+1\rangle$ for $\Delta \approx \omega - \omega_{eg} \approx \pm\nu$ with Rabi frequency $\eta\Omega$.

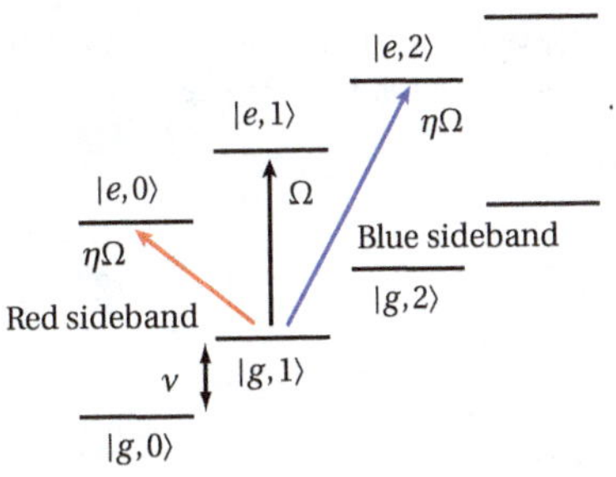

Fig. 23.3. The spectrum of energy levels of a trapped ion, consisting of two branches corresponding to the two electronic levels, each with a full set of associated harmonic oscillator levels.

23.2.2 Effective Hamiltonians Arising from Laser Induced Couplings

By tuning the laser as in Fig. 23.3 to the red or blue detuned sidebands, or to the direct transition with no change in sideband excitation, we can transfer the quantum state of the ion from any electronic ground state to the adjacent sidebands, or to the corresponding excited state. In particular, the three cases correspond to:

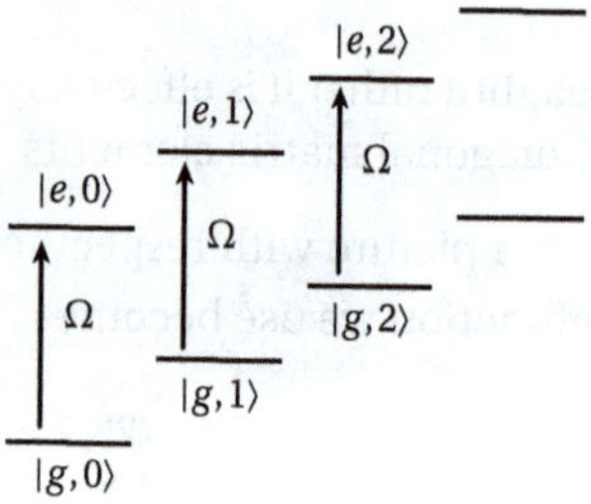

a) Laser on Resonance $|g,n\rangle \longrightarrow |e,n\rangle$: We choose $\Delta \approx 0$, and note that when we tune the laser close to the atomic transition frequency, while the transitions $|g,n\rangle \to |e,n\rangle$ will be excited, as long as $\eta \ll \nu$ and $\Omega \ll \nu$, excitation of the sidebands $|g,n\rangle \to |e,n\pm 1\rangle$ is suppressed, because they are off resonant. The Hamiltonian thus takes the approximate form (valid for $\eta \ll \nu$ and $\Omega \ll \nu$)

$$\tilde{H} \approx \hbar\nu\left(a^\dagger a + \tfrac{1}{2}\right) - \hbar\Delta|e\rangle\langle e| - \tfrac{1}{2}\hbar\Omega\Big(|e\rangle\langle g| + \text{h.c.}\Big). \qquad (23.2.8)$$

This approximate Hamiltonian describes motion decoupled from the electronic transitions, and we therefore have a two-level system.

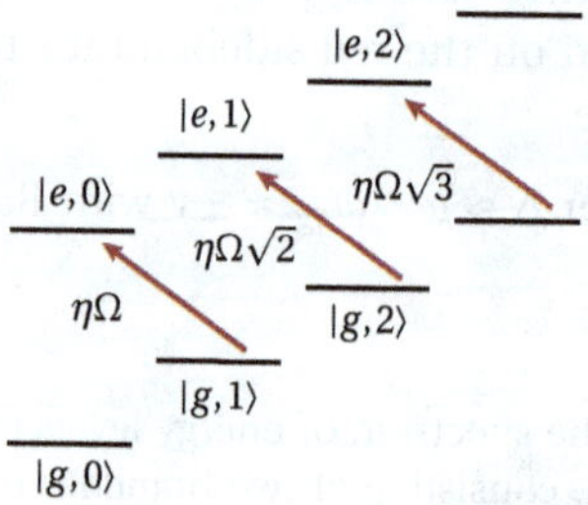

b) Emulation of the Jaynes–Cummings Model—Laser Tuned to the Lower Motional Sideband $|g,n\rangle \to |e,n-1\rangle$: Here we choose red detuning, $\Delta \approx -\nu$, and for $\Omega \ll \nu$ (in practice a rather strong condition) the bare atomic resonance is not excited. The resulting approximate Hamiltonian is a Jaynes–Cummings Hamiltonian with rotating-wave approximation, as introduced in *Bk. I: Sect. 12.4* (valid for $\Omega \ll \nu$)

$$\tilde{H} \approx \hbar\nu\left(a^\dagger a + \tfrac{1}{2}\right) - \hbar\Delta|e\rangle\langle e| - \tfrac{1}{2}i\hbar\eta\Omega\Big(|e\rangle\langle g|a + \text{h.c.}\Big). \qquad (23.2.9)$$

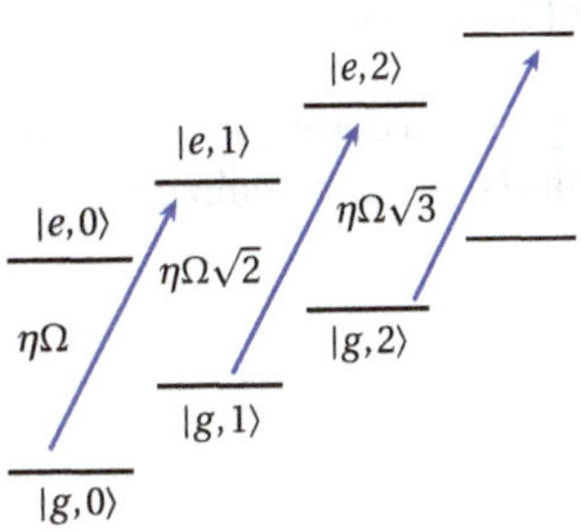

c) Laser Tuned to the Upper Motional Sideband $|g,n\rangle \to |e,n+1\rangle$: Here we choose blue detuning $\Delta \approx \nu$ and for $\Omega \ll \nu$ the bare atomic resonance is not excited. The approximate Hamiltonian is what might be called an "anti-Jaynes–Cummings" Hamiltonian, in which the transition of the atom from the ground state to the excited state is accompanied by the creation of a vibrational quantum (valid for $\Omega \ll \nu$)

$$\tilde{H} \approx \hbar\nu\left(a^\dagger a + \tfrac{1}{2}\right) - \hbar\Delta|e\rangle\langle e| - \tfrac{1}{2}\mathrm{i}\hbar\eta\Omega\left(|e\rangle\langle g|a^\dagger + \text{h.c.}\right). \quad (23.2.10)$$

23.3 Quantum State Engineering

S A Gardiner, Cirac and *Zoller* [23.3] showed that, using appropriate laser pulses, it is possible to use the different limiting Hamiltonians to prepare a wide variety of superposition states, including states in which the electronic and motional states are entangled.

23.3.1 Superpositions of Electronic States

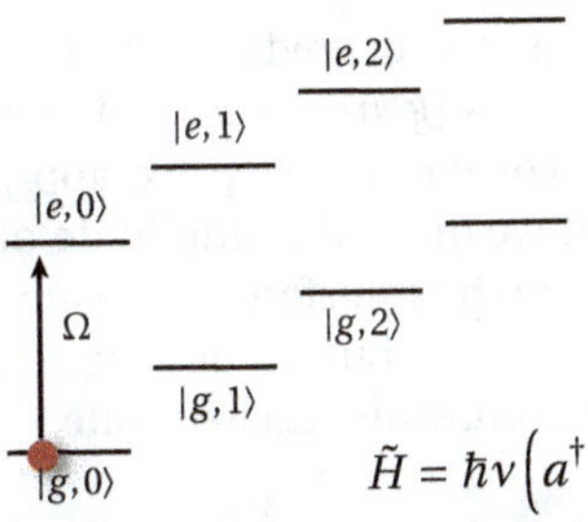

Starting from the ground state $|g,0\rangle$ we can prepare any superposition of the form

$$|g,0\rangle \longrightarrow \alpha|g,0\rangle + \beta|e,0\rangle, \quad (23.3.1)$$

by applying an appropriate on-resonance laser pulse characterized by a Rabi frequency $\Omega(t)$. This corresponds to the use of the Hamiltonian

$$\tilde{H} = \hbar\nu\left(a^\dagger a + \tfrac{1}{2}\right) - \hbar\Delta|e\rangle\langle e| - \tfrac{1}{2}\hbar\Omega(t)\left(|e\rangle\langle g| + \text{h.c.}\right). \quad (23.3.2)$$

This has no effect on the vibrational states.

23.3.2 Conversion of Electronic Superpositions to Motional Superpositions

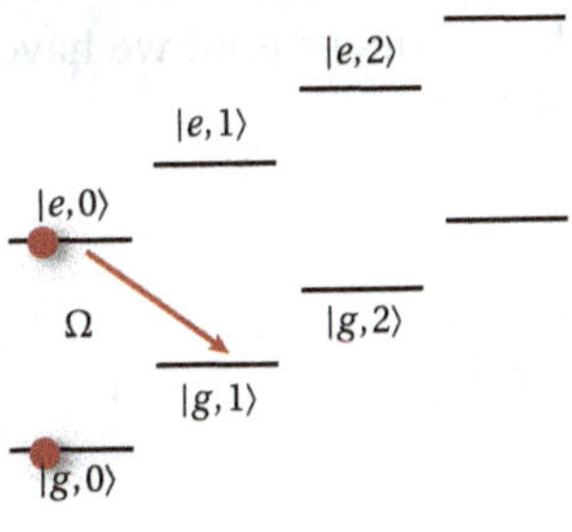

We can convert an atomic superposition to the same superposition of phonon states by applying a laser π-pulse on the red transition. The state $|g,0\rangle$ is not coupled to the laser light, so that $|g,0\rangle \longrightarrow |g,0\rangle$, while $|e,0\rangle \longrightarrow |g,1\rangle$.
The result is

$$\alpha|g,0\rangle + \beta|e,0\rangle \longrightarrow \alpha|g,0\rangle + \beta|g,1\rangle. \quad (23.3.3)$$

23.3.3 Generation of an Arbitrary Superposition of Motional States

We can engineer an arbitrary superposition state of phonon states by means of an iterative procedure, starting with the state $|g,0\rangle$, that is, we can realize the mapping

$$|g,0\rangle \longrightarrow |\Psi\rangle = \sum_{n=0}^{N} c_n |g,n\rangle. \tag{23.3.4}$$

for given coefficients c_n.

a) Procedure: Since we will do this using a sequence of Rabi pulses, this transformation is unitary, and has an inverse corresponding to applying the relevant pulses in the reverse order. Therefore, we will *start* with the state $|\Psi\rangle$, and show how to apply pulses which transfer all of the population to the state $|g,0\rangle$.

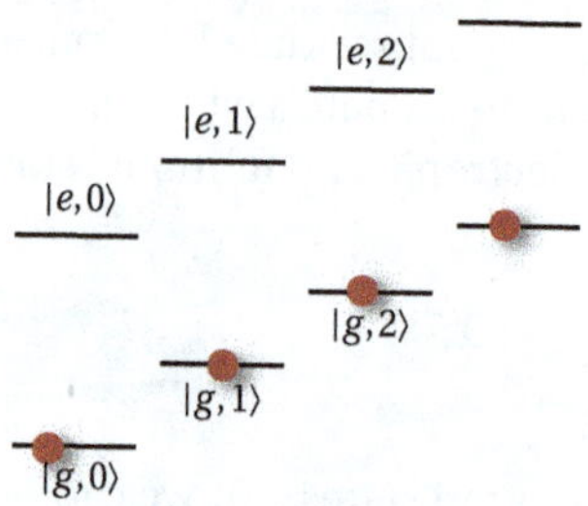

i) The procedure assumes that the initial state is, as in (23.3.4), composed of motional states only up to the value N.

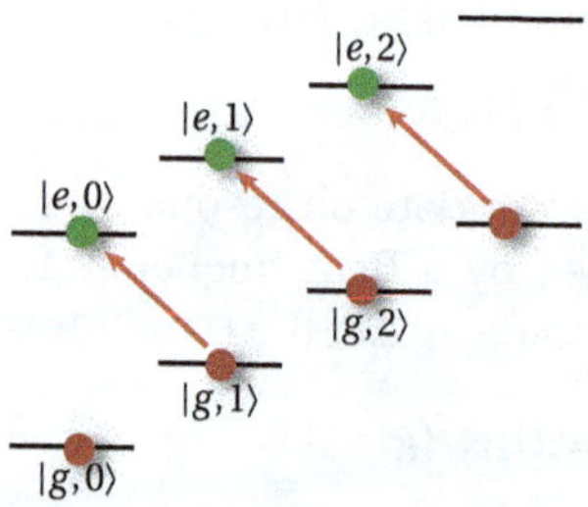

ii) By applying a laser on the red sideband we can couple the states $|g,n\rangle \longrightarrow |e\, n-1\rangle$ for all values of n. For the first step we apply a π-pulse so that we make the amplitude of $|g\rangle|N\rangle$ equal to zero by transferring the amplitude c_N to $|e\rangle|N-1\rangle$. But we now have a superposition of ground and excited state.

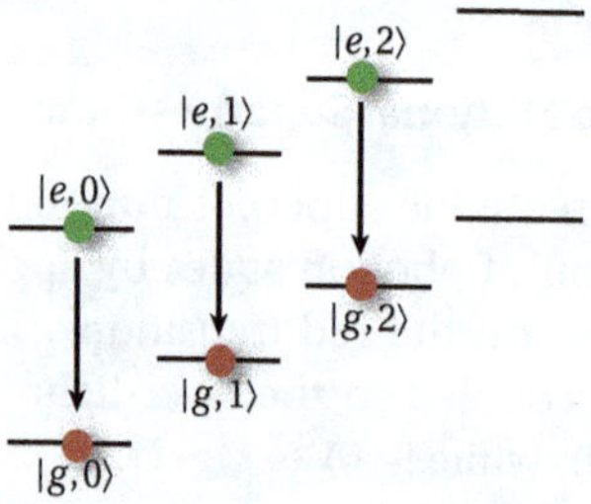

iii) In the second step, we apply a resonant laser, so that we transform the known superposition of $|g,N-1\rangle$ and $|e,N-1\rangle$ to $|g,N-1\rangle$, with no amplitude left in $|e,N-1\rangle$. Now we repeat the argument until we have transformed the state to $|g,0\rangle$.

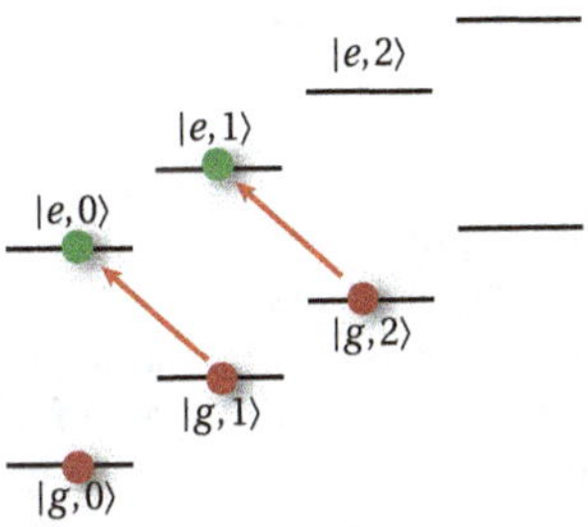

iv) We now have a state with phonon occupation only up to $N-1$, and no occupation in $|e, N-1\rangle$. We now start again with step one—although there is occupation in both the ground and excited electronic state of phonon level $N-1$, we can choose a pulse without detuning to put all the excitation into $|e, N-1\rangle$.

v) The process continues iteratively, until all occupation is in the state $|g, 0\rangle$. The inverse of the transformation resulting from all of the above steps produces the desired state (23.3.1), when applied to $|g, 0\rangle$.

23.A Appendix: Radio-Frequency Ion Traps

An ion interacts strongly with any applied electric field, but it is not possible to create a static electric field which will trap an ion, since this would require a potential minimum, which is forbidden by Earnshaw's theorem. Ion trapping is nevertheless possible using a *time-dependent* electric field, and to a good approximation, it is possible to create a system in which an ion behaves as if it were trapped in a three-dimensional harmonic potential.

Using such techniques, it is now possible to prepare and store a single ion in a trap for a very long time, and such an ion can be cooled to the vibrational ground state of the trap, where it can manipulated very precisely using appropriate laser pulses.

In this section, we base our description on the review by *Leibfried, Blatt, Monroe* and *Wineland* [23.4]. The most commonly used kind of trap is based on the use of a time-dependent electric quadrupole potential, which can be written

$$\Phi(\boldsymbol{x},t) = \tfrac{1}{2}U\left(\alpha x^2+\beta y^2+\gamma z^2\right)+\tfrac{1}{2}\tilde{U}\cos(\omega_{\rm rf}t)\left(\alpha' x^2+\beta' y^2+\gamma' z^2\right). \tag{23.A.1}$$

The frequency $\omega_{\rm rf}$ is typically in the MHz regime, and thus within the volume of the trap, the potentials satisfy Laplace's equation, leading to the requirements

$$\alpha+\beta+\gamma=0, \qquad \alpha'+\beta'+\gamma'=0. \tag{23.A.2}$$

From these requirements it is clear that both the static and the radio-frequency part of the potential must be repulsive in some directions, and attractive in others. However, the radio-frequency potential changes sign rapidly, and we shall see that for appropriate choices of the parameters this can lead to a net attractive potential in all directions.

23.A.1 Classical Equations of Motion

As in the case of a harmonic oscillator with a time-independent potential, the solutions of the classical equations of motion give all the information needed to construct the quantum solutions. For the potential considered in (23.A.1), the equations of motion for the x, y and z co-ordinates decouple from each other, and all have the same form.

The equation of motion (for a positively charged ion of charge $Z|e|$ and mass m) in the x-direction takes the form

$$\ddot{x} = -\frac{Z|e|}{m}\frac{\partial\Phi}{\partial x} = -\frac{Z|e|}{m}\left(U\alpha+\tilde{U}\alpha'\cos(\omega_{\rm rf}t)\right)x. \tag{23.A.3}$$

This is a very well known equation, known as *Mathieu's equation*, whose properties and solutions are well understood. Using the substitutions

$$\tau=\tfrac{1}{2}\omega_{\rm rf}t, \qquad a_x=\frac{4Z|e|U\alpha}{m\omega_{\rm rf}^2}, \qquad q_x=\frac{2Z|e|\tilde{U}\alpha'}{m\omega_{\rm rf}^2}, \tag{23.A.4}$$

it can be transformed into the standard form [23.5] of Mathieu's equation

$$\frac{d^2x}{d\tau^2}+\left(a_x-2q_x\cos(2\tau)\right)x=0. \tag{23.A.5}$$

a) Solutions of Mathieu's Equation: Since the coefficients in Mathieu's equation are periodic with period π, if $x(\tau)$ is a solution, then $x(\tau+n\pi)$ is also a solution. From this, and the fact that $x(-\tau)$ is obviously also a solution, it follows that a general solution can be written in the form

$$x(\tau) = Ae^{-i\beta_x\tau}F(\tau) + Be^{i\beta_x\tau}F(-\tau), \tag{23.A.6}$$

where

$$F(\tau) = \sum_{-\infty}^{\infty} C_n e^{-2ni\tau}. \tag{23.A.7}$$

In these equations A and B are constants determined by the initial conditions, while the coefficients C_n and the parameter β_x are determined by the parameters a_x and q_x of Mathieu's equation.

b) Floquet Solutions: A solution of the kind (23.A.6) in which only one of A, B is non-zero is known as a Floquet solution of Mathieu's equation, and $-\beta_x$ is called the *Floquet characteristic exponent.*

c) Expressions for the Coefficients: Substituting this expression for $x(\tau)$ into the Mathieu equation (23.A.5) yields the recursion relation for the coefficients:

$$C_{n+1} - D_n C_n + C_{n-1} = 0, \tag{23.A.8}$$

$$\text{in which} \quad D_n \equiv \left(a_x - (2n+\beta_x)^2\right)/q_x. \tag{23.A.9}$$

Recursive use of (23.A.8) yields two continued fraction expressions

$$C_{n+1} = \cfrac{C_n}{D_{n+1} - \cfrac{1}{D_{n+2} - \cfrac{1}{\cdots}}}, \tag{23.A.10}$$

$$C_{n-1} = \cfrac{C_n}{D_{n-1} - \cfrac{1}{D_{n-2} - \cfrac{1}{\cdots}}}. \tag{23.A.11}$$

From these two expressions and the equations (23.A.8, 23.A.9), we can derive an implicit equation from which the value of β_x can be determined, namely

$$\beta_x^2 \equiv a_x - q_x D_0 = a_x - q_x\left(\frac{C_1}{C_0} + \frac{C_{-1}}{C_0}\right), \tag{23.A.12}$$

$$= a_x - \cfrac{q_x}{D_1(\beta_x) - \cfrac{1}{D_2(\beta_x) - \cfrac{1}{\cdots}}} - \cfrac{q_x}{D_{-1}(\beta_x) - \cfrac{1}{D_{-2}(\beta_x) - \cfrac{1}{\cdots}}}. \tag{23.A.13}$$

Here we have explicitly written $D_n(\beta_x)$ to display the dependence on β_x, and thus to emphasize that (23.A.13) is the equation which must be solved to determine β_x—usually numerically.

d) Approximate Solutions: In practice the values of both $|a_x|$ and q_x are rather small compared to 1, and β_x is usually somewhat less than 1. This means that for $n \neq 0$ the values of D_n are quite large, and that the continued fraction expansion can be truncated after only a few terms.

For example, to a first approximation this means that, from (23.A.9), we can approximate $D_1 \approx -4/q_x$. Now truncate the continued fraction expression (23.A.13) by neglecting all terms of higher order than $D_{\pm 1}$, which gives

$$\beta_x^2 \approx a_x + \tfrac{1}{2}q_x^2. \tag{23.A.14}$$

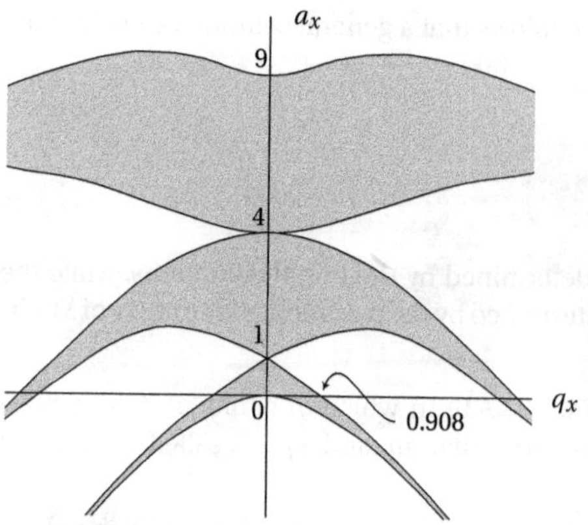

Fig. 23.4. Regions (shaded grey) where Mathieu's equation (23.A.5) has stable solutions, that is, where the characteristic exponent β_x defined in (23.A.6) is a real quantity. The figure shows only the three lowest regions. Ion trapping typically is implemented within the lowest region shown, and with both $0 < a_x \ll 1$ and $|q_x| \ll 1$.

Exercise 23.1 Approximation to Second Order: A more accurate result can be obtained by using the exact expression for $D_{\pm 1}(\beta_x)$, approximating $D_{\pm 2}(\beta_x) \approx 16/q_x$, and dropping terms of order higher than $D_{\pm 2}(\beta_x)$. Show that this yields a cubic equation in β_x^2 which in principle can be used to determine β_x. Find appropriate approximate solutions to this equation.

e) Stable Solutions of Mathieu's Equation: A *stable* solution of Mathieu's equation is one in which $x(\tau)$ is bounded for all τ, and corresponds to the trapping of the ion; solutions which are not stable do not trap the ion. For Mathieu's equation, the stability of the solution depends only on the values of the parameters q_x and a_x and not on the initial conditions.

The expression (23.A.6) for the solutions will be stable if β_x is real, and if the coefficients C_n are such as to yield convergence of the infinite series. For any value of β_x, we can see that $|D_{\pm n}| \sim 4n^2/q_x$, and that from (23.A.10) the coefficients C_n drop off very rapidly for $|n| \to \infty$. Thus, in this case the Fourier series certainly converges. The issue of stability depends therefore on the values of β_x for given values of a_x and q_x; these can be computed [23.5], and they follow the pattern illustrated in Fig. 23.4. The regions in which there are stable solutions, shaded grey, form a very interesting pattern, but this pattern is not easy to explain intuitively. The region usually used for ion trapping is that which contains the origin, and typically a_x is chosen to be very small, but positive, while $|q_x|$ is chosen to have a value between 0 and 0.908.

f) Behaviour of Solutions: Ion traps are normally operated in the first stable region, with very small values of a_x, and with $|q_x| < 0.908$, the stable range. In Fig. 23.5 we have shown two numerical solutions. One of these displays a motion very much like that of a particle moving in a harmonic potential, with only a small *micromotion* at the frequency the radio-frequency quadrupole potential. This kind of behaviour occurs for smaller q_x, and the resulting trap is rather weak, with a low oscillation frequency. Stronger trapping occurs at larger q_x, corresponding to a stronger radio-frequency quadrupole field, and this means that the trap frequency approaches much closer to the the quadrupole frequency. As seen from the figure, the motion of the ion is not very close to harmonic motion.

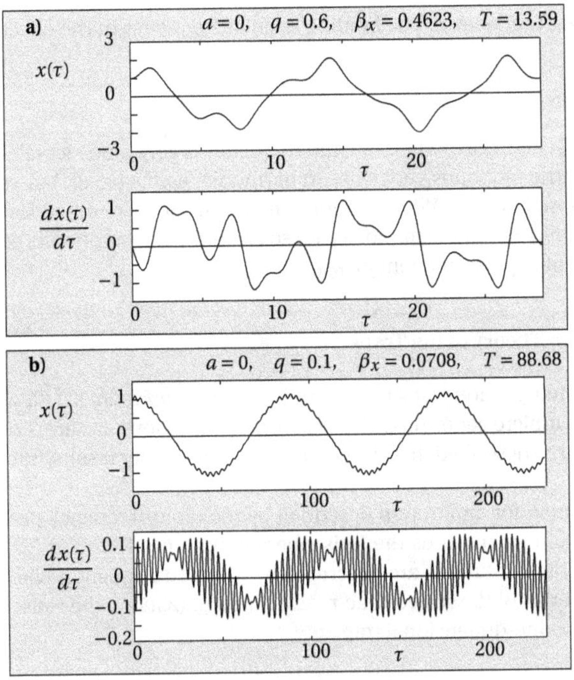

Fig. 23.5. Solutions of Mathieu's equation $d^2x/d\tau^2 + (a_x - 2q_x\cos(2\tau))\,x = 0$, for two sets of parameters: **a)** The micromotion frequency is about four times the effective trapping frequency, a realistic situation for a practical ion trap; **b)** The micromotion frequency is very much larger than the effective trap frequency, and the motion as revealed by the plot of $x(\tau)$ appears quite accurately harmonic. However in both cases, the micromotion velocity is of the same order of magnitude as the velocity expected from the effective trap.

g) Positive and Negative Frequency Solutions: For a stable solution, since β_x is real, the results of c) above show that coefficients C_n can also be chosen to be real, and $F(-\tau) = F^*(\tau)$. We can then write $x(\tau) \equiv G(\tau) + G^*(\tau)$, where the positive and negative frequency solutions are

$$G(\tau) = e^{-i\beta_x\tau}F(\tau), \tag{23.A.15}$$

$$G^*(\tau) = e^{i\beta_x\tau}F^*(\tau). \tag{23.A.16}$$

These solutions are Floquet solutions, and can be specified in terms of their initial conditions, namely

$$G(0) = u, \qquad \dot{G}(0) = -i(\beta_x u + v), \tag{23.A.17}$$

where the quantities u and v are defined by

$$u \equiv \sum_{-\infty}^{\infty} C_n = F(0), \qquad v \equiv \sum_{-\infty}^{\infty} 2nC_n = iF'(0). \tag{23.A.18}$$

The motion described by each of these solutions corresponds to a complex harmonic motion, given by the term $\exp(\pm i\beta_x\tau)$, modulated by the Floquet function $F(\tau)$ (or its complex conjugate) which is itself periodic with period π.

h) Trap Binding Frequency: The binding angular frequency of the effective trap is given by

$$\omega_{\text{trap}} = \tfrac{1}{2}\beta_x\omega_{\text{rf}}. \tag{23.A.19}$$

As can be seen in Fig. 23.5 b, when the trap frequency is very much less than the micromotion frequency, the motion is very close to harmonic, with a small, but very rapid micromotion superposed on this. When the micromotion is not very much faster than the trap frequency, as in Fig. 23.5 a, the motion is not so close to being harmonic, but the oscillation does take place at angular frequency ω_{trap}.

23.A.2 Quantum Theory of Ion Traps

The quantum theory of ion traps was formulated by *Combescure* [23.6], who showed how to describe a complete set of time-dependent wavefunctions, in terms of which the trap behaviour could be described. In this section we will derive her results, but using a different procedure.

The Hamiltonian for an ion trap described by the radio-frequency potential (23.A.1) is quadratic in $\boldsymbol{x}$ and $\boldsymbol{p}$, and thus the Heisenberg equations of motion for the trapped ion are linear, and their solutions are described by Mathieu's equation with operator initial conditions. In terms of the scaled time τ, as defined in (23.A.4), the Heisenberg equations of motion for the one-dimensional trap are

$$\frac{dx}{d\tau} = \frac{2}{m\omega_{\text{rf}}}\,p, \tag{23.A.20}$$

$$\frac{dp}{d\tau} = -\frac{m\omega_{\text{rf}}}{2}\big(a_x - q_x\cos(2\tau)\big)\,x. \tag{23.A.21}$$

a) Operator Initial Conditions: We consider the case of stable solutions, when β_x is real. In this case we can use the notation $G(\tau)$ introduced in (23.A.15, 23.A.16) and write the solution for the operator $x(\tau)$, corresponding to (23.A.6), as

$$x(\tau) = AG(\tau) + A^\dagger G^*(\tau). \tag{23.A.22}$$

These determine the operator A in terms of the initial value operators x_0, p_0; thus A is given by the equations

$$x_0 = u(A + A^\dagger), \qquad p_0 = -\frac{im\omega_{\text{rf}}(\beta_x u + v)}{2}(A - A^\dagger), \tag{23.A.23}$$

with u and v defined by (23.A.18).

From these we deduce that

$$A = \frac{x_0}{2u} + \frac{ip_0}{m\omega_{\text{rf}}(\beta_x u + v)}, \tag{23.A.24}$$

$$[A, A^\dagger] = \frac{\hbar}{m\omega_{\text{rf}}u(\beta_x u + v)}. \tag{23.A.25}$$

b) Commutation Relations and Normalization of C_n: The logical normalization of u and v (and hence the C_n) is given by the choice

$$m\omega_{\text{rf}}u(\beta_x u + v) = \hbar, \tag{23.A.26}$$

so that the commutation relations of A and $A^\dagger$ become the usual creation and destruction operator commutation relations

$$[A, A^\dagger] = 1, \tag{23.A.27}$$

and the expression in terms of initial momentum and position operators is

$$A = \frac{x_0}{2u} + \frac{iup_0}{\hbar}. \tag{23.A.28}$$

c) Commutation Relations for $x(\tau)$ and $p(\tau)$: We can check that the solutions we have derived are consistent with the canonical commutation relations as follows. From the solution (23.A.22) we can evaluate the commutator as

$$[x(\tau), p(\tau)] = \frac{m\omega_{\text{rf}}}{2}\left(G(\tau)\frac{dG^*(\tau)}{d\tau} - G^*(\tau)\frac{dG(\tau)}{d\tau}\right) \equiv \frac{m\omega_{\text{rf}}}{2}\mathcal{W}. \tag{23.A.29}$$

Since $G(\tau)$ and $G^*(\tau)$ are independent solutions of Mathieu's equation, $\mathcal{W}$, the right hand side of (23.A.29) is the Wronskian of the two solutions, which is time independent. Thus, $\mathcal{W}$ is determined by the value of the commutator at $\tau = 0$, that is $[x_0, p_0] = i\hbar$, so that

$$\mathcal{W} = \frac{2i\hbar}{m\omega_{\text{rf}}}. \tag{23.A.30}$$

The Wronskian and its time independence will play a significant role in the further analysis of the quantum theory of an ion trap.

d) Inverse Relations: From the expression (23.A.22), we can write

$$A = \frac{1}{\mathcal{W}}\left(x(\tau)\frac{dG^*(\tau)}{d\tau} - p(\tau)\frac{2G(\tau)}{m\omega_{\text{rf}}}\right). \tag{23.A.31}$$

If $U(\tau_1, \tau_0)$ is the time evolution operator, then the Heisenberg operator is

$$A(\tau) = U^{-1}(\tau, 0)AU(\tau, 0), \tag{23.A.32}$$

and we can therefore deduce that

$$A(-\tau) = \frac{1}{\mathcal{W}}\left(x(0)\frac{dG^*(\tau)}{d\tau} - p(0)\frac{2G^*(\tau)}{m\omega_{\text{rf}}}\right). \tag{23.A.33}$$

The Floquet periodicity means that $G(\tau)$ and $G'(\tau)$ have the periodicity property

$$G(\tau + N\pi) = e^{-iN\pi\beta_{\text{rf}}}G(\tau), \qquad G'(\tau + N\pi) = e^{-iN\pi\beta_{\text{rf}}}G'(\tau). \tag{23.A.34}$$

This property then leads to a similar property for $A(\tau)$

$$A(\tau + N\pi) = e^{-iN\pi\beta_{\text{rf}}}A(\tau). \tag{23.A.35}$$

e) Quantum States: Since A, $A^\dagger$ satisfy harmonic oscillator commutation relations, there exists a corresponding set of number states, which we shall write $|\xi, n\rangle$. These satisfy the usual relations

$$A|\xi, n\rangle = \sqrt{n}|\xi, n-1\rangle, \tag{23.A.36}$$

$$A^\dagger|\xi, n\rangle = \sqrt{n+1}|\xi, n+1\rangle, \tag{23.A.37}$$

$$A^\dagger A|\xi, n\rangle = n|\xi, n\rangle. \tag{23.A.38}$$

Furthermore, the time evolved state is then given by

$$|\xi, \tau, n\rangle = U(\tau, 0)|\xi, n\rangle, \tag{23.A.39}$$

and these mean that

$$A(-\tau)|\xi,\tau,n\rangle = \sqrt{n}|\xi,\tau,n-1\rangle, \tag{23.A.40}$$
$$A^\dagger(-\tau)|\xi,\tau,n\rangle = \sqrt{n+1}|\xi,\tau,n+1\rangle, \tag{23.A.41}$$
$$A^\dagger(-\tau)A(-\tau)|\xi,\tau,n\rangle = n|\xi,\tau,n\rangle. \tag{23.A.42}$$

The periodicity property of $A(\tau)$ then shows that

$$|\xi,\tau+N\pi,n\rangle = \lambda_{N,n}|\xi,\tau,n\rangle, \tag{23.A.43}$$

where $\lambda_{N,n}$ is a phase. It is important to note that we cannot determine this phase directly from the periodicity properties of $A(\tau)$; for that a detailed calculation of the wavefunctions is necessary. In (23.A.52) we show that $\lambda_{N,n} = \exp(\mathrm{i}(n+1/2)\beta_{\mathrm{rf}}N\pi)$.

f) Wavefunctions: The wavefunctions can be determined quite simply using solutions of the Heisenberg equations of motion, and the results are the same as those of Combescure [23.6].

g) Wavefunction for $n = 0$: From the explicit expression for $A(\tau)$ given in (23.A.33), and taking $x(0) \to x$, $p(0) \to -\mathrm{i}\hbar\partial_x$ as the Schrödinger picture operators, the $n = 0$ state wavefunction $\xi_0(x,\tau) \equiv \langle x|\xi,0,\tau\rangle$ is determined by the equation

$$\left(xG'(\tau)^* + 2\mathrm{i}\hbar\partial_x\frac{G(\tau)^*}{m\omega_{\mathrm{rf}}}\right)\xi_0(x,\tau) = 0, \tag{23.A.44}$$

with the solution

$$\xi_0(x,\tau) = \exp\left(\frac{\mathrm{i}m\omega_{\mathrm{rf}}}{4\hbar}\left(\frac{G(\tau)'}{G(\tau)}\right)^* x^2 + r(t)\right), \tag{23.A.45}$$

where $r(\tau)$ gives a normalization factor, which can be determined directly from the Schrödinger equation. In order to be consistent with the equations of motion (23.A.20, 23.A.21), the Schrödinger equation takes the form

$$\mathrm{i}\hbar\frac{\partial\xi_0(\tau)}{\partial\tau}\frac{\omega_{\mathrm{rf}}}{2} = \left(-\frac{\hbar^2}{2m}\frac{\partial^2}{\partial x^2} + \frac{m\omega_{\mathrm{rf}}^2x^2}{8}\left(a_x - 2q_x\cos(2\tau)\right)\right). \tag{23.A.46}$$

Direct substitution of the explicit form (23.A.45) into this Schrödinger equation, and use of the fact that $G(\tau)^*$ is a solution of Mathieu's equation, yields the explicit expression

$$r(\tau) = -\tfrac{1}{2}\log G(\tau)^* + c, \tag{23.A.47}$$

where c is a constant of integration.

h) Normalization of $\xi_0(x,\tau)$: Notice that

$$|\xi_0(x,\tau)|^2 = \exp\left(\frac{\mathrm{i}m\omega_{\mathrm{rf}}x^2}{4\hbar}\left[\frac{G(\tau)G'(\tau)^* - G(\tau)^*G'(\tau)}{|G(\tau)|^2}\right] - \log|G(\tau)| + 2\,\mathrm{Re}\,(c)\right). \tag{23.A.48}$$

Recognizing the Wronskian in this expression, and using its value, as given by (23.A.29), of $W = 2\mathrm{i}\hbar/m\omega_{\mathrm{rf}}$, we can choose an appropriate value of c to give, as the time-dependent normalized $n = 0$ probability density

$$|\xi_0(x,\tau)|^2 = \frac{1}{\sqrt{2\pi}|G(\tau)|}\exp\left(-\frac{x^2}{2|G(\tau)|^2}\right). \tag{23.A.49}$$

i) Normalized Wavefunction for $n = 0$: Using this normalization, the normalized time-dependent $n = 0$ wavefunction can be chosen to be

$$\xi_0(x,\tau) = \sqrt{\frac{1}{\sqrt{2\pi}G(\tau)^*}}\exp\left(\frac{im\omega_{\rm rf}}{4\hbar}\left(\frac{G(\tau)'}{G(\tau)}\right)^* x^2\right). \tag{23.A.50}$$

j) Wavefunctions for $n > 0$: These can be determined using the creation operator as in (23.A.37). It is clear that the resulting wavefunction will be a polynomial of order n multiplied by $\xi_0(x,\tau)$, and that the polynomials so defined will be orthogonal polynomials with weight function $|\xi_0(x,t)|^2$, as given in (23.A.49). They are therefore proportional to Hermite polynomials of argument $x/\sqrt{2}|G(\tau)|$. To determine the constant of proportionality it is only necessary to determine the coefficient of x^n in the polynomial, and using the explicit formula for the creation operator, this coefficient is $\left(G(\tau)^*\right)^{-n}/\sqrt{n!}$.

The wavefunction then can be determined to be

$$\begin{aligned}\xi_n(x,\tau) &= (2^n n!)^{-1/2}\left(\frac{G(\tau)}{G(\tau)^*}\right)^{n/2}\sqrt{\frac{1}{\sqrt{2\pi}G(\tau)^*}}\\ &\times \mathrm{H}_n\left(\frac{x}{\sqrt{2}|G(\tau)|}\right)\exp\left(\frac{im\omega_{\rm rf}}{4\hbar}\left(\frac{G(\tau)'}{G(\tau)}\right)^* x^2\right).\end{aligned} \tag{23.A.51}$$

k) Periodicity Properties of the Wavefunctions: The expression for the wavefunctions is composed of two parts with different periodicity properties as follows:

i) The second line is periodic in τ with period π, since the phase factors as in (23.A.34) cancel in the exponential, and do not occur in the modulus which appears in the Hermite polynomial.

ii) The first line has a periodicity property determined from (23.A.34) which does involve the phase factor $\exp(\mathrm{i}\beta_{\rm rf}\pi)$; explicitly we therefore find

$$\xi_n(x,\tau+N\pi) = \mathrm{e}^{-\mathrm{i}(n+1/2)\beta_{\rm rf}N\pi}\xi_n(x,\tau). \tag{23.A.52}$$

iii) This is the natural generalization of the harmonic oscillator periodicity,

$$\xi_{\mathrm{ho},n}(x,\tau+T) = \mathrm{e}^{-\mathrm{i}(n+1/2)\omega T}\xi_{\mathrm{ho},n}(x,\tau), \tag{23.A.53}$$

in which T may have any value, rather than the value $N\pi$ appropriate for the ion trap.

Exercise 23.2 The Harmonic Trap Limit: Consider the case where q_x is very small, and a_x is positive, so that the effect of the radio-frequency field is very weak. Mathieu's equation in this limit approaches a simple harmonic oscillator equation with angular frequency $\Omega = \frac{1}{2}\omega_{\rm rf}\sqrt{a_x}$. In this limit, show that

$$A = \sqrt{\frac{m\Omega}{2\hbar}}\,x_0 + \sqrt{\frac{1}{2m\hbar\Omega}}\,p_0\,. \tag{23.A.54}$$

This is the expression for the destruction operator for limiting harmonic trap.

24. The Ion Trap Quantum Computer

The ion trap quantum computer, devised by *Cirac* and *Zoller* in 1995 [24.1], uses ions, usually of calcium, beryllium or barium, trapped in a linear configuration ion trap of the form discussed in Chap. 23. The qubits are made using the electronic levels of the ions, and quantum information can be exchanged between ions using the Coulomb forces between them. The 1995 proposal used two major enabling techniques:

i) *Quantum Gates*: Using the techniques of quantum state engineering described in Sect. 23.3, it was shown how to implement a set of universal quantum gates, such as those described in Sect. 2.3. The implementation of a two-qubit gate relies on the transfer of quantum information to and from the electronic levels of an ion to the quantized centre of mass motion of the linear array of ions, in order to manipulate the quantum information encoded in one ion in a way which depends on the quantum information in another ion.

ii) *Initialization of the Qubits*: It is important to put the atoms used for the qubits into the initial state required. Normally this is the state in which the ion is in both the motional ground state, and in the chosen electronic state. The motional ground state is achieved by using sideband cooling, while the electronic ground state is reached by using techniques of optical pumping.

iii) *Efficient Detection*: For a quantum computer it is necessary to detect the quantum state of the qubit with 100% efficiency. This can be done using a three level system such as described in Sect. 19.4. In the Innsbruck quantum computer this is done with $^{40}Ca^{+}$, as illustrated in Fig. 24.1. For each ion, the internal electronic Zeeman levels $D_{5/2}(m=-1/2)$ and $S_{1/2}(m=-1/2)$ encode the logical states $|0\rangle$ and $|1\rangle$ of a qubit. For coherent operations, a laser at a wavelength of 729 nm excites the quadrupole transition $S_{1/2} \leftrightarrow D_{5/2}$ connecting the qubit states.

   Detection is achieved by illuminating the ion with light of wavelength 397 nm. If fluorescence is immediately visible, this is a measurement corresponding to the qubit being in the $S_{1/2}$ state; if no fluorescence is visible, the measurement corresponds to the qubit being in the $D_{5/2}$ state.

Since the original proposal different kinds of gates have been introduced—in particular the *Mølmer–Sørenson gate*, treated in Sect. 24.3, has been widely applied.

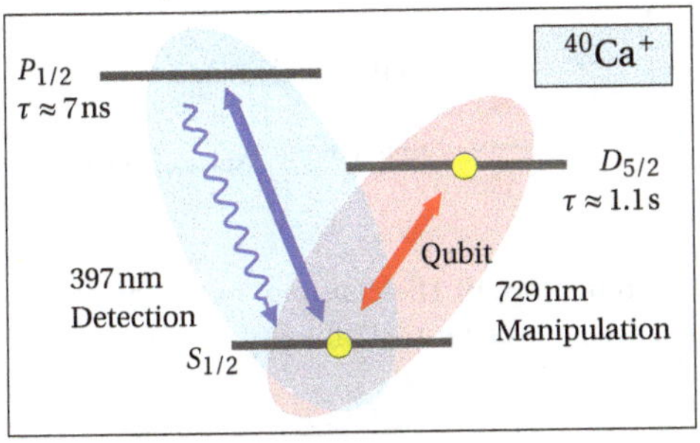

Fig. 24.1. The three-level system used in the Innsbruck quantum computer. The weak transition on the right provides the qubit, while the fluorescence detection occurs on the strong transition on the left.

24.1 Ions in a Linear Trap

We consider a string of N ions in a linear trap. A linear trap corresponds to a confinement of the motion along x-, y- and z-directions in an anisotropic harmonic potential with frequencies $\nu \equiv \nu_x \ll \nu_y, \nu_z$. The equilibrium positions of the ions will be given by the confining forces of the trapping potential balancing the Coulomb repulsion between the ions. If the ions have been previously laser cooled in all three dimensions, they undergo small oscillations around the equilibrium positions. In this case, the motion of the ions is described in terms of normal modes.

24.1.1 Two Ions in a Linear Trap

As an example, we will consider a one-dimensional model, consisting of two ions in a linear trap with internal levels $|g\rangle$ and $|e\rangle$, and coupled to laser light. In this case there are only two normal modes, the *centre of mass mode* and the *stretch mode*, as illustrated in Fig. 24.2 b. The Hamiltonian for this system is

$$\begin{aligned} H = {} & \frac{P_1^2}{2M} + \tfrac{1}{2}M\nu^2 X_1^2 + \frac{P_2^2}{2M} + \tfrac{1}{2}M\nu^2 X_2^2 + \frac{e^2}{4\pi\epsilon_0 |X_1 - X_2|} \\ & + \hbar\omega_{1eg}|e,1\rangle\langle e,1| + \hbar\omega_{2eg}|e,2\rangle\langle e,2| \\ & - \left(\tfrac{1}{2}\hbar\Omega_1(t)\mathrm{e}^{\mathrm{i}kX_1}\mathrm{e}^{-\mathrm{i}\omega_1 t}|e,1\rangle\langle g,1| + \mathrm{h.c.}\right) \\ & - \left(\tfrac{1}{2}\hbar\Omega_2(t)\mathrm{e}^{\mathrm{i}kX_2}\mathrm{e}^{-\mathrm{i}\omega_2 t}|e,2\rangle\langle g,2| + \mathrm{h.c.}\right). \end{aligned} \tag{24.1.1}$$

Introducing centre of mass and relative coordinates

$$X = \tfrac{1}{2}(X_1 + X_2), \qquad x = X_2 - X_1, \tag{24.1.2}$$

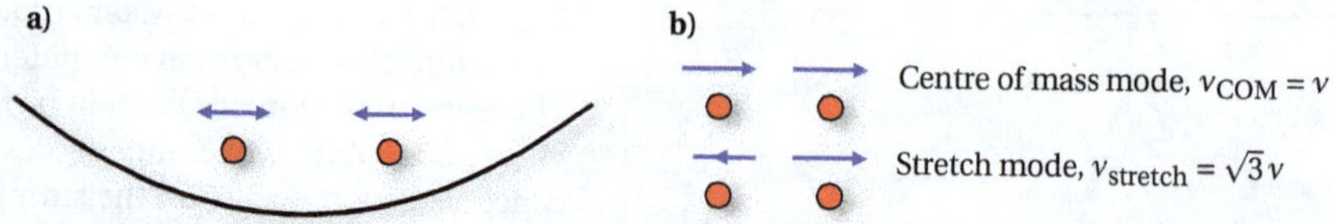

Fig. 24.2. a) Two ions held in the same harmonic trap; **b)** The centre of mass and stretch mode frequencies are always distinct, with a ratio of exactly $1:\sqrt{3}$ for any number of ions.

the Hamiltonian becomes

$$
\begin{aligned}
H = &\left(\frac{P^2}{2(2M)} + \tfrac{1}{2}(2M)\nu^2 X^2\right) + \left(\frac{p^2}{2(M/2)} + \tfrac{1}{2}(\tfrac{1}{2}M)\nu^2 x^2 + \frac{e^2}{4\pi\epsilon_0 |x|}\right) \\
&+\hbar\omega_{1eg}|e,1\rangle\langle e,1| + \hbar\omega_{2eg}|e,2\rangle\langle e,2| \\
&-\left(\tfrac{1}{2}\Omega_1(t)\mathrm{e}^{\mathrm{i}kX - \mathrm{i}kx/2 - \mathrm{i}\omega_1 t}|e,1\rangle\langle g,1| + \text{h.c.}\right) \\
&-\left(\tfrac{1}{2}\Omega_2(t)\mathrm{e}^{\mathrm{i}kX + \mathrm{i}kx/2 - \mathrm{i}\omega_2 t}|e,2\rangle\langle g,2| + \text{h.c.}\right).
\end{aligned}
\qquad (24.1.3)
$$

a) Quadratic Approximation about the Equilibrium Position: The equilibrium position is given by the value of the relative co-ordinate

$$
x_{\text{eq}} = \left(\frac{e^2}{2\pi\epsilon_0 M\nu^2}\right)^{1/3}, \qquad (24.1.4)
$$

and a quadratic approximation to the vibrational potential about this equilibrium point is

$$
V_{\text{vibrational}}(x) \approx \frac{3}{2}M\nu^2(x - x_{\text{eq}})^2. \qquad (24.1.5)
$$

This yields a stretch-mode angular frequency of

$$
\nu_{\text{stretch}} = \sqrt{3}\nu. \qquad (24.1.6)
$$

b) Hamiltonian in Terms of Phonon Operators and in a Rotating Frame: We can now write kX and kx in terms of creation and destruction operators, using the Lamb–Dicke parameter as in (23.1.20), and move in a frame rotating at the laser frequencies as in Sect. 23.1.6. This finally gives the form most useful for analysis of the problem

$$
\begin{aligned}
\tilde{H} = &\hbar\nu(a_{\text{cm}}^\dagger a_{\text{cm}} + \tfrac{1}{2}) + \hbar\sqrt{3}\nu(a_{\text{r}}^\dagger a_{\text{r}} + \tfrac{1}{2}) \\
&-\hbar\Delta_1|e,1\rangle\langle e,1| - \hbar\Delta_2|e,2\rangle\langle e,2| \\
&-\left(\tfrac{1}{2}\hbar\Omega_1(t)\mathrm{e}^{\mathrm{i}\eta_{\text{cm}}(a_{\text{cm}}+a_{\text{cm}}^\dagger)}\,\mathrm{e}^{-\mathrm{i}\eta_{\text{r}}(a_{\text{r}}+a_{\text{r}}^\dagger)}|e,1\rangle\langle g,1| + \text{h.c.}\right) \\
&-\left(\tfrac{1}{2}\hbar\Omega_2(t)\mathrm{e}^{\mathrm{i}\eta_{\text{cm}}(a_{\text{cm}}+a_{\text{cm}}^\dagger)}\,\mathrm{e}^{\mathrm{i}\eta_{\text{r}}(a_{\text{r}}+a_{\text{r}}^\dagger)}|e,2\rangle\langle g,2| + \text{h.c.}\right).
\end{aligned}
\qquad (24.1.7)
$$

The various parts of the Hamiltonian are:

i) The first line describes the harmonic collective oscillations of the centre of mass, and stretch mode with oscillation frequency ν and $\sqrt{3}\nu$ respectively.

ii) The second line is the bare atomic Hamiltonian for the first and second ion in the rotating frame, with $\Delta_{1,2}$ the laser detunings.

iii) The third and fourth lines are the laser couplings to the ions, with Rabi frequencies $\Omega_{1,2}$ of the laser acting on each ion, and $\eta_{\text{cm,r}}$ the Lamb-Dicke parameters for the centre of mass and stretch modes.

24.1.2 *N* Atoms in a Linear Trap

Our previous discussion of the Lamb-Dicke expansion of the Hamiltonian is readily extended to the general case of a string of N ions. In this case, there will be a set of N collective modes of ion motion, whose frequencies can be determined by using the same small amplitude approximation as for the case of two trapped ions.

a) Mode Frequencies: Remarkably, the frequencies of the first two modes are the same as for two trapped ions, that is:

i) The frequency of the centre of mass mode in the x-direction is $\nu_x \equiv \nu$.

ii) The next frequency is that of the stretch mode and has the exact value $\sqrt{3}\nu_x$.

iii) *James* [24.2] has numerically determined the eigenvalues and modes for all N up to 10. A feature of great significance for the quantum computer proposal is that the frequency spacing of the low lying modes is almost *independent* of the number of ions N in the trap. For $N = 10$ he gives the mode frequencies as approximately

$$\left\{1,\ \sqrt{3},\ \sqrt{5.8},\ \sqrt{9.4},\ \sqrt{13.6},\ \sqrt{18.5},\ \sqrt{23.9},\ \sqrt{29.8},\ \sqrt{36.3},\ \sqrt{43.2}\right\}\nu. \qquad (24.1.8)$$

For $N < 10$, the the first two frequencies are the same, and the remainder differ by no more that 10% from these values.

b) Laser Excitation of Modes: Because there is always a significant gap between the first two modes, it is possible to tune the frequency of a laser acting on *a single ion* in such a way as to *excite only the centre of mass mode.* This means that the excitation energy is shared by the translational motion of *all* of the ions. Thus, information stored in the electronic state of a single ion can be transferred to the centre of mass motion of the whole set of ions, and is therefore then known by all of the ions. This will be fundamental for the implementation of a quantum computer.

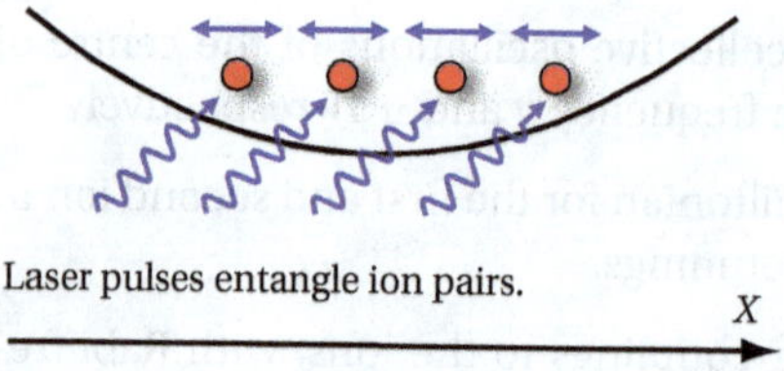
Laser pulses entangle ion pairs.
X

Fig. 24.3. Schematic view of an ion trap quantum computer, with N ions held in the same harmonic trap.

24.2 Implementation of a Two-Qubit Quantum Gate

As first proposed in [24.1], N cold ions interacting with laser light and moving in a linear trap provide a realistic physical system to implement a quantum computer. This gate, now known as the *Cirac–Zoller phase gate* is *conceptually* the simplest two-qubit gate so far devised.

The heart of this procedure, i.e. the dynamics presented in (24.2.16), was demonstrated by the NIST group in 1995 with a single beryllium ion [24.3]. In 2003 the Innsbruck group implemented the complete protocol with individual addressing of two calcium ions [24.4, 24.5].

a) Gate Procedure: The following procedure is used to perform a two-qubit gate between two ions in an ion string:

i) The string of ions is sideband-cooled to the motional ground state.

ii) The quantum information of the first qubit is transferred to the centre of mass mode of the motional degree of freedom (the *data bus mode*) of the ion string.

iii) The resulting motional state does not affect only the particular ion addressed, it affects the *ion string* as a whole.

iv) As a consequence, operations on any second ion, conditioned on the motional state of the ion string, can be performed.

v) Finally the quantum information of the motion is mapped back onto the first ion of the string, leaving the centre of mass degree of freedom in the ground state.

b) Distinctive Features: The distinctive features of this system are:

i) It allows the implementation of a complete pair of quantum gates between any set of ions. These ions need not be nearest neighbours.

ii) Decoherence is comparatively small.

iii) The final readout can be performed with essentially unit efficiency.

iv) It requires single ion addressing and ground-state cooling of the bus mode.

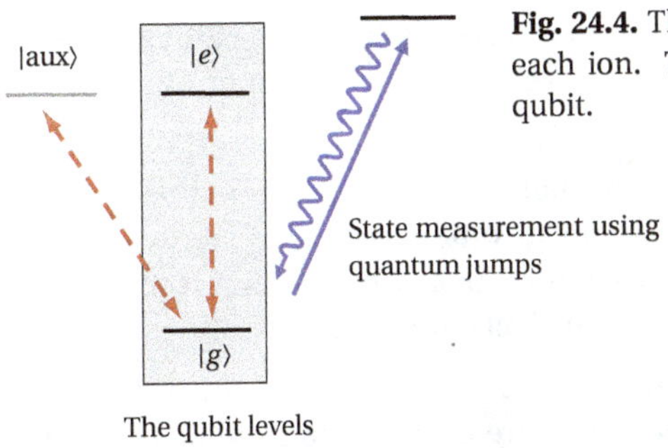

Fig. 24.4. The level scheme and transitions used for each ion. The two levels in the grey box form the qubit.

24.2.1 Representation of Qubits

Fig. 24.3 illustrates the arrangement of ions, laser beams and the motion of the ions; while Fig. 24.4 gives the level structure necessary for the device.

a) The Qubits: The quantum information is stored using long-lived internal states of the N ions, with

$$\left.\begin{array}{ll}\text{Ground state:} & |g,j\rangle \equiv |0,j\rangle, \\ \text{Excited state:} & |e,j\rangle \equiv |1,j\rangle,\end{array}\right\} \quad j=1,\dots,N. \tag{24.2.1}$$

The quantum information is thus stored in a superposition state of the form

$$|\text{Info}\rangle \equiv a_g|g,j\rangle + a_e|e,j\rangle. \tag{24.2.2}$$

Transitions between $|g,j\rangle$ and $|e,j\rangle$ are generated by on-resonant laser light with the particular polarization, which we will label as $q=0$.

b) Auxiliary State: In addition, we require that each ion should have another excited state

$$\text{Auxiliary state:} \quad |\text{aux},j\rangle, \qquad j=1,\dots,N. \tag{24.2.3}$$

This state is only occupied during the process of performing a gate operation, and is always unoccupied when no operation is being performed. Hence, it *does not participate in the storage of quantum information,* and for this reason we have given it the notation $|\text{aux}\rangle$, which is distinctly different from that used for the information storage states $|g\rangle$ and $|\text{aux}\rangle$.

Physically, the states $|\text{aux}\rangle$ and $|e\rangle$ would normally be very similar, perhaps being different Zeeman sublevels, and transitions between $|g,j\rangle$ and the auxiliary state $|\text{aux},j\rangle$ would be generated by on-resonant laser light with the particular polarization label $q=1$, different from that chosen for the state $|\text{aux}\rangle$, which we have denoted as $q=0$.

24.2.2 Manipulation of Qubits

In this system independent manipulation of each individual qubit is accomplished by addressing the ions with individual laser beams and inducing various kinds of Rabi rotation. In particular we make use of these Rabi rotations:

i) *Qubit Rotations*: These are the Rabi rotations described in Sect. 23.3.1, which produce superpositions of ground and excited states, without affecting the motional state of the ion;

ii) *Motional Conversion*: These are the Rabi rotations described Sect. 23.3.2, which convert *electronic superpositions* to *superpositions of motional states.* The information in the electronic excitation of a *particular* ion is transferred into the *collective centre of mass motion* of all of the ions.

The heart of the idea is the implementation of a two-qubit gate between *any* two ions in the trap by utilizing the collective centre of mass mode, which plays the role of a quantum data bus, making information stored in a given ion available to any other ion. For this, we must first ensure that the collective phonon modes have been cooled to the ground state.

24.2.3 Basic Gate Operations

Our aim is now to design an appropriate universal set of gates, in terms of which any quantum operation can be expressed, as we have explained in Sect. 2.3.

a) Single Qubit Rotations: These can be performed tuning a laser on resonance with the internal transition, with detuning $\Delta_j = 0$, and with polarization $q = 0$. In an interaction picture the corresponding Hamiltonian is that of Sect. 23.2.2 a

$$H_j = \tfrac{1}{2}\hbar\Omega\left(|e,j\rangle\langle g,j| + |g,j\rangle\langle e,j|\right). \tag{24.2.4}$$

For an interaction time $t = k\pi/\Omega$ (i.e. using a $k\pi$ pulse), this process is described by the following unitary evolution operator

$$V_j^k = \exp\Big(-\mathrm{i}(k\pi/2)\left(|g,j\rangle\langle e,j| + |e,j\rangle\langle g,j|\right)\Big), \tag{24.2.5}$$

so that we achieve a Rabi rotation

$$\begin{aligned}|g,j\rangle &\longrightarrow \cos(k\pi/2)\ |g,j\rangle - \mathrm{i}\sin(k\pi/2)|e,j\rangle\,,\\ |e,j\rangle &\longrightarrow \cos(k\pi/2)\ |e,j\rangle - \mathrm{i}\sin(k\pi/2)|g,j\rangle\,.\end{aligned} \tag{24.2.6}$$

b) Motional Conversion Hamiltonian: If the laser addressing the j-th ion is tuned to the lower motional sideband of the centre of mass mode, then in the interaction picture the Hamiltonian has the form given in Sect. 23.2.2 b. There are two cases, depending on whether or not the auxiliary state is being used.

Addressing the States Carrying the Quantum Information:

$$H_j^{eg} = \tfrac{1}{2}\Omega_N\left[|e,j\rangle\langle g,j|a + |g,j\rangle\langle e,j|a^\dagger\right]. \tag{24.2.7}$$

Addressing the Auxiliary State:

$$H_j^{\mathrm{aux}} = \tfrac{1}{2}\Omega_N\left[|\mathrm{aux},j\rangle\langle g,j|a + |g,j\rangle\langle \mathrm{aux},j|a^\dagger\right]. \tag{24.2.8}$$

Here:

i) $a^\dagger$ and a are the creation and annihilation operator of the centre of mass phonons.

ii) We have defined the *effective Rabi frequency* Ω_N in terms of the Lamb-Dicke parameter η and the number of ions N

$$\Omega_N = \frac{\eta\Omega}{\sqrt{N}} . \tag{24.2.9}$$

The factor $\sqrt{N}$ appears since the effective mass of the centre of mass motion is NM, and the amplitude of the mode scales like $1/\sqrt{NM}$.

iii) Note that, for simplicity, we have chosen the laser phase to be zero.

c) Notation: The qubits are made from the electronic states of the ions at locations j, and there will be a *qubit state* at each location j.

The motional state of the centre of mass will be described by quanta with values $0, 1, 2 \ldots$. This state is a property of the N ions as a whole, and we will call it the *data bus state.*

Because we will wish to consider sometimes one qubit, and sometimes two qubits, we will use two notations.

i) *One Qubit*: We will use a notation like

$$|e, j\rangle|0\rangle, \qquad \begin{cases} \text{Qubit } j \text{ is in the state } |e\rangle, \\ \text{Data bus is in the state is } |0\rangle. \end{cases} \tag{24.2.10}$$

ii) *Two Qubits*: When we wish to talk about two different qubits and the centre of mass motional state, the notation is

$$|e, j\rangle|g, j'\rangle|1\rangle, \qquad \begin{cases} \text{Qubit } j \text{ is in the state } |e\rangle, \\ \text{Qubit } j' \text{ is in the state } |g\rangle, \\ \text{Data bus is in the state is } |1\rangle. \end{cases} \tag{24.2.11}$$

d) Action of the Hamiltonian H_j^{eg}: As discussed in Sect. 23.2.2 b, the Hamiltonian H_j^{eg} couples states to each other as follows:

i) $|g, j\rangle|0\rangle$ is uncoupled from all states, and therefore does not change.

ii) Nontrivial couplings occur pair wise

$$\left.\begin{array}{llll} |g, j\rangle|1\rangle \longleftrightarrow |e, j\rangle|0\rangle, & \text{Rabi frequency} & \Omega_N, \\ |g, j\rangle|2\rangle \longleftrightarrow |e, j\rangle|1\rangle, & \text{Rabi frequency} & \sqrt{2}\Omega_N, \\ |g, j\rangle|3\rangle \longleftrightarrow |e, j\rangle|2\rangle, & \text{Rabi frequency} & \sqrt{3}\Omega_N, \\ \vdots \qquad\qquad \vdots & \vdots & \vdots \end{array}\right\} \tag{24.2.12}$$

These two transitions have different Rabi frequencies, and the second one involves two centre of mass quanta. It will be important to avoid any occurrence of such a transition, and for this the availability of the auxiliary state will be essential.

We can generate a $k\pi$ pulse for the first of these transitions by turning on a laser beam for the time $T_k = k\pi/\Omega_N$. The corresponding evolution of the system is described by the unitary operator

$$U_j^{eg}(k\pi) = \exp\left[-\tfrac{1}{2}\mathrm{i}k\pi\left(|e,j\rangle\langle g,j|a + |g,j\rangle\langle e,j|a^\dagger\right)\right]. \tag{24.2.13}$$

As noted above, this transformation does not change $|g,j\rangle|0\rangle$, whereas

$$\begin{aligned} |g,j\rangle|1\rangle &\longrightarrow \cos(k\pi/2)\ |g,j\rangle|1\rangle - \mathrm{i}\sin(k\pi/2)|e,j\rangle|0\rangle\,, \\ |e,j\rangle|0\rangle &\longrightarrow \cos(k\pi/2)\ |e,j\rangle|0\rangle - \mathrm{i}\sin(k\pi/2)|g,j\rangle|1\rangle\,. \end{aligned} \tag{24.2.14}$$

e) Action of the Hamiltonian H_j^{aux}: This is exactly the same as for H_j^{eg}, with the replacement $e \to \mathrm{aux}$.

24.2.4 The Controlled Phase Gate

Let us now show how a two-bit gate can be performed using these operators. We will utilize two qubits (as described in Sect. 2.3.2) namely:

i) *A Control Qubit*, which will be located on ion number m.

ii) *A Target Qubit*, which will be located on ion number n.

a) Component Operations: We consider the following three-step process (see Fig. 24.5):

i) *Control Qubit—Ion Number m*: A π laser pulse of the Hamiltonian H_m^{eg} is applied to the *control bit* at position m. The evolution corresponding to this step is given by $U_m^{eg}(\pi)$as seen in Fig. 24.5 a.

ii) *Target Qubit—Ion Number n*: We now apply a 2π pulse using the auxiliary state to the *target bit* at position n. The corresponding evolution operator $U_n^{\mathrm{aux}}(2\pi)$ changes the sign of the state $|g,n\rangle|1\rangle$ (without affecting the others) via a rotation through the auxiliary state $|\mathrm{aux},n\rangle|0\rangle$ (Fig. 24.5 b).

   It is essential to use the auxiliary state for this process. Since quantum information is coded on the state $|e,n\rangle|1\rangle$, choosing U_n^{eg} would invoke the transition $|g,n\rangle|2\rangle \leftrightarrow |e,n\rangle|1\rangle$ on that state, which we can see from (24.2.12), has the wrong Rabi frequency. However, no quantum information is carried on the auxiliary state $|\mathrm{aux},m\rangle|1\rangle$, and this means that the state of the system will have no component of this state.

iii) *Target Qubit—Ion Number m*: We repeat the transformation $U_m^{eg}(\pi)$ used in i), as in Fig. 24.5 c. This restores the target bit to its original state.

b) The Phase Gate: Thus, the unitary operation for the whole process is

$$\mathsf{U}_{m,n} \equiv U_m^{eg}(\pi)\, U_n^{\mathrm{aux}}(2\pi)\, U_m^{eg}(\pi). \tag{24.2.15}$$

The resulting effect on the relevant states is:

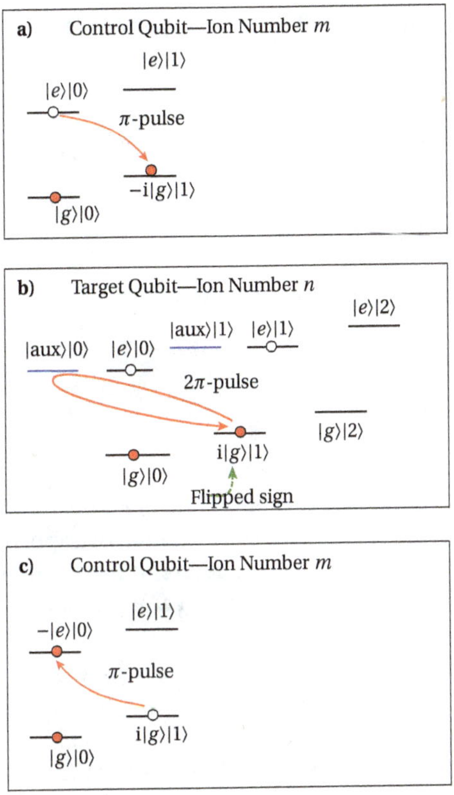

Fig. 24.5. The two-qubit quantum gate—see (24.2.16). **a)** First step—ion number m: The qubit of the first atom is swapped to the photonic data bus with a π-pulse on the lower motional sideband, with a phase change of $-\mathrm{i}$; **b)** Second step—ion number n: The state $|g,1\rangle$ acquires a minus sign due to a 2π-rotation via the auxiliary atomic level $|\mathrm{aux},1\rangle$ on the lower motional sideband; **c)** Third step—ion number m: The qubit of the first atom is swapped from the photonic data bus back to $|e,0\rangle$ with a π-pulse on the lower motional sideband, with a phase change of i.

$$
\begin{array}{lllllll}
 & \overset{U_m^{eg}(\pi)}{} & & \overset{U_n^{\mathrm{aux}}(2\pi)}{} & & \overset{U_m^{eg}(\pi)}{} & \\
|g,m\rangle|g,n\rangle|0\rangle & \longrightarrow & |g,m\rangle|g,n\rangle|0\rangle & \longrightarrow & |g,m\rangle|g,n\rangle|0\rangle & \longrightarrow & |g,m\rangle|g,n\rangle|0\rangle, \\
|g,m\rangle|e,n\rangle|0\rangle & \longrightarrow & |g,m\rangle|e,n\rangle|0\rangle & \longrightarrow & |g,m\rangle|e,n\rangle|0\rangle & \longrightarrow & |g,m\rangle|e,n\rangle|0\rangle, \\
|e,m\rangle|g,n\rangle|0\rangle & \longrightarrow & -\mathrm{i}\,|g,m\rangle|g,n\rangle|1\rangle & \longrightarrow & \mathrm{i}\,|g,m\rangle|g,n\rangle|1\rangle & \longrightarrow & |e,m\rangle|g,n\rangle|0\rangle, \\
|e,m\rangle|e,n\rangle|0\rangle & \longrightarrow & -\mathrm{i}\,|g,m\rangle|e,n\rangle|1\rangle & \longrightarrow & -\mathrm{i}\,|g,m\rangle|e,n\rangle|1\rangle & \longrightarrow & -\,|e,m\rangle|e,n\rangle|0\rangle.
\end{array}
\tag{24.2.16}
$$

The effect of this interaction is to change the sign of the state only when both ions are initially excited. Note that the state of the centre of mass mode is restored to the vacuum state $|0\rangle$ after the process.

This is exactly the the controlled-phase gate as defined in (2.3.8); this can be seen by mapping our notation, which is expressed in terms of the physical atomic states, to the qubit notation of Sect. 2.2. Thus we make the correspondence

$$\left.\begin{array}{lcl} |g,m\rangle\ |g,n\rangle\ |0\rangle & \longleftrightarrow & |0,0\rangle, \\ |g,m\rangle\ |e,n\rangle\ |0\rangle & \longleftrightarrow & |0,1\rangle, \\ |e,m\rangle\ |g,n\rangle\ |0\rangle & \longleftrightarrow & |1,0\rangle, \\ |e,m\rangle\ |e,n\rangle\ |0\rangle & \longleftrightarrow & |1,1\rangle. \end{array}\right\} \tag{24.2.17}$$

Hence equation (24.2.16) is indeed the controlled-phase gate $\mathcal{U}_\pi^{(2)}$ as defined in (2.3.8).

24.3 Mølmer–Sørensen Gate

An alternative implementation of a two-qubit gate was proposed by *Mølmer* and *Sørensen* [24.6,24.7] in 1999. The gate is used on the same configuration of trapped ions as in Sect. 24.1, in which the harmonic motion of the centre of mass mode provides a *data bus* which enables information transfer between any pair of ions, and the same approximations are made.

The difference arises from the use of a *bichromatic laser beam*, with two frequencies detuned from the qubit frequency ω_{eg} by a frequency close to $\pm\nu$, where ν is the frequency of the collective centre of mass mode. The detuning is *not* exactly ν. Virtual transitions can occur between pairs of ions via two-photon processes involving one photon of frequency $\omega_{eg}+\nu$, and the other of frequency $\omega_{eg}-\nu$. Thus, in contrast to the Cirac–Zoller gate, in which information is transferred into the centre of mass mode (the data bus) and then to another ion, in the Mølmer–Sørensen Gate the data bus is only virtually populated. The main features which arise are:

i) The vibrational degrees of freedom used for communication only enter virtually. Although they are crucial as intermediate states in the processes, population is not transferred to states with different vibrational excitation.

ii) Transition paths involving different unpopulated, vibrational states interfere destructively to eliminate the dependence of the gate parameters on vibrational quantum numbers. The gate is thus insensitive to thermal motion of the ions.

iii) Using this gate, one can construct states of the form

$$|\Psi\rangle = \frac{1}{\sqrt{2}}\left(e^{i\phi_g}|gg...g\rangle + e^{i\phi_e}|ee...e\rangle\right), \tag{24.3.1}$$

where $|gg...g\rangle$ and $|ee...e\rangle$ (N terms) are product states describing N ions which are all in the (same) internal state $|g\rangle$ or $|e\rangle$. States like (24.3.1) with three particles are known as GHZ states [24.8].

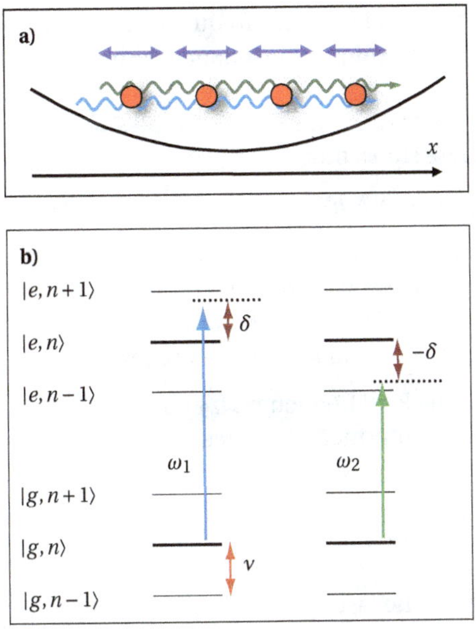

Fig. 24.6. The arrangements of energy levels and lasers for the Mølmer–Sørenson gate. **a)** Two lasers of *different* frequency illuminating *all* of the ions in the ion trap (compare with the arrangements in Fig. 24.3 for the Cirac–Zoller gate. **b)** Energy levels and laser detunings. Two ions with quantized vibrational motion are illuminated with lasers detuned close to the upper and lower sidebands.

24.3.1 Gate Hamiltonian

The Hamiltonian for the Mølmer–Sørensen gate is essentially the same as given in Sect. 24.1, but we omit the explicit reference to the stretch mode in (24.1.7), since this in practice plays no part in the operation of the device. In addition, the two laser fields are included, so that

$$H = H_0 + H_{\text{int}}\,, \tag{24.3.2}$$

$$H_0 = \hbar\nu\left(a^\dagger a + \tfrac{1}{2}\right) + \tfrac{1}{2}\hbar\omega_{eg}\sum_i \sigma_{z,i}\,, \tag{24.3.3}$$

$$H_{\text{int}} = \sum_{r=1}^{2}\sum_{i=1}^{N} \tfrac{1}{2}\hbar\Omega\left(\sigma_i^+ \mathrm{e}^{\mathrm{i}\eta_r(a+a^\dagger)-\mathrm{i}\omega_r t} + \text{h.c.}\right). \tag{24.3.4}$$

As in Sect. 24.1, the Pauli matrices σ_i represent the internal degrees of freedom for the i th ion. The laser frequencies are ω_r, where $r = 1$ or 2, and both lasers are, for simplicity, taken to have the same Rabi frequency Ω for all of the N ions.

The difference between the two frequencies in the bichromatic laser beam is very small, and therefore their k-vectors are essentially identical, so that we can set $\eta_1 = \eta_2 = \eta$. The ion trap is taken to be operating in the Lamb–Dicke limit, that is, $\eta\sqrt{n+1}$ is well below unity.

a) Laser Tunings: Let us first consider only two ions. The basic principle is as illustrated in Fig. 24.6, and is as follows: A bichromatic laser field with frequencies $\omega_1 = \omega_{eg} + \delta$ and $\omega_2 = \omega_{eg} - \delta$ illuminates both ions. One frequency is tuned near to the red sideband frequency $\omega_{eg} - \nu$, and the other frequency is tuned near to the blue sideband frequency $\omega_{eg} + \nu$. Thus, while the two frequencies sum up to twice the qubit frequency ω_{eg}, neither laser field is on resonance with the qubit frequency ω_{eg}.

Nevertheless, a *two-photon transition* is on resonance, and couples the states $|g,g,n\rangle$ and $|e,e,n\rangle$ via virtual intermediate states thus:

$$|g,g,n\rangle \longleftrightarrow \{|e,g,n+1\rangle,\ |g,e,n-1\rangle\} \longleftrightarrow |e,e,n\rangle, \tag{24.3.5}$$

where the first (second) letter denotes the internal state e or g of the first (second) ion and n is the quantum number for the relevant vibrational mode of the trap. The detuning from the sideband is chosen so large that the intermediate states $|e,g,n+1\rangle$ and $|g,e,n-1\rangle$ are not significantly populated in the process.

b) Elimination of Intermediate States: The Rabi frequency Ω_{eff} for the transition between the states $|g,g,n\rangle$ and $|e,e,n\rangle$, via intermediate states m, can be determined in second order perturbation theory,

$$\Omega_{\text{eff}}^2 = \frac{4}{\hbar^2}\left|\sum_m \frac{\langle e,e,n|H_{\text{int}}|m\rangle\langle m|H_{\text{int}}|g,g,n\rangle}{E_{g,g,n}+\hbar\omega_i - E_m}\right|^2, \tag{24.3.6}$$

where the laser energy $\hbar\omega_i$ is the energy of the laser addressing the ion which is excited in the intermediate state $|m\rangle$. If we restrict the sum to $|e,g,n+1\rangle$ and $|g,e,n-1\rangle$, we get

$$\Omega_{\text{eff}} = -\frac{\Omega^2\eta^2}{2(\nu-\delta)}, \tag{24.3.7}$$

where $\delta = \omega_1 - \omega_{eg}$ is the detuning of the laser addressing the first ion.

The states $|e,g,n\rangle$ and $|g,e,n\rangle$ are coupled to order zero in η, which may counteract the larger denominator in (24.3.6). These terms, however, cancel each other due to interference.

24.3.2 Transition Paths

We show in Fig. 24.7 a the "transition paths" between states $|g,g,n\rangle$ and $|e,e,n\rangle$ and in Fig. 24.7 b those between $|e,g,n\rangle$ and $|g,e,n\rangle$ mediated by the interaction with the laser fields within any pair of ions.

a) Destructive Interference between Paths: The interaction strengths of the different one-photon transitions in Fig. 24.7 depend on the value of the vibrational quantum number n. In the limit of $\eta\sqrt{n+1} \ll 1$, terms between states n and $n+1$ are proportional to $\sqrt{n+1}$, whereas terms between $n-1$ and n are proportional to $\sqrt{n}$. When we add the contributions from the four different transition paths shown in either panel of Fig. 24.7, the paths yield the coefficients

$$\left.\begin{aligned} &\frac{(n+1)\Omega^2\eta^2}{\delta-\nu}, && \text{paths with } n+1 \text{ vibrational quanta,} \\ &-\frac{n\Omega^2\eta^2}{\delta-\nu}, && \text{paths with } n \text{ vibrational quanta.} \end{aligned}\right\} \tag{24.3.8}$$

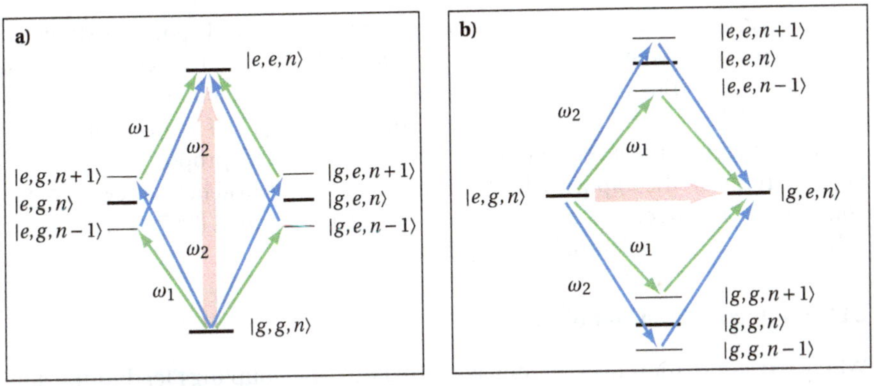

Fig. 24.7. Level scheme for a pair of ions sharing an oscillator degree of freedom.
a) By application of laser light with frequencies $\omega_{eg} \pm \delta$, where δ is somewhat smaller than the vibrational frequency ν, we identify four transition paths between the states $|g,g,n\rangle$ and $|e,e,n\rangle$, which interfere as described in the text. The overall result which occurs when the pulse length t_{gate} is appropriately chosen to satisfy (24.3.26), is represented by the broad rose-coloured arrow.
b) Four similar transition paths are identified between states $|e,g,n\rangle$ and $|g,e,n\rangle$, yielding the same effective coupling among these states as between the states in the left panel.

The coefficients have opposite signs, yielding a destructive interference which exactly removes the n-dependence from the total effective Rabi frequency of the two-photon transition. This implies that the electronic state evolves in a manner independent of the vibrational state—this result clearly permits the vibrational state to be even a thermal state.

b) Energy Shifts: The elegant results so far are slightly marred by the existence of energy shifts, which can also be calculated using perturbation theory. The most important terms which give energy shifts are

$$\Delta E_{ggn} = \Delta E_{een} = -\hbar\frac{(\eta\Omega)^2}{4}\frac{1}{\nu-\delta}, \tag{24.3.9}$$

$$\Delta E_{egn} = \hbar\left(\frac{(\eta\Omega)^2}{2}\frac{n}{\nu-\delta} - \frac{\Omega^2}{2\delta}\right), \tag{24.3.10}$$

$$\Delta E_{gen} = \hbar\left(-\frac{(\eta\Omega)^2}{2}\frac{n+1}{\nu-\delta} + \frac{\Omega^2}{2\delta}\right). \tag{24.3.11}$$

The energy shifts of the states $|e,e,n\rangle$ and $|g,g,n\rangle$ are identical and independent of n, but since the energy shifts of $|e,g,n\rangle$ and $|g,e,n\rangle$ depend on the vibrational quantum number, the time evolution introduces phase factors $\mathrm{e}^{-\mathrm{i}\Delta E_{egn}t/\hbar}$, which depend on n. This means, for example, that at the time

$$t = T_{\text{inv}} \equiv \frac{2\pi(\nu-\delta)}{\eta^2\Omega^2}, \tag{24.3.12}$$

where $|g,g,n\rangle$ and $|e,e,n\rangle$ are inverted, factors of $(-1)^n$ will tend to extinguish the coherence between internal states $|e,e,n\rangle$ and $|e,g,n\rangle$. This coherence can be restored by a trick resembling photon echoes [24.9]. Notice that the n dependent part of ΔE_{egn} is minus the n dependent part of ΔE_{gen}. If at any time $T/2$ we change the sign of the laser detuning δ, phase components proportional to n will begin to rotate in the opposite direction and at time T we will have a revival of the coherence. The effectiveness of this idea has been confirmed [24.7] by numerical solutions of the exact Schrödinger equation.

24.3.3 Effective Spin Hamiltonian

When the summation over all ions is carried out, the couplings lead to the spin Hamiltonian

$$H_{\text{spin}} = \frac{2\chi}{\hbar}\sum_{\substack{i,j \\ i\neq j}} \sigma_x^i \sigma_x^j, \qquad \chi \equiv \frac{\eta^2\Omega^2\nu}{2(\nu^2-\delta^2)}. \tag{24.3.13}$$

24.3.4 Use as a Two-Qubit Gate

When $N = 2$, the gate becomes a two-qubit gate, and since $(\sigma_x^1\sigma_x^2)^2 = 1$, the evolution operator for a laser pulse of duration t is

$$\mathcal{U}_{\text{pulse}}(t) = \cos\vartheta + \mathrm{i}\sin\vartheta\,\sigma_x^1\sigma_x^2, \qquad \vartheta \equiv \frac{2\chi t}{\hbar}. \tag{24.3.14}$$

The two-qubit Mølmer–Sørensen gate arises when we make the choice

$$t = t_{\text{MS}} \equiv \frac{\pi}{8\chi}, \tag{24.3.15}$$

leading to the form

$$\mathcal{U}_{\text{MS}} = \frac{1}{\sqrt{2}}\left(1 + \mathrm{i}\sigma_x^1\sigma_x^2\right). \tag{24.3.16}$$

When written out explicitly in terms of the states of the ions, the gate takes the form

$$\left.\begin{aligned}
|e,e,n\rangle &\to \tfrac{1}{\sqrt{2}}\big(|e,e,n\rangle + \mathrm{i}|g,g,n\rangle\big),\\
|e,g,n\rangle &\to \tfrac{1}{\sqrt{2}}\big(|e,g,n\rangle + \mathrm{i}|g,e,n\rangle\big),\\
|g,e,n\rangle &\to \tfrac{1}{\sqrt{2}}\big(|g,e,n\rangle + \mathrm{i}|e,g,n\rangle\big),\\
|g,g,n\rangle &\to \tfrac{1}{\sqrt{2}}\big(|g,g,n\rangle + \mathrm{i}|e,e,n\rangle\big).
\end{aligned}\right\} \tag{24.3.17}$$

We can define a qubit basis in terms of the eigenstates of σ_x by

$$\begin{aligned}
|1\rangle &= \tfrac{1}{\sqrt{2}}\big(|e\rangle + |g\rangle\big), & \sigma_x|1\rangle &= |1\rangle,\\
|0\rangle &= \tfrac{1}{\sqrt{2}}\big(|e\rangle - |g\rangle\big), & \sigma_x|0\rangle &= -|0\rangle.
\end{aligned} \tag{24.3.18}$$

These states are eigenstates of the unitary operation gate operation, and transform as

$$\mathcal{U}_{\rm MS}\begin{pmatrix}|1,1,n\rangle\\|1,0,n\rangle\\|0,1,n\rangle\\|0,0,n\rangle\end{pmatrix} = \exp\left(\frac{\mathrm{i}\pi}{4}\right)\begin{pmatrix}|1,1,n\rangle\\-\mathrm{i}\,|1,0,n\rangle\\-\mathrm{i}\,|0,1,n\rangle\\|0,0,n\rangle\end{pmatrix}. \tag{24.3.19}$$

This transformation can be transformed to the controlled phase gate $\mathcal{U}_\pi^{(2)}$ given in (2.3.8) by using and the single qubit S-gate given in (2.3.5) thus:

$$\mathcal{U}_\pi^{(2)} = \mathrm{e}^{-\mathrm{i}\pi/4}\, S^1 \otimes S^2\, \mathcal{U}_{\rm MS}\,, \tag{24.3.20}$$

where S^i is the S-gate acting on the i th ion.

24.3.5 Creation of GHZ-Like States

The GHZ-like states of the form

$$|\Psi\rangle = \tfrac{1}{\sqrt{2}}\left(\mathrm{e}^{\mathrm{i}\phi_g}|gg...g\rangle + \mathrm{e}^{\mathrm{i}\phi_e}|ee...e\rangle\right), \tag{24.3.21}$$

are of fundamental interest, and they can be created quite straightforwardly using the N-ion version of the Mølmer–Sørensen gate.

a) Angular Momentum Representation: The Hamiltonian $H_{\rm spin}$ is a symmetric function of all of the spins, and is best described by states of given total pseudo-angular momentum, represented by

$$\boldsymbol{J} \equiv \tfrac{1}{2}\hbar\sum_{i=1}^{N}\boldsymbol{\sigma}^i. \tag{24.3.22}$$

Any collective state of all ions is represented by a pseudo-angular momentum state $|J,M\rangle$, an eigenstate of the J_z operator. A state invariant under permutation of the ions, with a certain number N_e of excited ions is described as $|J = N/2, M = N_e - N/2\rangle$.

For example, the state (24.3.21) is a superposition of the states $|N/2,-N/2\rangle$ and $|N/2,N/2\rangle$.

If we begin with all ions in the ground state $|g\rangle$, the initial state of the ions is

$$|g,g,g,\dots g\rangle = \left|\tfrac{1}{2}N,-\tfrac{1}{2}N\right\rangle. \tag{24.3.23}$$

b) Spin Hamiltonian in the Angular Momentum Representation: Noting that $\left(\sigma_x^i\right)^2 = 1$, the spin Hamiltonian (24.3.13) is equivalent to (by discarding a constant)

$$H_{\rm spin} = \frac{\chi}{\hbar}(J_+^2 + J_-^2 + J_+J_- + J_-J_+) = 4\frac{\chi}{\hbar}J_x^2\,. \tag{24.3.24}$$

c) Time Evolution in the Angular Momentum Representation: The time evolution operator is

$$U(t) = \exp\big(-\mathrm{i}4\chi J_x^2 t/\hbar^2\big). \tag{24.3.25}$$

24.3.6 Creation of Many-Body Entanglement

Using this time evolution operator, we can show:

i) If we apply this to the initial state $|\tfrac{1}{2}N, -\tfrac{1}{2}N\rangle$, the population becomes distributed over all $|J, M\rangle$ states, with M differing from $-\tfrac{1}{2}N$ by an even number.

ii) If N is even, population is transferred all the way to the state $|\tfrac{1}{2}N, \tfrac{1}{2}N\rangle$, which is the second component of (24.3.21).

iii) At the instant

$$t = t_{\text{gate}} \equiv \frac{\pi}{8\chi}, \tag{24.3.26}$$

the state is precisely of the form (24.3.21), with

$$\phi_g = -\tfrac{1}{4}\pi, \qquad \phi_e = \tfrac{1}{4}\pi + \tfrac{1}{2}N\pi. \tag{24.3.27}$$

Proof: This result can be understood from the rotation properties of angular momenta:

i) The initial state can be expanded on eigenstates of J_x, namely

$$|\tfrac{1}{2}N, -\tfrac{1}{2}N\rangle = \sum_M c_M |\tfrac{1}{2}N, M\rangle_x. \tag{24.3.28}$$

ii) From the properties of the Wigner rotation functions [24.10], it follows that

$$|\tfrac{1}{2}N, \tfrac{1}{2}N\rangle = \sum_M c_M (-1)^{\frac{1}{2}N-M} |\tfrac{1}{2}N, M\rangle_x. \tag{24.3.29}$$

iii) Inserting the values for ϕ_g and ϕ_e mentioned above, we can therefore write the state (24.3.21) as

$$\sum_M c_M \frac{1}{\sqrt{2}}\big(\mathrm{e}^{-\mathrm{i}\pi/4} + (-1)^M \mathrm{e}^{\mathrm{i}\pi/4}\big)|\tfrac{1}{2}N, M\rangle_x. \tag{24.3.30}$$

The net factors multiplying the initial amplitudes c_M are unity for M even and $-\mathrm{i}$ for M odd. The action of $U(t)$ in the J_x basis amounts to a multiplication of each amplitude c_M by $\exp(-4\mathrm{i}\chi M^2 t)$, and for $t = \pi/(8\chi)$ this factor attains the desired value of unity for M even and $-\mathrm{i}$ for M odd.

iv) Note that the state (24.3.21) is not only a multi-particle entangled state; it is a superposition of two "mesoscopically distinguishable" states.

24.3.7 Summary

The advantage of the Mølmer–Sørensen gate compared to the Cirac–Zoller gate is the immunity to thermal motion of the ions, but there is the disadvantage that the dynamics of the gate depends on the use of virtual intermediate states. This means that he formulae such as (24.3.7) and (24.3.8) require a pulse of length of the order of $1/|\nu-\delta|$, which leads to a relatively slow gate operation. The Cirac–Zoller gate does not have this disadvantage—it can be fast, but requires the ions to be cooled to the ground state.

24.4 Geometric Phase Gates

The geometric phase gate is designed to provide a fast gate which has the same immunity to thermal motion of the ions as the Mølmer–Sørensen gate. The principle of operation relies on behaviour similar to that which gives rise to Berry's phase, which we treated in Sect. 5.4.1. *García-Ripoll, Zoller* and *Cirac* [24.11] developed a geometric phase gate, which we will now describe.

24.4.1 Geometric Phase in a Harmonic Oscillator

The idea of a geometric phase is best introduced by considering the controlled application of forces to a harmonic oscillator. We start with the driven oscillator Hamiltonian in the form

$$H(t) = \hbar\omega a^\dagger a - \frac{1}{\sqrt{2}} f(t)\big(a^\dagger + a\big). \tag{24.4.1}$$

a) Equations of Motion and their Solutions: The solution of the Schrödinger equation $\mathrm{i}\hbar\partial_t|\psi,t\rangle = H(t)|\psi,t\rangle$, in the case that the initial state is a coherent state, $|\psi,0\rangle = |\alpha_0\rangle$, is itself a coherent state *multiplied by a phase,*[1] of the form

$$|\psi,t\rangle = \mathrm{e}^{\mathrm{i}\phi(t)}|\alpha(t)\rangle. \tag{24.4.2}$$

The phase and coherent state amplitude satisfy

$$\frac{d\alpha(t)}{dt} = -\mathrm{i}\omega\alpha(t) + \mathrm{i}\frac{1}{\sqrt{2}\hbar} f(t), \tag{24.4.3}$$

$$\frac{d\phi(t)}{dt} = \frac{1}{2\sqrt{2}\hbar} f(t)\big(\alpha(t)^* + \alpha(t)\big). \tag{24.4.4}$$

The first equation has a solution,

$$\alpha(t) = \mathrm{e}^{\mathrm{i}\omega t}\left[\alpha_0 + \frac{\mathrm{i}}{\sqrt{2}\hbar}\int_0^t d\tau\,\mathrm{e}^{\mathrm{i}\omega\tau} f(\tau)\right], \tag{24.4.5}$$

[1] Note that the derivation of *Bk. I: Sect. 10.5.2* is incomplete, and does not give this phase.

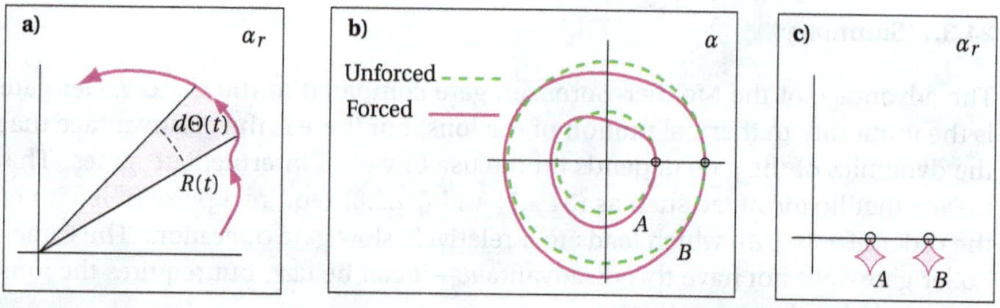

Fig. 24.8. a) Illustration of the relationship between the geometric phase and the area swept out by the trajectory of the coherent state variable $\alpha_r(t)$ in the rotating frame, as in (24.4.9). **b)** Trajectories in phase space of the coherent state variable $\alpha(t)$ of a forced harmonic oscillator with Hamiltonian $H = \hbar\omega a^\dagger a + F\sin(2\omega t)(a + a^\dagger)$. Trajectories of given by (24.4.5) without forcing (green dashed line) and with $F = 0.1\hbar\omega$ (solid red line), for two initial conditions. **c)** The same trajectories in a rotating frame of reference of the corresponding variable $\alpha_r(t)$ as given by (24.4.6).

b) Phase and Geometry: Let us move to a rotating frame, defined by

$$\alpha_r(t) \equiv e^{i\omega t}\alpha(t) \equiv R(t)e^{i\Theta(t)}. \tag{24.4.6}$$

In this rotating frame the only change in the coherent state amplitude $\alpha_r(t)$ arises from the interaction, thus we find

$$\frac{d\alpha_r(t)}{dt} = \frac{i}{\sqrt{2}\hbar}e^{i\omega\tau} f(\tau), \tag{24.4.7}$$

$$\frac{d\phi(t)}{dt} = \text{Im}\left(\alpha_r(t)^* \frac{d\alpha_r(t)}{dt}\right), \tag{24.4.8}$$

$$= R(t)^2 \frac{d\Theta(t)}{dt} = 2\frac{dA(t)}{dt}. \tag{24.4.9}$$

In the last equation $A(t)$ is the area swept out by the motion of $\alpha_R(t)$ in the phase space plane—see Fig. 24.8 a. The terminology *geometric* for this phase arises from this observation.

c) Phase Change around a Closed Loop: If the trajectory of $\alpha_r(t)$ forms a closed loop, the phase change around the loop is given by the area enclosed by the loop. Any trajectory with the same enclosed area will give the same phase. Therefore we will consider a driving force $f(t)$ which acts during a period $(0, T)$, and for which $\alpha_r(T) = \alpha_r(0)$, so that the quantum state of the system is unchanged, apart from the phase change. From (24.4.7), this requires

$$\int_0^T d\tau\, e^{i\omega\tau} f(\tau) = 0. \tag{24.4.10}$$

Using this condition, (24.4.7) and (24.4.8), the total phase is given by

$$\Phi(0,T) = \mathrm{Im}\left(\int_0^T d\tau \,\frac{\mathrm{i}}{\sqrt{2}\hbar}\,\mathrm{e}^{\mathrm{i}\omega\tau} f(\tau)\,\alpha_r^*(\tau)\right), \tag{24.4.11}$$

$$= \frac{1}{2\hbar^2}\,\mathrm{Im}\left(\int_0^T d\tau_1 \int_0^{\tau_1} d\tau_2\, \mathrm{e}^{\mathrm{i}\omega(\tau_1-\tau_2)} f(\tau_1) f(\tau_2)\right). \tag{24.4.12}$$

In Fig. 24.8 we show the phase space trajectories obtained by forcing two coherent states with a sinusoidal force at twice the oscillator frequency, choosing $f(t) = -\sqrt{2}F\sin(2\omega t)$.

d) The Geometric Phase is Independent of the Initial Quantum State: The most important aspect of the formula (24.4.12) is that $\Phi(0,T)$ depends only on the forcing function $f(t)$, and not on the initial coherent state amplitude. Since the coherent states form a complete set, this means that we can apply the force $f(t)$ to *any quantum state* and get the same phase change. This point is clear from Fig. 24.8. In Fig. 24.8 b the orbits arising from different initial conditions seem very different, but Fig. 24.8 c shows that in the rotating frame they have the same shape, and the same enclosed area A.

In other words, the total phase $\Phi(0,T)$ is insensitive to the initial motional state of the system and it is thus robust. This property is of crucial importance when we seek applications to real systems that cannot be cooled to the zero phonon limit.

24.4.2 Phase of Two Ions

Let us now consider the case of two trapped ions, as in Sect. 24.1.1. Here we will denote the two normal modes:

i) The centre of mass mode, $x_c = (x_1 + x_2)/2$, which oscillates with frequency ω_c.

ii) The stretch mode, $x_s = x_2 - x_1$, which oscillates with frequency ω_s.

If the ions are stored in the same harmonic trap with trap frequency ω, they have the incommensurate frequencies $\omega_c = \omega$ and $\omega_s = \omega\sqrt{3}$. (It would also be possible to use other trap configurations, which would permit $\omega_s = \omega$.)

a) State-Dependent Forces: We can impose the same state-dependent force on both ions, for example by means of an off-resonance laser that induces an AC Stark shift on only one of the internal states of the ions. The Hamiltonian will take the form

$$H = \hbar\omega_c a_c^\dagger a_c + \hbar\omega_s a_s^\dagger a_s - f(t)\mathbf{p}_1 x_1 - f(t)\mathbf{p}_2 x_2, \tag{24.4.13}$$

$$\begin{aligned} &= \hbar\omega_c a_c^\dagger a_c + \hbar\omega_s a_s^\dagger a_s + f(t)(\mathbf{p}_2 - \mathbf{p}_1)d \\ &\quad - f(t)(\mathbf{p}_1 + \mathbf{p}_2)\frac{\alpha_c}{\sqrt{2}}(a_c + a_c^\dagger) - f(t)(\mathbf{p}_2 - \mathbf{p}_1)\frac{\alpha_s}{2\sqrt{2}}(a_s + a_s^\dagger). \end{aligned} \tag{24.4.14}$$

Here:

i) d is the equilibrium distance between the ions.

ii) $\alpha_{c,s} = \sqrt{\hbar / m\omega_{c,s}}$ are the oscillator lengths.

iii) $\mathbf{p}_1$ and $\mathbf{p}_2$ are operators, which take on the value 1 or 0, depending on whether the ion is in internal state $|+1\rangle$ or $|-1\rangle$.

b) State-Dependent Geometric Phase: Since the modes are now decoupled, we can apply the formulae of Sect. 24.4.1 almost directly. In this case, corresponding to the fact that there are two modes, and these have different frequencies, the force must satisfy the two conditions

$$\int_0^T d\tau\, e^{i\omega_c \tau} f(\tau) = \int_0^T d\tau\, e^{i\omega_s \tau} f(\tau) = 0. \tag{24.4.15}$$

These provide the correct generalization of (24.4.10).

The total phase becomes an operator in the spin space, and can be written in terms of the coefficients

$$\mathcal{G}_{c,s} \equiv \int_0^T d\tau_1 \int_0^T d\tau_2 \frac{\sin\left(\omega_{c,s}|\tau_1 - \tau_2|\right)}{2m\hbar\omega_{c,s}}, \tag{24.4.16}$$

in the form

$$\Phi_{\text{gate}} = \tfrac{1}{2}\left(\mathbf{p}_1 + \mathbf{p}_2\right)\left(\mathcal{G}_c + \tfrac{1}{4}\mathcal{G}_s\right) + \mathbf{p}_1\mathbf{p}_2\left(\mathcal{G}_c - \tfrac{1}{4}\mathcal{G}_s\right). \tag{24.4.17}$$

The derivation follows the procedure which leads to the result (24.4.12).

c) The Geometric Phase Gate: The operator $\exp\left(i\Phi_{\text{gate}}\right)$ is the product of two single-qubit gates, and a two qubit gate. The two-qubit gate becomes a controlled-phase gate of the form (2.3.8) when one tunes the force function $f(t)$ so that:

i) The two conditions (24.4.15), one for the c-mode and one for the s-mode, are satisfied, ensuring that the the system returns to the same quantum state, apart from a phase.

ii) The coefficient of $\mathbf{p}_1\mathbf{p}_2$ becomes π, that is

$$\mathcal{G}_c - \tfrac{1}{4}\mathcal{G}_s = \pi. \tag{24.4.18}$$

d) Temperature Independence of the Geometric Phase Gate: Because the gate phase Φ_{gate} does not depend on the motional quantum states of the ions, the same phase will be found for any initial ion state, or any statistical mixture of initial ion states. Thus, for a factorizable density operator of the system, we can express the gate operation as

$$\rho_{\text{qubit}} \otimes \rho_{\text{motional}} \longrightarrow \left(e^{i\pi\mathbf{p}_1\mathbf{p}_2}\rho_{\text{qubit}}e^{-i\pi\mathbf{p}_1\mathbf{p}_2}\right) \otimes \rho_{\text{motional}}. \tag{24.4.19}$$

As in the Mølmer–Sørensen gate, this will only be true as long as the typical fluctuation time scale of the thermalized ion state is significantly longer than than the time to operate the gate.

e) Fast Geometric Gates: As we have constructed it, this gate can be made to operate very rapidly, since the forcing function $f(t)$ can be made quite strong. Using this freedom, *García-Ripoll, Zoller* and *Cirac* [24.11] have developed protocols which appear to be almost arbitrarily rapid.

f) Relationship to the Berry Phase: The Berry phase, which we have outlined in Sect. 5.4.1, applies to a phase developed as an energy eigenstate is adiabatically (and therefore slowly) moved by changing parameters of the Hamiltonian. Moving the parameter in a closed loop brings the state back to its original form, but with a phase change, which depends on the area swept out by the trajectory of the parameter. The geometric phase, as pointed out by *Aharonov* and *Anandan* [24.12], is the result of a dynamical evolution of a system in a cycle; the quantum state is taken around a trajectory, returning eventually to its original position in Hilbert space, but with the extra geometric phase. The two concepts are different, but closely related, and can overlap—a geometric phase evolved slowly may be the same thing as a Berry phase. The comprehensive review of the relationship between geometric and Berry phase by *Anandan, Christian* and *Wanelik* [24.13] gives a very clear view of the situation.

Gates which use the Berry phase, rather than the geometric phase, are called *adiabatic gates.* In principle, they are slow, because they must be operated adiabatically. However, as noted in [24.11], some gates of this kind are really geometric phase gates, and may be operated rapidly.

g) Relationship to the Mølmer–Sørensen Gate: *Sørensen* and *Mølmer* [24.14] noted that the Mølmer–Sørensen gate can in fact be viewed a geometric phase gate, and that it can in fact be implemented as a much faster gate than their original proposal.

24.5 The Ion Trap Quantum Computer in Practice

Currently, the most complete experimental implementation of an ion trap quantum computer is the Innsbruck quantum processor, whose basic principles we have discussed and illustrated in Sect. 1.4.1–Sect. 1.4.2, as well as in this chapter. While the device implements all of the necessary protocols, it is of course not constructed on a scale which can realize a practical and useful quantum processor. Even so, the actual device is enormously more complex than our theoretical outline—for a full description the reader is referred to [24.15].

The main points of note are:

i) Quantum algorithms can be implemented straightforwardly as *quantum circuits,* where the algorithm is built up from a standard universal set of operations containing single qubit operations and CNOT gates.

ii) The set of gates implemented allows the realization of any completely positive map, that is, any quantum Markov process can be implemented.

iii) In order to build a practical quantum information processor it is mandatory to minimize the error rate of the available operations. This is only possible if the relevant noise processes can be characterized, and if their physics is well understood.

There are many sources of noise and fluctuations. The major sources are fluctuations in the frequency, phase and amplitude of lasers, magnetic field fluctuations, and electric field fluctuations. These give rise to errors in the state initialization, in the coherent manipulations and in the state detection. As well, it has been found that the dominant source of errors depends on the actual sequence of operations performed.

iv) The available toolbox has been employed to realize various quantum algorithms, such as the quantum Fourier transform and the quantum order finding algorithm.

24.5.1 Quantum Computing Using Multiple Ion Traps on a Chip

Manipulating a large number of ions in a single trap presents immense technical difficulties, and both experience and scaling arguments suggest that this scheme is limited to computations on tens of ions. A possible solution to this problem [24.16, 24.17] involves the distribution of the ions in an array of multiple trap zones, fabricated microscopically on a silicon chip. In this architecture, multiqubit gate operations would involve a large number of processing zones, in each of which there would be a relatively small number of ions. Entanglement would be distributed between these zones by physically moving the ions or by optical means. This kind of approach is being pursued actively [24.18], and it should lead to a genuinely scalable ion trap quantum computer—one which would be of sufficient size to realize the full potential of quantum computing.

VIII Circuit Quantum Electrodynamics

PART VIII: CIRCUIT QUANTUM ELECTRODYNAMICS

Superconducting systems are naturally also electric circuits, and the quantum theory of electric circuits is quite well-established. However, it has only recently been possible to realize the full potential of quantum circuit theory as *circuit quantum electrodynamics*, for which a complete emulation of quantum optical systems is the aim. Quantum devices can now be realized, in which the harmonic oscillators are either resonant electric circuits of microwave cavities, and in which the two-level systems are the quantized levels of Josephson junctions. Using these devices, nearly all of the systems on quantum optics have been emulated. There are very real possibilities for quantum devices using this paradigm, and these appear to have very desirable features.

Here we give a brief survey of the main theoretical framework needed to describe such quantum devices. In Chap. 25 we cover the foundations of quantized electric circuit theory, in a form closely related to that of quantum optics. Here we include both Hamiltonian and dissipative dynamics.

In Chap. 26 we present the basic theory of Josephson junctions, and show how these can be used in quantum circuits to provide two-level systems, qubits, quantum amplifiers and efficient quantum measurement. Once all of the necessary approximations are made, the resulting theoretical structure differs little from that of quantum optics.

The main differences all arise from the longer wavelength of the electromagnetic radiation, for example:

i) Single photon counting is not feasible, on the other hand, microwave voltages can be amplified and measured with almost quantum-limited noise.

ii) Microwaves can be confined to waveguides, and the interaction between them and the Josephson junction-based two-level systems can be made very large, and confined to a few modes of the electromagnetic field.

iii) Everything takes place at temperatures significantly below 1K—this is less convenient than the room-temperature devices of quantum optics.

iv) There is no realistic way of transmitting quantum information using microwaves. The only realistic possibility is the optical fibres. This means that any quantum networking will necessarily involve quantum interfacing between optical and superconducting systems.

25. Quantum Circuit Theory

Quantum electronics was formulated before the establishment of quantum optics as the most significant field of quantum engineering, and its formulation as a form of quantized circuit theory was introduced by *Louisell* [25.1], and later put on a very solid foundation by *Yurke* and *Denker* [25.2]. With the development of quantum devices based on superconductors the field is achieving a new relevance, and a practical formulation of quantum network theory for application to superconducting quantum devices has now been put in place by *Devoret* [25.3] and applied [25.4] by the group of *Girvin* and *Schoelkopf*.

The main aim of the current experimental investigations is to make a superconducting device in which a two-level system is created in essentially the same way as in a real atom. This requires a potential in which the energy levels are not evenly spaced, and it is thus possible to tune to the transition between the particular pair of energy levels which we wish to be the two-level system. The necessary anharmonic potential is engineered by using the nonlinear properties of Josephson junctions.

Superconducting devices operate at microwave frequencies, and therefore can be coupled to the modes of a suitable transmission line. Thus an analogue of the ion trap quantum computer is in principle feasible, in which the superconducting two-level system takes the place of the energy levels of the ions, and the modes of the transmission line take the place of the vibrational modes of the ions.

In this chapter we will formulate quantum circuit theory using ordinary linear inductors, capacitors and transmission lines; in the next chapter we will show how to incorporate superconducting devices within this framework.

25.1 The *LC* Oscillator

The fundamental device underlying quantum circuit theory is the *LC* oscillator illustrated in Fig. 25.1. The circuit is characterized by six quantities:

i) *Device Parameters*: The inductance L of the inductor, and the capacitance C of the capacitor.

ii) *Dynamical Variables Associated with the Circuit*: Both the voltage V between the two connections linking the capacitor and the inductor, and the current I flowing through both the inductor and the capacitor are associated with the circuit as a whole.

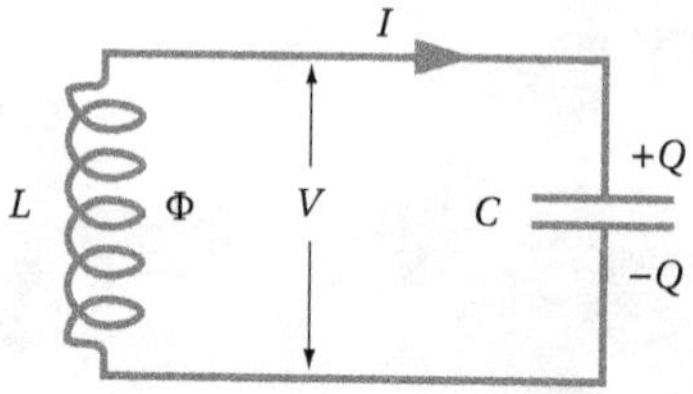

Fig. 25.1. Schematic of an *LC* oscillator, indicating the notation used for quantization.

iii) *Dynamical Variables Associated with the Devices*: The inductor is associated with the magnetic flux Φ, and the capacitor is associated with the charge Q.

These quantities are connected with each other by four equations:

Lenz's law	$V = -\partial_t \Phi\,,$	(25.1.1)
Conservation of charge	$I = \partial_t Q\,,$	(25.1.2)
Inductor flux–current law	$\Phi = L\,I,$	(25.1.3)
Capacitor charge-voltage law	$Q = C\,V.$	(25.1.4)

25.1.1 Lagrangian and Hamiltonian

To set up Lagrangian and Hamiltonian descriptions, we need to specify which of the four variables Q, I, Φ, V is to be interpreted as the canonical co-ordinate variable. Instead of the most common choice, which is the charge Q, when dealing with superconducting quantum devices the most convenient choice for the canonical co-ordinate is the flux Φ. This is because a Josephson junction behaves like a nonlinear inductor, whose energy is not a simply quadratic function of Φ—setting Φ to be the co-ordinate allows us to regard this as a potential energy, leaving the capacitative energy as the quadratic function of $\dot{\Phi}$ which can be regarded as kinetic energy as we shall now show.

a) Lagrangian: Therefore, making this choice, the energies are

Inductive energy	$E_L = \tfrac{1}{2}LI^2 = \dfrac{\Phi^2}{2L}\,,$	(25.1.5)
Capacitative energy	$E_C = \tfrac{1}{2}CV^2 = \dfrac{C\,\dot{\Phi}^2}{2}\,,$	(25.1.6)

and correspondingly the Lagrangian is

$$\mathcal{L}_{\text{osc}} = \frac{C\,\dot{\Phi}^2}{2} - \frac{\Phi^2}{2L}\,. \tag{25.1.7}$$

The equation of motion is thus

$$\ddot{\Phi} = -\omega_{\text{osc}}^2\Phi, \quad \text{where} \quad \omega_{\text{osc}} = \frac{1}{\sqrt{LC}}\,. \tag{25.1.8}$$

b) Hamiltonian: The canonical momentum corresponding to Φ is

$$\Pi_\Phi \equiv \frac{\partial \mathcal{L}_{\text{osc}}}{\partial \dot{\Phi}} = C\,\dot{\Phi} = -Q. \tag{25.1.9}$$

The Hamiltonian is then

$$\mathcal{H}_{\text{osc}} = \frac{\Pi_\Phi^2}{2C} + \frac{\Phi^2}{2L}. \tag{25.1.10}$$

This Hamiltonian is of course that of a harmonic oscillator.

25.1.2 Quantization of the Oscillator

The quantization can be achieved by using the generic procedure we have given in *Bk. I: Sect.16.3.* We will be able to write the destruction operator in terms of the *characteristic impedance* of the oscillator, defined by

$$Z_{\text{osc}} \equiv \sqrt{\frac{L}{C}} = \omega_{\text{osc}} L = \frac{1}{\omega_{\text{osc}} C}. \tag{25.1.11}$$

a) Operator Relationships: We use the definitions given in *Bk. I: (16.3.3)* and *Bk. I: (16.3.4)*; setting $m \to C$ and $\omega \to \omega_{\text{osc}}$, we find that mechanical quantity $\kappa \equiv \sqrt{\omega m}$ is mapped to $\kappa \to 1/\sqrt{Z_{\text{osc}}}$, and we find that:

i) *The Destruction Operator* is given by

$$a = \frac{1}{\sqrt{2\hbar}}\left(\frac{1}{\sqrt{Z_{\text{osc}}}}\Phi + \mathrm{i}\sqrt{Z_{\text{osc}}}\,\Pi\right) = \frac{1}{\sqrt{2\hbar}}\left(\frac{1}{\sqrt{Z_{\text{osc}}}}\Phi - \mathrm{i}\sqrt{Z_{\text{osc}}}\,Q\right). \tag{25.1.12}$$

ii) *The Hamiltonian* takes the usual harmonic form

$$\mathcal{H}_{\text{osc}} = \hbar\omega_{\text{osc}}\left(a^\dagger a + \tfrac{1}{2}\right). \tag{25.1.13}$$

iii) *The Charge and Flux Operators* are given by

$$Q = \mathrm{i}\sqrt{\frac{\hbar}{2Z_{\text{osc}}}}\left(a - a^\dagger\right), \tag{25.1.14}$$

$$\Phi = \sqrt{\frac{\hbar Z_{\text{osc}}}{2}}\left(a + a^\dagger\right). \tag{25.1.15}$$

iv) *The Operators for Voltage and Current* in the circuit are

$$V = \mathrm{i}\sqrt{\frac{\hbar}{2Z_{\text{osc}}}}\left(\frac{a - a^\dagger}{C}\right), \tag{25.1.16}$$

$$I = \sqrt{\frac{\hbar Z_{\text{osc}}}{2}}\left(\frac{a + a^\dagger}{L}\right). \tag{25.1.17}$$

b) Canonical Commutation Relations: The canonical commutation relations following from the quantization procedure take two forms, depending on our choice of operators

$$[Q,\Phi] = i\hbar, \qquad [V,I] = i\hbar. \tag{25.1.18}$$

c) Significance of the Characteristic Impedance: For an *LC* circuit as above, the characteristic impedance provides the bridge between the formulation in terms of quanta by the creation and destruction operators and the circuit quantities flux and charge. In a more complex circuit there may be a number of resonances, and these may be either series resonances or parallel resonances.

In AC circuit theory, one normally uses the the electrical engineer's j-operator rather than the physicist's imaginary i. Of course $\mathrm{i}^2 = \mathrm{j}^2 = -1$, but the electrical engineer writes an AC current as being proportional to $\mathrm{e}^{\mathrm{j}\omega t}$, whereas the Schrödinger equation sets the time dependence as $\mathrm{e}^{-\mathrm{i}\omega t}$.

i) *Parallel Resonance*: Bearing this in mind, we know the *frequency-dependent admittance* of the parallel *LC*-circuit is given by

$$Y(\omega) = \frac{1}{\mathrm{j}\omega L} + \mathrm{j}\omega C, \tag{25.1.19}$$

and resonance in this point of view is characterized by

$$Y(\omega_{\mathrm{osc}}) = 0. \tag{25.1.20}$$

The characteristic impedance is defined in terms of the slope of the function $Y(\omega)$ at $\omega = \omega_{\mathrm{osc}}$, explicitly

$$Z_{\mathrm{osc}} = \frac{2\mathrm{j}}{\omega_{\mathrm{osc}}\partial Y/\partial\omega|_{\omega=\omega_{\mathrm{osc}}}}. \tag{25.1.21}$$

ii) *Series Resonance*: The *frequency-dependent impedance* of a series *LC*-circuit is

$$Z(\omega) = \frac{1}{\mathrm{j}\omega C} + \mathrm{j}\omega L, \tag{25.1.22}$$

and resonance in this point of view is characterized by

$$Z(\omega_{\mathrm{osc}}) = 0. \tag{25.1.23}$$

The characteristic impedance is given by the slope of the function $Z(\omega)$ at $\omega = \omega_{\mathrm{osc}}$, explicitly

$$Z_{\mathrm{osc}} = -\mathrm{j}\frac{\omega_{\mathrm{osc}}}{2}\frac{\partial Z}{\partial\omega}\bigg|_{\omega=\omega_{\mathrm{osc}}}. \tag{25.1.24}$$

d) More Complex Circuits: For circuits other than a simple parallel *LC* circuit the the above formulation is generalized to the characterization of the parallel resonances, that is, those resonances defined by zeroes of the admittance. These can be determined experimentally, and the behaviour of the admittance as a function of ω measured, allowing the characteristic impedance to be determined.

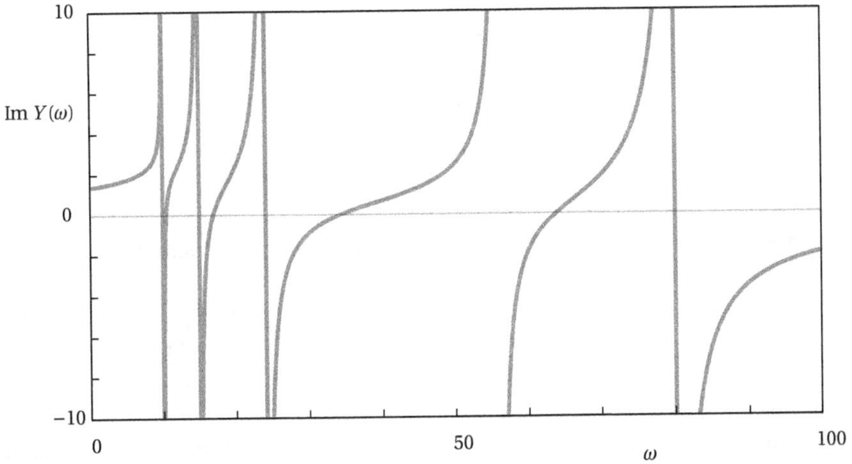

Fig. 25.2. Plot of the imaginary part of a typical admittance function, illustrating the alternating poles and zeroes corresponding to series and parallel resonances.

Foster's reactance theorem [25.5] states that the imaginary part of the admittance is always an increasing function of ω. This means that the zeroes, corresponding to the parallel resonance, must alternate with poles, which arise from the zeroes of the impedance of series resonances. A plot of the imaginary part of a typical admittance function is given in Fig. 25.2, which illustrates this behaviour.

A consequence of this increasing behaviour is that the slope of Im $Y(\omega)$ is positive, and thus the characteristic impedance as given by (25.1.21) is also positive. It also follows by a similar analysis that the characteristic impedance of a series resonance, as given by (25.1.24), is positive.

Exercise 25.1 Zero Point Fluctuations: Show that the zero point charge and flux fluctuations in the *LC* circuit are given by

$$\langle Q^2\rangle_{\text{ZPF}} = \frac{1}{2\hbar Z_{\text{osc}}}, \qquad \langle \Phi^2\rangle_{\text{ZPF}} = \frac{Z_{\text{osc}}}{2\hbar}. \tag{25.1.25}$$

Exercise 25.2 Formulation with Charge as the Canonical Co-ordinate: Formulate the more conventional Lagrangian and Hamiltonian theory of the *LC* oscillator, which uses charge Q as the canonical co-ordinate, so that the inductive energy is $E_L = \frac{1}{2}L\dot{Q}^2$, the capacitative energy is $E_C = \frac{1}{2}Q^2/C$, and the canonical momentum is $\Pi_Q \equiv L\dot{Q}$.

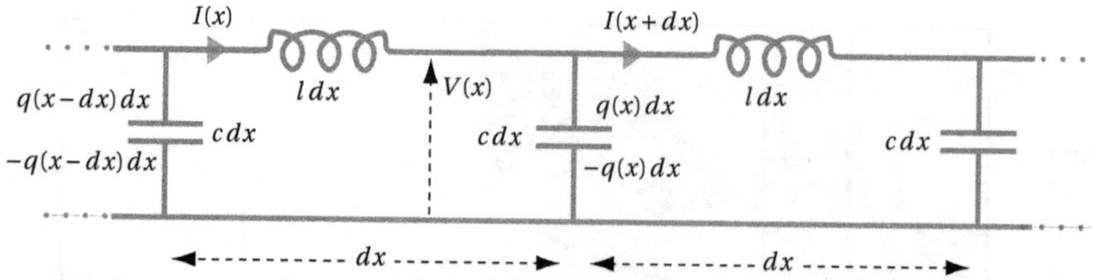

Fig. 25.3. Model of a transmission line composed of infinitesimal lumped circuit elements.

25.2 The Transmission Line

A transmission line can be made in a number of ways, such as coaxial cable, or a waveguide. Most commonly, however, in superconducting work it is made up of two conducting strips of metal separated by a very thin insulating layer. Any such a device has a capacitance c per unit length, and an inductance l per unit length, and can be thought of as a chain of LC circuits. The circuit elements are then infinitesimal capacitors $c\,dx$ and infinitesimal inductors $l\,dx$, connected as illustrated in Fig. 25.3.

25.2.1 Transmission Line Wave Equation

To derive the appropriate equations of motion we use Kirchhoff's laws as follows:

i) *Kirchhoff's Current Law*: We denote by $I(x,t)$ the current through the inductor at the point x. The inductance of the length dx of the inductor at x is $l\,dx$, so that the corresponding element of magnetic flux is given in terms of a flux per unit length $\varphi(x,t)$ by

$$\varphi(x,t)\,dx = I(x,t)\,l\,dx, \tag{25.2.1}$$

and thus

$$I(x,t) = \frac{\varphi(x,t)}{l}. \tag{25.2.2}$$

The charge on the capacitor at the point x is denoted by $q(x,t)\,dx$, so that Kirchhoff's current law requires

$$\partial_t q(x,t) = \frac{I(x,t) - I(x+dx,t)}{dx} = -\partial_x I(x,t)\,, \tag{25.2.3}$$

$$= -\frac{1}{l}\partial_x \varphi(x,t)\,. \tag{25.2.4}$$

ii) *Kirchhoff's Voltage Law*: We now apply Kirchhoff's voltage law around the left-hand loop in Fig. 25.3. The voltage differences across the capacitor at the points $x - dx$, x are

$$V(x,t) = \frac{q(x,t)\,dx}{c\,dx} = \frac{q(x,t)}{c}, \tag{25.2.5}$$

$$V(x-dx,t) = \frac{q(x-dx,t)\,dx}{c\,dx} = \frac{q(x-dx,t)}{c}, \tag{25.2.6}$$

while the voltage across the inductor is given by Lenz's law as

$$V_l(x,t) = -l\,dx\,\dot{I}(x,t) = -\dot{\varphi}(x,t)\,dx. \tag{25.2.7}$$

Kirchhoff's voltage law then requires

$$V(x,t) - V(x-dx,t) = V_l(x,t), \tag{25.2.8}$$

that is

$$\frac{1}{c}\partial_x q(x,t) = -\partial_t \varphi(x,t). \tag{25.2.9}$$

iii) *The Transmission Line Equations*: The equations we have derived are the fundamental equations governing the transmission line; we have shown that they are provided by Kirchhoff's circuit laws in the form (25.2.4, 25.2.9):

Kirchoff's current law:	$l\,\partial_t q(x,t) = -\partial_x \varphi(x,t),$	(25.2.10)
Kirchoff's voltage law:	$\partial_x q(x,t) = -c\,\partial_t \varphi(x,t).$	(25.2.11)

iv) *Transmission Line Wave Equation*: We can now eliminate either $q(x,t)$ or $\varphi(x,t)$ from the transmission line equations, to get two forms of the transmission line wave equation

$$\partial_t^2 \varphi(x,t) = v^2 \partial_x^2 \varphi(x,t), \tag{25.2.12}$$

$$\partial_t^2 q(x,t) = v^2 \partial_x^2 q(x,t), \tag{25.2.13}$$

where the propagation speed of the wave is given by

$$v \equiv \frac{1}{\sqrt{lc}}. \tag{25.2.14}$$

25.2.2 The Flux Potential and the Lagrangian Formulation

Because the charge per unit length $q(x,t)$ and the flux per unit length $\varphi(x,t)$ satisfy the second transmission line equation (25.2.11), we can introduce a *flux potential* $A(x,t)$ in terms of which we can write

$$\varphi(x,t) = \partial_x A(x,t), \tag{25.2.15}$$

$$q(x,t) = -c\,\partial_t A(x,t). \tag{25.2.16}$$

The flux potential $A(x,t)$ behaves like a one-dimensional version of the electromagnetic vector potential, providing voltage and current through the equations

$$
\begin{aligned}
I(x,t) &= \varphi(x,t)/l = \partial_x A(x,t)/l, && (25.2.17)\\
V(x,t) &= q(x,t)/c = -\partial_t A(x,t). && (25.2.18)
\end{aligned}
$$

Furthermore, substituting from (25.2.17, 25.2.18) into the first transmission line equation we find that $A(x,t)$ also satisfies the wave equation

$$\partial_t^2 A(x,t) = v^2 \partial_x^2 A(x,t). \tag{25.2.19}$$

a) Energy in Terms of the Flux Potential: The energy content of the system can then expressed in terms of the flux potential as

$$\text{Inductive energy:} \quad E_L = \tfrac{1}{2}\int dx\, l\, I(x,t)^2 = \frac{1}{2l}\int dx \left(\partial_x A(x,t)\right)^2, \tag{25.2.20}$$

$$\text{Capacitative energy:} \quad E_C = \tfrac{1}{2}\int dx\, c\, V(x,t)^2 = \frac{c}{2}\int dx \left(\partial_t A(x,t)\right)^2. \tag{25.2.21}$$

b) Lagrangian and Hamiltonian: Correspondingly we have

$$\mathcal{L}_{\mathrm{TL}} = \int \left(\frac{c}{2}\left(\partial_t A(x,t)\right)^2 - \frac{1}{2l}\left(\partial_x A(x,t)\right)^2\right) dx, \tag{25.2.22}$$

$$\mathcal{H}_{\mathrm{TL}} = \int \left(\frac{c}{2}\left(\partial_t A(x,t)\right)^2 + \frac{1}{2l}\left(\partial_x A(x,t)\right)^2\right) dx. \tag{25.2.23}$$

25.2.3 Boundary Conditions

When we couple a transmission line to an electrical circuit, we must make sure the voltage across the ends of the transmission line and the current through it match the corresponding circuit quantities. For the transmission line, the current and voltage are determined from $A(x,t)$ by (25.2.17, 25.2.18), and these must match whatever is imposed by the situation of the transmission line.

a) Isolated Transmission Lines: If the transmission line has a boundary at the point $x = A$, there are two possible boundary conditions corresponding to the idea that the transmission line is not connected to anything at $x = A$, namely:

i) *Zero Current Boundary Condition*: This corresponds to an *open-circuit configuration*, and from (25.2.17) this requires

$$\partial_x A(x,t)|_{x=A} = 0. \tag{25.2.24}$$

ii) *Zero Voltage Boundary Condition*: This corresponds to a *short-circuit configuration*, and from (25.2.18) this requires

$$\partial_t A(x,t)|_{x=A} = 0. \tag{25.2.25}$$

The most commonly used configuration experimentally is the open-circuit configuration.

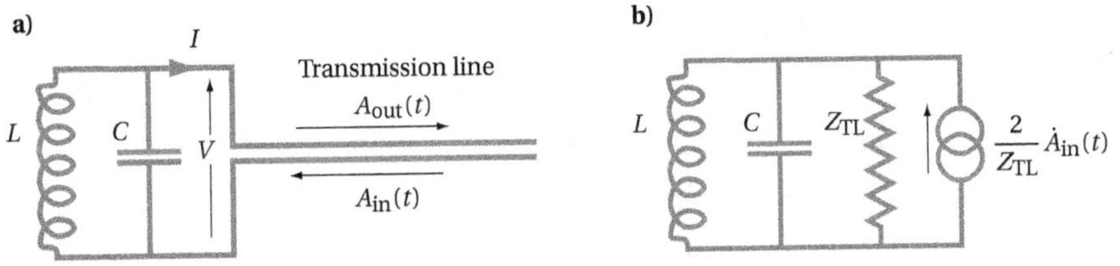

Fig. 25.4. a) Schematic of an *LC* oscillator coupled in parallel to a transmission line of characteristic impedance Z_{TL}; **b)** Equivalent circuit, with a parallel resistance of value Z_{TL}, and a current source arising from the incoming field from the transmission line.

Both of these boundary conditions are compatible with the conservation of energy, and give systems which can be directly quantized, as we shall show in Sect. 25.5.

b) Terminated Transmission Line: A related boundary condition, of great significance in the everyday use of transmission lines, is that corresponding to terminating the transmission line with a resistor R. Because the current through the transmission line must equal the current in the resistor, Ohm's law says that

$$\partial_t A(x,t) = -\frac{R}{l}\partial_x A(x,t). \qquad (25.2.26)$$

This does not correspond to a truly *isolated* transmission line, since a resistor is a dissipative element, composed of a very large number of modes. While this is a very simple setup classically, its quantization also requires the quantization of a resistor—in fact we will show in the next section that an ideal resistor R can be modelled as a transmission line. of characteristic impedance R.

25.3 Transmission Lines Coupled to Circuits

We will now consider a semi-infinite transmission line coupled to a resonant *LC* circuit, and will show how it comes about that the transmission line itself behaves like a resistor as far as the circuit is concerned. The resistive energy loss from the circuit arises from the radiation of energy down the transmission line to infinity. Considered as a whole, the coupled *LC* circuit and transmission line does conserve energy, but this is eventually all contained in an outgoing wave travelling in the transmission line.

25.3.1 Parallel Coupling of the *LC* Oscillator to a Transmission Line

Let us therefore couple a transmission line on the interval $0 \leqslant x < \infty$ to an *LC* circuit at the point $x = 0$; this will show how the boundary conditions are to be applied.

a) Circuit Equation of Motion: The circuit for parallel coupling is as illustrated in Fig. 25.4 a), and we will use the notation of the previous sections.

i) *Transmission Line Current and Voltage*: The voltage and current at $x = 0$ are given in terms of the transmission line quantities by (25.2.17, 25.2.18).

ii) *Circuit Current and Voltage*: From the point of view of the circuit, the voltage can be written in terms of the inductor as

$$V(t) = -\dot{\Phi}(t). \tag{25.3.1}$$

The current is then given by subtracting the current flowing into the capacitor from the current flowing out of the inductor. The capacitative current is given by

$$I_C(t) = C\dot{V}(t) = -C\ddot{\Phi}(t), \tag{25.3.2}$$

and thus the total current is

$$I(t) = \frac{\Phi(t)}{L} + C\ddot{\Phi}(t)\,. \tag{25.3.3}$$

iii) *Input and Output Fields*: The transmission line wave equation (25.2.19) has as its general solution the sum of an incoming travelling wave and an outgoing travelling wave

$$A(x,t) = A_{\text{in}}(t + x/v) + A_{\text{out}}(t - x/v). \tag{25.3.4}$$

The boundary conditions (25.2.17, 25.2.18) then give

$$V(t) = -\left(\dot{A}_{\text{in}}(t) + \dot{A}_{\text{out}}(t)\right), \tag{25.3.5}$$

$$I(t) = \frac{1}{Z_{\text{TL}}}\left(\dot{A}_{\text{in}}(t) - \dot{A}_{\text{out}}(t)\right). \tag{25.3.6}$$

Here we have introduced the *characteristic impedance of the transmission line*, defined by

$$Z_{\text{TL}} \equiv \sqrt{\frac{l}{c}} = lv, \tag{25.3.7}$$

and used (25.2.14) to get the second equality.

iv) *Driven and Damped Circuit Equation*: We can eliminate $\dot{A}_{\text{out}}(t)$, $V(t)$ and $I(t)$ from the two equations (25.3.5, 25.3.6) and (25.3.1, 25.3.3) to get

$$\boxed{C\ddot{\Phi}(t) + \frac{\Phi(t)}{L} + \frac{\dot{\Phi}(t)}{Z_{\text{TL}}} = \frac{2}{Z_{\text{TL}}}\dot{A}_{\text{in}}(t).} \tag{25.3.8}$$

b) Inputs, Outputs, Damping and Resistance: The equations of motion show that the transmission line, as seen by the *LC* circuit, behaves like a current source in parallel with a resistor of magnitude Z_{TL}. The magnitude of the current produced by the source is given by the right-hand side of the equation, namely $2\dot{A}_{\text{in}}(t)/Z_{\text{TL}}$. If the system is being considered classically, and at zero temperature, the incoming field $\dot{A}_{\text{in}}(t)$ can be set equal to zero, and then the transmission line behaves simply as an electric resistance.

The incoming field is however of interest in many ways.

i) Classically, at finite temperature the input field has thermal fluctuations. Comparing with the description of Johnson noise given in *Bk. I: Sect. 7.2*, it is apparent that the incoming field is the physical origin of the noise source.

ii) If the system is being treated quantum-mechanically, the input field becomes an operator, and can never be set equal to zero. At zero temperature this represents the vacuum fluctuations of the quantum field, which we can see drive the *LC* circuit. We will consider this in more detail after we deal fully with the quantization of the field in Sect. 25.5.

iii) The output field is given using (25.3.1, 25.3.5), namely

$$\dot{A}_{\text{out}}(t) = -\dot{A}_{\text{in}}(t) - \frac{1}{L}\dot{\Phi}(t). \tag{25.3.9}$$

The total field in the transmission line then follows from (25.3.12), and takes the form

$$A(x,t) = \dot{A}_{\text{in}}(t + x/v) - \dot{A}_{\text{in}}(t - x/v) - \frac{1}{L}\dot{\Phi}(t - x/v). \tag{25.3.10}$$

This field is then available to be measured by an appropriate instrument attached to the transmission line.

25.3.2 Series Coupling of the *LC* Oscillator to a Transmission Line

The treatment of the series circuit follows similar lines to those used for the parallel circuit, with an appropriate change in the use of the boundary conditions.

a) Circuit Equation of Motion: Referring now to Fig. 25.5, we note:

i) *Transmission Line Voltage and Current*: These are given by exactly the same formulae (25.2.17, 25.2.18) as for the parallel circuit.

ii) *Circuit Current and Voltage*: From the point of view of the circuit, these are given by

$$V(t) = -\dot{\Phi}(t) - \frac{1}{LC}\int_0^t dt'\,\Phi(t'), \tag{25.3.11}$$

$$I(t) = \frac{\Phi(t)}{L}, \tag{25.3.12}$$

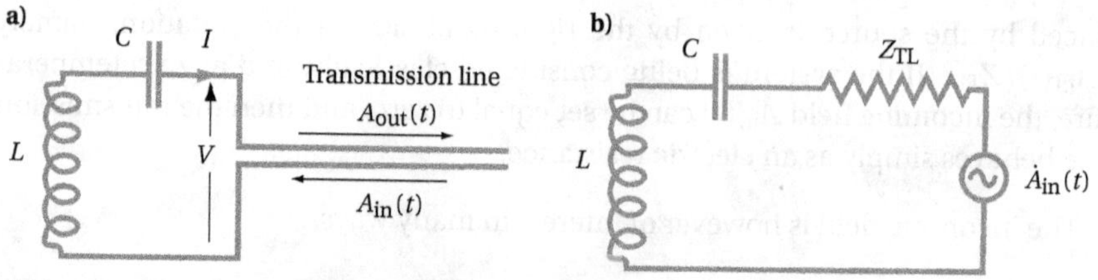

Fig. 25.5. a) Schematic of an LC oscillator coupled in series to a transmission line of characteristic impedance Z_{TL}; **b)** Equivalent circuit, with a series resistor of value Z_{TL}, and a voltage source arising from the incoming field from the transmission line.

iii) *Input and Output Fields:* We can now proceed as for the parallel circuit, and eliminate $\dot{A}_{out}(t)$, $V(t)$ and $I(t)$ from the equations (25.3.5, 25.3.6) and (25.3.11, 25.3.12) to get

$$L\dot{\Phi}(t) + \frac{1}{C}\int_0^t dt'\,\Phi(t') + Z_{TL}\Phi(t) = 2L\dot{A}_{in}(t). \qquad (25.3.13)$$

These equations look a little awkward, and in fact the series circuit is better formulated in using the charge on the capacitor as the canonical co-ordinate for the circuit, as in Ex. 25.2. From (25.1.2, 25.1.3), this means that we can write $Q(t) = \int_0^t \Phi(t')dt'/L$, so that (25.3.13) becomes

$$L\ddot{Q}(t) + \frac{Q(t)}{C} + Z_{TL}\dot{Q}(t) = 2\dot{A}_{in}(t). \qquad (25.3.14)$$

b) Inputs, Outputs, Damping and Resistance: The equations of motion show that the transmission line, as seen by the LC circuit, behaves like a voltage source in series with a resistor of magnitude Z_{TL}. The magnitude of the voltage is given by the right-hand side of the equation, namely $2\dot{A}_{in}(t)$. This interpretation as a voltage source in series with a resistor is of course exactly equivalent to the interpretation as a current source in parallel with the same resistance, as is well known in AC circuit theory.

The interpretation of the input and out fields is otherwise exactly the same as for the parallel circuit.

25.3.3 An Oscillator Embedded in a Transmission Line

The previous examples have involved a semi-infinite transmission line, which yields the simplest formalism. However, to consider a number of qubits coupled to each other via a transmission line, it will be necessary to consider a two-sided coupling. We will consider the setup of Fig. 25.6, in which we divide the transmission line into a left-hand and a right-hand part; thus effectively we have two

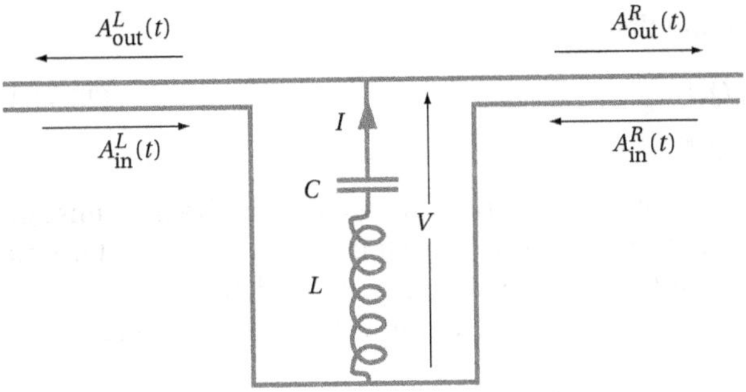

Fig. 25.6. Schematic of a series *LC* oscillator embedded in a transmission line.

transmission lines, and we need to apply the appropriate boundary conditions.

Taking the oscillator to be at $x = 0$, we must match to the two fields $A^R(x,t)$ and $A^L(x,t)$, whose solutions are

$$A^L(x,t) = A^L_{\mathrm{in}}(t - x/v) + A^L_{\mathrm{out}}(t + x/v), \tag{25.3.15}$$
$$A^R(x,t) = A^R_{\mathrm{in}}(t + x/v) + A^R_{\mathrm{out}}(t - x/v). \tag{25.3.16}$$

a) Boundary Conditions: These require that at $x = 0$

i) The voltage across both the left-hand and right-hand transmission lines matches that of the oscillator; that is

$$V(t) = -\partial_t A^L(0,t) = -\partial_t A^R(0,t), \tag{25.3.17}$$
$$= -\dot{A}^L_{\mathrm{in}}(t) - \dot{A}^L_{\mathrm{out}}(t), \tag{25.3.18}$$
$$= -\dot{A}^R_{\mathrm{in}}(t) - \dot{A}^R_{\mathrm{out}}(t). \tag{25.3.19}$$

ii) The current flowing *out of* the left-hand transmission line is equal to the sum of the oscillator current and the current flowing *into* the right-hand transmission line;

$$I(t) = \frac{\partial_x A^L(0,t) - \partial_x A^R(0,t)}{l}, \tag{25.3.20}$$
$$= \frac{1}{Z_{\mathrm{TL}}}\left(\dot{A}^L_{\mathrm{in}}(t) - \dot{A}^L_{\mathrm{out}}(t) + \dot{A}^R_{\mathrm{in}}(t) - \dot{A}^R_{\mathrm{out}}(t)\right). \tag{25.3.21}$$

b) Circuit Equation of Motion: The expressions for the circuit quantities are given by the same formulae (25.3.11, 25.3.12) as for the previous case, but we shall write them in terms of charge as we did in getting the final equation of motion (25.3.14). It is then only a matter of eliminating the output fields using (25.3.18, 25.3.19) to obtain the oscillator equation

$$L\ddot{Q}(t) + \frac{Q(t)}{C} + \tfrac{1}{2}Z_{\mathrm{TL}}\dot{Q}(t) = \dot{A}^L_{\mathrm{in}}(t) + \dot{A}^R_{\mathrm{in}}(t). \tag{25.3.22}$$

c) Output Fields: We also find that

$$\dot{A}^L_{\rm out}(t) = \dot{A}^R_{\rm in}(t) - \tfrac{1}{2} Z_{\rm TL}\dot{Q}(t), \tag{25.3.23}$$

$$\dot{A}^R_{\rm out}(t) = \dot{A}^L_{\rm in}(t) - \tfrac{1}{2} Z_{\rm TL}\dot{Q}(t). \tag{25.3.24}$$

These equations show that, as the characteristic impedance of the transmission line approaches zero, the incoming field from the left transfers directly to become the outgoing field on the right (and conversely), that is, the left and right transmission lines are united into a single two-sided infinite transmission line.

Exercise 25.3 The Asymmetric Configuration: Find the equations of motion for the case where the left and right transmission lines have different parameters l^L, l^R, c^L, c^R.

Exercise 25.4 Series Configurations: Work out the appropriate equations of motion for the two configurations below.

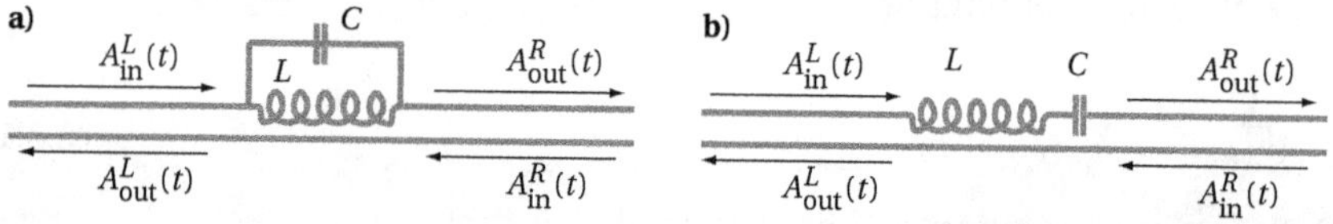

25.4 Lagrangian Formulation of Circuits Coupled to Transmission Lines

For application to quantum theory, a formulation of the coupling between the LC circuit and a transmission line in terms of the separate Lagrangians (25.1.7, 25.2.22) and an interaction Lagrangian is more appropriate for applications. There are two possible distinct couplings, which we identify as *voltage* and *current* coupling.

25.4.1 Voltage Coupling

In this case it is simplest to use the formulation in terms of the charge as canonical co-ordinate, as we did for the series circuit in Sect. 25.3.1 and Sect. 25.3.3. In practice it is common to use a finite transmission line, with endpoints at $x = A$ and $x = B$, and for convenience we choose $A < 0 < B$, and consider an almost localized interaction around the point $x = 0$.

a) Interaction Lagrangian: For voltage coupling, to the Lagrangian (25.2.22) we would most naturally add an interaction Lagrangian of the form

$$-\int_A^B \mu(x)\partial_t A(x,t)\, Q(t)\, dx. \tag{25.4.1}$$

Here $\mu(x)$ is a highly localized function, whose extent represents an approximate measure of the size of the LC circuit, but of course a true representation of a finite sized oscillator would require us to move beyond the lumped circuit model.

We will however not use the interaction in this form, but instead use the alternative form, obtained by adding a total time derivative to (25.4.1)

$$\mathcal{L}_V = \int_A^B \mu(x) A(x,t)\, \dot{Q}(t)\, dx \equiv \frac{1}{L}\int_A^B \mu(x) A(x,t)\, \Phi(t)\, dx, \tag{25.4.2}$$

which gives the same equations of motion. Since it does not depend on $\dot{\Phi}$ and therefore does not affect the definition of the canonical momentum, this form is better adapted to the LC oscillator Hamiltonian in the form we have chosen with Φ as the canonical co-ordinate.

The equations of motion which arise are

$$\partial_t^2 A(x,t) - v^2 \partial_x^2 A(x,t) = \frac{\mu(x)}{c} \dot{Q}(t), \tag{25.4.3}$$

$$\ddot{Q}(t) + \omega_{\text{osc}}^2 Q(t) = -\frac{1}{L}\int_A^B dx\, \mu(x) \partial_t A(x,t). \tag{25.4.4}$$

b) The Limit of Exact Locality: We will show that the limit that

$$\mu(x) \to -\delta(x), \tag{25.4.5}$$

corresponds to the embedded series circuit we have just treated in Sect. 25.3.3.

To make the comparison we consider only the case $-A = B \to \infty$. In the localized limit the equations of motion become

$$\partial_t^2 A(x,t) - v^2 \partial_x^2 A(x,t) = -\frac{\delta(x)}{c} \dot{Q}(t), \tag{25.4.6}$$

$$L\ddot{Q}(t) + \frac{Q(t)}{C} = \partial_t A(0,t). \tag{25.4.7}$$

The general solution of the wave equation (25.4.3) on the interval $-\infty < x < \infty$ is given by (25.A.13, 25.A.14); for this purpose we write this solution in the form

$$A(x,t) = -\tfrac{1}{2} Z_{\text{TL}} Q\left(t - \frac{|x|}{v}\right) + A_{\text{in}}^L\left(t + \frac{x}{v}\right) + A_{\text{in}}^R\left(t - \frac{x}{v}\right). \tag{25.4.8}$$

By substituting this solution into the circuit equation (25.4.7) we get

$$\boxed{L\ddot{Q}(t) + \frac{Q(t)}{C} + \tfrac{1}{2} Z_{\text{TL}} \dot{Q}(t) = \dot{A}_{\text{in}}^L(t) + \dot{A}_{\text{in}}^R(t),} \tag{25.4.9}$$

which is exactly the same as (25.3.22).

Exercise 25.5 Transformer Coupling: The Lagrangian formulation would also allow an exactly local limit like $\mu(x) \to \mu\delta(x)$, and this would give instead of (25.4.9), the modified equation

$$L\ddot{Q}(t) + \frac{Q(t)}{C} + \tfrac{1}{2}\mu^2 Z_{\text{TL}} \dot{Q}(t) = \mu\left(\dot{A}_{\text{in}}^L(t) + \dot{A}_{\text{in}}^R(t)\right), \tag{25.4.10}$$

Show that this corresponds to changing the boundary conditions (25.3.17, 25.3.20) to

$$\frac{V(t)}{\mu} = -\partial_t A^L(0,t) = -\partial_t A^R(0,t), \tag{25.4.11}$$

$$\mu I(t) = \frac{\partial_x A^L(0,t) - \partial_x A^R(0,t)}{l}. \tag{25.4.12}$$

Devise a circuit involving an ideal transformer which would achieve this coupling.

25.5 Quantization of the Transmission Line

The quantization procedure for a transmission line is similar to that used for quantum fields in *Bk. I: Ch.11.* We identify appropriate spatial modes which can be used to express the Lagrangian and Hamiltonian as a sum of corresponding independent individual harmonic oscillator quantities.

25.5.1 The Finite Transmission Line

We will consider a finite transmission line with $A < x < B$. Corresponding to the boundary conditions (25.2.24, 25.2.25), we can introduce two alternative sets of orthonormal eigenfunctions of the operator ∂_x^2 thus:

i) *Zero Current Boundary Condition at Both Ends*:

$$u_n(x) \equiv \sqrt{\frac{2}{B-A}}\cos\left(\frac{n\pi(x-A)}{B-A}\right), \qquad n = 0,1,2,3\ldots \tag{25.5.1}$$

ii) *Zero Voltage Boundary Condition at Both Ends*:

$$v_n(x) \equiv \sqrt{\frac{2}{B-A}}\sin\left(\frac{n\pi(x-A)}{B-A}\right), \qquad n = 1,2,3\ldots \tag{25.5.2}$$

The quantization procedure is similar in both cases, so here we will only treat the first case. We make the expansion

$$A(x,t) = \sum_{n=0}^{\infty} u_n(x) q_n(t), \tag{25.5.3}$$

and using the orthonormality of the eigenfunctions we find that the Lagrangian (25.2.22) and Hamiltonian (25.2.23) become

$$\mathcal{L}_{\text{TL}} = \frac{c}{2}\sum_{n=0}^{\infty}\left(\dot q_n(t)^2 - \omega_n^2 q_n(t)^2\right), \tag{25.5.4}$$

$$\mathcal{H}_{\text{TL}} = \sum_{n=0}^{\infty}\left(\frac{p_n(t)^2}{2c} + \frac{k_n^2 q_n(t)^2}{2l}\right), \tag{25.5.5}$$

$$\text{where}\quad k_n \equiv \frac{n\pi}{B-A}, \qquad \omega_n \equiv k_n v, \tag{25.5.6}$$

$$\text{and}\quad p_n(t) \equiv c\dot q_n(t). \tag{25.5.7}$$

The quantization follows the same procedure as for a single oscillator that was used in Sect. 25.1.1, and we find that

$$b_n \equiv \frac{1}{\sqrt{2\hbar}}\left(\sqrt{\frac{k_n}{Z_{\rm TL}}}\, q_n + i\sqrt{\frac{Z_{\rm TL}}{k_n}}\, p_n\right), \tag{25.5.8}$$

$$\mathcal{H}_{\rm TL} = \sum_{n=0}^{\infty} \hbar\omega_n \left(b_n^\dagger b_n + \tfrac{1}{2}\right). \tag{25.5.9}$$

25.5.2 Coupling to an *LC* Circuit

We will consider a finite transmission line with zero current boundary conditions at each end, and take the interaction in the form (25.4.2). We will also consider the symmetric situation in which $-A = B \equiv \frac{1}{2}\Lambda$, so that values of the eigenfunctions at the point $x = 0$ where the interaction takes place are

$$u_n(0) = \sqrt{\frac{2}{\Lambda}}\cos\left(\frac{n\pi}{2}\right) = \begin{cases} 0, & n \text{ odd}, \\ (-1)^{n/2}, & n \text{ even}. \end{cases} \tag{25.5.10}$$

Thus only the even n modes will be excited.

The interaction Hamiltonian follows from the interaction Lagrangian (25.4.2), and after writing the field in the form (25.5.3), this becomes

$$\mathcal{H}_{\rm Int} = \frac{1}{L}\sum_{n=0}^{\infty} \mu_n q_n(t)\Phi(t), \tag{25.5.11}$$

$$\text{where } \mu_n \equiv \int_{-A}^{A} u_n(x)\mu(x)\, dx. \tag{25.5.12}$$

In the case that the interaction is strictly local, we use $\mu(x) \to -\delta(x)$ as in (25.4.5), and find

$$\mathcal{H}_{\rm Int} = -\frac{2\hbar\sqrt{Z_{\rm TL}Z_{\rm osc}}}{L\sqrt{\pi}}\left(a^\dagger + a\right)\sum_{m=0}^{\infty}\frac{(-1)^m}{\sqrt{m}}\left(b_{2m} + b_{2m}^\dagger\right). \tag{25.5.13}$$

The full Hamiltonian is then

$$\mathcal{H} = \mathcal{H}_{\rm osc} + \mathcal{H}_{\rm TL} + \mathcal{H}_{\rm Int}, \tag{25.5.14}$$

where the oscillator and transmission line Hamiltonians are as in (25.1.13) and (25.5.9).

25.A Appendix: The Wave Equation in One Dimension

25.A.1 Green's Functions

All solutions of the one-dimensional wave equation with or without sources can be obtained from the solutions for the Green's function $G(x,t;x_0,t_0)$ which satisfies the equation

$$\partial_t^2 G(x,t;x_0,t_0) - v^2\partial_x^2 G(x,t;x_0,t_0) = \delta(x-x_0)\delta(t-t_0). \tag{25.A.1}$$

This equation can be integrated easily by changing variables to

$$y = t + x/v, \qquad z = t - x/v, \tag{25.A.2}$$

which transforms the equation to

$$\partial_y\partial_z G = 2v\,\delta(y)\delta(z). \tag{25.A.3}$$

By integrating with respect to y and z it a general solution of this equation is

$$G(x,t;x_0,t_0) = \frac{1}{2v}\Theta(z-z_0)\Theta(y-y_0) + P(y) + R(z), \tag{25.A.4}$$

$$= \frac{1}{2v}\Theta\left(t-t_0-\frac{|x-x_0|}{v}\right) + P\left(t+\frac{x}{v}\right) + R\left(t-\frac{x}{v}\right), \tag{25.A.5}$$

where $P(y)$ and $R(z)$ are arbitrary functions.

25.A.2 The Causal Green's Function

The *causal Green's function* is that solution for which the disturbance at (x_0,t_0) produces a signal only in the future of t_0. The first term (25.A.5) clearly is causal, and causality means that P and R must satisfy

$$P(t+x/v) = 0 \quad\text{and}\quad R(t-x/v) = 0, \qquad \text{for } t<t_0 \text{ and any value of } x, \tag{25.A.6}$$

$$\Longrightarrow \qquad P(z) \equiv R(z) \equiv 0. \tag{25.A.7}$$

Thus, the causal Green's function is

$$G_{\text{causal}}(x,t;x_0,t_0) = \frac{1}{2v}\Theta\left(t-t_0-\frac{|x-x_0|}{v}\right). \tag{25.A.8}$$

25.A.3 Time-Dependent Source of Fixed Shape

This is the problem which arises most commonly in transmission line work, and is written in the form

$$\partial_t^2 A(x,t) - v^2\partial_x^2 A(x,t) = g(x)f(t). \tag{25.A.9}$$

a) Causal Solution: This is written using the causal Green's function

$$A_{\text{causal}}(x,t) = \int_{-\infty}^{\infty} dx_0 \int_{-\infty}^{\infty} dt_0\, g(x_0)f(t_0)\,\frac{1}{2v}\Theta\left(t-t_0-\frac{|x-x_0|}{v}\right), \tag{25.A.10}$$

$$= \frac{1}{2v}\int_{-\infty}^{\infty} dx_0\, g(x_0)F\left(t-\frac{|x-x_0|}{v}\right), \tag{25.A.11}$$

where $F(t)$ is the time integral of $f(t)$,

$$F(t) = \int_{-\infty}^{t} f(t_0)\,dt_0. \tag{25.A.12}$$

b) Time-Dependent Point Source: In the case that $g(x) = g\delta(x)$, the solution becomes

$$A_{\text{causal}}(x,t) = \frac{g}{2v} F(t - |x|/v). \tag{25.A.13}$$

c) General Solution: This is given by adding possible solutions of the homogeneous equation to the causal solution, so that

$$A(x,t) = A_{\text{causal}}(x,t) + P(t + x/v) + R(t - x/v). \tag{25.A.14}$$

The functions $P(t+x/v)$ and $R(t-x/v)$ represent the initial conditions; these can represent thermal fluctuations, or quantum fluctuations in the case that the system is quantized.

26. Superconducting Quantum Devices

The application of superconductors to quantum information relies on a phenomenological model of quantum tunnelling in a Josephson junction, in which few physical details are explicitly included. It is essentially the same kind of modelling as is used in conventional electronics, with discrete devices connected together in an electrical circuit. There is a significant amount of freedom associated with the quantization of superconducting devices, and the full many-body formulation of superconductivity is certainly not necessary. Josephson junction devices always consist of pieces of superconductor which are sufficiently small that the only degree of freedom is the number of charge quanta present. The most significant other feature is that the piece of superconductor can be described by a pure state wavefunction.

The reader is referred to the paper by *Clarke* and *Wilhelm* [26.1] for a detailed review of the application of superconducting electronics to quantum devices, and to *Blais, Huang, Wallraff, Girvin* and *Schoelkopf* [26.2] for the original ideas on the architectures for superconducting quantum computing.

26.1 The Josephson junction

Let us consider the simplest possible model of a Josephson junction, consisting of two pieces of superconductor separated by a very thin insulating layer. The structure is illustrated schematically in Fig. 26.1, with a more realistic picture of the actual fabrication of a Josephson junction in Fig. 26.2. We will characterize the state of the system in terms of the numbers n_L and n_R of charge quanta on either side of the junction. These charge quanta are electron *pairs*, known as *Cooper pairs*, each carrying charge $2e$. The quantum state of the system can therefore be described in terms of the states $|n_L, n_R\rangle$.

In writing the Hamiltonian for this system, we need to consider the following issues:

i) *The Electrical Potential Energy*: The junction has a capacitance C_J, leading to a corresponding energy term. This means that the capacitative energy can be written

$$H_C = E_C \sum_{n_L,n_R} (n_L - n_R - n_g)^2 |n_L, n_R\rangle\langle n_L, n_R| , \tag{26.1.1}$$

$$\text{where} \quad E_C = \frac{e^2}{2C_J} . \tag{26.1.2}$$

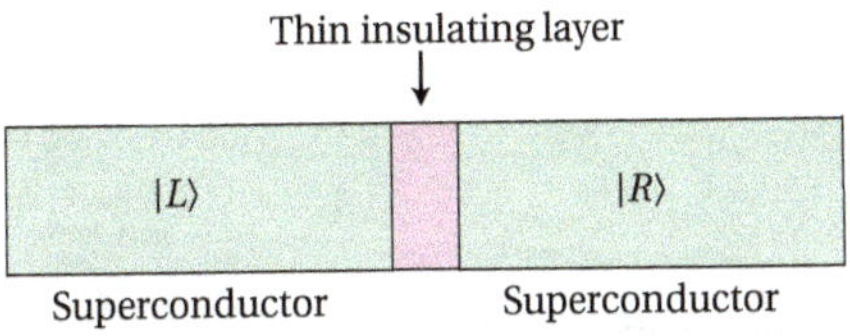

Fig. 26.1. Schematic of a Josephson junction.

Here n_g is an *offset charge*, which accounts for the possibility that, in addition to the Cooper pairs, there may also be charge arising from normal electrons, or an imposed voltage. The quantity n_g is not necessarily an integer, unlike n_L and n_R.

ii) *The Tunnelling Hamiltonian*: We assume that tunnelling of individual Cooper pairs takes place, and is described by

$$H_{\text{Tnl}} = -\tfrac{1}{2}E_J \sum_{n_L,n_R} \Big(|n_L,n_R\rangle\langle n_L-1,n_R+1| + |n_L,n_R\rangle\langle n_L+1,n_R-1|\Big). \quad (26.1.3)$$

The parameter E_J is known as the *Josephson energy*.

26.1.1 Analysis of the Hamiltonian

Both of the terms in the Hamiltonian leave the total $n_L + n_R$ unchanged, so it is convenient to use a simpler notation involving only one variable

$$|m\rangle \equiv |N_L + m, N_R - m\rangle, \quad (26.1.4)$$

in which:

i) The variable $-m$ is number of Cooper pairs *transferred* from left to right, starting from the state $|N_L, N_R\rangle$.

ii) The charge transferred from left to right is $2e\,m$ (where the electron charge has the value $-e$).

iii) The reference state $|N_L, N_R\rangle$ is arbitrary, and we will take N_L and N_R to be fixed quantities, so that the only degree of freedom is m.

iv) Here the range of m will treated as infinite, corresponding to fact that the number of electrons involved is macroscopic.

The Josephson Hamiltonian can then be rewritten in the subspace of fixed N_L and N_R in the form

$$H_J \equiv H_C + H_{\text{Tnl}} \quad (26.1.5)$$

$$= 4E_C \sum_m (m-n_g)^2 |m\rangle\langle m| - \tfrac{1}{2}E_J \sum \big(|m\rangle\langle m+1| + |m+1\rangle\langle m|\big). \quad (26.1.6)$$

Here $\quad n_g = \tfrac{1}{2}(N_L + N_R - n_0). \quad (26.1.7)$

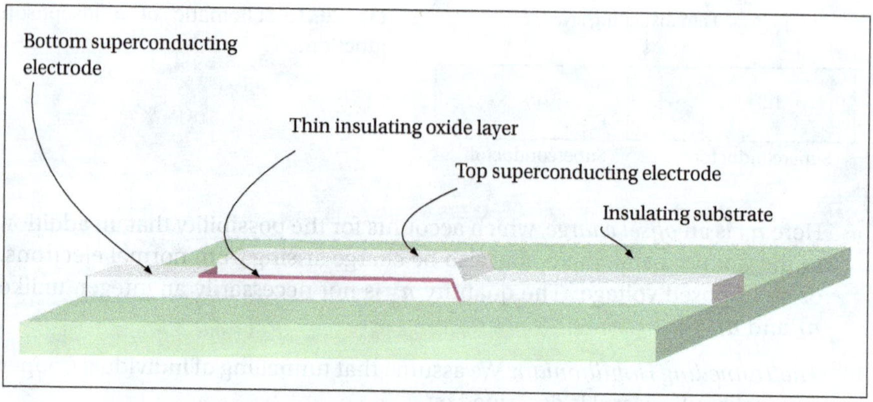

Fig. 26.2. Fabrication of a Josephson junction by two evaporated superconducting layers separated by a thin insulating oxide layer.

a) **Phase States:** Let us introduce *phase states*, defined by

$$|\delta\rangle \equiv \sum_{m=-\infty}^{\infty} e^{im\delta}|m\rangle. \tag{26.1.8}$$

If the Josephson junction is in the quantum state $|\Psi, t\rangle$, then there will a wave-function

$$\psi(\delta, t) \equiv \langle\delta|\Psi, t\rangle\,. \tag{26.1.9}$$

b) **Operators:** Correspondingly there is a *phase operator* Δ such that

$$e^{i\Delta} = \sum_{m=-\infty}^{\infty} |m\rangle\langle m+1|, \tag{26.1.10}$$

$$e^{i\Delta}|\delta\rangle = e^{i\delta}|\delta\rangle\,. \tag{26.1.11}$$

There is also a *number operator*

$$N \equiv \sum_{m=-\infty}^{\infty} m|m\rangle\langle m|. \tag{26.1.12}$$

These operators have the commutation relations

$$[N, e^{i\Delta}] = -e^{i\Delta}, \tag{26.1.13}$$

$$[N, e^{-i\Delta}] = e^{-i\Delta}, \tag{26.1.14}$$

$$[N, \cos\Delta] = -i\sin\Delta, \tag{26.1.15}$$

$$[N, \sin\Delta] = i\cos\Delta. \tag{26.1.16}$$

These are all equivalent to the canonical commutation relation

$$[\Delta, N] = -i. \tag{26.1.17}$$

26.1.2 Flux and Charge as Canonical Co-ordinate and Momentum

The charge operator (that is, the operator for the charge transferred from left to right)

$$Q \equiv 2eN, \tag{26.1.18}$$

and if we define the *flux operator*

$$\Phi \equiv \frac{\hbar}{2e}\Delta, \tag{26.1.19}$$

we see that the commutation relation is

$$[\Phi, -Q] = i\hbar, \tag{26.1.20}$$

that is, exactly as given in (25.1.18) for flux and charge as defined in Sect. 25.1.1, where we found that $-Q$ is the momentum canonically conjugate to Φ.

a) The Flux Quantum Φ_0: The coefficient relating flux and phase in (26.1.19) is related to the *flux quantum* Φ_0; precisely the flux quantum is defined by

$$\Phi_0 \equiv \frac{h}{2e}. \tag{26.1.21}$$

b) The Reduced Flux Quantum φ_0: The actual coefficient in (26.1.19) is a quantity we will call the *reduced flux quantum,* defined by

$$\varphi_0 \equiv \frac{\Phi_0}{2\pi} = \frac{\hbar}{2e}, \quad \text{so that} \quad \Phi = \varphi_0\Delta. \tag{26.1.22}$$

This is a more convenient quantity in Josephson junction work, because of its direct relationship between the phase operator and the flux operator.

26.1.3 Operator Form of the Hamiltonian

We can write the Hamiltonian (26.1.5–26.1.7) directly in terms of the number and phase operators $N(t)$ and $\Delta(t)$ as

$$H_J = 4E_C(N - n_g)^2 - E_J\cos\Delta. \tag{26.1.23}$$

Alternatively, we can write the Hamiltonian in terms of flux $\Phi(t)$ and charge $Q(t)$ as

$$H_J = \frac{(Q-q_g)^2}{2C_J} - E_J\cos\left(\frac{\Phi}{\varphi_0}\right), \tag{26.1.24}$$

$$q_g \equiv 2en_g\,. \tag{26.1.25}$$

Here q_g is known as the *residual charge* on the capacitor.

26.1.4 Equations of Motion for a Josephson Junction

Let us now investigate the Heisenberg equations of motion and the Schrödinger equation arising from the Josephson Hamiltonian.

a) Schrödinger Equation: The action of N on the phase state is $N|\delta\rangle = \mathrm{i}\partial_\delta|\delta\rangle$, so that in the Schrödinger equation for the wavefunction $\psi(\delta,t)$ we make the replacement $N \to -\mathrm{i}\partial_\delta$; the Schrödinger equation corresponding to the Josephson Hamiltonian in the form (26.1.23) is thus

$$\mathrm{i}\hbar\frac{\partial\psi(\delta,t)}{\partial t} = \left\{4E_C\left(-\mathrm{i}\frac{\partial}{\partial\delta} - n_g\right)^2 - E_J\cos\delta\right\}\psi(\delta,t). \tag{26.1.26}$$

b) Heisenberg Equations: The Heisenberg equations arising from the Josephson Hamiltonian (26.1.23) and the commutation relations (26.1.20) are

$$\dot{Q}(t) = \frac{E_J}{\varphi_0}\sin\left(\frac{\Phi(t)}{\varphi_0}\right), \tag{26.1.27}$$

$$\dot{\Phi}(t) = \frac{q_g - Q(t)}{C_J}\,. \tag{26.1.28}$$

Although these equations are fully quantum-mechanical, practical Josephson junctions often work in a regime where the equations of motion are almost classical. In fact it is more usual in the superconducting literature to treat these equations as classical equations, which are subsequently quantized.

c) The Critical Current: Up to now, we have characterized the Josephson junction by its *Josephson energy*, but it is more common to characterize the junction by the *critical current* I_0, defined by

$$I_0 \equiv \frac{E_J}{\varphi_0}\,. \tag{26.1.29}$$

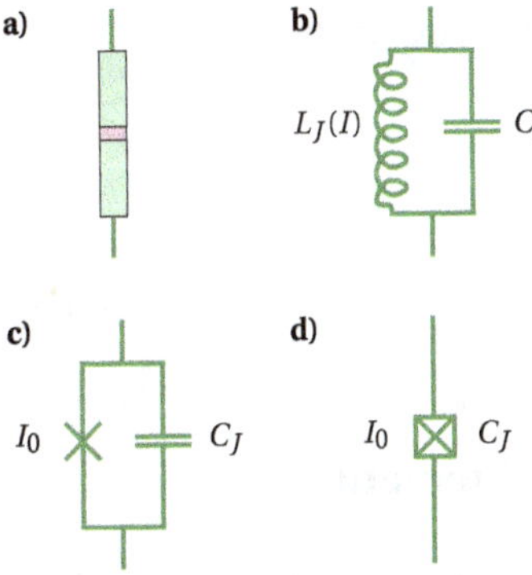

Fig. 26.3. Depictions of a Josephson junction.
a) Representation of the physical structure;
b) Equivalence to a nonlinear inductor $L_J(I)$ in parallel with a capacitor representing the intrinsic capacitance of the junction;
c) The cross is the conventional symbol for the ideal junction, here again in parallel with its intrinsic capacitance;
d) The junction and the capacitance always occur together, and this is represented in a circuit by the symbol of a cross contained in a square.

From the first Heisenberg equation (26.1.27), it follows that the tunnelling current through the junction is bounded in absolute value by I_0; it is not possible to achieve a higher supercurrent. (A higher current can be forced through the junction only by breaking the electron pairing, but this is a normal current.)

26.1.5 The Josephson Junction as a Non-Linear Inductor

When Δ can be considered to be small the Hamiltonian (26.1.23) can be written

$$H_J \approx \frac{(Q-q_g)^2}{2C_J} + \frac{\Phi^2 E_J}{2{\varphi_0}^2} + \text{constant}, \tag{26.1.30}$$

and the behaviour is that of an inductor of size ${\varphi_0}^2/E_J$.

For larger values of Δ we can interpret the flux dependent energy term as that of a nonlinear inductor, as long as some care is taken in defining this concept, since the definition of a nonlinear inductance depends on what physical aspect of inductance we wish to focus on. For a linear inductance the quantity L, in the expression $E = \frac{1}{2}LI^2$ for the inductive energy, in the expression $V = L\dot{I}$ corresponding to Lenz's law, and in the expression $\Phi = LI$ for magnetic flux arising from a current, has the same meaning in each case. When the inductor is nonlinear, however, these are all different quantities.

For the Josephson junction we know that the magnetic energy is given by the tunnelling Hamiltonian $H_{\text{Tnl}} = -E_J\cos(\Phi/\varphi_0)$, and the most useful definition of nonlinear inductance is in practice that corresponding to Lenz's law, which in this case is given by (26.1.28). This equation is to taken in conjunction with its partner, (26.1.27) from which we can see that

$$\Phi(t) = \varphi_0 \arcsin\left(\frac{I(t)}{I_0}\right). \tag{26.1.31}$$

The second equation, being Lenz's law, is also a law for the voltage,

$$V(t) = -\dot{\Phi}(t) \tag{26.1.32}$$

We now take the time derivative of (26.1.31), and eliminate $\dot{\Phi}(t)$, getting

$$V(t) = \dot{I}(t)\, L_J\big(I(t)\big), \tag{26.1.33}$$

where $L_J(I)$ is defined to be the *nonlinear inductance*, explicitly

$$L_J(I) \equiv \frac{\varphi_0}{\sqrt{I_0^2 - I^2}}. \tag{26.1.34}$$

When expressed in terms of flux, we find

$$L_J^{-1} = \frac{E_J}{\varphi_0{}^2}\cos\left(\frac{\Phi}{\varphi_0}\right). \tag{26.1.35}$$

26.2 The Josephson Junction as a Circuit Element

The description of the Josephson junction we have given so far is of an isolated device, which is invariably used as a circuit element. The notations used for the Josephson junction as a circuit element are shown in Fig. 26.3. The two notations Fig. 26.3 c and Fig. 26.3 d, are the most common, and are used interchangeably.

26.2.1 Josephson Oscillations

When we insert a Josephson junction in a circuit, the current and voltage given by the Josephson equations must conform to Kirchhoff's circuit laws, and we can set up the two elementary circuit configurations given in Fig. 26.4.

a) The Fixed Voltage Configuration: The configuration Fig. 26.4 a sets the voltage across the Josephson junction to the value $V(t)$. The equation (26.1.28) gives the voltage across the junction, so that we deduce that

$$\Phi(t) = \Phi(0) + \int_0^t V(t')\,dt', \tag{26.2.1}$$

$$q_g(t) = Q(t) + C_J V(t), \tag{26.2.2}$$

$$\dot{Q}(t) = I_0 \sin\left(\frac{2e\left(\Phi(0) + \int_0^t V(t')\,dt'\right)}{\hbar}\right). \tag{26.2.3}$$

In the case that the applied voltage has the time-independent value V_0, the last equation corresponds to an oscillating current through the junction, of magnitude I_0 and with the *Josephson frequency*

$$f_J = \frac{2eV_0}{h} = \frac{V_0}{\varphi_0} = 483.6 \times 10^{12}\, V_0\,\text{Hz}. \tag{26.2.4}$$

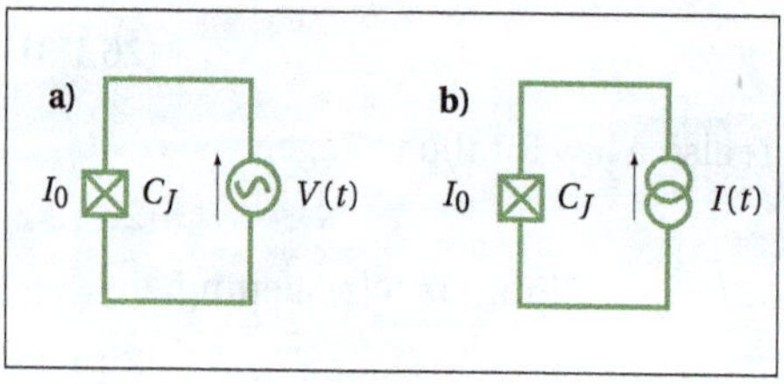

Fig. 26.4. Fundamental circuit configurations for a Josephson junction. **a)** The fixed voltage configuration; **b)** The fixed current configuration.

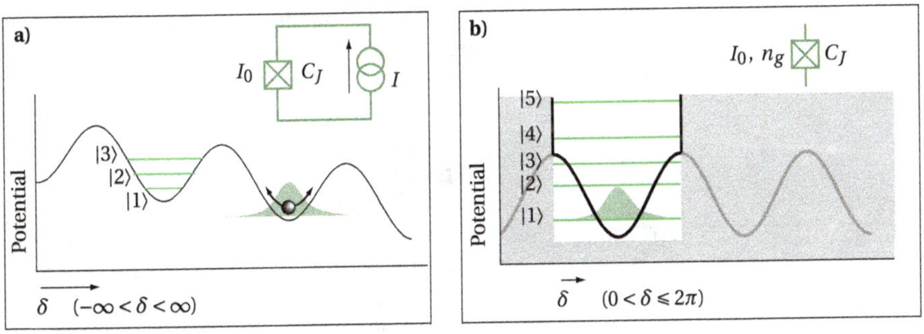

Fig. 26.5. Josephson potentials for different circuit configurations.
a) When inserted in a circuit with a constant bias current, the potential has the "washboard form" (the sum of a linear potential and a sinusoidal potential), and the range of δ is infinite.The potential is characterized by two quantities, the critical current I_0 and the capacitance C.
b) For an isolated junction, as used in the *transmon qubit*, the potential is sinusoidal, and the range of δ is $(0, 2\pi]$, with periodic boundary conditions. The potential is characterized by three quantities, the critical current I_0, the residual charge n_g and the capacitance C.

b) The Fixed Current Configuration: The circuit equations for the configuration Fig. 26.4 b follow from the equations of motion (26.1.27, 26.1.28) by treating the residual charge q_g as a dynamical variable, given in terms of the imposed current by

$$\dot{q}_g = I(t). \tag{26.2.5}$$

We can differentiate (26.1.28) and combine these equations to get

$$I(t) = C_J\ddot{\Phi}(t) + I_0 \sin\left(\frac{\Phi(t)}{\varphi_0}\right). \tag{26.2.6}$$

i) *The Washboard Potential*: For the case of a constant input current I, it is more conventional to express this equation in terms of the phase variable $\Delta(t) = \Phi(t)/\varphi_0$, in the form

$$\frac{\hbar C_J}{2e}\ddot{\Delta}(t) = I - I_0 \sin\Delta(t) = -\frac{2e}{\hbar}U'\big(\Delta(t)\big), \tag{26.2.7}$$

where (using the definitions (26.1.29) of I_0 and (26.1.21) of φ_0) we have introduced the *washboard potential*, defined by

$$U(\delta) \equiv -\varphi_0 I\delta - E_J\cos\delta. \tag{26.2.8}$$

This is the potential of a tilted washboard for a particle of mass $\hbar C_J/2e$, as illustrated in Fig. 26.5 a.

ii) *Classical Motion*: The equation of motion is a Heisenberg operator equation, but the behaviour is often well approximated by the classical equations, since a Josephson junction is a macroscopic device. For the classical equations, we

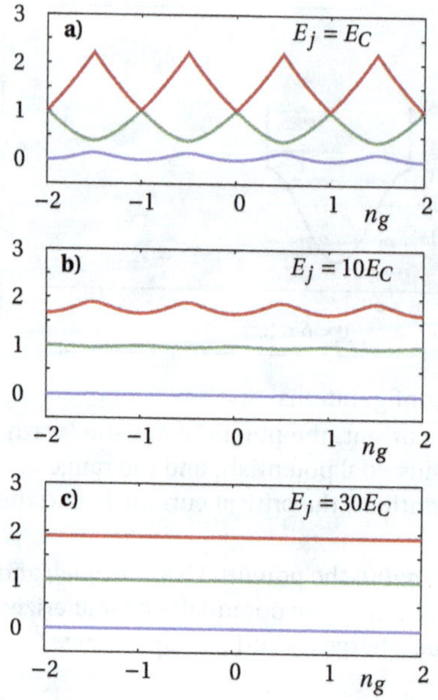

Fig. 26.6. The qualitative behaviour of the first three energy levels of the Josephson Hamiltonian (26.1.6) for three different values of the ratio E_J/E_C, and for a range of values of n_g. For each plot, the ground-state eigenvalue for $n_g = 0$ has been set equal to zero, and the eigenvalues normalized by dividing by the excitation energy of the second eigenvalue when $n_g = 0$. (The Hamiltonian was truncated to a 201 × 201 matrix to obtain these results. No significant alteration is found using a larger truncation.)

will use the notation $\delta(t)$ for the phase, and in that case, the equation of motion (26.2.7) is the same as that of a particle at position $\delta(t)$ moving in the washboard potential.

The nature of the motion is very different, depending on whether I is greater than or less than I_0. For $I < I_0$ the particle remains trapped in one of the potential wells, and executes oscillations in that well—the average value of $\dot{\delta}(t)$ over these oscillations is zero. However, when $I > I_0$, the particle surmounts the barrier, and runs down the washboard—in this case the average value of $\dot{\delta}(t)$ becomes non-zero.

The voltage across the junction is given by Lenz's law, that is (26.1.32), so we can write

$$V(t) = -\varphi_0 \dot{\delta}(t). \tag{26.2.9}$$

The averaged voltage will therefore be non-zero when $I > I_0$, that is, a mean DC voltage will appear when the Josephson junction is in the running state.

iii) *Quantization Effects*: For a sufficiently small junction, the energy in each well is in fact quantized, as indicated in Fig. 26.5 a, and this provides the basis for the use of the Josephson junction as a qubit.

26.2.2 The Open Circuit Configuration

The open circuit configuration is of interest because of its application to the *transmon qubit*, which we will consider in Sect. 26.3.1.

a) Effect of the Residual Charge: Because it is electrically isolated, the system is described by the Josephson Hamiltonian (26.1.6), or equivalently by the Schrödinger equation (26.1.26). The range of the phase variable is $(0, 2\pi]$, giving a potential of the form illustrated in Fig. 26.5 b, in which *periodic boundary conditions* are used. The residual charge affects the kinetic term, not the potential, and plays a very significant role. If we write

$$\psi(\delta, t) = e^{in_g\delta}\bar{\psi}(\delta, t), \tag{26.2.10}$$

the Schrödinger equation (26.1.26) takes on the conventional form

$$i\hbar\frac{\partial\bar{\psi}(\delta, t)}{\partial t} = \left\{4E_C\frac{\partial^2}{\partial\delta^2} - E_J\cos\delta\right\}\bar{\psi}(\delta, t), \tag{26.2.11}$$

but the periodic boundary condition takes on the form

$$\bar{\psi}(0, t) = e^{2\pi i n_g}\bar{\psi}(2\pi, t). \tag{26.2.12}$$

The effect of n_g on the energy levels arises entirely from this changed boundary condition, a special case of a *twisted boundary condition.*

b) Eigenvalues of the Josephson Hamiltonian: The simplest numerical way to determine the eigenvalues of the Josephson Hamiltonian is to diagonalize the matrix form of (26.1.6), using a truncation procedure. The results of this procedure are given in Fig. 26.6, in which the first few energy levels are plotted as a function of the residual charge number n_g. Unless $E_J \gg E_C$ the dependence on n_g is very strong—the addition of a single electron can change the eigenvalue from its maximum value to its minimum value. Since stray charges are hard to avoid in any experimental situation, there will be unavoidable fluctuations in the energy levels, unless one uses the device in the region $E_J \gg E_C$.

The eigenfunctions of the Schrödinger equation for this model can also be obtained exactly in terms of Mathieu functions [26.3, 26.4]. The crucial feature that emerges from this solution is that the charge dispersion, i.e. the maximal variation of the eigenenergies with n_g, is *exponentially* suppressed with E_J/E_C while the relative anharmonicity decreases only algebraically with a slow power law in E_J/E_C. As a consequence, there exists a regime with $E_J \gg E_C$—the *transmon regime*—where the anharmonicity is much larger than the linewidth arising from fluctuations of the offset charge n_g, thus satisfying the operability condition of a qubit [26.5].

This result can also be seen from the nature of the potential as illustrated in Fig. 26.5 b—for sufficiently large E_J the wavefunction becomes exponentially small at the ends $\delta = 0$ and $\delta = 2\pi$ of the range of δ, making the wavefunction very insensitive to the boundary condition.

26.3 Qubit Architectures

In order to be useful for quantum information processing tasks, several Josephson qubits must be made to controllably interact with each other and interactions with environmental degrees of freedom must be minimized. In circuit quantum electrodynamics, this is achieved by coupling the Josephson junctions to a common microwave environment with a desired discrete mode structure.

There are many different proposals for qubits based in Josephson junctions, and these are nicely reviewed by *Devoret* and *Schoelkopf* [26.6], where they note that

> A Josephson junction transforms an electric circuit into a true artificial atom, for which the transition from the ground state to the excited state $|g\rangle \leftrightarrow |e\rangle$ can be selectively excited and used as a qubit. The Josephson junction can be placed in parallel with an inductor, or can even replace the inductor completely, as in the case of the so-called "charge qubits". Potential energy functions of various shapes can be obtained by varying the relative strengths of three characteristic circuit energies associated with the inductance, capacitance, and tunnel.
>
> When several of these qubits, which are non-linear oscillators behaving as artificial atoms, are coupled to true oscillators (photons in a microwave cavity), one obtains, for low-lying excitations, an effective multiqubit, multicavity system Hamiltonian of the form
>
> $$H_{\text{eff}} = \hbar\left\{\sum_j \left(\omega_j^q b_j^\dagger b_j + \tfrac{1}{2}\alpha_j (b_j^\dagger b_j)^2\right) + \sum_m \omega_m^r a_m^\dagger a_m + \sum_{j,m} \chi_{j,m} b_j^\dagger b_j a_m^\dagger a_m\right\}, \tag{26.3.1}$$
>
> describing anharmonic qubit mode amplitudes indexed by j coupled to harmonic cavity modes indexed by m. The symbols a, b, and ω refer to the mode amplitudes and frequency, respectively. When driven with appropriate microwave signals, this system can perform arbitrary quantum operations at speeds determined by the non-linear interaction strengths α and χ, typically resulting in single-qubit gate times within 5 to 50 ns ($\alpha/2\pi \approx 200$ MHz) and two-qubit entangling gate times within 50 to 500 ns ($\chi/2\pi \approx 20$ MHz). There is also a weak induced anharmonicity of the cavity modes which we have neglected here.

In this section, we will limit our considerations to the *transmon qubit*, which displays all of the essential features mentioned above.

26.3.1 The Transmon Qubit

The *transmon qubit*, introduced by the Yale group [26.4], consists of an isolated Josephson junction, and is thus an application of the open circuit configuration of Sect. 26.2.2. The transmon was originally created using resonant LC circuits, but more recently has been constructed using a microwave cavity [26.7]. This is the configuration we shall describe, and is illustrated in Fig. 26.7 and Fig. 26.8. The energy levels of the system are not evenly spaced, and an effective two-level system can be isolated by appropriate tuning of the microwave field.

The transmon is designed to operate in a regime of significantly increased ratio of Josephson energy and charging energy E_J/E_C. It benefits from the fact that its charge dispersion decreases exponentially with E_J/E_C, while its loss in anharmonicity is described by a weak power law. This results in a dramatic insensitivity to charge noise, and an increase in the qubit-photon coupling, while maintaining sufficient anharmonicity for selective qubit control.

It usually employs an *antenna*, consisting of a length of superconductor sufficiently long to generate a significant dipole coupling to a microwave radiation field, and the microwave field employed is normally confined to a high Q superconducting cavity [26.8]. In essence, this makes it equivalent to an artificial atom, coupled to the cavity electromagnetic field by its dipole moment—in effect, a synthetic microwave version of cavity QED. This geometry presents the qubit with a well-controlled electromagnetic environment, limiting the possibility of relaxation through spontaneous emission into the multiple modes that may be possible with a complicated chip and its associated wiring. The qubit is placed in the center of the cavity, maximizing the coupling to the lowest frequency TE101 mode at $\omega_{\text{cav}} \approx 2\pi \times 8\,\text{GHz}$, which is used for readout and control.

26.3.2 Formulation as Cavity QED

Looking at the Josephson Hamiltonian in the form (26.1.23), we can make two approximations:

i) In the *transmon regime* we know that the effect of n_g becomes exponentially negligible, so we will set $n_g = 0$.

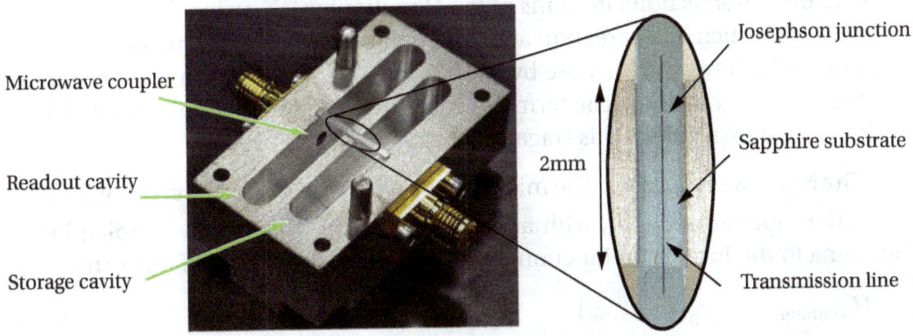

Fig. 26.7. Photograph of one half of two aluminum waveguide cavities coupled to a transmon qubit, as used by the Yale group in [26.7]. The right-hand side of the figure shows a detail of the qubit, fabricated on a sapphire substrate 1.4 mm wide, 15 mm long and 430 microns thick. The coupling strength of the qubit is determined by the length of the coupling antenna which extends into each cavity. The upper cavity, with a resonant frequency of $\omega_m = 2\pi \times 8.2564\,\text{GHz}$, is used for qubit readout and the lower cavity, with a frequency of $\omega_c = 2\pi \times 9.2747\,\text{GHz}$, is used to store and manipulate quantum states. (*Image kindly supplied by Prof G Kirchmair and Dr B Vlastakis.*)

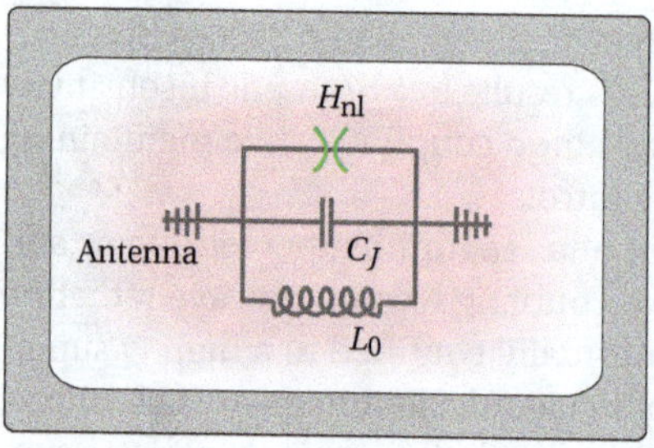

Fig. 26.8. Schematic diagram of a transmon. The Josephson junction is represented by the circuit diagram, in which the non-linear inductance is separated into the "spider" symbol, representing $H_{\rm nl}$, and the linear part L_0. The antennas are purely schematic, and often are simply linear conductors. They provide the interaction with the electromagnetic cavity mode, represented by the pink shading.

ii) The anharmonicity becomes small, but not negligible. This means that we can expand the cosine potential in powers of Δ, and truncate at the fourth order.

Using these approximations, we get an approximate Hamiltonian, which we write in terms of flux and charge operators as

$$H_J \approx H_0 + H_{\rm nl} - E_J, \tag{26.3.2}$$

$$H_0 \equiv \frac{Q^2}{2C_J} + \frac{\Phi^2}{2L_0}, \tag{26.3.3}$$

$$H_{\rm nl} \equiv -\frac{\varphi_0^2\Phi^4}{24L_0}, \tag{26.3.4}$$

$$L_0 \equiv \frac{\varphi_0^2}{E_J} = L_J(0). \tag{26.3.5}$$

The quantization is done in terms of the Hamiltonian H_0, using the procedure of Sect. 25.1.2, which gives creation and destruction operators for the bar Josephson junction which we shall denote by b_0, $b_0^\dagger$, in the same way as in (25.1.12), and a resonant frequency $\omega_{J,0}$. The term $H_{\rm nl}$ remains as an interaction term, which we will not treat explicitly at this stage.

a) The Microwave Cavity: The microwave cavity is described by creation and destruction operators a_0, $a_0^\dagger$, with a frequency $\omega_{\rm cav,0}$. This is now coupled by the antenna to the Josephson junction, by a coupling Hamiltonian of the form

$$H_{\rm coupling} = g\big(a_0^\dagger b_0 + b_0^\dagger a_0\big). \tag{26.3.6}$$

This coupling between the microwave cavity and the Josephson junction gives a non-diagonal quadratic part of the Hamiltonian, which can be diagonalized to give the Hamiltonian in terms of normal modes. The coupling is sufficiently small for us to be able to distinguish:

i) *The Josephson Junction Mode*: The mode operators b, $b^\dagger$, which describe the mode predominantly located on the Josephson junction, with frequency $\omega_{J,1}$.

ii) *The Cavity Mode*: The mode operators a, $a^\dagger$, which describe the mode predominantly located in the microwave cavity, with frequency $\omega_{\rm cav,1}$.

Following (25.1.12), the flux operator Φ is proportional to $b_0 + b_0^\dagger$, and the diagonalization procedure means that this will be proportional to $b + b^\dagger + \kappa(a + a^\dagger)$, where κ is determined by the diagonalization process, and normally is small. We then insert Φ into H_{nl}, and make the rotating wave approximation, so that only terms involving $a^\dagger a$ and $b^\dagger b$ remain. The result is the Hamiltonian of the form given in (26.3.1)

$$\mathcal{H} = \hbar\left\{\omega_J b^\dagger b + \omega_{\text{cav}} a^\dagger a - \tfrac{1}{2}K b^\dagger b^\dagger bb - \tfrac{1}{2}K_{\text{cav}} a^\dagger a^\dagger aa - \chi b^\dagger b a^\dagger a\right\}. \tag{26.3.7}$$

Here, ω_J and ω_{cav} differ from $\omega_{J,1}$ and $\omega_{\text{cav},1}$ by amounts proportional to κ, which arise when normally ordering the fourth-order terms.

In experimental situations, the frequencies ω_J and ω_{cav} are of the order of $2\pi \times 10^{10}$ Hz, while the nonlinear coefficients are about a factor 10^{-3} smaller.

b) The Qubit Hamiltonian: The energy levels of the Josephson junction are not evenly spaced, being given by

$$E_n = \hbar\left(\omega + \tfrac{1}{2}K\right)n - \tfrac{1}{2}\hbar K n^2, \qquad n = 0, 1, 2, \ldots \tag{26.3.8}$$

and the transmon is normally used in the regime where only the levels $n = 0$ and $n = 1$ are excited. We can therefore treat these as a two-level system described by Pauli matrices, and make the replacements

$$b \longrightarrow \sigma^-, \qquad b^\dagger \longrightarrow \sigma^+, \qquad b^\dagger b \longrightarrow \tfrac{1}{2}(1 + \sigma_z). \tag{26.3.9}$$

The Hamiltonian takes the form

$$\mathcal{H}_{\text{Transmon}} = \hbar\left\{\tfrac{1}{2}\omega_J(1 + \sigma_z) + \omega_{\text{cav}} a^\dagger a - \tfrac{1}{2}K_{\text{cav}} a^\dagger a^\dagger aa - \tfrac{1}{2}\chi a^\dagger a \sigma_z\right\}. \tag{26.3.10}$$

This is the same form as the dressed cavity QED Hamiltonian (22.1.32) discussed in Sect. 22.1.5. There is one additional feature, namely the term $\frac{1}{2}K_{\text{cav}} a^\dagger a^\dagger aa$, which is fourth-order, and is a kind of *Kerr nonlinearity* induced by the presence of the superconductor.

c) Energy Levels: The energy levels are written $E_{e,n}$ and $E_{g,n}$, and are

$$E_{g,n} = \hbar\left(\omega_{\text{cav}} + \tfrac{1}{2}\chi\right)n - \tfrac{1}{2}\hbar K_{\text{cav}} n(n-1), \tag{26.3.11}$$

$$E_{e,n} = \hbar\omega_J + \hbar\left(\omega_{\text{cav}} - \tfrac{1}{2}\chi\right)n - \tfrac{1}{2}\hbar K_{\text{cav}} n(n-1). \tag{26.3.12}$$

d) Use as a Qubit: The cavity mode can play the part of the data bus, in the same way as the phononic modes in the ion trap quantum computer. Normally, the number of microwave photons will correspond to $n = 0$ or $n = 1$, and in both cases the anharmonicity term $a^\dagger a^\dagger aa$ will be zero. The states $|g\rangle$ and $|e\rangle$ are the qubit states, and

$$E_{g,0} = 0, \qquad E_{e,0} = \hbar\omega_J, \tag{26.3.13}$$

$$E_{g,1} = \hbar\left(\omega_{\text{cav}} + \tfrac{1}{2}\chi\right), \qquad E_{e,1} = \hbar\omega_J + \hbar\left(\omega_{\text{cav}} - \tfrac{1}{2}\chi\right). \tag{26.3.14}$$

It is quite straightforward to put a number of transmon qubits in the same cavity, but the method chosen for the ion trap quantum computer, which uses individual focused laser beams to address the individual ions, cannot be applied at microwave wavelengths. Instead, these can be made individually addressable by choosing a different value of the Josephson frequency ω_J for each qubit—for example see [26.9] and [26.10].

In principle, the same gates as used for the ion trap quantum computer can be implemented.

26.3.3 Measurement of the Output

Because there are no microwave analogues of the optical photodetectors used to implement the measurement protocols for the ion trap quantum computer, completely different detection strategies are needed. A very effective measurement procedure for this kind of system has been devised [26.11]. It uses the coupling term $\frac{1}{2}\chi a^\dagger a \sigma_z$ in the transmon Hamiltonian, which makes the time evolution of the quantum state of the cavity dependent on the quantum state of the qubit.

The procedure is in principle very simple, and depends on the fact that the transmon Hamiltonian is composed of the commuting operators $a^\dagger a$ and σ_z:

i) *Initial State*: Take the system initially in the product state

$$|0\rangle \otimes \big(c_e|e\rangle + c_g|g\rangle\big), \tag{26.3.15}$$

that is, as a vacuum state of the cavity, and a superposition $c_e|e\rangle + c_g|g\rangle$ of the excited and ground states of the qubit.

ii) *Creation of a Coherent Cavity State*: Apply a strong short coherent microwave pulse to the cavity. During this pulse, the transmon Hamiltonian has a negligible effect, and the quantum state of the cavity becomes a coherent state $|\alpha\rangle$. In practice $|\alpha|^2$ attains a value corresponding to about five photons.

iii) *Entanglement Resulting from Time Evolution*: Let the system evolve freely under the transmon Hamiltonian. The evolution of coherent state variable will now depend on the qubit state. For simplicity, let us firstly neglect the anharmonic term proportional to $K_{\rm cav}$, since in practice the effect of this term is relatively small. In this case, if the system evolves for a time τ, then the state will evolve into an entangled state

$$|\alpha\rangle \otimes \big(c_e|e\rangle + c_g|g\rangle\big) \longrightarrow c_e|\alpha_e(\tau)\rangle \otimes |e\rangle + c_g|\alpha_g(\tau)\rangle|g\rangle, \tag{26.3.16}$$

in which

$$\alpha_g(\tau) \equiv \alpha e^{i(\omega_{\rm cav}+\chi/2)\tau}, \qquad \alpha_e(\tau) \equiv \alpha e^{i(\omega_{\rm cav}-\chi/2)\tau}. \tag{26.3.17}$$

iv) *Distinguishable Coherent Cavity States*: The amplitudes of the coherent states thus produced, for which $|\alpha|^2 \approx 5$, are not macroscopic, but are readily distinguishable. If τ is chosen so that $\chi\tau \approx \pi$, the overlap between the two states in the superposition is (see *Bk. I: (10.5.11)*)

$$\left|\langle \alpha e^{i(\omega_{\rm cav}-\chi/2)\tau} \,|\, \alpha e^{i(\omega_{\rm cav}+\chi/2)\tau}\rangle\right| = \exp\left(-2|\alpha|^2 \sin^2\left(\tfrac{1}{2}\chi\tau\right)\right) \approx 0.000045, \tag{26.3.18}$$

so that these states are essentially orthogonal.

v) *Quantum-Limited Amplification*: To measure these amplitudes, a *quantum-limited amplifier* is used. In *Bk. I: Sect.14.3.1*, a harmonic oscillator system with both *gain* and *loss* is described—the quantum-limited amplifier is the special case in which the loss term is zero, and we have called this an *ideal amplifier* in *Bk. I: Sect.16.2.1* d. The amplifier used in practice is the *Josephson ring modulator*, which we describe in detail in Sect. 26.4.1. At milliKelvin temperatures, this amplifier is a very accurate implementation of a quantum limited amplifier.

   The procedure is in fact exactly that described in *Bk. I: Sect.19.5.1* c, where we showed that noise is added, but the amount of noise is equivalent to half a quantum. This means that result of amplification on $N \equiv a^\dagger a$, for gain G, is equivalent to

$$\langle N\rangle \longrightarrow G\langle N\rangle, \qquad \mathrm{var}[N] \longrightarrow G^2\left(\mathrm{var}[N]+\tfrac{1}{2}\right). \tag{26.3.19}$$

   For $|\alpha|^2 \approx 5$ the variance is $\mathrm{var}[N] \approx 5$, and the extra half quantum does not affect the distinguishability of the two values of the coherent state amplitude.

vi) *Decoherence as a Result of Amplification*: As a result of the amplification, all coherence between the two coherent states is lost, as explained in *Bk. I: Sect.19.4*. Thus, the entangled pure state on the right of (26.3.16) becomes a mixed state, with a density operator of the form

$$\rho_{\mathrm{meas}} = \rho_{\mathrm{cav},e} \otimes |e\rangle\langle e| + \rho_{\mathrm{cav},g} \otimes |g\rangle\langle g|, \tag{26.3.20}$$

   where the density operators $\rho_{\mathrm{cav},e}$ and $\rho_{\mathrm{cav},g}$ represent the mixed states of the electromagnetic field which result from the amplification process.

26.3.4 Other Qubit Architectures

There is a range of of other architectures less closely aligned with the quantum optics viewpoint, which we shall not cover here. The reviews by *Clarke* and *Wilhelm* [26.1] and by *Martinis* [26.12] provide a good coverage of the range of possible architectures under consideration.

26.4 Josephson Junction Amplification

Because the Josephson junction is a non-linear device it can be used to generate parametric amplification, just as non-linear dielectric crystals can be used to produce optical parametric amplification, as noted in Sect. 21.2.1. This capability was used in 1988 to produce squeezing of both thermal microwaves [26.13] and in 1990 to squeeze microwave vacuum noise [26.14]. In the same way as we have noted in Chap. 21, it is possible to create both phase-sensitive and phase-preserving amplifiers. In this section we will not consider the full range of possible methods of

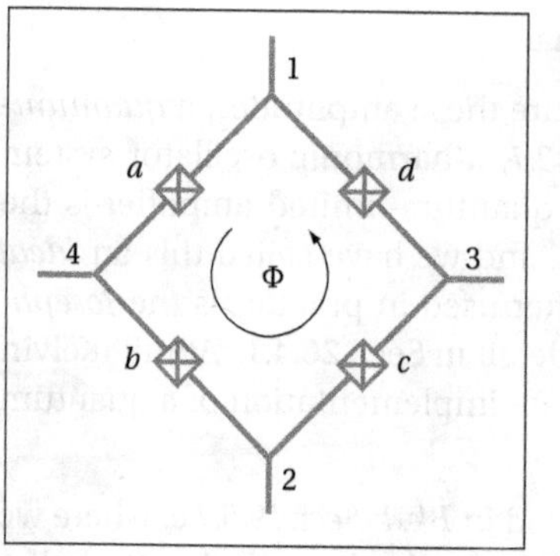

Fig. 26.9. The Josephson Ring modulator. Four Josephson junctions are connected in a Wheatstone bridge configuration.

constructing such amplifiers. Rather, because of its application to the measurement process used in the previous section, we shall consider only the Josephson ring modulator.

26.4.1 The Josephson Ring Modulator

The Josephson ring modulator was devised by the Yale group [26.15,26.16] to provide a quantum-limited phase-preserving amplifier of microwaves. It consists of four Josephson junctions in a Wheatstone bridge configuration. This gives a symmetric device, which greatly enhances the purity of the amplification process and simplifies both its operation and its analysis. As noted in [26.15]

> The device consists of four nominally identical Josephson junctions forming a ring threaded by a magnetic flux Φ. This magnetic flux induces a fixed, circulating current around the ring. When operated with a bias current lower than their critical current I_0, Josephson junctions behave as pure nonlinear inductors with inductance $L_J = \varphi_0/(I_0 \cos\delta)$, where δ is the gauge-invariant phase of the junction and $\varphi_0 = \hbar/2e$ is the reduced flux quantum. They are the only known non-linear and non-dissipative circuit elements working at microwave frequencies. The ring has three orthogonal electrical modes coupled to the junctions: two differential ones, X and Y, and a common one, Z. They provide the minimum number of modes for three-wave mixing.

The configuration is shown in Fig. 26.9; if V_i are the potentials at the nodes 1, 2, 3, 4, then we can define *node fluxes* Φ_i by

$$\dot{\Phi}_i = -V_i, \qquad i = 1,2,3,4. \tag{26.4.1}$$

The voltages across the Josephson junctions a, b, c, d, and their corresponding fluxes, are given by

$$\left.\begin{aligned} V_a &= -\dot{\Phi}_a = V_1 - V_4, \\ V_b &= -\dot{\Phi}_b = V_4 - V_2, \\ V_c &= -\dot{\Phi}_c = V_2 - V_3, \\ V_d &= -\dot{\Phi}_d = V_3 - V_1. \end{aligned}\right\} \tag{26.4.2}$$

Since the sum $V_a + V_b + V_c + V_d = 0$, we can deduce that

$$\Phi_a + \Phi_b + \Phi_c + \Phi_d = \Phi, \qquad \text{a constant.} \tag{26.4.3}$$

a) Flux Quantization: The phase change across junction a is given $\delta_a = \Phi_a/\varphi_0$, and similarly for junctions b, c, d, so the total phase change around the loop must be a multiple of 2π, giving the *flux quantization condition for a superconducting loop*

$$\Phi = 2n\pi\varphi_0. \tag{26.4.4}$$

b) Flux Modes as Dynamical Variables: Consistent with this constraint, we can define independent *flux modes* Φ_X, Φ_Y, Φ_Z, by setting

$$\left.\begin{aligned}
\Phi_a &= \tfrac{1}{4}(\ \ 2\Phi_X - 2\Phi_Y + 2\Phi_Z + \Phi\ ),\\
\Phi_b &= \tfrac{1}{4}(\ \ 2\Phi_X + 2\Phi_Y - 2\Phi_Z + \Phi\ ),\\
\Phi_c &= \tfrac{1}{4}(-2\Phi_X + 2\Phi_Y + 2\Phi_Z + \Phi\ ),\\
\Phi_a &= \tfrac{1}{4}(-2\Phi_X - 2\Phi_Y - 2\Phi_Z + \Phi\ ).
\end{aligned}\right\} \tag{26.4.5}$$

The Hamiltonian for the system is the sum of the Hamiltonian for each of the Josephson junctions, each of which is of the form (26.1.24). The junctions are designed to have a large area, and therefore a high capacitance, so that capacitative term can be neglected. Summing the four terms, and using the appropriate trigonometric identities, the Hamiltonian can be written

$$\begin{aligned}
H_{\text{Ring}} = -4E_J\Bigg\{ &\cos\left(\frac{\Phi_X}{2\varphi_0}\right)\cos\left(\frac{\Phi_Y}{2\varphi_0}\right)\cos\left(\frac{\Phi_Z}{2\varphi_0}\right)\cos\left(\frac{\Phi}{4\varphi_0}\right)\\
&+\sin\left(\frac{\Phi_X}{2\varphi_0}\right)\sin\left(\frac{\Phi_Y}{2\varphi_0}\right)\sin\left(\frac{\Phi_Z}{2\varphi_0}\right)\sin\left(\frac{\Phi}{4\varphi_0}\right)\Bigg\}.
\end{aligned} \tag{26.4.6}$$

c) The Parametric Hamiltonian Approximation: The loop flux Φ can be adjusted, and a particular value which provides the most natural parametric amplifier is the choice $\Phi = \frac{1}{2}\pi\varphi_0$. Making this choice, we can investigate small flux approximation in which Φ_X, Φ_Y, Φ_Z are all much smaller than φ_0, and expand the sines and cosines up to third order (and drop a constant term) to get the approximate Hamiltonian

$$H_{\text{Ring}} \approx \lambda\Phi_X\Phi_Y\Phi_Z + \mu\left(\Phi_X^2 + \Phi_Y^2 + \Phi_Z^2\right), \tag{26.4.7}$$

$$\lambda \equiv -\frac{E_J}{2\sqrt{2}\varphi_0^3}, \qquad \mu = -\frac{E_J}{\sqrt{2}\varphi_0^2}. \tag{26.4.8}$$

The net effect is the sum of three inductive terms, one for each mode flux, and the interesting trilinear term, which can be used to provide parametric amplification.

There operating points other than $\Phi = \frac{1}{2}\pi\varphi_0$ available, which we shall not treat here. For these, and a detailed description of the circuitry need, the reader should consult [26.15].

d) Use as a Parametric Amplifier: The Hamiltonian treats all of the fields on an equal basis, and we now want to show what must be done to get a Hamiltonian of the form (21.3.1).

i) *Pump Mode*: We set $\Phi_Z \to \mathcal{E}(t)$, a large classical field, which we will write in the form

$$\mathcal{E}(t) = \mathrm{i}E\big(\mathrm{e}^{-\mathrm{i}\omega_p t} - \mathrm{e}^{\mathrm{i}\omega_p t}\big). \tag{26.4.9}$$

ii) *Signal Mode*: Let us choose Φ_X to be the signal field. We assume that the Hamiltonian for the signal field includes a capacitative part, and an inductive part which includes that which arises in the approximate Hamiltonian (26.4.7), thus giving a harmonic oscillator with resonant frequency ω_s. We will represent the signal by a single mode at this frequency thus

$$\Phi_X = u\big(a\mathrm{e}^{-\mathrm{i}\omega_s t} + a^\dagger \mathrm{e}^{\mathrm{i}\omega_s t}\big), \tag{26.4.10}$$

where u is an appropriate mode normalization.

iii) *Idler Mode*: We choose Φ_Y to be the idler field, and this is treated similarly to the signal field, thus it is written

$$\Phi_Y = v\big(b\mathrm{e}^{-\mathrm{i}\omega_i t} + b^\dagger \mathrm{e}^{\mathrm{i}\omega_i t}\big), \tag{26.4.11}$$

where v is an appropriate mode normalization.

iv) *Frequency Matching*: For operation as a nondegenerate parametric amplifier we choose the frequencies to satisfy

$$\omega_p = \omega_s + \omega_i\,, \qquad \omega_s \neq \omega_i\,. \tag{26.4.12}$$

We now make a rotating wave approximation, yielding a parametric Hamiltonian of the form (21.3.1), namely

$$H_{\text{Ring}} \longrightarrow \mathrm{i}\lambda E u v\big(a^\dagger b^\dagger - ab\big). \tag{26.4.13}$$

v) *More Detailed Treatment*: In [26.15] a more detailed treatment than our simple rotating-wave method is given, but the essential conclusions are the same.

IX INTERFACING QUANTUM NETWORKS

PART IX: INTERFACING QUANTUM NETWORKS

The concept of a *quantum network* involves the following principal components:

i) *The Quantum Processor*: The long-lived internal states of atoms and ions currently provide the best method of processing quantum information, as exemplified by the ion trap quantum computer of Chap. 24.

ii) *The Quantum Memory*: Here too, the long-lived internal states of atoms and ions currently provide the best method of storing quantum information. There are a number of options currently under experimental investigation for such quantum memories.

iii) *The Quantum Node*: It is convenient to use the terminology *quantum node* to describe any local device which contains quantum processors or quantum memory.

iv) *The Quantum Channel*: It is possible to transfer quantum information with high fidelity using a propagating electromagnetic field. At present an optical fibre represents the best qubit carrier for fast and reliable communication over long distances. We will call any such carrier a *quantum channel.*

v) *The Quantum Interface*: The concept of a *quantum interface* represents not so much a device in itself, but a device and a choice of design of both a quantum node and a quantum channel by which an efficient transfer of quantum information can be effected.

This part of the book is devoted to the design of quantum memories, and their corresponding quantum interfaces. There are several schemes under development, but we shall limit ourselves to schemes based on the use of a three-level atom in the Λ-configuration of Sect. 6.1.1. The fundamental tool in all of this part of the book is the use of a control field to effect a change of the coupling between the qubit level and the external optical field.

In Chap. 27 we show how to interface a single atom of this kind confined in a cavity to the propagating optical field incident on the cavity. We then consider in Chap. 28 an *ensemble* of such atoms in a cavity, which provides many advantages compared to a single atom. Finally, in Chap. 29 treat the case of a quantized electromagnetic field freely propagating through an atomic vapour—here no cavity is involved.

27. Cavity Quantum Electrodynamics Networks

The three-level atom in the Λ-configuration of Chap. 6 provides a flexible basis for a quantum memory, which can be implemented in a number of ways. The essential features are illustrated in Fig. 27.1, namely:

i) The two ground-state levels, $|g\rangle$ and $|s\rangle$, are approximately degenerate, and very long lived.

ii) The transition $g \longleftrightarrow e$ couples to a quantized mode of the electromagnetic field, which carries the quantum information from the associated quantum channels.

iii) The transition $s \longleftrightarrow e$ couples to a strong classical dressing field with Rabi frequency $\Omega(t)$.

27.1 The Cavity QED Quantum Memory

In Chap. 22, we considered a two-level atom coupled to an optical cavity. Here we wish to study the three-level atom in the Λ-configuration, as treated in Sect. 6.1.1. In conjunction with an optical cavity, a single such atom can provide a one qubit

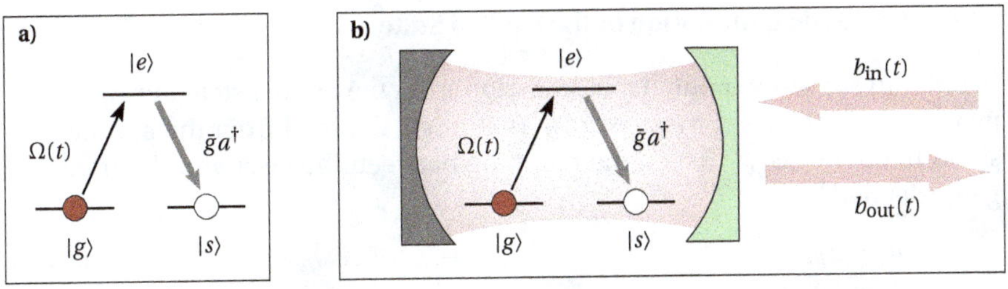

Fig. 27.1. a) The abstract three level atom in a Λ-configuration using a classical dressing field for the $g \leftrightarrow e$ transition, and a quantized field for the $s \leftrightarrow e$ transition. **b)** Fuller representation, illustrating the cavity and its mode. The mirror on the left is totally reflecting, whereas the right-hand mirror has a non-zero transmission, permitting the cavity mode to couple to input and output fields.

memory, in which the qubit is encoded in the two degenerate ground states, so that

$$|0\rangle \equiv |g\rangle, \qquad |1\rangle \equiv |s\rangle. \tag{27.1.1}$$

The dressing field $\Omega(t)$ can be used to control the storage and transmission of the quantum information. When $\Omega(t) \to 0$, the two states $|g\rangle$ and $|s\rangle$ are both stable, and provide a long lived quantum memory. However, when $\Omega(t) \neq 0$, a cavity mode quantum can be created, which in turn communicates with the external electromagnetic field. We will show that this process can faithfully transfer quantum information between the qubit and the external electromagnetic field. Cascading two atom-cavity systems together, we can then transfer quantum information from one such system to the other.

27.1.1 Hamiltonian for the Atom-Cavity System

We first set up the description of the atom-cavity system without any coupling to input and output. To describe the interaction of the atom with the cavity mode and the dressing field, we use the Hamiltonian (where, for simplicity, we choose $\Omega(t)$ and $\bar{g}$ to be real)

$$\begin{aligned} H_{\text{atom-cavity}} &= \hbar\omega_C a^\dagger a + \hbar\omega_{ge}|e\rangle\langle e| + \hbar\bar{g}\Big(|e\rangle\langle s|a + a^\dagger|s\rangle\langle e|\Big) \\ &\quad + \tfrac{1}{2}\hbar\Omega(t)\Big(\mathrm{e}^{-\mathrm{i}\omega_L t}|e\rangle\langle g| + \mathrm{e}^{\mathrm{i}\omega_L t}|g\rangle\langle e|\Big). \end{aligned} \tag{27.1.2}$$

In this Hamiltonian:

i) The operators a and $a^\dagger$ are the destruction and creation operators for cavity mode of frequency ω_C.

ii) The atomic states $|s\rangle$, $|e\rangle$ and $|g\rangle$ refer to the three-level system with excitation frequency ω_0 inside the cavity as in Fig. 27.1. The qubit is stored in a superposition of the two degenerate ground states $|s\rangle$ and $|g\rangle$.

27.1.2 Adiabatic Elimination of the Excited State

In order to suppress spontaneous emission from the excited state during the Raman process, the laser frequency ω_L is strongly detuned from the atomic transition frequency ω_{ge}; that is, the detuning between the laser and the transition $g \leftrightarrow e$ defined by

$$\Delta \equiv \omega_L - \omega_{ge}, \tag{27.1.3}$$

satisfies the conditions

$$|\Delta| \gg |\Omega(t)|, \qquad |\Delta| \gg |\bar{g}|. \tag{27.1.4}$$

In such a case, one can adiabatically eliminate the excited state $|e\rangle$, as we have shown in Sect. 6.2. The new Hamiltonian for the dynamics of the two ground states becomes, in a rotating frame for the cavity modes at the laser frequency,

$$H_{gs} = -\hbar\delta a^\dagger a + \hbar g(t)\Big(|g\rangle\langle s|a + |s\rangle\langle g|a^\dagger\Big), \tag{27.1.5}$$

$$g(t) \equiv \frac{\bar{g}\Omega(t)}{2\Delta}. \tag{27.1.6}$$

Here $\delta = \omega_L - \omega_c$ is the Raman detuning between the laser and the cavity mode frequencies.

a) Cavity Quantum Electrodynamics Hamiltonian: This Hamiltonian is of the same form as the cavity quantum electrodynamics Hamiltonians of Chap. 22, when these are written in a similar rotating frame. It has the form of a two-level atom with "ground" state $|g\rangle$ and "excited" state $|s\rangle$, coupled to a quantized cavity mode with an effective coupling constant $g(t)$, whose time dependence can be specified arbitrarily by adjusting the intensity of the dressing laser.

Exercise 27.1 Derivation of the Adiabatic Elimination: Note that the Hamiltonian (27.1.2) connects the states with n photons $|g,n\rangle$, $|e,n\rangle$, and the state with $n+1$ photons $|s,n+1\rangle$. By considering states of the form

$$|\Psi,n,t\rangle = a_g(t)|g,n\rangle e^{-i(\omega_L+n\omega_c)t} + a_e(t)|e,n\rangle e^{-in\omega_c t} + a_s(t)|s,n+1\rangle e^{-in\omega_c t}, \tag{27.1.7}$$

use the procedure of Sect. 6.2 to show that, when the conditions (27.1.4) are satisfied, this wavefunction satisfies a Schrödinger equation with a Hamiltonian

$$\bar{H}_{gs} = \frac{\hbar\Omega(t)^2}{4\Delta}|g\rangle\langle g| + \hbar\left(\frac{\bar{g}^2}{\Delta}a^\dagger a - \delta\right)|s\rangle\langle s| + \frac{\hbar\bar{g}\Omega(t)}{2\Delta}\Big(a|g\rangle\langle s| + a^\dagger|s\rangle\langle g|\Big). \tag{27.1.8}$$

Exercise 27.2 Validity of the Approximate Hamiltonian: The adiabatically eliminated Hamiltonian H_{gs}, as given by (27.1.5), can be derived from the form (27.1.8) as follows:

i) Neglect the first two terms inversely proportional to Δ, which are light shift terms. However the final term, which is also inversely proportional to Δ, is not neglected.

   Justify this procedure.

ii) Transform to an appropriate rotating frame. Show that the unitary transformation which produces H_{gs} is

$$|\bar{\Psi},n,t\rangle = U(t)|\Psi,n,t\rangle, \tag{27.1.9}$$

$$U(t) = \exp\Big(i\delta\big(a^\dagger a - |s\rangle\langle s|\big)t\Big). \tag{27.1.10}$$

27.1.3 Interaction with Input and Output Fields

The input and output couple directly to the cavity mode operator a. For the case that the input is the vacuum, we can, following Sect. 9.4.2, write a quantum stochastic Schrödinger equation

$$d|\Psi,t\rangle = \left(-\frac{\mathrm{i}}{\hbar}H_{\text{eff}}\,dt + \sqrt{\kappa}\,a\,dB^\dagger(t)\right)|\Psi,t\rangle, \tag{27.1.11}$$

$$H_{\text{eff}} \equiv H_{\text{atom-cavity}} + \hbar\,\delta\omega\,a^\dagger a - \tfrac{1}{2}\mathrm{i}\hbar\kappa\,a^\dagger a. \tag{27.1.12}$$

For most practical purposes, we will use the Hamiltonian (27.1.5) from which the excited state has been adiabatically eliminated, and absorb the frequency shift $\delta\omega$ into the definition of δ, leading to

$$H_{\text{eff}} \longrightarrow -\hbar\left(\delta + \tfrac{1}{2}\mathrm{i}\kappa\right)a^\dagger a + \hbar g(t)\Big(|g\rangle\langle s|a + |s\rangle\langle g|a^\dagger\Big). \tag{27.1.13}$$

We will solve the quantum stochastic Schrödinger equation (27.1.11) by an extension of the methods used in Ex. 9.3 and Ex. 9.4.

a) Basis States and Initial Conditions: To describe the possible states of the system in this degree of approximation, we need a notation for the combined states of the atom, the cavity mode, and the electromagnetic field, which we shall adapt from that used for Ex. 9.4.

The relevant situation is one in which, initially, we have a qubit stored in the system, so that:

i) The incoming electromagnetic field is in the vacuum state: $|0\rangle$;

ii) The atom is in a superposition state: $c_g|g\rangle + c_s|s\rangle$;

iii) The cavity mode is initially a zero-photon state state: $|\text{cav},0\rangle$.

b) Solutions of the Quantum Stochastic Schrödinger Equation: Using this set of basis states, we can write the time-dependent wavefunction as

$$\begin{aligned}|\Psi,t\rangle = {} & |g\rangle\otimes|\text{cav},0\rangle\otimes|\Phi,g,0,t\rangle + |g\rangle\otimes|\text{cav},1\rangle\otimes|\Phi,g,1,t\rangle \\ & + |s\rangle\otimes|\text{cav},0\rangle\otimes|\Phi,s,0,t\rangle + |s\rangle\otimes|\text{cav},1\rangle\otimes|\Phi,s,1,t\rangle,\end{aligned} \tag{27.1.14}$$

with the initial conditions for the coefficients (which are quantum states of the external electromagnetic field)

$$|\Phi,g,0,t=0\rangle = c_g|0\rangle, \tag{27.1.15}$$

$$|\Phi,s,0,t=0\rangle = c_s|0\rangle, \tag{27.1.16}$$

$$|\Phi,g,1,t=0\rangle = |\Phi,s,1,t=0\rangle = 0. \tag{27.1.17}$$

i) The quantum stochastic Schrödinger equation corresponding to the equations (27.1.11–27.1.13) yields the equations of motion

$$d|\Phi,g,0,t\rangle = \sqrt{\kappa}\,dB^\dagger(t)|\Phi,g,1,t\rangle - \mathrm{i}g(t)|\Phi,s,1,t\rangle\,dt, \tag{27.1.18}$$

$$d|\Phi,g,1,t\rangle = -\left(\mathrm{i}\delta + \tfrac{1}{2}\kappa\right)|\Phi,g,1,t\rangle\,dt, \tag{27.1.19}$$

$$d|\Phi,s,0,t\rangle = \sqrt{\kappa}\,dB^\dagger(t)|\Phi,s,1,t\rangle, \tag{27.1.20}$$

$$d|\Phi,s,1,t\rangle = -\mathrm{i}g(t)|\Phi,g,0,t\rangle\,dt - \left(\mathrm{i}\delta + \tfrac{1}{2}\kappa\right)|\Phi,s,1,t\rangle\,dt. \tag{27.1.21}$$

ii) The initial condition (27.1.17) for $|\Phi, g, 1, t\rangle$ and (27.1.19) show that

$$|\Phi, g, 1, t\rangle = 0. \tag{27.1.22}$$

iii) The equations (27.1.18, 27.1.21) then become two coupled ordinary differential equations

$$\frac{d}{dt}|\Phi, g, 0, t\rangle = -\mathrm{i}g(t)|\Phi, s, 1, t\rangle, \tag{27.1.23}$$

$$\frac{d}{dt}|\Phi, s, 1, t\rangle = -\mathrm{i}g(t)|\Phi, g, 0, t\rangle - \left(\mathrm{i}\delta + \tfrac{1}{2}\kappa\right)|\Phi, s, 1, t\rangle. \tag{27.1.24}$$

The solutions of these equations can be written

$$|\Phi, g, 0, t\rangle = c_g F_g(t)|0\rangle, \tag{27.1.25}$$

$$|\Phi, s, 1, t\rangle = c_g F_s(t)|0\rangle, \tag{27.1.26}$$

where $F_g(t)$ and $F_s(t)$ are the solutions of

$$\dot{F}_g(t) = -\mathrm{i}g(t)F_s(t), \tag{27.1.27}$$

$$\dot{F}_s(t) = -\mathrm{i}g(t)F_g(t) - \left(\mathrm{i}\delta + \tfrac{1}{2}\kappa\right)F_s(t), \tag{27.1.28}$$

$$F_g(0) = 1, \qquad F_s(0) = 0. \tag{27.1.29}$$

iv) The equation (27.1.20) has a solution which can be written in terms of this as

$$|\Phi, s, 0, t\rangle = c_s|0\rangle + c_g\sqrt{\kappa}\int_0^t F_s(t')\,dB^\dagger(t')\,|0\rangle. \tag{27.1.30}$$

c) Interpretation of the Solutions: The set of equations (27.1.22, 27.1.25, 27.1.26, 27.1.30), substituted into (27.1.14) to give the complete solution for $|\Psi, t\rangle$.

27.1.4 Quantum Information Transfer to the Electromagnetic Field

Provided $g(t) \neq 0$, the solutions of the differential equations (27.1.27–27.1.29) approach zero as $t \to \infty$—see Ex. 27.3. In this limit the only one of the four coefficients in (27.1.4) which does not vanish is $|\Phi, s, 0, t \to \infty\rangle$, and the corresponding solution for the wavefunction then becomes

$$|\Psi, t \to \infty\rangle = |s\rangle \otimes |\mathrm{cav}, 0\rangle \otimes \left(c_g|F_s, 1\rangle + c_s|0\rangle\right), \tag{27.1.31}$$

where we have defined a *state of one photon, with amplitude* $F_s(t)$ as

$$|F_s, 1\rangle \equiv \sqrt{\kappa}\int_0^\infty F_s(t)\,dB^\dagger(t)\,|0\rangle. \tag{27.1.32}$$

Therefore, the decay process transfers the quantum information from the atomic qubit, where it is held in the superposition $c_g|g\rangle + c_s|s\rangle$, to the electromagnetic field, where it is held in the form $c_g|0\rangle + c_s|F_s, 1\rangle$.

27.1.5 A Programmable Single Photon Source

If $c_s = 0$, this procedure gives a single photon state with amplitude $F_s(t)$. To the extent that it is possible to choose a form of $g(t)$ to produce a given form for $F_s(t)$,

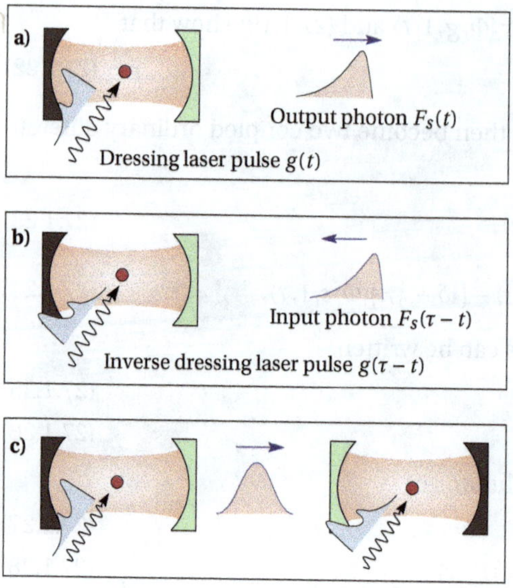

Fig. 27.2. a) Cavity decay produces a photon wave packet. The photon inside the atom has been produced by laser excitation of an atom. **b)** A "time reversed" cavity decay restores the atom in its original state. **c)** Transfer of the quantum state from one atom to the other using a time-symmetric pulse.

this can be considered as a programmable single photon source. Physically, this means that the atom is put into the state $|g\rangle$, and held there until the laser field is applied. The photon can therefore be produced on demand.

Exercise 27.3 Solving the Quantum Stochastic Schrödinger Equation: Explicitly show:

i) For solutions of the differential equation (27.1.27, 27.1.28), that

$$\frac{d}{dt}\left(|F_g(t)|^2 + |F_S(t)|^2\right) = -\kappa|F_S(t)|^2, \tag{27.1.33}$$

and this means that $F_g(t \to \infty) \to 0$ and $F_S(t \to \infty) \to 0$ provided $g(t) \neq 0$.

ii) From (27.1.33) it follows that the one photon state (27.1.32) is correctly normalized; that is

$$\langle F_S, 1|F_S, 1\rangle = \kappa \int_0^\infty |F_S(t)|^2 \, dt = 1. \tag{27.1.34}$$

Exercise 27.4 The Quantum Trajectory Picture: It is also possible to use the stochastic wavefunction simulation methods of Chap. 19, and the notation of Sect. 18.4, for which in this case

$$\mathbf{K} \equiv -\frac{i}{\hbar} H_{\text{eff}}, \qquad \mathbf{J} \equiv \sqrt{\kappa}\, a, \tag{27.1.35}$$

where H_{eff} is as given in (27.1.13).

Use the algorithm of Sect. 19.2 to construct quantum trajectory solutions in terms of the solutions $F_g(t)$ and $F_S(t)$ of the equations (27.1.27) and (27.1.28).

27.2 Quantum Information Transfer between Nodes

The procedure we have described in the previous section can effect the quantum state transfer from the *atom* to *electromagnetic field*

$$\text{Atom:} \qquad c_g|g\rangle + c_s|s\rangle \longrightarrow |s\rangle, \tag{27.2.1}$$

$$\text{Electromagnetic field:} \qquad |0\rangle \longrightarrow c_g|F_s, 1\rangle + c_s|0\rangle. \tag{27.2.2}$$

The function $F_s(t)$ determines the shape of the electromagnetic field pulse output by the cavity, is as defined by the equations (27.1.27–27.1.29). Its shape and magnitude depend on the detuning δ, the damping κ, and on the particular choice of coupling function $g(t)$, itself proportional to the time-dependent dressing field $\Omega(t)$, whose form we can choose as we wish. This means that we have considerable freedom to engineer the shape and the timing of the output pulse.

In particular, the quantum information will remain on the atom as long as $g(t) = 0$, that is, until we apply the dressing field. We can therefore transfer the information at will.

27.2.1 The Problem of Photon Reflection from the Second Cavity

However, to be useful, this method must be complemented by a method of transferring the quantum information out of the electromagnetic field and into another quantum node. The central problem in such transmission is that in general the photon is partially reflected from the second cavity, preventing accurate transfer of the qubit into the second cavity.

The solution to this problem is to apply a dressing field to the second cavity which allows for perfect absorption of the pulse without reflections. The following argument shows that this is indeed possible.

Consider a photon which leaks out of an optical cavity and propagates away as a wavepacket as in Fig. 27.2a. Imagine that we were able to time reverse this wavepacket and send it back into the cavity as in Fig. 27.2b. Then this would restore the original superposition state of the atom, provided we apply the corresponding time-reversed dressing pulse.

In particular, we can drive the atom in a transmitting cavity in such a way that the outgoing pulse is already symmetric in time, the wavepacket entering a receiving cavity would mimic this time reversed process, thus generating the state of the first atom in the second one as in Fig. 27.2c.

27.2.2 The Cavity QED Model of Quantum Information Transmission

The simplest possible configuration of quantum transmission between two nodes consists of two identical three-level atoms 1 and 2, each with degenerate ground states of the form considered in the previous section, contained in identical cavities, as illustrated in Fig. 27.3, with the output from the first cavity coupled unidirectionally into the input of the second cavity.

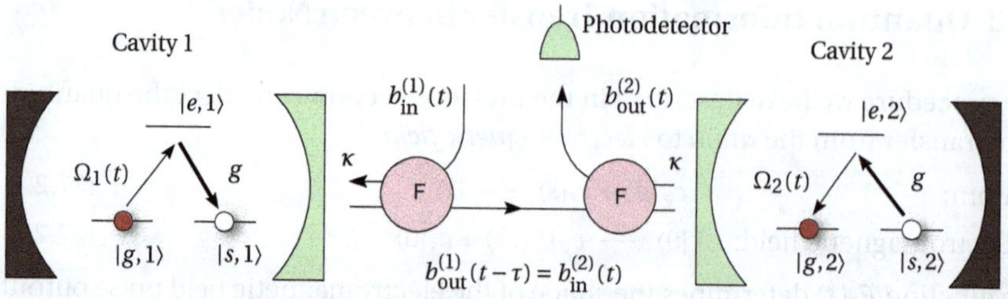

Fig. 27.3. Representation of unidirectional quantum transmission between two atoms in optical cavities (the quantum nodes) connected by an optical fibre (the quantum channel). For ideal quantum transmission, the photodetector should never detect any photons

a) Atom-Cavity Hamiltonians: We will adiabatically eliminate the excited states, as in Sect. 27.1.2, so that there are two atom-cavity Hamiltonians as in (27.1.5)

$$H_i = -\hbar\delta a_i^\dagger a_i + \hbar g_i(t)\Big(|g,i\rangle\langle s,i|a_i + a_i^\dagger|s,i\rangle\langle g,i|\Big). \quad (i=1,2) \tag{27.2.3}$$

b) Formulation as a Cascaded Quantum System: The cascaded quantum systems formalism of Chap. 12 provides the description of the unidirectional coupling. We follow Sect. 12.2.3, with $c_i \to a_i$, $\Gamma_i \to \kappa$, so that the quantum stochastic Schrödinger equation for this system is written using

$$H_{\rm eff} = H_1 + H_2 + \frac{i\hbar\kappa}{2}\left(a_1^\dagger a_1 + a_2^\dagger a_2 + 2a_2^\dagger a_1\right), \tag{27.2.4}$$

and the two operators

$$\mathbf{K} \equiv -\frac{i}{\hbar}H_{\rm eff}, \qquad \mathbf{J} \equiv \sqrt{\kappa}(a_1 + a_2). \tag{27.2.5}$$

Following (18.4.1), the quantum stochastic Schrödinger equation is then written as

$$d|\Psi,t\rangle = \Big(\mathbf{K}\,dt + \mathbf{J}\,dB^\dagger(t)\Big)|\Psi,t\rangle. \tag{27.2.6}$$

27.2.3 Achieving Ideal Quantum Transmission

The essential aim must be to effect *ideal quantum transmission* from the first atom to the second. There is a degree of arbitrariness about what transformations happen to the the other degrees of freedom, and we will choose these as follows:

$$\text{First atom:} \quad c_g|g,1\rangle + c_s|s,1\rangle \longrightarrow |s,1\rangle, \tag{27.2.7}$$

$$\text{First cavity:} \quad |\text{cav}_1,0\rangle \longrightarrow |\text{cav}_1,0\rangle, \tag{27.2.8}$$

$$\text{Second atom:} \quad |s,2\rangle \longrightarrow c_g|g,2\rangle + c_s|s,2\rangle, \tag{27.2.9}$$

$$\text{Second cavity:} \quad |\text{cav}_2,0\rangle \longrightarrow |\text{cav}_2,0\rangle. \tag{27.2.10}$$

The state of the incoming electromagnetic field is the vacuum, and the final state of the system must not radiate any information into the output field, which must also be therefore a vacuum. From the quantum stochastic Schrödinger equation (27.2.6), we can see that this can be achieved if we can find a solution for $|\Psi, t\rangle$ such that

$$\mathbf{J}|\Psi, t\rangle = \sqrt{\kappa}\big(a_1 + a_2\big)|\Psi, t\rangle = 0. \tag{27.2.11}$$

27.2.4 Use of the Quantum Trajectory Picture

The quantum trajectory picture, which is discussed in this context for a single atom-cavity system in Ex. 27.4, provides the most convenient way of treating this system. The quantum jumps which characterize this method, can be thought of as resulting from the continuous monitoring of the output field from the second cavity by a photodetector, as illustrated in Fig. 27.3.

a) Evolution Equations: The evolution of the system under continuous observation is described by the wavefunction $|\tilde{\psi}, t\rangle$ in the system Hilbert space, as formulated in Sect. 19.2, and is determined by the operators **K** and **J** given in (27.2.5). During the time intervals when no count is detected, the evolution of this wavefunction is given by

$$\frac{d}{dt}|\tilde{\psi}, t\rangle = \mathbf{K}|\tilde{\psi}, t\rangle. \tag{27.2.12}$$

The detection of a count at time t is associated with a quantum jump with jump operator **J** according to

$$|\tilde{\psi}, t + dt)\rangle \propto \mathbf{J}|\tilde{\psi}, t\rangle. \tag{27.2.13}$$

The probability density for a jump (detector click) to occur during the time interval from $(t, t + dt)$ is given by $\langle\tilde{\psi}, t|\mathbf{J}^\dagger\mathbf{J}|\tilde{\psi}, t\rangle\, dt$.

If we can find solutions such that $\mathbf{J}|\tilde{\psi}, t\rangle = 0$, then:

i) The probability of a quantum jump becomes zero, so that no jumps occur.

ii) If $\mathbf{J}|\tilde{\psi}, t\rangle = 0$, the action of the non-Hermitian part of the Hamiltonian becomes Hermitian, that is, we can set $a_1 \to -a_2$ and $a_2 \to -a_1$, yielding

$$\left(a_1^\dagger a_1 + a_2^\dagger a_2 + 2a_2^\dagger a_1\right)|\tilde{\psi}, t\rangle \longrightarrow -\left(a_1^\dagger a_2 - a_2^\dagger a_1\right)|\tilde{\psi}, t\rangle. \tag{27.2.14}$$

This means that the evolution in the absence of jumps is unitary. The system remains in a *dark state* of the cascaded quantum system, which cannot emit any light into the outside world. Physically, this means that the wavepacket is not reflected from the second cavity.

b) Designing the Dressing Laser Pulses: We can now formulate the quantum control problem: We wish to design the laser pulses in both cavities in such a way that ideal quantum transmission condition (27.2.11) is satisfied.

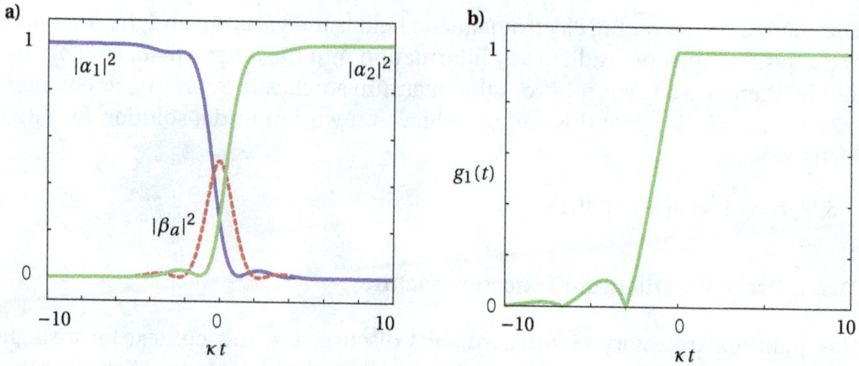

Fig. 27.4. a) Populations $\alpha_{1,2}(t)^2$ and $\beta_a(t)^2$ for a specific ideal transmission pulse; **b)** The coupling function $g_1(t) = g_2(-t)$; see [27.1] for details.

We expand the state of the system

$$|\tilde{\psi}, t\rangle = c_g|ss\rangle|00\rangle \\ +c_e\left(\alpha_1(t)|gs\rangle|00\rangle + \alpha_2(t)|sg\rangle|00\rangle + \beta_1(t)|ss\rangle|10\rangle + \beta_2(t)|ss\rangle|01\rangle\right), \tag{27.2.15}$$

where we use the abbreviated notation $|ss\rangle|00\rangle \equiv |s,1\rangle|s,1\rangle|0\rangle_{c_1}|0\rangle_{c_2}$ etc. Ideal quantum transmission will occur for

$$\alpha_1(-\infty) = \alpha_2(+\infty) = 1, \tag{27.2.16}$$

with all of the other coefficients being zero at the initial and final time. The first term on the right-hand side of (27.2.15) does not change under the time evolution generated by H_{eff}.

Defining symmetric and antisymmetric coefficients $\beta_{1,2} = (\beta_s \mp \beta_a)/\sqrt{2}$, we find the following *evolution equations*

$$\dot{\alpha}_1(t) = g_1(t)\beta_a(t)/\sqrt{2}, \tag{27.2.17}$$

$$\dot{\alpha}_2(t) = -g_2(t)\beta_a(t)/\sqrt{2}, \tag{27.2.18}$$

$$\dot{\beta}_a(t) = -g_1(t)\alpha_1(t)/\sqrt{2} + g_2(t)\alpha_2(t)/\sqrt{2}. \tag{27.2.19}$$

The condition for ideal quantum transmission (27.2.11) implies $\beta_s(t) = 0$, and therefore

$$\dot{\beta}_s(t) = g_1(t)\alpha_1(t)/\sqrt{2} + g_2(t)\alpha_2(t)/\sqrt{2} + \kappa\beta_a(t) \equiv 0. \tag{27.2.20}$$

As well this, there is the normalization condition

$$|\alpha_1(t)|^2 + |\alpha_2(t)|^2 + |\beta_a(t)|^2 = 1. \tag{27.2.21}$$

The mathematical problem is now to find pulse shapes $\Omega_{1,2}(t) \propto g_{1,2}(t)$ such that the conditions (27.2.16, 27.2.20) are fulfilled.

c) Solution: Solving these equations is a difficult problem, as imposing the conditions (27.2.16, 27.2.20) on the solutions of the differential equations (27.2.17–27.2.19) gives functional relations for the pulse shape whose solution are not obvious. In [27.1] a class of solutions was constructed which satisfy the above conditions, based on the physical expectation (which we noted in Sect. 27.2.1) that the time evolution in the second cavity should reverse the time evolution in the first one. Thus one looks for solutions satisfying the *symmetric pulse condition*:

$$g_2(t) = g_1(-t) \quad \text{for all } t. \tag{27.2.22}$$

As an illustration, Fig. 27.4a shows results of numerically integrating the full time-dependent Schrödinger equation with the effective Hamiltonian (27.2.4) for one of the laser pulses, shown in Fig. 27.4b, constructed in this way. As the figure shows, the quantum transmission is indeed very close to ideal.

Exercise 27.5 Constructing Time-Symmetric Pulses: Although the exact construction of a reversible output pulse function $F_S(t)$ is not straightforward, a trial and error method can quickly give a good approximation.

Solve the equations (27.1.27–27.1.29) numerically, and adjust $g(t)$ and the other parameters to get a reversible pulse—for example, the choice

$$g(t) = \min(0.011t + 0.016t^2, 0.7), \quad \kappa = 0.5, \quad \delta = 0, \tag{27.2.23}$$

provides a very acceptable result for a reversible $F_S(t)$.

28. The Dark-State Ensemble Quantum Memory

This chapter will present a model for a quantum memory which is quite different from that presented in the previous chapter. This model, introduced by *Fleischhauer* and *Lukin* [28.1, 28.2], accepts that, although a quantum memory could be set up using a single atom in a Λ-configuration interacting with a quantized optical cavity mode, this is experimentally difficult, and probably rather fragile. A more robust possibility uses an ensemble of a macroscopic number of atoms in the Λ-configuration of a three-level system, and stores the information in collective excitations of the many-atom system. Thus, instead of a *single atom* in a cavity, this model uses an *ensemble* of N such atoms, where N is typically the number of atoms inside a gas cell, and is a macroscopic number.

Both kinds of model use the concept of the concept of a *dark state* of the three-level system as introduced in Chap. 6, but in the case of ensembles there are families of dark states, and it is these families that are utilized for the quantum memory. As in the previous chapter, in using the dark state in a quantum memory, we replace the laser field which generates the coupling $e \longleftrightarrow g$ with a quantized cavity mode of the electromagnetic field, in which the quantum information to be stored can be encoded. We then use an ensemble of $N \gg 1$ identical three-level systems. The dark states then involve all of the atoms coherently, so that any individual atom has only a very small amplitude to be in any state other than the ground state $|g\rangle$.

Furthermore, the dark state becomes a linear combination of:

i) A state $|\boldsymbol{g}\rangle|n\rangle$, with all atoms in the ground state, and n quanta in the cavity mode.

ii) Symmetrized superposition states in which $k \leqslant n$ of the N atoms are in the state $|\boldsymbol{s}\rangle$, the remaining $N-k$ atoms are in the state $|\boldsymbol{g}\rangle$, and there are $N-k$ photons in the cavity mode. We denote these states by $|\boldsymbol{s}^k\rangle|0\rangle$.

The process of adiabatic transfer can be implemented to pass from the first kind of state, in which the information is encoded in the photons, to the second kind of state, where the information is encoded isomorphically in the coherent excitations of the atomic ensemble.

In this simplified version of a quantum memory, the cavity is considered to be perfectly reflecting, giving an isolated system composed of atoms and pho-

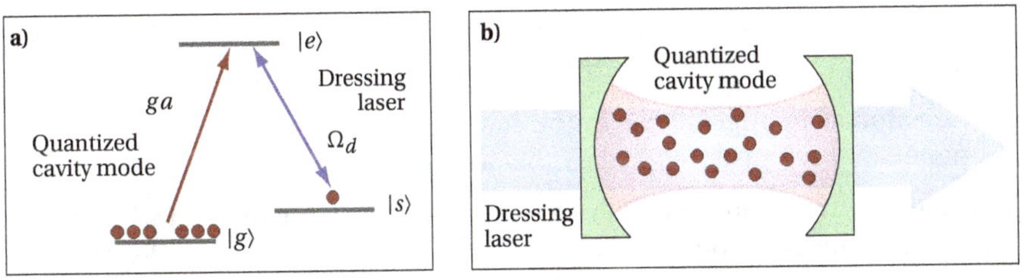

Fig. 28.1. The dark-state ensemble quantum memory. **a)** The level scheme and optical fields involved. The dressing laser is a classical control field, while the operator a represents a quantized cavity mode, coupled to the transition with coupling constant g according to the Hamiltonian (28.1.1). **b)** The cavity is transparent to the dressing field, but totally reflecting to the quantized mode a.

tons. However, the main advantage which arises from the use of large numbers of weakly excited atoms is that cavities are not really necessary. A quantized light beam passing through an ensemble of macroscopic dimensions can be attenuated over a finite distance, and produce no reflections. The process of attenuation is generated by adiabatic transfer, which transfers the quantum information from the light beam to the atoms in the ensemble. This provides an alternative to the cascaded systems method of preventing reflections given in the previous chapter.

This chapter will establish the basic principles of this kind of quantum memory; the task of the following chapter will be to formulate a method of transferring the information from a *propagating* quantized light field into the ensemble of atoms, in a way that does not involve the use of a cavity.

28.1 Quantum Memory Using an Ensemble of Λ-Systems

In this section we shall introduce the idea of *quantum state mapping*, in which the quantum state of a light field is mapped onto a corresponding state of an ensemble of atoms. The essential concepts of quantum state mapping are most easily understood for the case of a single mode light field, for example, the field inside a cavity. We will show in Chap. 29 how to treat the more realistic scenario involving a propagating light field in an atomic medium.

28.1.1 Hamiltonian for N Three-Level Atoms

We consider a collection of N three-level atoms in a Λ-configuration interacting with two optical modes, as illustrated in Fig. 28.1, with the following features:

i) A quantized radiation mode of frequency $\omega_p = \omega_{ge}$, chosen on resonance with the transition $|g\rangle \longleftrightarrow |e\rangle$, characterized by the harmonic oscillator operators $a, a^\dagger$.

ii) A strong *dressing field*, which we treat classically, and which is resonant with the transition $|s\rangle \longleftrightarrow |e\rangle$, so that $\omega_d = \omega_{se}$. This classical control field has Rabi frequency $\Omega_d(t)$.

iii) The atoms are originally all in the state $|g\rangle$, and only a small fraction of the population is transferred to the other levels.

The system corresponds to that described for a single atom by Fig. 13.4, but most importantly, the probe field is now chosen to be a quantized mode. We can move to a rotating frame in which the appropriate Hamiltonian is

$$H = \sum_{i=1}^{N} \hbar\left(-\Delta_d\sigma_{ee}^i - \delta\sigma_{ss}^i + ga\sigma_{eg}^i + ga^\dagger\sigma_{ge}^i + \tfrac{1}{2}\Omega_d\sigma_{es}^i + \tfrac{1}{2}\Omega_d^*\sigma_{se}^i\right). \tag{28.1.1}$$

The condition for the existence of a dark state is $\delta = 0$, but for future reference, we will retain δ in the Hamiltonian during our development, setting it equal to zero only when we wish to consider the dark state. This means that our Hamiltonian corresponds to that in (13.3.18) with $\Gamma_e \to 0$, that is, in the absence of loss from the cavity.

a) Collective Operators: The system is conveniently described in terms of three sets of operators with angular momentum commutation relations, namely (using the notation of (8.A.1))

$$J_+ = \sum_{i=1}^{N}\sigma_{sg}^i, \qquad J_- = \sum_{i=1}^{N}\sigma_{gs}^i, \qquad J_0 = \tfrac{1}{2}\sum_{i=1}^{N}(\sigma_{ss}^i - \sigma_{gg}^i), \tag{28.1.2}$$

$$K_+ = \sum_{i=1}^{N}\sigma_{eg}^i, \qquad K_- = \sum_{i=1}^{N}\sigma_{ge}^i, \qquad K_0 = \tfrac{1}{2}\sum_{i=1}^{N}(\sigma_{ee}^i - \sigma_{gg}^i), \tag{28.1.3}$$

$$L_+ = \sum_{i=1}^{N}\sigma_{es}^i, \qquad L_- = \sum_{i=1}^{N}\sigma_{se}^i, \qquad L_0 = \tfrac{1}{2}\sum_{i=1}^{N}(\sigma_{ee}^i - \sigma_{ss}^i). \tag{28.1.4}$$

Within each of these sets, the operators have angular momentum commutation relations, as in Appendix 8.A. The commutation relations between the different sets are easy to work out, for example, some which we will need shortly are

$$[J_-, L_-] = K_-, \qquad [J_-, L_+] = 0, \tag{28.1.5}$$

$$[J_-, K_-] = 0, \qquad [J_-, K_+] = -L_+, \tag{28.1.6}$$

$$[K_-, L_-] = 0, \qquad [K_-, L_+] = J_-. \tag{28.1.7}$$

Others can be similarly computed.

b) Dicke-Like States of Atoms and Photons: In the case we are considering, all of the atoms are initially in the ground state g, and we will write this state as

$$|\mathbf{g}\rangle \equiv |g_1, g_2, \ldots, g_N\rangle. \tag{28.1.8}$$

The state $|\mathbf{g}\rangle$ is simultaneously:

i) An eigenstate of the angular momentum $\boldsymbol{J}$ of the form $|j, m\rangle$ in which $j = \tfrac{1}{2}N$ and $m = -\tfrac{1}{2}N$.

ii) An eigenstate of the angular momentum $\boldsymbol{K}$ of the form $|k,q\rangle$ in which $k = \frac{1}{2}N$ and $q = -\frac{1}{2}N$.

The only states that can be created from $|\boldsymbol{g}\rangle$ by the Hamiltonian (28.1.1) are the totally symmetric states of the kind introduced by *Dicke* [28.3], of which some are

$$|\boldsymbol{g}\rangle = |g_1, g_2, \dots, g_N\rangle, \tag{28.1.9}$$

$$|\boldsymbol{e}\rangle = \frac{1}{\sqrt{N}}\sum_{j=1}^{N} |g_1,\dots,e_j,\dots,g_N\rangle, \tag{28.1.10}$$

$$|\boldsymbol{s}\rangle = \frac{1}{\sqrt{N}}\sum_{j=1}^{N} |g_1,\dots,s_j,\dots,g_N\rangle, \tag{28.1.11}$$

$$|\boldsymbol{ee}\rangle = \frac{1}{\sqrt{2N(N-1)}}\sum_{i\neq j=1}^{N} |g_1,\dots,e_i,\dots,e_j,\dots,g_N\rangle. \tag{28.1.12}$$

If the field is initially in a state with one photon, the relevant eigenstates are $|\boldsymbol{g},1\rangle$, $|\boldsymbol{e},0\rangle$ and $|\boldsymbol{s},0\rangle$, and for two excitations, the interaction involves three or more states, etc.

28.1.2 The Family of Dark States

In terms of these operators, the Hamiltonian (28.1.1) becomes

$$H = \hbar\Delta_d Q_e + \hbar\delta Q_s + \hbar g\left(aK_+ + a^\dagger K_-\right) + \tfrac{1}{2}\hbar\left(\Omega_d L_+ + \Omega_d^* L_-\right), \tag{28.1.13}$$

$$Q_e \equiv -\sum_{i=1}^{N}\sigma_{ee}^i, \qquad Q_s \equiv -\sum_{i=1}^{N}\sigma_{ss}^i. \tag{28.1.14}$$

Although we shall later want to consider cases in which $\delta \neq 0$, in this section we will set the detuning $\delta \to 0$, since this is necessary to get the dark states.

As in the case when both driving fields are treated classically, this Hamiltonian has dark states, which have no component of the excited state of any of the atoms. Hence, if $|D\rangle$ is a dark state, then $\sigma_{ee}^i|D\rangle = \sigma_{ge}^i|D\rangle = \sigma_{se}^i|D\rangle = 0$, and the action of the Hamiltonian on it will lead to the *dark state condition*

$$\bar{H}|D\rangle = 0, \tag{28.1.15}$$

in which we have defined

$$\bar{H} \equiv \sum_{i=1}^{N}\left(-\Delta_d\sigma_{ee}^i + ga\sigma_{eg}^i + \tfrac{1}{2}\Omega_d\sigma_{es}^i\right) \equiv gaK_+ + \tfrac{1}{2}\Omega_d L_+. \tag{28.1.16}$$

a) Construction of the Family of Dark-States: This proceeds inductively:

i) Note that the state $|\boldsymbol{g}\rangle$ defined in (28.1.8), with all atoms in the state g, and with no photons, is a dark state, since $a|\boldsymbol{g}\rangle = \sigma_{es}^i|\boldsymbol{g}\rangle = \sigma_{ee}^i|\boldsymbol{g}\rangle = 0$.

ii) Note that Q_e obviously commutes with $a^\dagger$ and J_+, so that, using the commutation relations (28.1.5, 28.1.6), and noting that $\delta = 0$,

$$[\bar{H}, a^\dagger] = gK_+, \qquad [\bar{H}, J_+] = \tfrac{1}{2}\Omega_d K_+. \tag{28.1.17}$$

iii) Using these it follows that

$$[\bar{H}, a^\dagger - \alpha J_+] = 0 \quad \text{provided} \quad \alpha = \frac{2g}{\Omega_d}. \tag{28.1.18}$$

iv) We therefore conclude that, since $|\mathbf{g}\rangle$ is a dark state, the family of states

$$|D, n\rangle \equiv \mathcal{N}\left(a^\dagger - \alpha J_+\right)^n |\mathbf{g}\rangle, \quad \text{(where } \mathcal{N} \text{ is a normalization factor)}, \tag{28.1.19}$$

is a family of dark states.

Exercise 28.1 Explicit Expression for the Dark States:

i) Explicitly show that the state $|\mathbf{g}\rangle$ is an eigenstate of the angular momentum $\boldsymbol{J}$ of the form $|j, m\rangle$ in which $j = \frac{1}{2}N$ and $m = -\frac{1}{2}N$.

ii) Use the angular momentum relations (8.A.4) to show that

$$|D, n\rangle = \mathcal{N}\sum_{k=0}^{n}\left(\sqrt{\frac{n!}{(n-k)!k!}}\sqrt{\frac{N!n!}{(N-k)!}}(-\alpha)^k\right)|N, \mathbf{s}^{(k)}, n-k\rangle. \tag{28.1.20}$$

Here, the notation $|N, \mathbf{s}^{(k)}, n-k\rangle$ means the symmetrized and normalized state with $n-k$ photons, and N atoms, of which k atoms are in the s state, and $N-k$ are in the g state.

Exercise 28.2 Asymptotic Form: For $N \gg k$, we can use Stirling's approximation to write $N!/(N-k)! \approx N^k$. Writing $\Omega_d = |\Omega_d|e^{i\varphi_d}$ as in (13.3.12), show that in this case, we can write

$$|D, n\rangle \approx \sum_{k=0}^{n}\sqrt{\frac{n!}{k!(n-k)!}}\left(-\sin\theta e^{i\varphi_d}\right)^k(\cos\theta)^{n-k}|\mathbf{s}^k, n-k\rangle, \tag{28.1.21}$$

in which

$$\tan\theta = \frac{2g\sqrt{N}}{|\Omega_d|}, \qquad |\mathcal{N}|^2 \longrightarrow \frac{1}{n!\left(1+|\alpha|^2\right)^n}. \tag{28.1.22}$$

b) Adiabatic Transfer Using Multi-Atom Dark States: The dark states $|D, n\rangle$, like all dark states, do not contain the excited state and are thus immune to spontaneous emission. The formula (28.1.22) which determines the angle θ characterizing the superposition, is rather different from the formula (13.3.11) for the corresponding situation with a single atom and two coherent driving fields. The denominator is the same, but in the numerator there is the replacement $|\Omega_p| \longrightarrow 2g\sqrt{N}$. This can be understood as the vacuum Rabi frequency (see Sect. 22.1.6) for this cavity, where the coupling constant is enhanced by $g \to g\sqrt{N}$, because N atoms contribute equally and coherently.

28.1.3 The Principle of the Dark State Quantum Memory

The utility of these dark states comes from the fact that they are a *family of dark states*, characterized by the parameters n, θ and φ_d defined in Ex. 28.2. The dependence of φ_d is trivial, and can be absorbed into the definition of the state $|\mathbf{s}^k\rangle$, and in what follows we will set $\varphi_d \longrightarrow 0$.

i) For an initial photonic number state, the adiabatic transfer is, for any given n

$$|D, n, \theta = 0\rangle \equiv |\mathbf{g}\rangle|n\rangle \longleftrightarrow |D, n, \theta = \tfrac{1}{2}\pi\rangle \equiv |\mathbf{s}^n\rangle|0\rangle. \tag{28.1.23}$$

ii) If the initial photonic state is characterized by density matrix ρ^{Phot}, the transfer takes the form

$$\left(\sum_{n,m} \rho_{nm}^{\text{Phot}} |n\rangle\langle m|\right) \otimes |\mathbf{g}\rangle\langle\mathbf{g}| \longleftrightarrow |0\rangle\langle 0| \otimes \left(\sum_{n,m} \rho_{nm}^{\text{Phot}} |\mathbf{s}^n\rangle\langle\mathbf{s}^m|\right). \tag{28.1.24}$$

The last result is highly significant—using adiabatic transfer, *it is possible to transfer all of the quantum information in the photonic mode to equivalent information in the ensemble of three-level atoms.*

This amounts to a *quantum memory*, which transfers information encoded in light into atomic information. While it is very hard to store photons for a reasonably long time, the coherence of atomic internal states can be maintained for a quite long time using current technology. Quantum information can be stored by transferring the photonic information to information encoded in atomic internal states. Using a reversed process of adiabatic transfer, it will be possible to transfer the information back into photonic form at a time which can be chosen at will.

28.2 Approximate Equations of Motion

In the situation where the quantized field is weak, and therefore the number of photons involved is small, and when relatively few atoms are excited from the state $|\mathbf{g}\rangle$, we can make approximations which lead to relatively simple linear equations of motion. We will use these in Sect. 28.3 to give a relatively simple description of the process of information storage and retrieval in the dark state quantum memory.

28.2.1 The Harmonic Approximation to an Ensemble of Atoms

A single two-level system is quite different from a harmonic oscillator, but when a large number of atoms interact weakly with an electromagnetic field, the states in which the mean number of excited atoms is small do behave almost harmonically.

a) Harmonic Oscillator Approximation for $\boldsymbol{J}$: Considering the operators $\boldsymbol{J}$, defined by (28.1.2). The actions of the operators on angular momentum eigenstates with $j = \frac{1}{2}N$, and $m = -\frac{1}{2}N + n$, in the case that $n \ll N$, are

$$\begin{aligned} J_0 \left|\tfrac{1}{2}N, -\tfrac{1}{2}N+n\right\rangle &= \left(-\tfrac{1}{2}N+n\right)\left|\tfrac{1}{2}N, -\tfrac{1}{2}N+n\right\rangle \\ &\approx -\tfrac{1}{2}N\left|\tfrac{1}{2}N, -\tfrac{1}{2}N+n\right\rangle, \end{aligned} \tag{28.2.1}$$

$$\begin{aligned} J_+ \left|\tfrac{1}{2}N, -\tfrac{1}{2}N+n\right\rangle &= \sqrt{(N-n)(n+1)}\left|\tfrac{1}{2}N, -\tfrac{1}{2}N+n+1\right\rangle \\ &\approx \sqrt{N(n+1)}\left|\tfrac{1}{2}N, -\tfrac{1}{2}N+n+1\right\rangle, \end{aligned} \tag{28.2.2}$$

$$\begin{aligned} J_- \left|\tfrac{1}{2}N, -\tfrac{1}{2}N+n\right\rangle &= \sqrt{n(N-n+1)}\left|\tfrac{1}{2}N, -\tfrac{1}{2}N+n-1\right\rangle \\ &\approx \sqrt{Nn}\left|\tfrac{1}{2}N, -\tfrac{1}{2}N+n\right\rangle. \end{aligned} \tag{28.2.3}$$

The harmonic approximation is made by introducing harmonic oscillator operators $S^\dagger$ and S, and making the mappings from the operators and states for atoms to those for a harmonic oscillator:

$$J_+ \longrightarrow \sqrt{N}S^\dagger, \tag{28.2.4}$$

$$J_- \longrightarrow \sqrt{N}S, \tag{28.2.5}$$

$$J_0 \longrightarrow -\tfrac{1}{2}N + S^\dagger S, \tag{28.2.6}$$

$$\left|\tfrac{1}{2}N, -\tfrac{1}{2}N+n\right\rangle_{\boldsymbol{J}} \longrightarrow |n\rangle_S. \tag{28.2.7}$$

These yield the angular momentum commutation relations (8.A.2) to lowest non-vanishing order in an expansion in powers of N^{-1}. The subscript $\boldsymbol{J}$ and S are added to avoid confusion with the similar states corresponding to the $g \leftrightarrow e$ operators, which we will introduce next.

b) Harmonic Oscillator Approximation for $\boldsymbol{K}$: Because $\boldsymbol{K}$ involves excitation from the highly occupied state $|\boldsymbol{g}\rangle$, we can write a harmonic oscillator approximation similar to that for $\boldsymbol{J}$, namely

$$K_+ \longrightarrow \sqrt{N}E^\dagger, \tag{28.2.8}$$

$$K_- \longrightarrow \sqrt{N}E, \tag{28.2.9}$$

$$K_0 \longrightarrow -\tfrac{1}{2}N + E^\dagger E, \tag{28.2.10}$$

$$\left|\tfrac{1}{2}N, -\tfrac{1}{2}N+m\right\rangle_{\boldsymbol{K}} \longrightarrow |m\rangle_E. \tag{28.2.11}$$

c) Physical Basis for the Approximations: The restriction to only those states which are close to the state $|\boldsymbol{g}\rangle$ is the basis for the harmonic oscillator approximations we have introduced. This restriction means that in the states we wish to consider, with $0 \leqslant n, m \ll N$, the e and s excitations are very sparsely spread over the full range of atoms.

It is important to note that there is no similar harmonic oscillator approximation for $\boldsymbol{L}$, since neither $|e\rangle$ nor $|s\rangle$ is highly occupied.

28.2.2 Equations of Motion in the Harmonic Oscillator Approximation

From the Hamiltonian (28.1.13) written in terms of the $\boldsymbol{J}$, $\boldsymbol{K}$, and $\boldsymbol{L}$ operators, the equations of motion can be written

$$\dot{K}_- = -\mathrm{i}\Delta_d K_- + \mathrm{i}gaK_0 + \tfrac{1}{2}\mathrm{i}\Omega_d J_- , \tag{28.2.12}$$

$$\dot{J}_- = -\mathrm{i}\delta J_- + \mathrm{i}ga[J_-, K_+] + \mathrm{i}\Omega_d^* K_- , \tag{28.2.13}$$

$$\dot{a} = \mathrm{i}gK_- . \tag{28.2.14}$$

These equations and their Hermitian conjugates form a closed set of equations. Since all other operators can be obtained using the commutation relations between the $\boldsymbol{J}$, $\boldsymbol{K}$, and $\boldsymbol{L}$ operators, their solutions contain all of the physical information about the system.

The case of interest, for which we want to consider solutions, is that in which:

i) The atoms are initially all in the ground state, that is the initial state is $|\mathbf{g}\rangle$.

ii) The quantized probe field a is weak, and therefore only a very small proportion of the atoms is transferred out of the ground state.

a) Harmonic and Perturbative Approximation: In this case, the harmonic approximations of the previous section can be made. We work to the lowest order, so that we set $K_0 \to -N/2$.

At the same time we make a perturbation approximation in terms of the parameter $g\sqrt{N}/|\Omega_d|$. Naively, we can say that the initial values of the operators S_- and E_- are zero, corresponding to assumed initial state $|\mathbf{g}\rangle$. This means that the time evolution yields solutions for S_- and E_- of order of magnitude $g\sqrt{N}$, from which we can see that the commutator term in (28.2.13) is of third order in the perturbation parameter, and can therefore be neglected. This results in the linear equations of motion

$$\dot{E} = -\mathrm{i}\Delta_d E + \mathrm{i}g\sqrt{N}\,a + \tfrac{1}{2}\mathrm{i}\Omega_d S, \tag{28.2.15}$$

$$\dot{S} = -\mathrm{i}\delta S + \tfrac{1}{2}\mathrm{i}\Omega_d^* E, \tag{28.2.16}$$

$$\dot{a} = \mathrm{i}g\sqrt{N}E. \tag{28.2.17}$$

b) Approximate Effective Hamiltonian: These equations of motion can be regarded as arising from the approximate effective Hamiltonian

$$H_{\text{Harmonic}} = \hbar\Big(\Delta_d E^\dagger E + \delta S^\dagger S - g\sqrt{N}\big(a^\dagger E + E^\dagger a\big) - \tfrac{1}{2}\Omega_d E^\dagger S - \tfrac{1}{2}\Omega_d^* S^\dagger E\Big). \tag{28.2.18}$$

c) Adiabatic Elimination of the Excited State: In the case that the system is far-detuned, so that $\Delta \gg \Omega_d$ and $\Delta \gg g\sqrt{N}$, we an adiabatically eliminate the excited state operator E to get the two-level equations

$$\dot{S} = \mathrm{i}a\left(\frac{g\sqrt{N}\Omega_d^*}{2\Delta_d}\right) + \mathrm{i}S\left(-\delta + \frac{\Omega_d^2}{4\Delta_d}\right), \qquad (28.2.19)$$

$$\dot{a} = \mathrm{i}S\left(\frac{g\sqrt{N}\Omega_d}{2\Delta_d}\right) + \mathrm{i}a\left(\frac{g^2 N}{\Delta_d}\right). \qquad (28.2.20)$$

The second terms on the right-hand side of these equations can normally be omitted. They result from the detuning δ and light-shifts; the light-shifts are normally very small, and δ is normally chosen equal to zero.

Exercise 28.3 Elimination of Light-Shifts: Show that, if we choose

$$\delta = \frac{\Omega_d^2 - 4g^2 N}{4\Delta_d}, \qquad (28.2.21)$$

we can choose a rotating frame in which the second terms on the right-hand side of (28.2.19, 28.2.20) are rigorously eliminated.

28.3 Quantum State Transfer and Quantum Memory

The approximate description in terms of the harmonic oscillator operators E and S provides an essential and realistic simplified description of the system. Combined with the methods of input-output theory described in Chap. 11, a description in terms of linear quantum Langevin equations is possible, and this makes it possible to follow the progress of the quantum information, through as many stages as necessary, and with relative ease. There are two key aspect aspects that must be achieved at every stage:

i) *Perfect Transmission of Quantum Information between a Light Field and an Atomic Ensemble*: This must be achieved at every stage in the network. It requires the use of adiabatic transfer techniques with time-dependent optical pulses carefully chosen to allow the perfect absorption or emission of optical pulses with a known time dependence.

ii) *Minimization of Coupling of the Atoms to Extraneous Optical Modes*: This arises as a natural consequence of the use of an atomic ensemble. The information is distributed only in the Dicke-like modes of the atomic ensemble, and the coupling of these to the relevant optical fields is enhanced in proportion to the number of atoms involved.

28.3.1 The Λ-I Configuration

The dark-state quantum memory we have set up now needs to be connected with incoming and outgoing quantum information, which will be carried on light signals. The configuration we will choose is shown in Fig. 28.2, and we will call it the Λ-I configuration. It is characterized by the following features:

i) All of the atoms are initially in the $|g\rangle$ state, as in Fig. 28.2 b.

ii) The transition $|g\rangle \leftrightarrow |e\rangle$ is coupled with a coupling coefficient g to a quantized ring cavity mode $a(t)$, which constitutes the quantum signal, and which is itself connected to the input and output quantum signals $b_{\text{in}}(t)$ and $b_{\text{out}}(t)$.

iii) The driving laser is coupled to the transition $|s\rangle \leftrightarrow |e\rangle$. The driving laser and the ring cavity mode are chosen to be copropagating, so as to satisfy the collective coupling condition

$$(k_d - k_a)\,L_{\text{Ensemble}} \ll \pi. \tag{28.3.1}$$

Here k_d and k_a are respectively the wave vectors of the dressing laser and the quantum signal, and L_{Ensemble} is the length of the atomic ensemble.

iv) We assume off-resonant coupling with a large detuning Δ_d as illustrated in Fig. 28.2 b. The detuning δ is of course set to zero to satisfy the dark-state condition.

v) In the related Λ-II configuration—discussed in Sect. 28.3.5 and illustrated in Fig. 28.3—the roles of the quantized cavity mode and the driving laser are interchanged.

a) Quantum Langevin Equations for Coupling to the Input and Output Signals: Corresponding to the equations of motion as given in (28.2.19) and (28.2.20), we can neglect the second terms on the right-hand side as in Ex. 28.3. We then use the methods of Chap. 11, in particular adapting Ex. 11.1, to add an input and damping

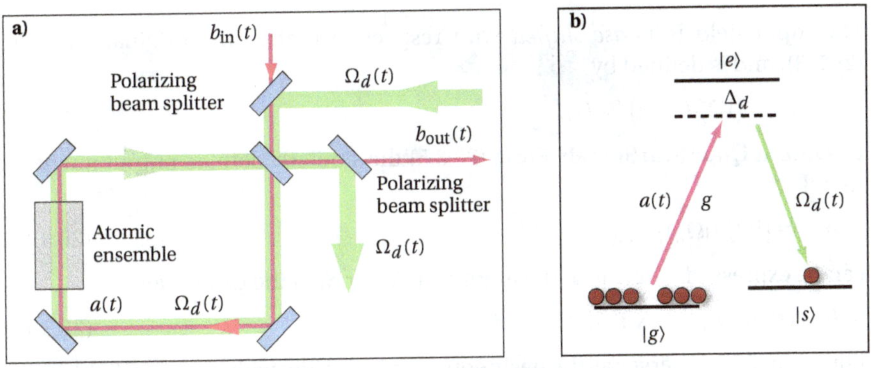

Fig. 28.2. a) An atomic ensemble in a weak coupling cavity. **b)** The Λ-I level configuration.

to the cavity mode, so that the operators satisfy the quantum Langevin equations

$$\dot{S} = \mathrm{i}\left(\frac{g\sqrt{N}\Omega_d^*}{2\Delta_d}\right) a, \tag{28.3.2}$$

$$\dot{a} = \mathrm{i}\left(\frac{g\sqrt{N}\Omega_d}{2\Delta_d}\right) S - \tfrac{1}{2}\kappa a - \sqrt{\kappa}\, b_{\mathrm{in}}(t). \tag{28.3.3}$$

Here:

i) κ is the cavity decay rate;

ii) As in Sect. 11.1.1, $b_{\mathrm{in}}(t)$ is the input quantum signal with the properties

$$\left[b_{\mathrm{in}}(t), b_{\mathrm{in}}^{\dagger}(t')\right] = \delta(t-t'). \tag{28.3.4}$$

iii) The output quantum signal $b_{\mathrm{out}}(t)$ from the cavity is connected to the input by the fundamental input-output relation (11.2.6)

$$b_{\mathrm{out}}(t) = b_{\mathrm{in}}(t) + \sqrt{\kappa}a. \tag{28.3.5}$$

b) Approximate Equations: We will consider the bad-cavity limit, in which the cavity decay rate is much larger than the coupling rate, that is

$$\kappa \gg \frac{|\Omega_d| g\sqrt{N}}{2\Delta_d}. \tag{28.3.6}$$

In this situation we can adiabatically eliminate the cavity mode a, and obtain an effective quantum Langevin equation for the collective atomic mode operator $S(t)$

$$\dot{S}(t) = -\tfrac{1}{2}\bar{\kappa} S(t) - \sqrt{\bar{\kappa}}\, \bar{b}_{\mathrm{in}}(t), \tag{28.3.7}$$

where the effective coupling rate is

$$\bar{\kappa} = \frac{N|\Omega_d g|^2}{\Delta_d^2 \kappa}. \tag{28.3.8}$$

The input field is *phase shifted* with respect to that in the original equation (28.3.3), and is defined by

$$\bar{b}_{\mathrm{in}}(t) \equiv \left(\mathrm{i}\Omega_d^*/|\Omega_d|\right) b_{\mathrm{in}}(t). \tag{28.3.9}$$

c) Output Quantum Signal: From (28.3.5) the similarly defined output quantum signal

$$\bar{b}_{\mathrm{out}} \equiv \left(\mathrm{i}\Omega_d^*/|\Omega_d|\right) b_{\mathrm{out}}, \tag{28.3.10}$$

can be expressed in terms of the atomic operator S, in the usual way

$$\bar{b}_{\mathrm{out}}(t) = \bar{b}_{\mathrm{in}}(t) + \sqrt{\bar{\kappa}}\, S(t). \tag{28.3.11}$$

Equation (28.3.7) serves as the basic equation in dealing with quantum light memory.

Exercise 28.4 Adiabatic Elimination of the Cavity Field: Explicitly check the adiabatic elimination procedure which leads to (28.3.7).

28.3.2 The Information Transfer Process

The aim of this section is to formulate methods for quantifying the *mapping efficiency* of quantum information between the atomic ensemble and the electromagnetic field.

a) Input Pulse Shape and Single-Mode Bosonic Operators: The input signal must consist of a pulse of light in a definite kind of quantum state, and with a definite pulse shape $F_{\text{in}}(t)$. This pulse itself will normally be produced by the output of some definite quantum process—for example, in Ex. 9.3, we saw that the pulse shape from the output of a free cavity mode has the form

$$F_{\text{in}}(t) = \sqrt{\frac{\kappa}{1-e^{-\kappa T}}}\, e^{-\kappa t/2}, \qquad (0 \leqslant t \leqslant T) \tag{28.3.12}$$

where κ is the cavity decay rate and T denotes the pulse duration. In a more relevant example in Sect. 27.1.3 we saw how to compute the pulse shape from a cavity QED model of a quantum memory.

For such an input pulse with a definite shape, we can define a single-mode Bosonic operator in terms a normalized mode function $F_{\text{in}}(t)$; thus

$$B_{\text{in}} \equiv \int_0^T F_{\text{in}}(t)\bar{b}_{\text{in}}(t)\,dt, \qquad \text{provided} \quad \int_0^T |F_{\text{in}}(t)|^2\,dt = 1. \tag{28.3.13}$$

The operators thus defined obey the Bosonic commutation relation

$$\left[B_{\text{in}}\,,\,B_{\text{in}}^\dagger\right] = 1. \tag{28.3.14}$$

Similarly, for the normalized output optical signal $B_{\text{out}}(t)$ with a definite shape $F_{\text{out}}(t)$, we can also define a single-mode operator

$$B_{\text{out}} \equiv \int_0^T F_{\text{out}}(t)b_{\text{out}}(t)dt. \tag{28.3.15}$$

b) Functioning of the Quantum Memory: The quantum light memory has the following functions:

i) The faithful transfer of the quantum state of the input optical mode B_{in}, with a chosen pulse shape $F_{\text{in}}(t)$, to the state of the collective atomic mode described by $S(t)$.

ii) To store the state in the collective atomic mode $S(t)$.

iii) To read out the stored state again into an output optical mode B_{out} with a chosen pulse shape $F_{\text{out}}(t)$.

c) Transferring the Input into the Collective Atomic Mode: The time-dependent light-atom interaction is described by the quantum Langevin equation (28.3.7), with the effective coupling rate

$$\bar{\kappa}(t) = \frac{N|\Omega_d(t)g|^2}{\Delta_d^2\kappa}. \tag{28.3.16}$$

Here we are assuming that the Rabi frequency $\Omega_d(t)$ changes sufficiently slowly for the adiabatic approximation to be made. Subject to this constraint, $\bar{\kappa}(t)$ can be adjusted by controlling the time dependence of the Rabi frequency $\Omega_d(t)$, that is, by changing the classical laser intensity.

The quantum Langevin equation (28.3.7) is linear and has the solution

$$\boxed{S(T) = S(0)\mathrm{e}^{-\int_0^T \bar{\kappa}(t)\,dt/2} - \int_0^T \mathrm{e}^{-\int_t^T \bar{\kappa}(\tau)\,d\tau/2}\sqrt{\bar{\kappa}(t)}\,\bar{b}_{\rm in}(t)\,dt.} \tag{28.3.17}$$

The second term on the right represents the mode of the input which is written into the quantum memory. It is most convenient to start by writing equations for the un-normalized mode function thus indicated, which we define as

$$\bar{F}_{\rm in}(t) \equiv \begin{cases} \mathrm{e}^{-\int_t^T \bar{\kappa}(\tau)\,d\tau/2}\sqrt{\bar{\kappa}(t)}, & 0 < t < T, \\ 0, & \text{otherwise}. \end{cases} \tag{28.3.18}$$

For a given form of $\bar{F}_{\rm in}(t)$, these equations implicitly specify the form of $\bar{\kappa}(t)$, and in fact they mean that $\bar{\kappa}(t)$ satisfies the differential equation

$$\frac{d\bar{\kappa}(t)}{dt} = \frac{2}{\bar{F}_{\rm in}(t)}\frac{d\bar{F}_{\rm in}(t)}{dt}\bar{\kappa}(t) - \bar{\kappa}(t)^2, \qquad 0 < t < T. \tag{28.3.19}$$

This differential equation can be solved by writing $\bar{\mu}(t) = 1/\bar{\kappa}(t)$, in terms of which it becomes

$$\frac{d\bar{\mu}(t)}{dt} = -\frac{2}{\bar{F}_{\rm in}(t)}\frac{d\bar{F}_{\rm in}(t)}{dt}\bar{\mu}(t) + 1, \tag{28.3.20}$$

and this linear equation is easily solved. Using $\bar{F}_{\rm in}(T) = \sqrt{\bar{\kappa}(T)}$, which follows from the definition (28.3.18), we find

$$\bar{\kappa}(t) = \begin{cases} \dfrac{\bar{F}_{\rm in}(t)^2}{1 - \int_t^T \bar{F}_{\rm in}(s)^2\,ds}, & 0 < t < T, \\ 0, & \text{otherwise}. \end{cases} \tag{28.3.21}$$

We conclude that we can specify the mode function $\bar{F}_{\rm in}(t)$, and using this equation, we can then determine $\bar{\kappa}(t)$. Therefore through (28.3.8), this determines the Rabi frequency we need as

$$|\Omega_d(t)| = \frac{\Delta_d}{g}\sqrt{\frac{\kappa\bar{\kappa}(t)}{N}}. \tag{28.3.22}$$

d) The Mapping Efficiency: To relate this expression to the definition (28.3.13) of the input mode operator, we can define the *normalized mode function* by

$$F_{\text{in}}(t) \equiv \frac{1}{\sqrt{\eta(T)}} \bar{F}_{\text{in}}(t), \tag{28.3.23}$$

where, to ensure that the mode function is normalized,

$$\eta(T) = \int_0^T e^{-\int_t^T \bar{\kappa}(\tau)\,d\tau} \bar{\kappa}(t)\,dt, \tag{28.3.24}$$

$$= 1 - e^{-\int_0^T \bar{\kappa}(\tau)\,d\tau}. \tag{28.3.25}$$

The mode operator B_{in} is then defined as in (28.3.13), and it can be seen from (28.3.17) that the collective atomic operator at time T is given in terms of $\eta(T)$, which we shall call the *mapping efficiency*, by the mapping

$$S(T) = \sqrt{1-\eta(T)}\, S(0) - \sqrt{\eta(T)}\, B_{\text{in}}. \tag{28.3.26}$$

Since $\bar{\kappa}(t)$ is positive, the mapping efficiency $\eta(T)$ tends to 1 after a sufficiently long time T—more precisely, provided $\int_0^\infty \bar{\kappa}(\tau)\,d\tau \to \infty$, the mapping efficiency $\eta(\infty) \to 1$.

When the mapping efficiency $\eta(T) = 1$, the photonic state is faithfully mapped to the atomic state with $S(T) = -B_{\text{in}}$.

e) Added Noise—Beam Splitter Analogy: For $\eta(T) < 1$, the state mapping has an inherent loss component, arising from the term proportional to $S(0)$. The definition (28.3.13) of B_{in} shows that it depends on the input $b_{\text{in}}(t)$ only for $0 \leqslant t \leqslant T$, and as we have seen in Sect. 11.2.3, for this time range (apart from $t = 0$) $b_{\text{in}}(t)$ commutes with both $S(0)$ and $S(0)^\dagger$. The commutator is in fact given by (11.2.12), from which it follows that

$$[S(0), B_{\text{in}}] = [S^\dagger(0), B_{\text{in}}] = 0. \tag{28.3.27}$$

This means the $S(0)$ behaves like a kind of vacuum noise; it is quantum mechanically independent of B_{in}, and its quantum state is the ground state, corresponding to all atoms being in the ground state at time 0.

The mapping process obeys the same kinds of equations as does a beam splitter, such as we have discussed in Sect. 21.5.2. Here, however, the two modes mixed by the process are completely different from each other physically, one being the travelling quantized electromagnetic pulse, and the other a collective excitation in the ensemble of atoms.

Exercise 28.5 Commutation Relation: Explicitly show from the solution (28.3.17) that $[S(T), S^\dagger(T)] = 1$.

Exercise 28.6 The Impedance Matching Condition: The differential equation given by (28.3.19) is sometimes called an "impedance matching condition", since it is equivalent to there being no reflection from the system input, which in microwave and similar systems is arranged by impedance matching. A better terminology would be a "mode matching" condition, since it arises here from the requirement to transfer all information to the correct mode.

Check that the differential equation (28.3.19) is equivalent to the explicit expression (28.3.18).

Exercise 28.7 A Square Input Pulse:

i) Consider a square pulse for the normalized mode function

$$F_{\text{in}}(t) = \begin{cases} \dfrac{1}{\sqrt{T}}, & 0 < t < T, \\ 0, & \text{otherwise.} \end{cases} \tag{28.3.28}$$

Use the solution (28.3.21) to show that

$$\bar{\kappa}(t) = \begin{cases} \dfrac{1}{t}, & 0 < t < T, \\ 0, & \text{otherwise.} \end{cases} \tag{28.3.29}$$

Notice that this solution requires $\bar{\kappa}(t) \to \infty$ as $t \to 0$, which means that this solution is physically attainable for early times.

ii) Compare with the case of a square pulse for $\bar{\kappa}(t)$. What kind of $\bar{F}_{\text{in}}(t)$ results?

28.3.3 Storing the Quantum Information

After mapping the photonic state to the atomic state, the classical laser is turned off, and $\bar{\kappa}(t) \to 0$ for all later times. The quantum Langevin equation (28.3.7) then shows that $S(t)$ becomes a constant, and the information therefore remains stored in the internal atomic mode.

28.3.4 Transferring the Stored Input to the Output

We now need a way of reading out this state to an output optical pulse with a chosen pulse shape $F_{\text{out}}(t)$; that is, we would like to map the state of the atomic mode $S(T)$ to the output optical mode defined by

$$B_{\text{out}} = \int_0^T F_{\text{out}}(t) b_{\text{out}}(t)\, dt. \tag{28.3.30}$$

a) Output Field Quantum Langevin Equation: For this we need the quantum Langevin equation written in terms of the *output field* $\bar{b}_{\text{out}}(t)$, which, as shown in Sect. 11.2.2, is obtained from the quantum Langevin equation (28.3.7) by substituting for $\bar{b}_{\text{in}}(t)$ using the input-output relationship (28.3.11). The result is

$$\dot{S}(t) = \tfrac{1}{2}\bar{\kappa}(t) S(t) - \sqrt{\bar{\kappa}(t)}\, \bar{b}_{\text{out}}(t). \tag{28.3.31}$$

It is appropriate to solve for the initial value $S(0)$ in terms of the final value $S(T)$, which can be written as

$$S(0) = S(T)\mathrm{e}^{-\int_0^T \bar\kappa(t)\,dt/2} + \int_0^T \mathrm{e}^{-\int_0^t \bar\kappa(\tau)\,d\tau/2}\sqrt{\bar\kappa(t)}\,\bar b_{\mathrm{out}}(t)\,dt. \tag{28.3.32}$$

The two terms on the right hand side are independent of each other, since the output field commutes with all system operators in the future, as we noted in Sect. 11.2.3. Notice also that this equation, while very similar to (28.3.17), differs from it in a number of details.

b) Output Field and Mapping Efficiency: As previously, we write a mode function in an un-normalized form

$$\bar F_{\mathrm{out}}(t) \equiv \mathrm{e}^{-\int_0^t \bar\kappa(\tau)\,d\tau/2}\sqrt{\bar\kappa(t)}\,, \tag{28.3.33}$$

and in this case the effective coupling constant $\bar\kappa(t)$ satisfies the equation

$$\frac{d\bar\kappa(t)}{dt} = \frac{2}{F_{\mathrm{out}}(t)}\frac{dF_{\mathrm{out}}(t)}{dt}\bar\kappa(t) + \bar\kappa(t)^2, \tag{28.3.34}$$

which is the time reverse of (28.3.19), the corresponding equation for write-in of the photonic state. With this condition, one can deduce from (28.3.32) and the input-output relation (28.3.11) that after time T, the outgoing optical mode is given by the equation

$$S(0) = \sqrt{1-\eta(T)}\,S(T) + \sqrt{\eta(T)}\,B_{\mathrm{out}}. \tag{28.3.35}$$

Here $\eta(T)$ is given by the same equation (28.3.25) as for the mapping into the memory—however the function $\bar\kappa(t)$, in terms of which $\eta(T)$ is defined, is not the same in both cases.

This equation is not yet in the form needed, since the operator $S(t)$ occurs at the times $t=0$ and $t=T$. We can eliminate the dependence of $S(T)$ by using the solution (28.3.17) of the quantum Langevin equation in terms of the input field $b_{\mathrm{in}}(t)$, which (after some manipulation) gives

$$B_{\mathrm{out}} = \sqrt{\eta(T)}\,S(0) + \sqrt{1-\eta(T)}\,\bar n_{\mathrm{out}}, \tag{28.3.36}$$

$$\bar n_{\mathrm{out}} \equiv \int_0^T F_{\mathrm{out}}(t)\,b_{\mathrm{in}}(t)\,dt. \tag{28.3.37}$$

In (28.3.36) we have the desired relationship between the output B_{out} and information stored in the atomic mode $S(0)$. Unless the mapping efficiency $\eta(T)$ is 100%, there is also the added term $\bar n_{\mathrm{out}}$.

As with the process of writing the signal into the quantum memory, the read-out process has the form of beam-splitter, as in Sect. 21.5.2, with the signal going in one port, and the noise in the other port.

c) Added Noise in the Output Process: Since $\bar{n}_{\rm out}$ depends only on $b_{\rm in}(t)$ for $t > 0$, it commutes with and is independent of $S(0)$ and $S^\dagger(0)$. It also has the commutation relation $[\bar{n}_{\rm out}, \bar{n}^\dagger_{\rm out}] = 1$. It thus plays the role of an added vacuum noise. The expression for $\bar{n}_{\rm out}$ has a "hybrid" form, involving the output mode and the input noise. It represents the quantum noise input into $S(t)$ during the write-out process.

28.3.5 The Λ-II Configuration

In the Λ-II configuration, illustrated in Fig. 28.3, each atom has the same three levels $|g\rangle$, $|e\rangle$ and $|s\rangle$, as in the Λ-I configuration described in Sect. 28.3.1. In this case the classical driving laser with Rabi frequency $\Omega_d(t)$ couples to the transition $|g\rangle \leftrightarrow |e\rangle$, while the quantum signal $a(t)$ couples to the transition $|s\rangle \leftrightarrow |e\rangle$. The notation and experimental layout are essentially the same as for the Λ-I configuration, but the physical behaviour is very different.

a) Effective Hamiltonian: In this level configuration, the effective Hamiltonian, after adiabatic elimination of the upper level $|e\rangle$, is

$$H = \hbar\left(\frac{\sqrt{N}\Omega_d}{2\Delta_d}\right) S^\dagger a^\dagger + \hbar\left(\frac{\sqrt{N}\Omega^*_d}{2\Delta_d}\right) Sa, \tag{28.3.38}$$

where S is the collective atomic annihilation operator defined as before, and we have neglected the trivial light shift terms.

This Hamiltonian is that of two-mode squeezing, which has been treated in Sect. 21.3. Here only the cavity mode operators a, $a^\dagger$ are connected to an output field, which is the same situation as the case considered in Sect. 21.4.2. Consequently, this Hamiltonian does not create correlated beams of light. In contrast, here we will see that the correlation is between the output field and the state of the atomic ensemble, and therefore measurement of the output light field gives information about the quantum state of the atomic ensemble.

b) Quantum Langevin Equation: In the same way as for the Λ-I configuration, we can write the quantum Langevin equations corresponding to the Hamiltonian

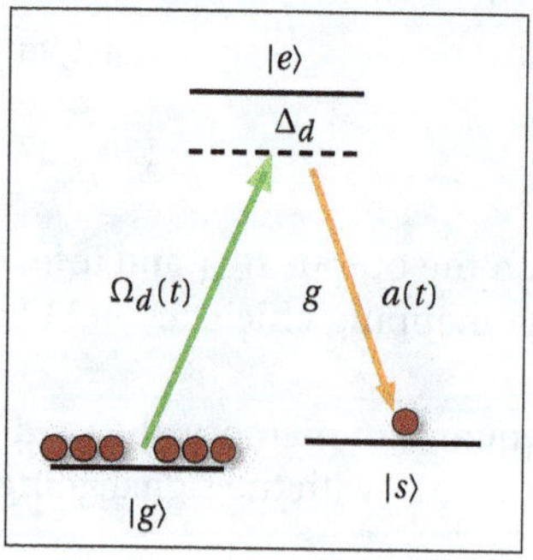

Fig. 28.3. The Λ-II level configuration.

(28.3.38), and then adiabatically eliminate the cavity mode $a(t)$ in the bad-cavity limit to obtain a quantum Langevin equation for the collective atomic mode, which in this case is $S^\dagger(t)$

$$\dot{S}^\dagger(t) = \tfrac{1}{2}\bar{\kappa}(t)S^\dagger(t) - \sqrt{\bar{\kappa}(t)}\,\bar{b}_{\text{in}}(t), \tag{28.3.39}$$

where the effective coupling rate is given by

$$\bar{\kappa}(t) = \frac{N|\Omega_d(t)g|^2}{\Delta_d^2\kappa}. \tag{28.3.40}$$

As in the case of the readout of a stored signal, it is advantageous to consider the solutions of both the quantum Langevin equation (28.3.39) and the corresponding equation written in terms of the output

$$\dot{S}^\dagger(t) = -\tfrac{1}{2}\bar{\kappa}(t)S^\dagger(t) - \sqrt{\bar{\kappa}(t)}\,\bar{b}_{\text{out}}(t). \tag{28.3.41}$$

c) Solutions of Quantum Langevin Equations: The solutions of the two equations are

$$S^\dagger(T) = S(0)e^{\int_0^T \bar{\kappa}(t)\,dt/2} - \int_0^T e^{\int_t^T \bar{\kappa}(\tau)\,d\tau/2}\sqrt{\bar{\kappa}(t)}\,\bar{b}_{\text{in}}(t)\,dt, \tag{28.3.42}$$

$$S^\dagger(0) = S(T)e^{\int_0^T \bar{\kappa}(t)\,dt/2} + \int_0^T e^{\int_0^t \bar{\kappa}(\tau)\,d\tau/2}\sqrt{\bar{\kappa}(t)}\,\bar{b}_{\text{out}}(t)\,dt. \tag{28.3.43}$$

d) Mode Function and Operators for Input and Output: For these equations we define the appropriate un-normalized mode function by

$$\bar{G}(t) \equiv \sqrt{\bar{\kappa}(t)}\,e^{\int_t^T \bar{\kappa}(\tau)\,d\tau/2}, \tag{28.3.44}$$

and this satisfies the differential equation

$$\frac{d\bar{\kappa}(t)}{dt} = \frac{2}{\bar{G}(t)}\frac{d\bar{G}(t)}{dt}\bar{\kappa}(t) - \bar{\kappa}(t)^2. \tag{28.3.45}$$

The normalized mode is defined by

$$G(t) = \frac{1}{\sqrt{\eta(T)}}\bar{G}(t), \qquad \eta(T) = e^{\int_0^T \bar{\kappa}(\tau)\,d\tau} - 1. \tag{28.3.46}$$

Using the mode function, we define the input and output mode operators

$$B_{\text{in}} \equiv \frac{1}{\sqrt{\eta(T)}}\int_0^T \bar{G}(t)b_{\text{in}}(t)\,dt, \tag{28.3.47}$$

$$B_{\text{out}} \equiv \frac{1}{\sqrt{\eta(T)}}\int_0^T \bar{G}(t)b_{\text{out}}(t)\,dt. \tag{28.3.48}$$

e) Input-Output Transfer Matrix: In terms of these definitions, the solutions (28.3.42, 28.3.43) become

$$S^\dagger(T) = \sqrt{1+\eta(T)}\,S^\dagger(0) - \sqrt{\eta(T)}\,B_{\text{in}}, \tag{28.3.49}$$

$$S^\dagger(0) = \sqrt{1+\eta(T)}\,S^\dagger(T) + \sqrt{\eta(T)}\,B_{\text{out}}. \tag{28.3.50}$$

These equations are best arranged to express the outputs in terms of the inputs, that is, by eliminating $S^\dagger(T)$ from the second equation we get the set

$$\begin{pmatrix} S^\dagger(T) \\ B_{\text{out}} \end{pmatrix} = \begin{pmatrix} \sqrt{1+\eta(T)} & -\sqrt{\eta(T)} \\ -\sqrt{\eta(T)} & \sqrt{1+\eta(T)} \end{pmatrix} \begin{pmatrix} S^\dagger(0) \\ B_{\text{in}} \end{pmatrix}. \tag{28.3.51}$$

The matrix in this equation is a transfer matrix from inputs to outputs. However, the two inputs are of quite a different kind from each other.

i) *The Ensemble Operators* $S^\dagger(0)$, $S^\dagger(0)$ represent the initial and final information about the ensemble.

ii) *The Mode Operators* B_{in}, B_{out} represent the information which streams into and out of the ensemble—the "flying qubits".

f) Bogoliubov Transformation: If we write

$$r_B \equiv \frac{1}{2}\int_0^T \bar{\kappa}(\tau)\,d\tau, \tag{28.3.52}$$

then the transfer equations take the form of a *Bogoliubov transformation*

$$\begin{pmatrix} S^\dagger(T) \\ B_{\text{out}} \end{pmatrix} = \begin{pmatrix} \cosh r_B & -\sinh r_B \\ -\sinh r_B & \cosh r_B \end{pmatrix} \begin{pmatrix} S^\dagger(0) \\ B_{\text{in}} \end{pmatrix}. \tag{28.3.53}$$

We will want to consider the case that both of the modes described by destruction operators $S(0)$ and B_{in} are initially the vacuum state, which will be denoted by $|0\rangle_a$ and $|0\rangle_p$ respectively in the following (the subscripts "a" for atoms, and "p" for photons). Transferring the solution to the Schrödinger picture, we conclude (as in Sect. 21.3) that the collective atomic mode and the effective single mode for the output quantum signal are in a two-mode squeezed state

$$|\phi\rangle = \operatorname{sech} r_B \sum_n \tanh^n r_B\,|n\rangle_a |n\rangle_p\,. \tag{28.3.54}$$

This is a two-mode squeezed state, in which the photons emerging from the cavity are entangled with the state of the atomic ensemble. It forms the basis for the quantum repeater, as introduced by *Duan, Lukin, Cirac* and *Zoller* [28.4] in 2001.

29. Spatially Extended Atomic Ensembles

In the previous two chapters we have concentrated on atoms, either singly or in an ensemble, in conjunction with an optical cavity. The cavity is used to ensure good coupling to the external optical fields which carry the quantum information into and out of the cavity. However, the function of the cavity can also be achieved in an extended ensemble in a vapour cell, in which the light beams propagate while exchanging information with the atoms, regarded as being essentially at rest.

The major change is the absence of a cavity mode, whose place is taken by an electromagnetic field which propagates through the vapour, with which it interacts. In this chapter we will extend the treatment of the previous chapter, and in doing so will make use of the *harmonic approximation* of the ensemble operators, introduced in Sect. 28.2.2. This approximation is essentially equivalent to treating the atomic vapour as a linear polarizable optical medium.

We will therefore introduce a rather standard treatment of light propagation in a polarizable vapour, in which linearity can be assumed because the individual atoms are only weakly coupled to the electromagnetic field. This treatment will be semiclassical. We will then consider how propagation occurs, depending on the detuning of the probe from resonance.

29.1 Light Propagation in an Atomic Vapour—Semiclassical Theory

We can write Maxwell's equation in a dielectric medium, such as is provided by an atomic vapour,

$$\nabla\cdot\boldsymbol{D} = 0, \qquad \nabla\cdot\boldsymbol{B} = 0, \tag{29.1.1}$$

$$\nabla\times\boldsymbol{E} = -\frac{\partial\boldsymbol{B}}{\partial t}, \qquad \nabla\times\boldsymbol{H} = \frac{\partial\boldsymbol{D}}{\partial t}. \tag{29.1.2}$$

The kind of medium we are considering is non-magnetic, has no free charges, and no conductivity. The constitutive equations can then be written

$$\boldsymbol{D} = \epsilon_0\boldsymbol{E}+\boldsymbol{P}, \tag{29.1.3}$$

$$\boldsymbol{B} = \mu_0\boldsymbol{H}, \tag{29.1.4}$$

using which we can derive the wave equation

$$\nabla\times(\nabla\times\boldsymbol{E})+\mu_0\epsilon_0\frac{\partial^2\boldsymbol{E}}{\partial t^2} = -\mu_0\frac{\partial^2\boldsymbol{P}}{\partial t^2}. \tag{29.1.5}$$

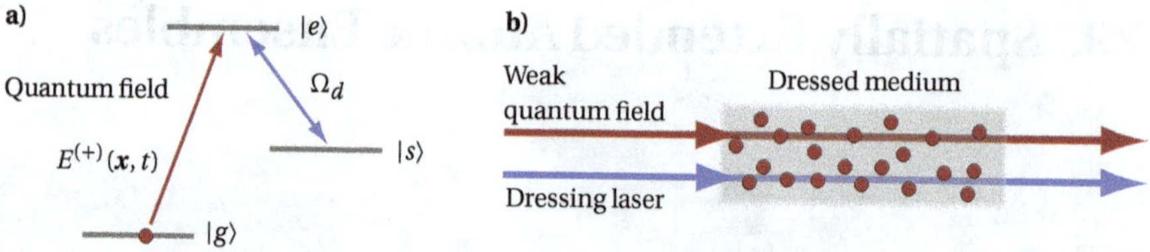

Fig. 29.1. The setup for light propagation through a vapour of three-level atoms. **a)** The energy levels, transitions and optical fields; **b)** The atoms are relatively densely spread over a finite volume, which is illuminated by the pump and probe fields. There is no optical cavity.

Here $\boldsymbol{P}$ is determined by the macroscopic polarization in the vapour, and this arises as a result of the optical electric field. Classically, the polarization is taken to be a linear function of the electric field, of the form $\boldsymbol{P}(t) = \kappa \boldsymbol{E}(t)$, but at optical frequencies this simple phenomenological formula is obviously inappropriate, for two reasons:

i) The relationship between $\boldsymbol{P}(t)$ and $\boldsymbol{E}(t)$ can easily be non-linear at achievable laser intensities.

ii) In the limit that this relationship can be taken to be linear, at optical frequencies the polarization would depend on the electric field at all times up to t, giving a relationship more like $\boldsymbol{P}(t) = \int_0^\infty \kappa(\tau)\boldsymbol{E}(t-\tau)\,d\tau$.

Nevertheless the case when the polarization can be taken as a linear functional of the electric field is in practice very important, even when non-linear effects are important. For example, in the case of the three-level system of Sect. 13.3, there are two optical fields to consider, of which only the dressing field is sufficiently strong to cause non-linear effects. The probe field is weak, and one can expect a linear relationship between it and the polarization.

29.1.1 Electric Polarization and Susceptibility

Let us now consider the kind of situation illustrated in Fig. 29.1, in which a vapour of three-level atoms is illuminated with a weak quantum field and a strong dressing field. The electric polarization $\boldsymbol{P}(\boldsymbol{x}, t)$ is the dipole moment per unit volume of the medium under consideration. In the case of an atomic vapour, this arises from the sum of the individual dipole moments of the atoms inside a sufficiently small volume element ΔV located at the position $\boldsymbol{x}$. The polarization generated in the atomic medium by the applied fields is of primary interest, since it acts as a source term in Maxwell's equations and determines the electromagnetic field dynamics. From our formulation of the electric interaction given in Chap. 4, the dipole moment operator for an individual atom interacting with an optical field is decomposable into positive and negative frequency parts, and the part of Hamiltonian (13.3.18) corresponding to the interaction of the atom with the probe field can be written as

$$H_p^{\rm dip} = \tfrac{1}{2}\hbar\left(\Omega_p\sigma_{eg}+\Omega_p^*\sigma_{ge}\right), \tag{29.1.6}$$

$$= -\mathcal{E}_p \boldsymbol{d}_{eg}\cdot\boldsymbol{\epsilon}\,\sigma_{eg} - (\mathcal{E}_p\boldsymbol{d}_{eg}\cdot\boldsymbol{\epsilon})^*\,\sigma_{ge}. \tag{29.1.7}$$

This expression is written in a rotating frame, and the two Hermitian conjugate terms represent the contributions to the Hamiltonian from the electric field at frequencies $\pm\omega_p$. Thus, the mean dipole moments at each frequency are

$$\left.\begin{aligned} \bar{\boldsymbol{D}}(\omega_p) &\equiv \boldsymbol{d}_{ge}\rho_{eg}, & \bar{\boldsymbol{D}}(-\omega_p) &\equiv \boldsymbol{d}_{eg}\rho_{ge},\\ \bar{\boldsymbol{D}}(\omega_d) &\equiv \boldsymbol{d}_{se}\rho_{es}, & \bar{\boldsymbol{D}}(-\omega_d) &\equiv \boldsymbol{d}_{es}\rho_{se}, \end{aligned}\right\} \tag{29.1.8}$$

and the polarizations in the rotating frame at the various frequencies are

$$\boldsymbol{P}(\boldsymbol{x},\pm\omega_p) = \frac{N}{\Delta V}\bar{\boldsymbol{D}}(\pm\omega_p), \tag{29.1.9}$$

$$\boldsymbol{P}(\boldsymbol{x},\pm\omega_d) = \frac{N}{\Delta V}\bar{\boldsymbol{D}}(\pm\omega_d). \tag{29.1.10}$$

To compute the polarization, we need to compute the stationary density operator, and this can be done perturbatively as in the following exercise.

Exercise 29.1 Perturbative Calculation of the Susceptibility: The master equation (13.3.19)—where for simplicity we choose phases so that Ω_p and Ω_d are real—can be written in the form

$$\dot\rho = \left(L_0+\Omega_p L_1\right)\rho, \tag{29.1.11}$$

and the stationary solution for small Ω_p can be computed peturbatively to first order as

$$\rho_s \approx \rho^0+\Omega_p\rho^1, \tag{29.1.12}$$

in which ρ^0 and ρ^1 satisfy the equations

$$L_0\rho^0 = 0, \tag{29.1.13}$$

$$L_0\rho^1 = -L_1\rho^0. \tag{29.1.14}$$

i) Show that $\rho^0 = |g\rangle\langle g|$, and interpret this result.

ii) By taking matrix elements of (29.1.14) show that

$$\rho^1_{ee} = \rho^1_{es} = \rho^1_{se} = \rho^1_{ss} = 0, \tag{29.1.15}$$

$$(\Delta_p-\Delta_d)\rho^1_{sg} = \tfrac{1}{2}\Omega_d\rho^1_{eg}, \tag{29.1.16}$$

$$\left(\Delta_p+\tfrac{1}{2}\mathrm{i}\Gamma_e\right)\rho^1_{eg}-\tfrac{1}{2}\Omega_d\rho^1_{sg} = \tfrac{1}{2}. \tag{29.1.17}$$

iii) Hence show that

$$\rho^1_{eg} = \frac{2\delta}{2\delta\left(2\Delta_p+\mathrm{i}\Gamma_e\right)-|\Omega_d|^2}, \tag{29.1.18}$$

$$\rho^1_{sg} = \frac{\Omega_d}{2\delta\left(2\Delta_p+\mathrm{i}\Gamma_e\right)-|\Omega_d|^2}. \tag{29.1.19}$$

29.1.2 Perturbative Regime in the Probe Field—Susceptibility

The *linear susceptibility* $\chi(\omega)$ is the linear response of the polarization to an infinitesimal electric displacement field of frequency ω_p. From (29.1.18) and (29.1.8), it follows that $\rho_{eg} \approx \Omega_p \rho^1_{eg}$, so that the dipole moment to this order of approximation is

$$\bar{\boldsymbol{D}}(\omega_p) \approx \boldsymbol{d}_{ge}\Omega_p\rho^1_{eg} = -\mathcal{E}_p \boldsymbol{d}_{ge}\frac{2\boldsymbol{d}_{eg}\cdot\boldsymbol{\epsilon}}{\hbar}\rho^1_{eg}. \tag{29.1.20}$$

The resultant polarization is not necessarily parallel with the direction of the electric field, that is, with $\boldsymbol{\epsilon}$, and therefore the linear response to an optical field must be described in general by a tensor. However, a convenient scalar measure is given by the component of the polarization in the direction of the electric field, namely

$$\boldsymbol{\epsilon}^*\cdot\boldsymbol{P}(\boldsymbol{x},\omega_p) = -\frac{2|\boldsymbol{d}_{eg}\cdot\boldsymbol{\epsilon}|^2}{\hbar}\mathcal{E}_p\rho^1_{eg} \equiv \varepsilon_0\chi^{(1)}(\omega_p)\mathcal{E}_p, \tag{29.1.21}$$

and this equation gives the definition of the *scalar linear susceptibility* $\chi^{(1)}(\omega_p)$. Explicitly substituting the value (29.1.18) for ρ^1_{eg}, we find

$$\chi^{(1)}(\omega_p) = \left(\frac{|\boldsymbol{d}_{ge}\cdot\boldsymbol{\epsilon}|^2 n_{\text{Atoms}}}{\varepsilon_0\hbar}\right)\frac{4\delta}{|\Omega_d|^2 - 2\delta\left(2\Delta_p + \mathrm{i}\Gamma_e\right)}. \tag{29.1.22}$$

Here $n_{\text{Atoms}} \equiv N/\Delta V$ is the number density of the atomic vapour.

a) Comparison with the Susceptibility for a Two-Level Atom: The coherence for the damped two-level atom given in (13.2.8) can be written in the limit that both $\Omega_R \ll |\Delta|$ and $\Omega_R \ll \Gamma$

$$\bar{\rho}_{eg}(t) = -\frac{\Omega_R}{2\Delta - \mathrm{i}\Gamma}, \tag{29.1.23}$$

and this is compared with the result for the three-level system in Fig. 29.2.

b) Interpretation of the Susceptibility: The expression (29.1.22) obscures the dependence on the actual laser frequencies, and becomes more understandable when expressed directly in terms of these frequencies thus:

$$\chi^{(1)}(\omega_p) = \left(\frac{|\boldsymbol{d}_{ge}\cdot\boldsymbol{\epsilon}|^2 n_{\text{Atoms}}}{\varepsilon_0\hbar}\right)\frac{4(\omega_p - \omega_d - \omega_{sg})}{|\Omega_d|^2 - 2(\omega_p - \omega_d - \omega_{sg})\left(2(\omega_p - \omega_{eg}) + \mathrm{i}\Gamma_e\right)}. \tag{29.1.24}$$

c) Autler–Townes Splitting: As can be seen in Fig. 29.2, the resonance is split in two for non-zero Ω_d, which is a reflection of the existence of the dark state. The existence of the splitting is obvious from (29.1.24), since the susceptibility—both real and imaginary part—is zero when $\omega_p = \omega_d + \omega_{sg}$, unless $\Omega_d \to 0$. In the latter case the susceptibility is simply that of the two-level system corresponding to the transition $e \leftrightarrow g$.

The splitting is therefore of the same order of magnitude as Ω_d, and this can be made very small. Since the linewidth can be much narrower than Γ_e, this yields a very sharp feature in the spectrum, which was first noted by *Autler* and *Townes* [29.1].

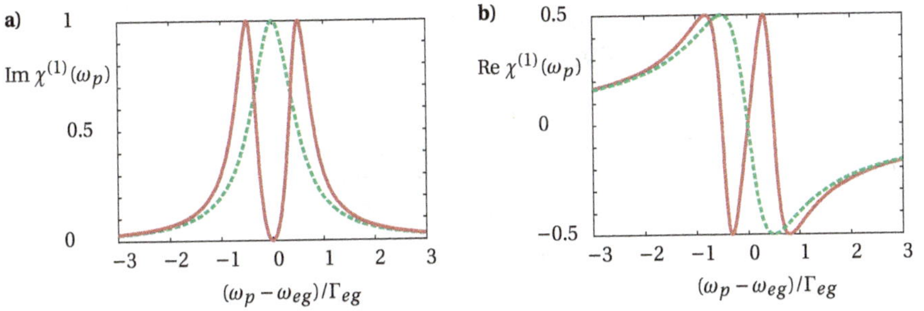

Fig. 29.2. The susceptibility $\chi^{(1)}(\omega_p)$, as given in (29.1.24), plotted as a function of $\omega_p - \omega_{eg}$ for fixed values of ω_d, ω_{eg} and ω_{es} is plotted red, while, for comparison, the susceptibility (29.1.23) for a similar two-level system is plotted as a green dashed line. **a)** The imaginary part represents absorption, which for the three-level system exhibits a zero—corresponding to the existence of the dark state—when $\Omega_d \neq 0$. The splitting into two peaks becomes arbitrarily narrow as $\Omega_d \to 0$, and is known as the *Autler–Townes* splitting. **b)** The real part also crosses zero, with a slope that can become arbitrarily large as $\Omega_d \to 0$, giving rise to an arbitrarily large refractive index in the bandwidth between the two absorption peaks.

29.1.3 Electromagnetically Induced Transparency

The vanishing of the absorption under these conditions is now known as *electromagnetically induced transparency*, corresponding to a picture in which applying the electromagnetic field Ω_d produces the transparency. This phenomenon was first observed by *Boller, Imamoğlu* and *Harris* [29.2].

29.2 Propagation in One Dimension

We want to consider the propagation of a beam of light, of frequency ω_0 and wavenumber $k_0 = \omega_0/c$, through an atomic vapour, under the conditions:

i) *One-Dimensional Description*: We will simplify the description by writing a one dimensional model in the z-direction and using a scalar electric field $E(z,t)$.

ii) *Slowly Varying Amplitude Approximation*: We will write the field in the form

$$E(z,t) = \mathcal{E}(z,t)e^{-i(\omega_0 t - k_0 z)} + \mathcal{E}^*(z,t)e^{i(\omega_0 t - k_0 z)}. \tag{29.2.1}$$

The space and time variation of the electric field amplitude $\mathcal{E}(z,t)$ is assumed to be very weak compared with the optical scales k_0 and ω_0. This means that there can be no genuinely sharp boundaries in the spatial layout of any system under consideration. However, there can be boundaries which are sharp on a coarse grained spatial scale, but smooth on the scale of the wavelengths involved.

iii) *Polarization*: Similarly, we will write the polarization of the medium in a form using a slowly varying amplitude

$$P(z,t) = \mathcal{P}(z,t)e^{-i(\omega_0 t - k_0 z)} + \mathcal{P}^*(z,t)e^{i(\omega_0 t - k_0 z)}. \tag{29.2.2}$$

Since the polarization arises from the atomic vapour we can also expect the density operator matrix elements to exhibit the same kind of space and time dependence, that is, for an atom at the position z_i we will write

$$\rho_{eg}^{(i)}(t) = \tilde{\rho}_{eg}(z_i,t)e^{-i(\omega_0 t - k_0 z_i)}. \tag{29.2.3}$$

iv) *Atomic and Polarization Densities*: This now lets us write the polarization in the form

$$\mathcal{P}(z,t) = d_{eg}\sum_{i=1}^{N}\delta(z-z_i)\tilde{\rho}_{eg}^{(i)}(t) \longrightarrow d_{eg}\tilde{\rho}_{eg}(z,t). \tag{29.2.4}$$

The aggregation of pointlike atoms is smoothed over the coarse spatial scale to give the equivalent smooth function $\tilde{\rho}_{eg}(z,t)$. We can similarly define quantities $\rho_{ee}(z,t)$ and $\rho_{gg}(z,t)$ as *atomic densities* at z in the excited and ground states.

29.2.1 The Wave Equation

Corresponding to the full three-dimensional wave equation (29.1.5), we can write

$$-\frac{\partial^2 E}{\partial z^2} + \frac{1}{c^2}\frac{\partial^2 E}{\partial t^2} = -\mu_0\frac{\partial^2 P}{\partial t^2}. \tag{29.2.5}$$

Here E is the amplitude of the electric field corresponding to a definite optical polarization.

a) Use of the Slowly Varying Amplitude Approximation: We will write the wave equation in the form

$$\left(\frac{\partial}{\partial z} + \frac{1}{c}\frac{\partial}{\partial t}\right)\left(-\frac{\partial}{\partial z} + \frac{1}{c}\frac{\partial}{\partial t}\right)E = -\mu_0\frac{\partial^2 P}{\partial t^2}. \tag{29.2.6}$$

We now insert the slowly varying amplitude forms (29.2.2, 29.2.3) into the wave equation, and proceed as follows:

i) Since $\mathcal{E}(z,t)$ is slowly varying, we can write approximately

$$\left(-\frac{\partial}{\partial z} + \frac{1}{c}\frac{\partial}{\partial t}\right)\left(\mathcal{E}(z,t)e^{-i(\omega_0 t - k_0 z)}\right) \approx -2ik_0\mathcal{E}(z,t)e^{-i(\omega_0 t - k_0 z)}. \tag{29.2.7}$$

ii) Similarly, we can write

$$\frac{\partial^2}{\partial t^2}\left(\mathcal{P}(z,t)e^{-i(\omega_0 t - k_0 z)}\right) \approx -\omega_0^2\mathcal{P}(z,t)e^{-i(\omega_0 t - k_0 z)}. \tag{29.2.8}$$

iii) We separately equate the coefficients of the positive and negative frequency terms $e^{-i(\omega_0 t - k_0 z)}$, since these are of very high frequency compared to the slowly varying amplitudes.

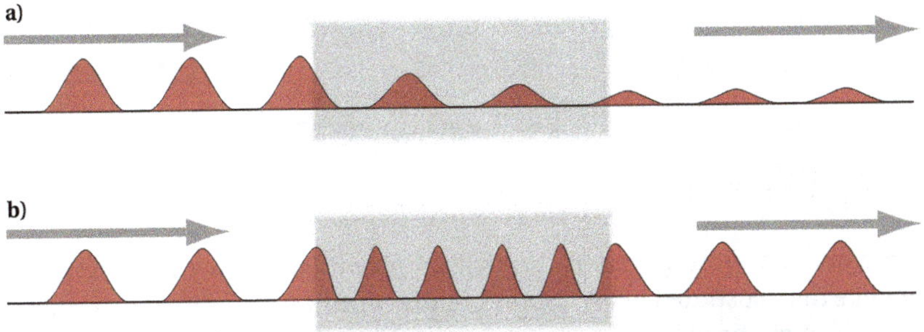

Fig. 29.3. a) A pulse train with a frequency far off resonance is absorbed as it traverses the atomic vapour. **b)** On resonance the speed of light is slowed, the pulses are spatially shortened, and traverse the medium at reduced speed. On leaving the medium they regain their original speed, height and length.

iv) Then using $\omega_0 = k_0 c$ and $c^2 = 1/\mu_0\varepsilon_0$, we get the equation of motion for the slowly varying amplitude

$$\left(\frac{\partial}{\partial t} + c\frac{\partial}{\partial z}\right)\mathcal{E}(z,t) = \frac{i\omega_0}{\varepsilon_0}\mathcal{P}(z,t). \tag{29.2.9}$$

In this form it is clear that, in the slowly varying amplitude approximation, only waves propagating from left to right are supported, and reflections are forbidden.

b) Frequency of the Slowly Varying Amplitudes: The slowly varying amplitude and polarization can be written in terms of the Fourier components of the corresponding positive frequency amplitudes as

$$\mathcal{E}(z,t) = \int_{\mathcal{B}} d\omega\, e^{-i\omega t}\,\tilde{\mathcal{E}}(z,\omega+\omega_0), \qquad \mathcal{P}(z,t) = \int_{\mathcal{B}} d\omega\, e^{-i\omega t}\,\tilde{\mathcal{P}}(z,\omega+\omega_0), \tag{29.2.10}$$

where $\mathcal{B}$ represents a narrow bandwidth around $\omega = 0$.

29.2.2 Absorptive and Dispersive Behaviour

The behaviour of the solutions of these equations depends strongly on the frequency dependence of the susceptibility, and, as can be seen from (29.1.22) and Fig. 29.2, there is a strong frequency dependence near the value $\delta = 0$, that is, when $\omega_p \approx \omega_d + \omega_{sg}$. The susceptibility (29.1.22) relates these through

$$\tilde{\mathcal{P}}(z,\omega+\omega_0) = \varepsilon_0\chi^{(1)}(\omega+\omega_0)\tilde{\mathcal{E}}(z,\omega+\omega_0). \tag{29.2.11}$$

The behaviour thus described depends on the choice of ω_0.

a) Far off Resonance Behaviour: If we choose ω_0 to be very different from $\omega_d + \omega_{sg}$, then δ is almost constant over the bandwidth $\mathcal{B}$, so that within $\mathcal{B}$ we can approximate $\chi^{(1)}(\omega+\omega_0) \approx \chi^{(1)}(\omega_0)$ and thus we can write

$$\mathcal{P}(z,t) \approx \varepsilon_0 \chi^{(1)}(\omega_0)\mathcal{E}(z,t). \tag{29.2.12}$$

Substituting into the equation (29.2.9), we get a propagation equation

$$\left(\frac{\partial}{\partial t} + c\frac{\partial}{\partial z}\right)\mathcal{E}(z,t) = \mathrm{i}\omega_0 \chi^{(1)}(\omega_0)\mathcal{E}(z,t). \tag{29.2.13}$$

This corresponds to propagation through an absorbing medium, giving a behaviour like that illustrated in Fig. 29.3 a, where an incident pulse train is absorbed as it goes through the medium, but is not slowed.

b) On-Resonance Behaviour: If we tune so that $\omega_0 = \omega_d + \omega_{sg}$ exactly, then $\delta \longrightarrow (\omega+\omega_0) - (\omega_d + \omega_{sg}) \longrightarrow \omega$. If, in addition, $\mathcal{B}$ is sufficiently narrow, then we can approximate the susceptibility formula (29.1.22) as

$$\chi^{(1)}(\omega+\omega_0) \approx \left(\frac{|\boldsymbol{d}_{ge}\cdot\boldsymbol{\epsilon}|^2 n_{\text{Atoms}}}{\varepsilon_0\hbar}\right)\frac{4\omega}{|\Omega_d|^2}. \tag{29.2.14}$$

Fourier transformed, this gives

$$\mathcal{P}(z,t) \approx \left(\frac{4|\boldsymbol{d}_{ge}\cdot\boldsymbol{\epsilon}|^2 n_{\text{Atoms}}}{\hbar|\Omega_d|^2}\right)\mathrm{i}\frac{\partial\mathcal{E}(z,t)}{\partial t}. \tag{29.2.15}$$

Substituting into the equation (29.2.9), we get a propagation equation

$$\left(\frac{\partial}{\partial t} + c\frac{\partial}{\partial z}\right)\mathcal{E}(z,t) = -\left(\frac{4\omega_0|\boldsymbol{d}_{ge}\cdot\boldsymbol{\epsilon}|^2 n_{\text{Atoms}}}{\varepsilon_0\hbar|\Omega_d|^2}\right)\frac{\partial\mathcal{E}(z,t)}{\partial t}. \tag{29.2.16}$$

The behaviour in this case is purely dispersive, since the resulting equation of motion corresponds to a propagation equation

$$\left(\frac{\partial}{\partial t} + v_g\frac{\partial}{\partial z}\right) = 0, \tag{29.2.17}$$

in which the group velocity is

$$v_g(z) = \frac{c}{1+n_g}, \tag{29.2.18}$$

where $1+n_g$ is a refractive index given by

$$n_g = \frac{4\omega_0|\boldsymbol{d}_{ge}\cdot\boldsymbol{\epsilon}|^2 n_{\text{Atoms}}}{\varepsilon_0\hbar|\Omega_d|^2}. \tag{29.2.19}$$

In this case a train of light pulses is slowed down and compressed as it passes through the vapour, as shown in Fig. 29.3 b. By making the atomic density n_g sufficiently large, the group velocity v_g can be made very small, giving rise to the concept of *slow light*, where v_g can be only metres per second. This was dramatically demonstrated by *Hau et al.* in 1999, using the very high atomic density in a Bose–Einstein condensate of about a million sodium atoms [29.3].

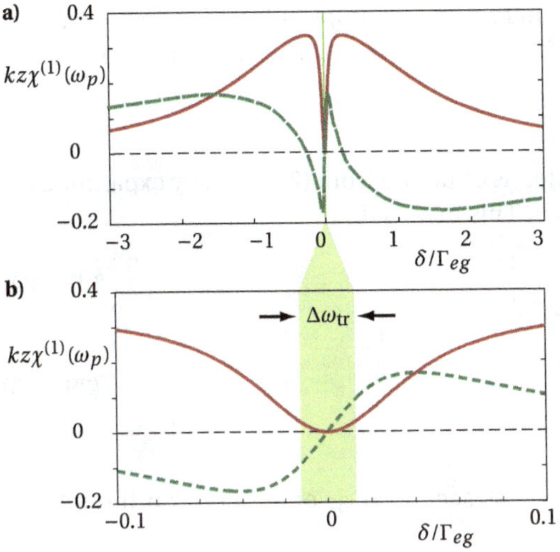

Fig. 29.4. Illustration of the idea of the transmission window. The real part of $kz\chi^{(1)}(\omega_p)$ (red solid line) and the imaginary part of $kz\chi^{(1)}(\omega_p)$ (green dashed line) are plotted on different scales, for the case where $\Omega_d = \Gamma_{eg}/6$, making the transparency feature considerably narrower than the linewidth. The green shading represents the range of ω_p in which absorption would be less than a few percent.

29.2.3 The Transparency Window

If there is information coded in an incoming light pulse, this can be regarded as being stored in the slow light for the reasonably long time it takes a pulse to propagate through the vapour. However, to assess how well quantum information could be so stored, a more detailed treatment of the quantum dynamics of the system is needed.

The restriction to a narrow bandwidth $\mathcal{B}$ in deriving the classical equation, and introduced in (29.2.10) is equivalent to the slowly varying amplitude approximation used by both the classical and the quantum formulation. It is important to quantify the range of validity of this kind of approximation, and we will do this using the concept of the *transparency window*.

Let us consider a stationary solution of the off-resonance equation of motion (29.2.13) evaluated at $\omega_0 \longrightarrow \omega_p$, which has the general form

$$\mathcal{E}(z,t) = e^{-k_0\chi^{(1)}(\omega_p)z} f(z-ct), \tag{29.2.20}$$

where $f(x)$ is an arbitrary function of x. This solution corresponds to a wave propagating from left to right, modulated by an exponential function of z.

The *transmission function* measures the phase and amplitude of the light as it traverses the medium, and is most conveniently expressed as a function of the detuning (as is also defined in (13.3.4)

$$\delta = \omega_p - \omega_d - \omega_{sd}, \tag{29.2.21}$$

and is given by

$$T(\delta,z) = \exp\big(ikz\chi^{(1)}(\omega_p)\big). \tag{29.2.22}$$

Since $\chi^{(1)}(\omega_p)$ has a positive imaginary part, the amplitude will decrease with increasing z—the light is thus absorbed by the medium. The attenuation of the amplitude of the transmitted light is given by

$$|T(\delta,z)| \approx \exp\left(-\delta^2/\Delta\omega_{\text{tr}}^2\right), \tag{29.2.23}$$

in which the approximate expression is obtained from (29.1.22), by expanding the imaginary part to second order in δ. This means that

$$\Delta\omega_{\text{tr}} = \frac{\varepsilon_0\hbar}{2|\boldsymbol{d}\cdot\boldsymbol{\epsilon}|^2}\frac{|\Omega_d|^2}{\sqrt{n_{\text{Atoms}}\Gamma_e k_{eg} z}} \tag{29.2.24}$$

$$= \frac{|\Omega_d|^2}{4\sqrt{\Gamma_{eg}\Gamma_e}}\frac{1}{\sqrt{\sigma n_{\text{Atoms}} z}}. \tag{29.2.25}$$

In this result:

i) We have used the absorption cross section σ_{abs} of an atom on the $g \longrightarrow e$ transition

$$\sigma_{\text{abs}} = \frac{3\lambda_{ge}^2}{2\pi}. \tag{29.2.26}$$

ii) In (29.2.25) we have used the result (8.5.29) to write the result in terms of Γ_{eg} instead of the dipole matrix element.

The quantity $\Delta\Omega_{\text{tr}}$ defines the *transparency window*—this is the range of ω_p such that

$$|\delta| \equiv |\omega_p - \omega_d - \omega_{sg}| < \Delta\omega_{\text{tr}}. \tag{29.2.27}$$

As can be seen from Fig. 29.4, this is also the range of ω_p in which the dispersion, that is Re $\chi^{(1)}(\omega_p)$, can be taken as linear in δ, and hence it represents the range of frequencies in which the dispersive propagation equation as given by (29.2.16), is a valid approximation.

29.2.4 Slow Light

In the case a spatially dependent Rabi frequency, the term on the right-hand-side of the propagation equation (29.2.16) simply leads to a modification of the group velocity of the electric field, leading to a propagation equation

$$\left(\frac{\partial}{\partial t} + v_g(z)\frac{\partial}{\partial z}\right)\mathcal{E}(z) = 0, \tag{29.2.28}$$

in which the group velocity is

$$v_g(z) = \frac{c}{1+n_g(z)}, \tag{29.2.29}$$

where $1+n_g(z)$ is a refractive index, in which

$$n_g(z) = \frac{4g^2 N(z)}{|\Omega_d(z)|^2}. \tag{29.2.30}$$

a) Solution of the Wave Equation: The wave equation (29.2.28) has the exact solution describing a propagation with a spatially varying velocity $v_g(z)$

$$\mathcal{E}(z,t) = \mathcal{E}\left(0, t - \int_0^z dz' \frac{1}{v_g(z')}\right), \tag{29.2.31}$$

where $\mathcal{E}(0,t')$ denotes the field entering the interaction region at $z = 0$.

b) Temporal Profile of the Pulse: The solution (29.2.31) gives the same time dependence at any position, with a delay corresponding to the propagation time. Consequently, all aspects of the electromagnetic field at a single time are unaffected. For example:

i) *The Electromagnetic Energy Flux*: This is the same at any position.

ii) *The Spectrum of the Pulse*: This also remains unchanged

$$S(z,\omega) \equiv \int_{-\infty}^{\infty} d\tau\, e^{-i\omega\tau} \left\langle \mathcal{E}^*(z,t)\mathcal{E}(z,t-\tau)\right\rangle \tag{29.2.32}$$

$$= S(0,\omega). \tag{29.2.33}$$

In particular the spectral width stays constant

$$\Delta\omega_p(z) = \Delta\omega_p(0). \tag{29.2.34}$$

c) Compression of the Spatial Pulse Profile: On the other hand a spatial change of the group velocity leads to a compression of the spatial pulse length, so that the pulse length L_{pulse} inside the vapour medium, is given by

$$L_{\text{pulse}} = \frac{v_g}{c} L_0, \tag{29.2.35}$$

where L_0 is the free-space value of the spatial pulse length.

d) Delay or Storage Time: Let us consider the possible storage of a pulse of temporal length τ_{pulse} in the electromagnetically induced transparency vapour medium, which is contained within a length l. The *pulse delay time* is defined as the time the pulse is delayed on exiting the medium as a result of introducing the vapour, and is given by

$$\tau_{\text{delay}} \equiv \frac{n_g l}{c}. \tag{29.2.36}$$

If this is a sufficiently long time, it is possible to consider this as the storage of information encoded in a pulse. Let us consider some criteria for this to be a realistic storage technique.

i) *Opacity*: It is convenient to introduce the *opacity* α of the medium, contained within a length l, and in the absence of electromagnetically induced transparency

$$\alpha \equiv \frac{3}{8\pi^2} n_{\text{atoms}} \lambda^3 k l = \tfrac{1}{2}\sigma_{\text{abs}} l. \tag{29.2.37}$$

So defined, the *amplitude* $\mathcal{E}$ decays by a factor $\exp(-\alpha)$ as the field traverses the length l of the medium.

ii) *Opacity, Transparency Width and Pulse Length*: Using the opacity, we can write the transparency width of a medium of length l defined by (29.2.24, 29.2.25) in the form

$$\Delta\omega_{\rm tr} = \frac{|\Omega|^2}{4\Gamma_{eg}\sqrt{\alpha}} = \frac{\sqrt{\alpha}}{\tau_{\rm delay}}. \tag{29.2.38}$$

Hence, for a given value of l, large delay times imply a narrow transparency window. To avoid absorption, the pulse cannot be permitted to change any faster than on a time scale $\sim 1/\Delta\omega_{\rm tr}$, so this means that the pulse length itself must increase as the delay time increases.

iii) *Figure of Merit for a Memory Device*: Hence there is an upper bound for the ratio of achievable delay or storage time to the initial pulse length of a photon, namely

$$\frac{\tau_{\rm delay}}{\tau_{\rm pulse}} \leqslant \sqrt{\alpha}\,. \tag{29.2.39}$$

The ratio $\tau_{\rm delay}/\tau_{\rm pulse}$ is the figure of merit for a memory device. The larger this ratio the better suited is the system for a temporal storage.

iv) *Medium length*: There is another quantity which is important for the storage capacity, namely the ratio of medium length l to the length $L_{\rm pulse}$ of an individual pulse inside the medium. Following similar arguments as above one finds that this quantity is also limited by the square root of the opacity

$$\frac{l}{L_{\rm pulse}} \leqslant \sqrt{\alpha}, \tag{29.2.40}$$

where we have assumed $n_g \gg 1$.

In practice, the achievable opacity α of atomic vapor systems is limited to values below 10^4 resulting in upper bounds for the ratio of time delay to pulse length of the order of 100.

29.3 Quantum Theory of Light Propagation in an Atomic Vapour

We now consider the propagation of a quantized light probe field in a medium of three-level Λ-atoms in the presence of a strong classical dressing field. Since we will always work within the weak probe field limit, it will be appropriate to use linearized (or harmonic) formalism of Sect. 28.2.2, and to generalize this to the case of a vapour which is distributed in space.

29.3.1 One-Dimensional Electromagnetic Field Operators

We want to introduce the idea of a quantized version of the one-dimensional slowly varying amplitude approximation introduced in (29.2.1), and we do this by writing

$$E^{(+)}(\boldsymbol{x}, t) = \sqrt{\frac{\hbar\omega_p}{2\varepsilon_0}} f(x, y) \mathcal{A}(z, t)\, \mathrm{e}^{\mathrm{i}k_p(z-ct)}. \tag{29.3.1}$$

In this equation the electric field operator is chosen to represent excitations such that

i) All modes have a transverse profile given by the real function $f(x, y)$, normalized so that $\int dx\,dy\, f(x, y)^2 = 1$. This profile is very broad compared to the width of the volume containing the atoms, so that the atomic interaction does not depend on the atom's transverse location.

ii) We will introduce the operator $\mathrm{e}^{\mathrm{i}k_p(z-ct)}\mathcal{A}(z, t)$ to represent modes with excitations in in the z direction, and with wavenumbers k in a very narrow range $|k_z| < \eta$. This means we can write

$$\mathcal{A}(z, t) = \frac{1}{\sqrt{L}} \sum_{|k|<\eta} \mathrm{e}^{\mathrm{i}kz} a_k(t), \tag{29.3.2}$$

where $a_k(t)$ represents the destruction operator for the z wavenumber $k + k_p$. The commutation relation for $\mathcal{A}(z, t)$ is then

$$\left[\mathcal{A}(z, t), \mathcal{A}^\dagger(z', t)\right] = \delta(z - z'). \tag{29.3.3}$$

The operator $\mathcal{A}(z, t)$ is the quantized version of the slowly varying electric field $\mathcal{E}(z, t)$ defined in (29.2.1).

iii) Here $L = V^{1/3}$, where V is the quantization volume.

29.3.2 Spatially Dependent Collective Atomic Operators

We consider the atomic vapour to be contained within a tube of cross-sectional area ΔA, and we want to consider atomic operators corresponding to the atoms contained within an interval $(z, z + \Delta z)$, whose volume is therefore $\Delta A\, \Delta z$. The operators E and S introduced in Sect. 28.2.1 now become z-dependent, and we can write, in the volume element at the point z,

$$E(z, t) = \int_z^{z+\Delta z} \tilde{E}(z', t)\, dz', \tag{29.3.4}$$

$$S(z, t) = \int_z^{z+\Delta z} \tilde{S}(z', t)\, dz'. \tag{29.3.5}$$

The operators $\tilde{E}(z, t)$ and $\tilde{S}(z, t)$ behave as independent commuting quantum fields, with the non-zero commutation relations

$$\left[\tilde{S}(z, t), \tilde{S}^\dagger(z, t)\right] = \delta(z - z'), \tag{29.3.6}$$

$$\left[\tilde{E}(z, t), \tilde{E}^\dagger(z, t)\right] = \delta(z - z'), \tag{29.3.7}$$

With these commutation relations, the operators $E(z, t)$ and $S(z, t)$ have harmonic oscillator commutation relations at each point z, and are independent of the corresponding operators at different points.

29.3.3 Hamiltonian and Equations of Motion

Let us now develop an approximate Hamiltonian description, which will involve the operators $\tilde{S}(z,t)$, $\tilde{E}(z,t)$ and $\mathcal{A}(z,t)$.

a) Approximate Hamiltonian: We generalize the Hamiltonian (28.2.18) to include the linear density of atoms $N(z)$, a Hamiltonian to give one dimensional electrodynamics, and also to include spontaneous emission arising the the $g \leftrightarrow e$ transition; it takes the form

$$\begin{aligned} H_{\text{Eff}} = & -\tfrac{1}{2}\hbar \int dz \left(\Omega_d^*(z)\tilde{E}(z)\tilde{S}^\dagger(z) + \Omega_d(z)\tilde{S}(z)\tilde{E}^\dagger(z)\right) \\ & -\hbar g \int dz \sqrt{N(z)} \left(\mathcal{A}(z)\tilde{E}^\dagger(z) + \mathcal{A}^\dagger(z)\tilde{E}(z)\right) \\ & +\tfrac{1}{2}i\hbar c \int dz \left(\mathcal{A}^\dagger(z)\frac{\partial \mathcal{A}(z)}{\partial z} - \mathcal{A}(z)\frac{\partial \mathcal{A}^\dagger(z)}{\partial z}\right) \\ & + H_{\text{Spontaneous emission}} . \end{aligned} \tag{29.3.8}$$

Here, the final term is the kind of Hamiltonian introduced in Part III to account for interaction with the radiation field, and here it gives rise to a damping term proportional to $-\frac{1}{2}\Gamma_{eg}\tilde{E}$, and an appropriate quantum noise term.

b) Quantum Stochastic Differential Equations: When the noise is included we get quantum stochastic differential equations

$$d\tilde{S} = \tfrac{1}{2}i\Omega_d^*\tilde{E}\,dt, \tag{29.3.9}$$

$$d\tilde{E} = \left\{-\tfrac{1}{2}\Gamma_{ge}\tilde{E} + \tfrac{1}{2}i\Omega_d\tilde{S} + ig\mathcal{A}\sqrt{N(z)}\right\}dt + dZ(z,t), \tag{29.3.10}$$

$$d\mathcal{A} = \left\{-c\frac{\partial \mathcal{A}}{\partial z} + ig\sqrt{N(z)}\,\tilde{E}\right\}dt, \tag{29.3.11}$$

where the non-zero product of the noise terms is

$$dZ(z,t)dZ^\dagger(z',t) = \Gamma_{ge}\delta(z-z')\,dt. \tag{29.3.12}$$

Because they are derivable directly from a Hamiltonian, these equations of motion preserve commutation relations and all other quantum-mechanical properties.

c) Interpretation of the Approximations: The approximation corresponds to assuming an non-depletable collection of Bosonic atoms in the ground state, with excitation to the states $|e\rangle$ and $|s\rangle$ generated by the Bosonic creation operators $\tilde{E}^\dagger$ and $\tilde{S}^\dagger$. The Bosonic nature of the excitations corresponds to the symmetry of the wavefunctions of the Dicke states, as described in Sect. 28.1.1b.

d) Representation Using the P-Function: The approximate Hamiltonian is bilinear in the creation and destruction operators, and has a straightforward linear damping and noise from a zero-temperature heat bath. Using the rule and methods for the P-function as given in *Bk. I: Sect.17.2*, we can associate a P-function

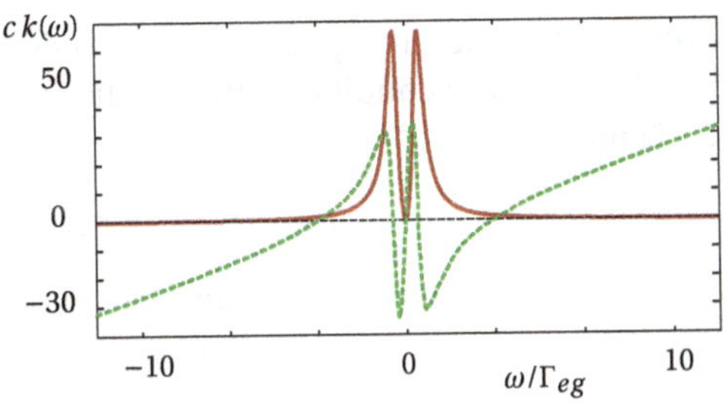

Fig. 29.5. The dispersion relation which arises from (29.3.13, 29.3.14). The real part is plotted as the dashed green line, and the imaginary part is the solid red line. The approximate equations (29.3.16, 29.3.17) are valid in the region of positive slope of the real part near $\omega = 0$.

of three z-dependent coherent state variables, $e(z)$, $s(z)$ and $a(z)$. Because the noise is at temperature zero, the equation of motion for the P-function will be a Liouville equation, corresponding to classical equations of motion for the phase space variables of exactly the same form as (29.3.9–29.3.11), except that the term $dZ(z,t)$ is not present—there is no noise.

e) Conclusion: Since the P-function equations have no stochastic term, we can eliminate the variable $e(z)$, and get the equations for the variables $s(z)$ and $a(z)$

$$\left(\frac{\partial}{\partial t}+c\frac{\partial}{\partial z}\right)a(z,t) = \frac{2g\sqrt{N(z)}}{\Omega_d^*}\frac{\partial s(z,t)}{\partial t}, \tag{29.3.13}$$

$$\left(\frac{\partial^2}{\partial t^2}+\tfrac{1}{2}\Gamma_{eg}\frac{\partial}{\partial t}+\tfrac{1}{2}|\Omega_d|^2\right)s(z,t) = -g\sqrt{N(z)}\Omega_d^* a(z,t). \tag{29.3.14}$$

For $N(z) = N$ independent of position, we can derive a dispersion relation by considering solutions of these equations proportional to $\exp(ikz - i\omega t)$, from which we find that

$$ck(\omega) = \omega + \frac{2g^2 N\omega}{\tfrac{1}{2}|\Omega_d|^2 - \tfrac{1}{2}i\Gamma_{eg}\omega - \omega^2}. \tag{29.3.15}$$

This is plotted in Fig. 29.5.

29.3.4 Solutions of the Equations of Motion

Provided time dependences involved are on a time scale much slower than given by the Rabi frequency $\Omega_d(t)$, we can make the adiabatic approximation and neglect the time derivative of $\Omega_d(t)$, and use instead the simplified equations

$$s(z,t) = -\frac{2g\sqrt{N(z)}}{\Omega_d(z,t)}a(z,t), \tag{29.3.16}$$

$$\left(\frac{\partial}{\partial t}+c\frac{\partial}{\partial z}\right)a(z,t) = -\frac{4g^2N(z)}{|\Omega_d(z,t)|^2}\frac{\partial a(z,t)}{\partial t}. \tag{29.3.17}$$

a) Polarization and Electric Field: The first of these equations gives the polarization $s(z,t)$ in terms of the electric field $a(z,t)$, and the second corresponds to the propagation with a reduced group velocity of (29.2.29), $v_g(z,t)$, although in this case we are introducing a time-dependent group velocity. As we reduce $\Omega_d(z,t)$, the group velocity decreases, ultimately reaching zero when $\Omega_d(z,t) = 0$.

b) Stopping and Re-Accelerating Wavepackets: Thus, it appears possible to slow down a pulse of light to a complete halt. Under these conditions, the solutions of the equations of motion (29.3.16, 29.3.17) are

$$a(z,t) = 0, \qquad \frac{\partial a(z,t)}{\partial t} = 0. \tag{29.3.18}$$

In any procedure by which $\Omega_d(z,t) \longrightarrow 0$ from some non-zero value, we can determine $s(z,t)$ by

$$s(z,t) = -\lim_{\Omega_d \to 0}\left(\frac{2g}{\Omega_d(z,t)}\, a(z,t)\right). \tag{29.3.19}$$

In this limit, the information in the electromagnetic field has been transferred completely into the polarization $s(z,t)$.

29.4 Quantum Memory Using Dark State Polaritons

In this section we will use the concept of a dark state more explicitly to show how to make a quantum memory based on the dark state formulation of Sect. 28.1. The use of adiabatic transfer techniques in the case of light propagating through an vapour provides a very advantageous alternative to the method of the previous section, which simply uses the existence of a slow light speed to give a delay to the propagation of a pulse.

29.4.1 Interpretation as Dark-State Polaritons

In this section, we shall for simplicity choose phases so that $\Omega_d(z,t)$ is real and positive. We will define a pair of mixed field operators,

$$\Psi(z,t) = \cos\theta(t)\, a(z,t) - \sin\theta(t)\, s(z,t), \tag{29.4.1}$$

$$\Phi(z,t) = \sin\theta(t)\, a(z,t) + \cos\theta(t)\, s(z,t), \tag{29.4.2}$$

with the mixing angle given by

$$\tan^2\theta(t) = \frac{4g^2 N}{\Omega_d(z,t)^2} = n_g(t). \tag{29.4.3}$$

$\Psi(z,t)$ and $\Phi(z,t)$ are superpositions of the electromagnetic field operator $a(z,t)$ and the collective atomic operator $s(z,t)$, whose proportions are determined by $\theta(t)$, which can be controlled by changing the magnitude of $\Omega_d(z,t)$, that is, the strength of the external driving field.

a) Adiabatic Approximation: In the adiabatic limit, corresponding to the validity of (29.3.16), one finds

$$\Phi(z,t) \longrightarrow 0, \tag{29.4.4}$$

and consequently

$$a(z,t) \longrightarrow \cos\theta(t)\, \Psi(z,t), \tag{29.4.5}$$

$$s(z,t) \longrightarrow -\sin\theta(t)\, \Psi(z,t). \tag{29.4.6}$$

Furthermore, the equation of motion (29.3.17) for $a(z,t)$ corresponds to the equation of motion

$$\left(\frac{\partial}{\partial t} + c\cos^2\theta(t)\frac{\partial}{\partial z}\right)\Psi(z,t) = 0. \tag{29.4.7}$$

29.4.2 Stopping and Re-Accelerating Photon Wavepackets

Corresponding to what we have already seen in Sect. 29.2.4, the solutions of the wave equation in the form (29.4.7) can be written in terms of the group velocity $v_g(t) = c\cos^2\theta(t)$ in the form

$$\Psi(z,t) = \Psi\left(z - c\int_0^t d\tau\, \cos^2\theta(\tau), 0\right). \tag{29.4.8}$$

These are shape preserving, and because the representation is in terms of a P-function, they also preserve the quantum state of the the initial wavepacket. Note that:

i) For $\theta \longrightarrow 0$, corresponding to a strong external driving field $\Omega_d^2 \gg g^2 N$, the polariton has a purely photonic character, that is

$$\Psi(z,t) \longrightarrow a(z,t), \tag{29.4.9}$$

and the propagation velocity is c, that of the vacuum speed of light.

ii) For $\theta \to \pi/2$, that is, in the opposite limit of a weak driving field, $\Omega_d^2 \ll g^2 N$, the polariton becomes a spin-wave,

$$\Psi(z,t) \longrightarrow -s(z,t), \tag{29.4.10}$$

and its propagation velocity approaches zero.

iii) Thus the following mapping can be realized

$$a(z) \longleftrightarrow s(z) = \frac{\tilde{\sigma}_{gs}(z')e^{i\Delta k z'}}{\sqrt{N(z)}}, \tag{29.4.11}$$

where

$$z' = z + z_0 = z + c\int_0^\infty d\tau\, \cos^2\theta(\tau). \tag{29.4.12}$$

This is the essence of the transfer technique of quantum states from photon wave-packets propagating at the speed of light to stationary atomic excitations, or stationary spin waves.

iv) Adiabatically rotating the mixing angle from $\theta = 0$ to $\theta = \pi/2$ decelerates the polariton to a complete halt, changing its character from purely electromagnetic to purely atomic. Because of the linearity of the equation of motion and the conservation of the spatial shape, the quantum state of the polariton is not changed during this process.

v) Similarly, the polariton can be re-accelerated to the vacuum speed of light; in this process the stored quantum state is transferred back to the field.

References

Chapter 1

[1.1] A. Einstein, B. Podolsky, and N. Rosen, *Can Quantum-Mechanical Description of Physical Reality Be Considered Complete?*, Phys. Rev. **47**, 777 (1935). 3

[1.2] J. S. Bell, *On the Einstein–Podolsky–Rosen Paradox*, Physics **1**, 195 (1964). 3, 34, 35

[1.3] J. F. Clauser, M. A. Horne, A. Shimony, and R. A. Holt, *Proposed Experiment to Test Local Hidden-Variable Theories*, Phys. Rev. Lett. **23**, 880 (1969). 3

[1.4] A. Aspect, P. Grangier, and G. Roger, *Experimental Tests of Realistic Local Theories via Bell's Theorem*, Phys. Rev. Lett. **47**, 460 (1981). 3, 4

[1.5] R. P. Feynman, *Simulating Physics with Computers*, International Journal of Theoretical Physics **21**, 467 (1982). 4, 18, 29, 31

[1.6] R. J. Glauber, *The Quantum Theory of Optical Coherence*, Phys. Rev. **130**, 2530 (1963). 4, 251

[1.7] R. P. Feynman, *Quantum Mechanical Computers*, Foundations of Physics **16**, 507 (1986). 6, 18

[1.8] P. W. Shor, in *Proceedings of the 35th Annual Symposium on the Foundations of Computer Science* (IEEE Computer Society Press, New York, 1994), p. 124. 6, 22, 27

[1.9] J. I. Cirac and P. Zoller, *Quantum Computations with Cold Trapped Ions*, Phys. Rev. Lett. **74**, 4091 (1995). 6, 14, 18, 358, 362

[1.10] F. Schmidt-Kaler, H. Häffner, S. Gulde, M. Riebe, G. P. T. Lancaster, T. Deuschle, C. Becher, W. Hänsel, J. Eschner, C. F. Roos, and R. Blatt, *How to Realize a Universal Quantum Gate with Trapped Ions*, Appl. Phys. B **77**, 789 (2003). 6, 362

[1.11] D. Leibfried, B. DeMarco, V. Meyer, D. Lucas, M. Barrett, J. Britton, W. M. Itano, B. Jelenkovic, C. Langer, T. Rosenband, and D. J. Wineland, *Experimental Demonstration of a Robust, High-Fidelity Geometric Two Ion-Qubit Phase Gate*, Nature **422**, 412 (2003). 6

[1.12] M. A. Nielsen and I. I. Chuang, *Quantum Computation and Quantum Information*, 1st ed. (Cambridge University Press, Cambridge, 2000, 2010). 6, 25, 34, 252

[1.13] M. J. Collett and C. W. Gardiner, *Squeezing of Intracavity and Travelling Wave Light Fields Produced in Parametric Amplification*, Phys. Rev. A **30**, 1386 (1984). 8, 121, 154

[1.14] C. W. Gardiner and M. J. Collett, *Input and Output in Damped Quantum Systems—Quantum Stochastic Differential Equations and the Master Equation*, Phys. Rev. A **31**, 3761 (1985). 8, 121, 154, 161

[1.15] T. Pellizzari, S. A. Gardiner, J. I. Cirac, and P. Zoller, *Decoherence, Continuous Observation, and Quantum Computing: A Cavity QED Model,* Phys. Rev. Lett. **75**, 3788 (1995). 15

[1.16] H. J. Kimble, *The Quantum Internet,* Nature **453**, 1023 (2008). 15, 32, 337

[1.17] D. Loss and D. P. DiVincenzo, *Quantum Computation with Quantum Dots,* Phys. Rev. A **57**, 120 (1998). 16

[1.18] R. Hanson, L. P. Kouwenhoven, J. R. Petta, S. Tarucha, and L. M. K. Vandersypen, *Spins in Few-Electron Quantum Dots,* Rev. Mod. Phys. **79**, 1217 (2007). 16

[1.19] K. M. Weiss, J. M. Elzerman, Y. L. Delley, J. Miguel-Sanchez, and A. Imamoğlu, *Coherent Two-Electron Spin Qubits in an Optically Active Pair of Coupled InGaAs Quantum Dots,* Phys. Rev. Lett. **109**, 107401 (2012). 16

[1.20] M. P. Hedges, J. J. Longdell, Y. Li, and M. J. Sellars, *Efficient Quantum Memory for Light,* Nature **465**, 1052 (2010). 17

[1.21] M. Zhong, M. P. Hedges, R. L. Ahlefeldt, J. G. Bartholomew, S. E. Beavan, S. M. Wittig, J. J. Longdell, and M. J. Sellars, *Optically Addressable Nuclear Spins in a Solid with a Six-Hour Coherence Time,* Nature **517**, 177 (2015). 17

[1.22] E. Knill, R. Laflamme, and G. J. Milburn, *A Scheme for Efficient Quantum Computation with Linear Optics,* Nature **409**, 46 (2001). 17

[1.23] T. B. Pittman, M. J. Fitch, B. C. Jacobs, and J. D. Franson, *Experimental Controlled-NOT Logic Gate for Single Photons in the Coincidence Basis,* Phys. Rev. A **68**, 032316 (2003). 17

[1.24] E. Martin-Lopez, A. Laing, T. Lawson, R. Alvarez, X.-Q. Zhou, and J. L. O'Brien, *Experimental Realization of Shor's Quantum Factoring Algorithm Using Qubit Recycling,* Nature Photonics **6**, 773 (2012). 17

[1.25] R. Hanson and D. D. Awschalom, *Coherent Manipulation of Single Spins in Semiconductors,* Nature **453**, 1043 (2008). 17

Chapter 2

[2.1] R. P. Feynman, *Simulating Physics with Computers,* International Journal of Theoretical Physics **21**, 467 (1982). 4, 18, 29, 31

[2.2] R. P. Feynman, *Quantum Mechanical Computers,* Foundations of Physics **16**, 507 (1986). 6, 18

[2.3] J. I. Cirac and P. Zoller, *Quantum Computations with Cold Trapped Ions,* Phys. Rev. Lett. **74**, 4091 (1995). 6, 14, 18, 358, 362

[2.4] P. W. Shor, in *Proceedings of the 35th Annual Symposium on the Foundations of Computer Science* (IEEE Computer Society Press, New York, 1994), p. 124. 6, 22, 27

[2.5] A. Ekert and R. Jozsa, *Quantum Computation and Shor's Factoring Algorithm,* Rev. Mod. Phys. **68**, 733 (1996). 22, 26, 27, 28, 29

[2.6] D. P. DiVincenzo, *The Physical Implementation of Quantum Computation,* Fortschritte der Physik **48**, 771 (2000). 22

[2.7] D. Deutsch, *Quantum Computational Networks*, Proc. Roy. Soc. London Ser. A **425**, 73 (1989). 23

[2.8] T. Sleator and H. Weinfurter, *Realizable Universal Quantum Logic Gates*, Phys. Rev. Lett. **74**, 4087 (1995). 23

[2.9] A. Peres, *Quantum Theory: Concepts and Methods* (Kluwer, Dordrecht, Boston, 1995). 25

[2.10] M. A. Nielsen and I. I. Chuang, *Quantum Computation and Quantum Information*, 1st ed. (Cambridge University Press, Cambridge, 2000, 2010). 6, 25, 34, 252

[2.11] V. Vedral, A. Barenco, and A. Ekert, *Quantum Networks for Elementary Arithmetic Operations*, Phys. Rev. A **54**, 147 (1996). 26

[2.12] S. Lloyd, *Universal Quantum Simulators*, Science **273**, 1073 (1996). 29

[2.13] H. F. Trotter, *On the Product of Semigroups of Operators*, Proceedings of the American Mathematical Society **10**, 545 (1959). 30

[2.14] T. Kato, in Trotter's Product Formula for an Arbitrary Pair of Self-Adjoint Contraction Semigroups, *Topics in Functional Analysis (Essays Dedicated to M. G. Krein on the Occasion of his 70th Birthday)* (Academic Press, Boston, 1978). 30

Chapter 3

[3.1] H. J. Kimble, *The Quantum Internet*, Nature **453**, 1023 (2008). 15, 32, 337

[3.2] M. A. Nielsen and I. I. Chuang, *Quantum Computation and Quantum Information*, 1st ed. (Cambridge University Press, Cambridge, 2000, 2010). 6, 25, 34, 252

[3.3] R. Horodecki, P. Horodecki, M. Horodecki, and K. Horodecki, *Quantum Entanglement*, Rev. Mod. Phys. **81**, 865 (2009). 34, 39

[3.4] J. S. Bell, *On the Einstein–Podolsky–Rosen Paradox*, Physics **1**, 195 (1964). 3, 34, 35

[3.5] C. H. Bennett, H. J. Bernstein, S. Popescu, and B. Schumacher, *Concentrating Partial Entanglement by Local Operations*, Phys. Rev. A **53**, 2046 (1996). 39

[3.6] S. Popescu and D. Rohrlich, *Thermodynamics and the Measure of Entanglement*, Phys. Rev. A **56**, 3319 (1997). 39

[3.7] L. Amico, R. Fazio, A. Osterloh, and V. Vedral, *Entanglement in Many-Body Systems*, Rev. Mod. Phys. **80**, 517 (2008). 39

[3.8] W. Dür, G. Vidal, and J. I. Cirac, *Three Qubits Can be Entangled in Two Inequivalent Ways*, Phys. Rev. A **62**, 062314 (2000). 40

[3.9] D. M. Greenberger, M. Horne, and A. Zeilinger, in *Bell's Theorem, Quantum Theory and Conceptions of the Universe*, edited by M. Kafatos (Kluwer Academic (Also available as arXiv:0712.0921), Dordrecht, the Netherlands, 1989), p. 69. 40

[3.10] A. Peres, *Separability Criterion for Density Matrices*, Phys. Rev. Lett **77**, 1413 (1996). 41

[3.11] R. Horodecki, P. Horodecki, and M. Horodecki, *Quantum α-Entropy Inequalities: Independent Condition for Local Realism?*, Phys. Lett A **210**, 377 (1996). 41

[3.12] B. P. Lanyon, C. Hempel, D. Nigg, M. Müller, R. Gerritsma, F. Zähringer, P. Schindler, J. T. Barreiro, M. Rambach, G. Kirchmair, M. Hennrich, P. Zoller, R. Blatt, and C. F. Roos, *Universal Digital Quantum Simulation with Trapped Ions*, Science **334**, 57 (2011). 42

[3.13] W. K. Wooters and W. H. Zurek, *A Single Quantum Cannot be Cloned*, Nature **299**, 802 (1982). 42

[3.14] H. Barnum, C. M. Caves, C. A. Fuchs, R. Jozsa, and B. Schumacher, *Noncommuting Mixed States Cannot Be Broadcast*, Phys. Rev. Lett. **76**, 2818 (1996). 42, 43

[3.15] C. H. Bennett, G. Brassard, C. Crepeau, R. Jozsa, A. Peres, and W. K. Wootters, *Teleporting an Unknown Quantum State via Dual Classical and Einstein–Podolsky–Rosen Channels*, Phys. Rev. Lett. **70**, 1895 (1993). 43

[3.16] D. Bouwmeester, J.-W. Pan, K. Mattle, M. Eibl, H. Weinfurter, and A. Zeilinger, *Experimental Quantum Teleportation*, Nature **390**, 575 (1997). 43

[3.17] D. Boschi, S. Branca, F. De Martini, L. Hardy, and S. Popescu, *Experimental Realization of Teleporting an Unknown Pure Quantum State via Dual Classical and Einstein-Podolsky-Rosen Channels*, Phys. Rev. Lett. **80**, 1121 (1998). 43

[3.18] A. Furusawa, J. L. Sørensen, S. L. Braunstein, C. A. Fuchs, H. J. Kimble, and E. S. Polzik, *Unconditional Quantum Teleportation*, Science **282**, 706 (1998). 43

Chapter 4

[4.1] L. Allen and J. H. Eberly, *Optical Resonance and Two-Level Atoms* (Dover, Mineola, N. Y., 1975). 47, 372

[4.2] B. R. Mollow, *Pure-State Analysis of Resonant Light Scattering: Radiative Damping, Saturation, and Multiphoton Effects*, Phys. Rev. A **12**, 1919 (1975). 47, 138, 140

[4.3] T. A. Savard, K. M. O'Hara, and J. E. Thomas, *Laser-Noise-Induced Heating in Far-Off Resonance Optical Traps*, Phys. Rev. A **56**, R1095 (1997). 50

[4.4] C. W. Gardiner, J. Ye, H. C. Nagerl, and H. J. Kimble, *Evaluation of Heating Effects on Atoms Trapped in an Optical Trap*, Phys. Rev. A **61**, 045801 (2000). 50

Chapter 5

[5.1] F. Bloch, *Nuclear Induction*, Phys. Rev. **70**, 460 (1946). 54, 59

[5.2] M. V. Berry, *Quantal Phase Factors Accompanying Adiabatic Changes*, Proceedings of the Royal Society of London. A. Mathematical and Physical Sciences **392**, 45 (1984). 68

Chapter 6

[6.1] S. Schiemann, A. Kuhn, S. Steuerwald, and K. Bergmann, *Efficient Coherent Population Transfer in NO Molecules Using Pulsed Lasers*, Phys. Rev. Lett. **71**, 3637 (1993). 77, 78, 180

[6.2] T. A. Laine and S. Stenholm, *Adiabatic Processes in Three-Level Systems*, Phys. Rev. A **53**, 2501 (1996). 77, 180

[6.3] K. Bergmann, H. Theuer, and B. W. Shore, *Coherent Population Transfer among Quantum States of Atoms and Molecules*, Rev. Mod. Phys. **70**, 1003 (1998). 77, 180

Chapter 7

[7.1] P. L. Kapitza and P. A. M. Dirac, *The Reflection of Electrons from Standing Light Waves*, Mathematical Proceedings of the Cambridge Philosophical Society **29**, 297 (1933). 80, 84

[7.2] P. E. Moskowitz, P. L. Gould, and D. E. Pritchard, *Deflection of Atoms by Standing-Wave Radiation*, J. Opt. Soc. Am. B **2**, 1784 (1985). 80, 84

[7.3] P. L. Gould, G. A. Ruff, and D. E. Pritchard, *Diffraction of Atoms by Light: The Near-Resonant Kapitza–Dirac Effect*, Phys. Rev. Lett. **56**, 827 (1986). 80, 84

[7.4] A. D. Cronin, J. Schmiedmayer, and D. E. Pritchard, *Optics and Interferometry with Atoms and Molecules*, Rev. Mod. Phys. **81**, 1051 (2009). 83, 87

[7.5] P. J. Martin, B. G. Oldaker, A. H. Miklich, and D. E. Pritchard, *Bragg Scattering of Atoms from a Standing Light Wave*, Phys. Rev. Lett. **60**, 515 (1988). 84

[7.6] M. Kozuma, L. Deng, E. W. Hagley, J. Wen, R. Lutwak, K. Helmerson, S. L. Rolston, and W. D. Phillips, *Coherent Splitting of Bose–Einstein Condensed Atoms with Optically Induced Bragg Diffraction*, Phys. Rev. Lett. **82**, 871 (1999). 84

[7.7] H. Müller, S.-W. Chiow, and S. Chu, *Atom-Wave Diffraction between the Raman–Nath and the Bragg Regime: Effective Rabi Frequency, Losses, and Phase Shifts*, Phys. Rev. A **77**, 023609 (2008). 87

[7.8] P. Marte, P. Zoller, and J. L. Hall, *Coherent Atomic Mirrors and Beam Splitters by Adiabatic Passage in Multilevel Systems*, Phys. Rev. A **44**, R4118 (1991). 87, 88

Chapter 8

[8.1] I. I. Sobelman, *Introduction to the Theory of Atomic Spectra* (Pergamon, Oxford, 1972). 90, 94, 95, 110

[8.2] I. I. Sobelman, *Atomic Spectra and Radiative Transitions* (Springer, Berlin, New York, 1979). 90, 94, 95

[8.3] M. J. Seaton, *Quantum Defect Theory*, Rep. Prog. Phys. **46**, 167 (1983). 93

[8.4] E. Arimondo, M. Inguscio, and P. Violino, *Experimental Determinations of the Hyperfine Structure in the Alkali Atoms*, Rev. Mod. Phys. **49**, 31 (1977). 100, 101

[8.5] D. M. Brink and G. R. Satchler, *Angular Momentum* (Oxford University Press, Oxford, 1975). 111, 112, 374

Chapter 9

[9.1] M. J. Collett and C. W. Gardiner, *Squeezing of Intracavity and Travelling Wave Light Fields Produced in Parametric Amplification*, Phys. Rev. A **30**, 1386 (1984). 8, 121, 154

[9.2] C. W. Gardiner and M. J. Collett, *Input and Output in Damped Quantum Systems—Quantum Stochastic Differential Equations and the Master Equation*, Phys. Rev. A **31**, 3761 (1985). 8, 121, 154, 161

[9.3] B. R. Mollow, *Pure-State Analysis of Resonant Light Scattering: Radiative Damping, Saturation, and Multiphoton Effects*, Phys. Rev. A **12**, 1919 (1975). 47, 138, 140

Chapter 10

[10.1] W. J. Munro and C. W. Gardiner, *Non-Rotating-Wave Master Equation*, Phys. Rev. A **53**, 2633 (1996). 149

Chapter 11

[11.1] M. J. Collett and C. W. Gardiner, *Squeezing of Intracavity and Travelling Wave Light Fields Produced in Parametric Amplification*, Phys. Rev. A **30**, 1386 (1984). 8, 121, 154

[11.2] C. W. Gardiner and M. J. Collett, *Input and Output in Damped Quantum Systems—Quantum Stochastic Differential Equations and the Master Equation*, Phys. Rev. A **31**, 3761 (1985). 8, 121, 154, 161

[11.3] B. Yurke and J. S. Denker, *Quantum Network Theory*, Phys. Rev. A **29**, 1419 (1984). 154, 383

[11.4] C. W. Gardiner, *Input and Output in Damped Quantum Systems III: Formulation of Damped Systems Driven by Fermion Fields*, Opt. Comm. **243**, 57 (2004). 168

Chapter 12

[12.1] K. S. Choi, H. Deng, J. Laurat, and H. J. Kimble, *Mapping Photonic Entanglement into and out of a Quantum Memory*, Nature **452**, 67 (2008). 169

[12.2] S. Ritter, C. Nolleke, C. Hahn, A. Reiserer, A. Neuzner, M. Uphoff, M. Mucke, E. Figueroa, J. Bochmann, and G. Rempe, *An Elementary Quantum Network of Single Atoms in Optical Cavities*, Nature **484**, 195 (2012). 169

[12.3] M. I. Kolobov and I. V. Sokolov, *Quantum Theory of Light Interaction with an Optical Amplifier*, Opt. Spektrosk. **62**, 112 (1987). 169

[12.4] H. J. Carmichael, *Quantum Trajectory Theory for Cascaded Open Systems*, Phys. Rev. Lett. **70**, 2273 (1993). 169

[12.5] C. W. Gardiner, *Driving a Quantum System with the Output Field from Another Driven Quantum System*, Phys. Rev. Lett. **70**, 2269 (1993). 169

[12.6] C. W. Gardiner and A. S. Parkins, *Driving Atoms with Light of Arbitrary Statistics*, Phys Rev A **50**, 1792 (1994). 169, 173, 174

[12.7] H. J. Carmichael, *Statistical Methods in Quantum Optics 2: Non-Classical Fields* (Springer, Berlin, Heidelberg, New York, 2008). 173, 337

[12.8] K. Stannigel, P. Rabl, and P. Zoller, *Driven-Dissipative Preparation of Entangled States in Cascaded Quantum-Optical Networks*, New Journal of Physics **14**, 063014 (2012). 174

[12.9] T. Ramos, H. Pichler, A. J. Daley, and P. Zoller, *Quantum Spin Dimers from Chiral Dissipation in Cold-Atom Chains*, Phys. Rev. Lett. **113**, 237203 (2014). 174

[12.10] H. Pichler, T. Ramos, A. J. Daley, and P. Zoller, Quantum Optics of Chiral Spin Networks, arXiv:1411.2963 [quant-ph], (2014). 174

Chapter 13

[13.1] S. Schiemann, A. Kuhn, S. Steuerwald, and K. Bergmann, *Efficient Coherent Population Transfer in NO Molecules Using Pulsed Lasers*, Phys. Rev. Lett. **71**, 3637 (1993). 77, 78, 180

[13.2] T. A. Laine and S. Stenholm, *Adiabatic Processes in Three-Level Systems*, Phys. Rev. A **53**, 2501 (1996). 77, 180

[13.3] K. Bergmann, H. Theuer, and B. W. Shore, *Coherent Population Transfer among Quantum States of Atoms and Molecules*, Rev. Mod. Phys. **70**, 1003 (1998). 77, 180

Chapter 14

[14.1] R. H. Lehmberg, *Radiation from an N-Atom System. I. General Formalism*, Phys. Rev. A **2**, 883 (1970). 185

[14.2] R. H. Lehmberg, *Radiation from an N-Atom System. II. Spontaneous Emission from a Pair of Atoms*, Phys. Rev. A **2**, 889 (1970). 185

[14.3] M. Abramowitz and I. A. Stegun, *Handbook of Mathematical Functions* (Dover, New York, 1965). 188, 189, 198, 350, 352

[14.4] R. H. Dicke, *Coherence in Spontaneous Radiation Processes*, Phys. Rev. **93**, 99 (1954). 194, 437

[14.5] M. Gross and S. Haroche, *Superradiance: An Essay on the Theory of Collective Spontaneous Emission*, Physics Reports **93**, 301 (1982). 194

Chapter 15

[15.1] P. Marte, R. Dum, R. Taieb, and P. Zoller, *Resonance Fluorescence From Quantized One-Dimensional Molasses*, Phys. Rev. A **47**, 1378 (1993). 201

Chapter 16

[16.1] S. Stenholm, *The Semiclassical Theory of Laser Cooling*, Rev. Mod. Phys. **58**, 699 (1986). 212, 228

[16.2] J. Dalibard and C. Cohen-Tannoudji, *Laser Cooling below the Doppler Limit by Polarization Gradients: Simple Theoretical Models*, J. Opt. Soc. Am. B **6**, 2023 (1989). 212

[16.3] C. N. Cohen-Tannoudji and W. D. Phillips, *New Mechanisms for Laser Cooling*, Physics Today **44 (10)**, 33 (1990). 212

[16.4] J. Dalibard and C. Cohen-Tannoudji, *Atomic Motion in Laser Light: Connection Between Semiclassical and Quantum Descriptions*, Journal of Physics B: Atomic and Molecular Physics **18**, 1661 (1985). 212

[16.5] H. J. Metcalf and P. van der Straten, *Laser Cooling and Trapping* (Springer, New York, Berlin, Heidelberg, 1999). 212

[16.6] P. D. Lett, W. D. Phillips, S. L. Rolston, C. E. Tanner, R. N. Watts, and C. I. Westbrook, *Optical Molasses*, J. Opt. Soc. Am. B Soc. **6**, 2084 (1989). 227

Chapter 17

[17.1] M. Lindberg and S. Stenholm, *The Master Equation for Laser Cooling of Trapped Particles*, Journal of Physics B: Atomic and Molecular Physics **17**, 3375 (1984). 228

[17.2] S. Stenholm, *The Semiclassical Theory of Laser Cooling*, Rev. Mod. Phys. **58**, 699 (1986). 212, 228

[17.3] J. I. Cirac, R. Blatt, P. Zoller, and W. D. Phillips, *Laser Cooling of Trapped Ions in a Standing Wave*, Phys. Rev. A **46**, 2668 (1992). 228

Chapter 18

[18.1] L. Mandel, *Fluctuations of Photon Beams and their Correlations*, Proceedings of the Physical Society **72**, 1037 (1958). 243

[18.2] P. L. Kelley and W. H. Kleiner, *Theory of Electromagnetic Field Measurement and Photoelectron Counting*, Phys. Rev. **136**, A316 (1964). 243

[18.3] R. J. Glauber, *The Quantum Theory of Optical Coherence*, Phys. Rev. **130**, 2530 (1963). 4, 251

[18.4] R. J. Glauber, *Coherent and Incoherent States of the Radiation Field*, Phys. Rev. **131**, 2766 (1963). 251

[18.5] M. Corti and V. Degiorgio, *Third-Order Intensity Correlations in Laser Light*, Optics Communications **11**, 1 (1974). 251

[18.6] M. A. Nielsen and I. I. Chuang, *Quantum Computation and Quantum Information*, 1st ed. (Cambridge University Press, Cambridge, 2000, 2010). 6, 25, 34, 252

[18.7] D. F. Walls and G. J. Milburn, *Quantum Optics*, 2nd ed. (Springer, Berlin, Heidelberg, 2008). 255, 337

Chapter 19

[19.1] P. Alsing and H. J. Carmichael, *Spontaneous Dressed-State Polarization of a Coupled Atom and Cavity Mode*, Quantum Optics: Journal of the European Optical Society Part B **3**, 13 (1991). 267

[19.2] N. G. van Kampen, *Stochastic Processes in Physics and Chemistry* (North Holland, Amsterdam, New York, Oxford, 1981, 1992). 274

Chapter 20

[20.1] H. P. Yuen and J. Shapiro, *Optical Communication with Two-Photon Coherent States—Part I: Quantum-State Propagation and Quantum Noise*, IEEE Transactions on Information Theory **24**, 657 (1978). 285

[20.2] J. Shapiro, H. P. Yuen, and A. Mata, *Optical Communication with Two-Photon Coherent States—Part II: Photoemissive Detection and Structured Receiver Performance*, IEEE Transactions on Information Theory **25**, 179 (1979). 285

[20.3] H. P. Yuen and J. Shapiro, *Optical Communication with Two-Photon Coherent States—Part III: Quantum Measurements Realizable with Photoemissive Detectors*, IEEE Transactions on Information Theory **26**, 78 (1980). 285

Chapter 21

[21.1] J. N. Hollenhorst, *Quantum Limits on Resonant-Mass Gravitational-Radiation Detectors*, Phys. Rev. D **19**, 1669 (1979). 301

[21.2] C. M. Caves, *Quantum-Mechanical Noise in an Interferometer*, Phys. Rev. D **23**, 1693 (1981). 302

[21.3] R. E. Slusher, L. W. Hollberg, B. Yurke, J. C. Mertz, and J. F. Valley, *Observation of Squeezed States Generated by Four-Wave Mixing in an Optical Cavity*, Physical Review Letters **55**, 2409 (1985). 308

[21.4] L.-A. Wu, H. J. Kimble, J. L. Hall, and H. Wu, *Generation of Squeezed States by Parametric Down Conversion*, Phys. Rev. Lett. **57**, 2520 (1986). 308, 311

[21.5] R. Movshovich, B. Yurke, P. G. Kaminsky, A. D. Smith, A. H. Silver, R. W. Simon, and M. V. Schneider, *Observation of Zero-Point Noise Squeezing via a Josephson-Parametric Amplifier*, Phys. Rev. Lett. **65**, 1419 (1990). 308, 417

[21.6] N. Bloembergen, *Nonlinear Optics: a Lecture Note and Reprint Volume* (Benjamin, New York, 1965). 309

[21.7] C. M. Caves, *Quantum Limits on Noise in Linear Amplifiers*, Phys. Rev. D **26**, 1817 (1982). 321

Chapter 22

[22.1] R. Miller, T. E. Northup, K. M. Birnbaum, A. Boca, A. D. Boozer, and H. J. Kimble, *Trapped Atoms in Cavity QED: Coupling Quantized Light and Matter*, Journal of Physics B: Atomic, Molecular and Optical Physics **38**, S551 (2005). 327, 328, 337

[22.2] E. Jaynes and F. Cummings, *Comparison of Quantum and Semiclassical Radiation Theories with Application to the Beam Maser,* Proc IEEE **51**, 89 (1963). 327

[22.3] G. Rempe, H. Walther, and N. Klein, *Observation of Quantum Collapse and Revival in a One-Atom Maser,* Phys. Rev. Lett. **58**, 353 (1987). 328, 336

[22.4] J. M. Raimond, M. Brune, and S. Haroche, *Manipulating Quantum Entanglement with Atoms and Photons in a Cavity,* Rev. Mod. Phys. **73**, 565 (2001). 328

[22.5] A. Blais, R.-S. Huang, A. Wallraff, S. M. Girvin, and R. J. Schoelkopf, *Cavity Quantum Electrodynamics for Superconducting Electrical circuits: An Architecture for Quantum Computation,* Phys. Rev. A **69**, 062320 (2004). 334, 383, 402

[22.6] S. Haroche and J.-M. Raimond, *Exploring the Quantum* (Oxford, New York, 2006). 334

[22.7] H. J. Kimble, *The Quantum Internet,* Nature **453**, 1023 (2008). 15, 32, 337

[22.8] D. F. Walls and G. J. Milburn, *Quantum Optics,* 2nd ed. (Springer, Berlin, Heidelberg, 2008). 255, 337

[22.9] H. J. Carmichael, *Statistical Methods in Quantum Optics 2: Non-Classical Fields* (Springer, Berlin, Heidelberg, New York, 2008). 173, 337

Chapter 23

[23.1] C. A. Blockley, D. F. Walls, and H. Risken, *Quantum Collapses and Revivals in a Quantized Trap,* EPL (Europhysics Letters) **17**, 509 (1992). 343

[23.2] C. A. Blockley and D. F. Walls, *Cooling of a Trapped Ion in the Strong-Sideband Regime,* Phys. Rev. A **47**, 2115 (1993). 343

[23.3] S. A. Gardiner, J. I. Cirac, and P. Zoller, *Nonclassical States and Measurement of General Motional Observables of a Trapped Ion,* Phys. Rev. A **55**, 1683 (1997). 347

[23.4] D. Leibfried, R. Blatt, C. Monroe, and D. Wineland, *Quantum Dynamics of Single Trapped Ions,* Rev. Mod. Phys. **75**, 281 (2003). 350

[23.5] M. Abramowitz and I. A. Stegun, *Handbook of Mathematical Functions* (Dover, New York, 1965). 188, 189, 198, 350, 352

[23.6] M. Combescure, *Trapping of Quantum Particles for a Class of Time-Periodic Potentials: A Semi-Classical Approach,* Annals of Physics **173**, 210 (1987). 354, 356

Chapter 24

[24.1] J. I. Cirac and P. Zoller, *Quantum Computations with Cold Trapped Ions,* Phys. Rev. Lett. **74**, 4091 (1995). 6, 14, 18, 358, 362

[24.2] D. James, *Quantum Dynamics of Cold Trapped Ions with Application to Quantum Computation,* Applied Physics B: Lasers and Optics **66**, 181 (1998). 361

[24.3] C. Monroe, D. M. Meekhof, B. E. King, W. M. Itano, and D. J. Wineland, *Demonstration of a Fundamental Quantum Logic Gate,* Phys. Rev. Lett. **75**, 4714 (1995). 362

[24.4] F. Schmidt-Kaler, H. Haffner, M. Riebe, S. Gulde, G. P. T. Lancaster, T. Deuschle, C. Becher, C. F. Roos, J. Eschner, and R. Blatt, *Realization of the Cirac–Zoller Controlled-NOT Quantum Gate,* Nature **422**, 408 (2003). 362

[24.5] F. Schmidt-Kaler, H. Häffner, S. Gulde, M. Riebe, G. P. T. Lancaster, T. Deuschle, C. Becher, W. Hänsel, J. Eschner, C. F. Roos, and R. Blatt, *How to Realize a Universal Quantum Gate with Trapped Ions,* Appl. Phys. B **77**, 789 (2003). 6, 362

[24.6] A. Sørensen and K. Mølmer, *Quantum Computation with Ions in Thermal Motion,* Phys. Rev. Lett. **82**, 1971 (1999). 368

[24.7] K. Mølmer and A. Sørensen, *Multiparticle Entanglement of Hot Trapped Ions,* Phys. Rev. Lett. **82**, 1835 (1999). 368, 372

[24.8] D. M. Greenberger, M. A. Horne, A. Shimony, and A. Zeilinger, *Bell's Theorem without Inequalities,* American Journal of Physics **58**, 1131 (1990). 368

[24.9] L. Allen and J. H. Eberly, *Optical Resonance and Two-Level Atoms* (Dover, Mineola, N. Y., 1975). 47, 372

[24.10] D. M. Brink and G. R. Satchler, *Angular Momentum* (Oxford University Press, Oxford, 1975). 111, 112, 374

[24.11] J. J. García-Ripoll, P. Zoller, and J. I. Cirac, *Coherent Control of Trapped Ions Using Off-Resonant Lasers,* Phys. Rev. A **71**, 062309 (2005). 375, 378, 379

[24.12] Y. Aharonov and J. Anandan, *Phase Change During a Cyclic Quantum Evolution,* Phys. Rev. Lett. **58**, 1593 (1987). 379

[24.13] J. Anandan, J. Christian, and K. Wanelik, *Resource Letter GPP-1: Geometric Phases in Physics,* American Journal of Physics **65**, 180 (1997). 379

[24.14] A. Sørensen and K. Mølmer, *Entanglement and Quantum Computation with Ions in Thermal Motion,* Phys. Rev. A **62**, 022311 (2000). 379

[24.15] P. Schindler, D. Nigg, T. Monz, J. T. Barreiro, E. Martinez, S. X. Wang, S. Quint, M. F. Brandl, V. Nebendahl, C. F. Roos, M. Chwalla, M. Hennrich, and R. Blatt, *A Quantum Information Processor with Trapped Ions,* New Journal of Physics **15**, 123012 (2013). 379

[24.16] D. Kielpinski, C. Monroe, and D. J. Wineland, *Architecture for a Large-Scale Ion-Trap Quantum Computer,* Nature **417**, 709 (2002). 380

[24.17] R. Blatt and D. Wineland, *Entangled States of Trapped Atomic Ions,* Nature **453**, 1008 (2008). 380

[24.18] C. Monroe, R. Raussendorf, A. Ruthven, K. R. Brown, P. Maunz, L.-M. Duan, and J. Kim, *Large-Scale Modular Quantum Computer Architecture with Atomic Memory and Photonic Interconnects,* Phys. Rev. A **89**, 022317 (2014). 380

Chapter 25

[25.1] W. H. Louisell, *Radiation and Noise in Quantum Electronics* (McGraw–Hill, New York, 1964). 383

[25.2] B. Yurke and J. S. Denker, *Quantum Network Theory,* Phys. Rev. A **29**, 1419 (1984). 154, 383

[25.3] M. H. Devoret, in *Quantum Fluctuations (Les Houches, Session LXIII 1995)*, edited by S. Reynaud, E. Giacobino, and J. Zinn-Justin (Elsevier, Amsterdam, 1997), Vol. LXIII, p. 351. 383

[25.4] A. Blais, R.-S. Huang, A. Wallraff, S. M. Girvin, and R. J. Schoelkopf, *Cavity Quantum Electrodynamics for Superconducting Electrical circuits: An Architecture for Quantum Computation*, Phys. Rev. A **69**, 062320 (2004). 334, 383, 402

[25.5] R. M. Foster, *A Reactance Theorem*, Bell Systems Technical Journal **3**, 259 (1924). 387

Chapter 26

[26.1] J. Clarke and F. K. Wilhelm, *Superconducting Quantum Bits*, Nature **453**, 1031 (2008). 402, 417

[26.2] A. Blais, R.-S. Huang, A. Wallraff, S. M. Girvin, and R. J. Schoelkopf, *Cavity Quantum Electrodynamics for Superconducting Electrical circuits: An Architecture for Quantum Computation*, Phys. Rev. A **69**, 062320 (2004). 334, 383, 402

[26.3] A. Cottet, *Implementation of a Quantum Bit in a Superconducting Circuit*, Ph. D. thesis, Université Paris VI (2002). 411

[26.4] J. Koch, T. M. Yu, J. Gambetta, A. A. Houck, D. I. Schuster, J. Majer, A. Blais, M. H. Devoret, S. M. Girvin, and R. J. Schoelkopf, *Charge-Insensitive Qubit Design Derived from the Cooper Pair Box*, Phys. Rev. A **76**, 042319 (2007). 411, 412

[26.5] J. A. Schreier, A. A. Houck, J. Koch, D. I. Schuster, B. R. Johnson, J. M. Chow, J. M. Gambetta, J. Majer, L. Frunzio, M. H. Devoret, S. M. Girvin, and R. J. Schoelkopf, *Suppressing Charge Noise Decoherence in Superconducting Charge Qubits*, Phys. Rev. B **77**, 180502 (2008). 411

[26.6] M. H. Devoret and R. J. Schoelkopf, *Superconducting Circuits for Quantum Information: An Outlook*, Science **339**, 1169 (2013). 412

[26.7] G. Kirchmair, B. Vlastakis, Z. Leghtas, S. E. Nigg, H. Paik, E. Ginossar, M. Mirrahimi, L. Frunzio, S. Girvin, and R. Schoelkopf, *Observation of Quantum State Collapse and Revival Due to the Single-Photon Kerr Effect*, Nature **495**, 205 (2013). 412, 413

[26.8] H. Paik, D. I. Schuster, L. S. Bishop, G. Kirchmair, G. Catelani, A. P. Sears, B. R. Johnson, M. J. Reagor, L. Frunzio, L. I. Glazman, S. M. Girvin, M. H. Devoret, and R. J. Schoelkopf, *Observation of High Coherence in Josephson Junction Qubits Measured in a Three-Dimensional Circuit QED Architecture*, Phys. Rev. Lett. **107**, 240501 (2011). 413

[26.9] M. D. Reed, L. DiCarlo, B. R. Johnson, L. Sun, D. I. Schuster, L. Frunzio, and R. J. Schoelkopf, *High-Fidelity Readout in Circuit Quantum Electrodynamics Using the Jaynes-Cummings Nonlinearity*, Phys. Rev. Lett. **105**, 173601 (2010). 416

[26.10] L. DiCarlo, M. D. Reed, L. Sun, B. R. Johnson, J. M. Chow, J. M. Gambetta, L. Frunzio, S. M. Girvin, M. H. Devoret, and R. J. Schoelkopf, *Preparation and Measurement of Three-Qubit Entanglement in a Superconducting Circuit*, Nature **467**, 574 (2010). 416

[26.11] M. Hatridge, S. Shankar, M. Mirrahimi, F. Schackert, K. Geerlings, T. Brecht, K. M. Sliwa, B. Abdo, L. Frunzio, S. M. Girvin, R. J. Schoelkopf, and M. H. Devoret, *Quantum Back-Action of an Individual Variable-Strength Measurement*, Science **339**, 178 (2013). 416

[26.12] J. Martinis, *Superconducting Phase Qubits*, Quantum Information Processing **8**, 81 (2009). 417

[26.13] B. Yurke, P. Kaminsky, R. Miller, E. Whittaker, A. Smith, A. Silver, and R. Simon, *Observation of 4.2 K Equilibrium-Noise Squeezing via a Josephson-Parametric Amplifier*, Physical Review Letters **60**, 764 (1988). 417

[26.14] R. Movshovich, B. Yurke, P. G. Kaminsky, A. D. Smith, A. H. Silver, R. W. Simon, and M. V. Schneider, *Observation of Zero-Point Noise Squeezing via a Josephson-Parametric Amplifier*, Phys. Rev. Lett. **65**, 1419 (1990). 308, 417

[26.15] N. Bergeal, R. Vijay, V. E. Manucharyan, I. Siddiqi, R. J. Schoelkopf, S. M. Girvin, and M. H. Devoret, *Analog Information Processing at the Quantum Limit with a Josephson Ring Modulator*, Nat Phys **6**, 296 (2010). 418, 419, 420

[26.16] N. Bergeal, F. Schackert, M. Metcalfe, R. Vijay, V. E. Manucharyan, L. Frunzio, D. E. Prober, R. J. Schoelkopf, S. M. Girvin, and M. H. Devoret, *Phase-Preserving Amplification Near the Quantum Limit with a Josephson Ring Modulator*, Nature **465**, 64 (2010). 418

Chapter 27

[27.1] J. I. Cirac, P. Zoller, H. J. Kimble, and H. Mabuchi, *Quantum State Transfer and Entanglement Distribution among Distant Nodes in a Quantum Network*, Phys. Rev. Lett. **78**, 3221 (1997). 432, 433

Chapter 28

[28.1] M. Fleischhauer and M. D. Lukin, *Quantum Memory for Photons: Dark-State Polaritons*, Phys. Rev. A **65**, 022314 (2002). 434

[28.2] M. Fleischhauer, A. Imamoğlu, and J. P. Marangos, *Electromagnetically Induced Transparency: Optics in Coherent Media*, Rev. Mod. Phys. **77**, 633 (2005). 434

[28.3] R. H. Dicke, *Coherence in Spontaneous Radiation Processes*, Phys. Rev. **93**, 99 (1954). 194, 437

[28.4] L. M. Duan, M. D. Lukin, J. I. Cirac, and P. Zoller, *Long-Distance Quantum Communication with Atomic Ensembles and Linear Optics*, Nature **414**, 413 (2001). 452

Chapter 29

[29.1] S. H. Autler and C. H. Townes, *Stark Effect in Rapidly Varying Fields*, Phys. Rev. **100**, 703 (1955). 456

[29.2] K.-J. Boller, A. Imamoğlu, and S. E. Harris, *Observation of Electromagnetically Induced Transparency*, Phys. Rev. Lett. **66**, 2593 (1991). 457

[29.3] L. V. Hau, S. E. Harris, Z. Dutton, and C. H. Behroozi, *Light Speed Reduction to 17 Metres per Second in an Ultracold Atomic Gas*, Nature **397**, 594 (1999). 460

Author Index

Subject Index

U

V

W

Z

www.ingramcontent.com/pod-product-compliance
Lightning Source LLC
Chambersburg PA
CBHW070740220326
41598CB00026B/3713